Plant Pathology for Sustainable Agriculture

Series Editors:
Ravinder Kumar · Govind P. Rao

Plant Pathology for Sustainable Agriculture is a comprehensive book series that presents current research, tools, and practices that strengthen plant health and support sustainable crop production. The series brings together scientific knowledge, field experience, and technological advances to create a steady body of reliable and accessible resources. Each book adds to a growing collection that highlights practical solutions, modern diagnostics, and integrated approaches used in plant disease management across diverse crops and environments.

The series covers a wide range of topics, reflecting the depth of plant pathology research. It includes studies on viruses, bacteria, fungi, phytoplasmas, nematodes, seed-borne pathogens, point-of-care diagnostics, molecular and genomic tools, disease forecasting, climate-linked risks, crop resistance strategies, and sustainable management practices. It also includes emerging areas such as digital agriculture, remote sensing, and microbiome-based solutions.

The series serves researchers, faculty members, extension workers, and students who seek dependable scientific content. It also supports plant health professionals, diagnostic laboratories, breeding programs, policy groups, and anyone involved in crop protection and sustainable agricultural systems. It acts as an essential resource for readers who need clear, evidence-based information that supports plant disease research and practical decision-making.

Ravinder Kumar • Rahul Kumar Tiwari
Milan Kumar Lal
Virendra Kumar Baranwal • Brajesh Singh
Editors

Fruit Crops and Viral Pathogens Volume II

Molecular Insights and Sustainable Control

Springer

Editors
Ravinder Kumar
Division of Plant Pathology
ICAR-Indian Agricultural Research Institute
New Delhi, Delhi, India

Rahul Kumar Tiwari
Division of Crop Protection
ICAR-Indian Institute of Sugarcane Research
Lucknow, Uttar Pradesh, India

Milan Kumar Lal
Division of Crop Physiology & Biochemistry
ICAR-National Rice Research Institute
Cuttack, Odisha, India

Virendra Kumar Baranwal
Division of Plant Pathology
ICAR-Indian Agricultural Research Institute
New Delhi, Delhi, India

Brajesh Singh
Physiology, Biochemistry & Post-Harvest Technology
ICAR-Central Potato Research Institute
Shimla, Himachal Pradesh, India

ISSN 3120-5976 ISSN 3120-5984 (electronic)
Plant Pathology for Sustainable Agriculture
ISBN 978-981-95-7260-1 ISBN 978-981-95-7261-8 (eBook)
https://doi.org/10.1007/978-981-95-7261-8

This Springer imprint is published by the registered company Springer Nature Singapore Pte Ltd.
The registered company address is: 152 Beach Road, #21-01/04 Gateway East, Singapore 189721, Singapore

Preface

Fruit crops are indispensable to global agriculture, nutrition, and rural economies, contributing substantially to dietary diversity, livelihood security, and international trade. However, viral diseases remain one of the most persistent and economically devastating constraints to fruit crop production. The perennial nature of many fruit crops, coupled with vegetative propagation, long production cycles, complex virus–vector relationships, and increasing climatic variability, makes virus management in fruit systems particularly challenging. In recent decades, the emergence of new viral pathogens, mixed infections, and rapidly evolving virus populations has further intensified these challenges, demanding innovative and integrated solutions. The book *Fruit Crops and Viral Pathogens: Molecular Insights and Sustainable Control* has been developed to address these pressing concerns through a comprehensive and multidisciplinary lens. This volume brings together contemporary knowledge on the biology, ecology, molecular mechanisms, and management of fruit-infecting viruses, integrating classical plant virology with modern advances in omics technologies, diagnostics, biotechnology, and regulatory frameworks. It is intended as a valuable reference for plant virologists, molecular biologists, horticulturists, breeders, biotechnologists, extension professionals, and postgraduate students engaged in fruit crop protection and sustainable horticulture. The initial chapters set the foundation by examining the economic losses and quality deterioration caused by viral diseases in fruit crops, followed by detailed accounts of viral threats affecting major tropical and subtropical fruits. The influence of climate change on virus emergence, geographic expansion, and outbreak dynamics is critically analyzed, highlighting how altered temperature regimes, vector behavior, and host physiology are reshaping fruit virus epidemiology worldwide. The complexity of viral disease complexes, including synergistic and mixed infections, is explored to underscore their implications for diagnosis, symptom expression, and control strategies. A strong emphasis is placed on understanding virus biology at the molecular and biochemical levels. Chapters addressing viral pathogenesis, virus–vector interactions, suppressors of RNA silencing, epigenetic regulation, and genome characterization provide mechanistic insights into how viruses manipulate host and vector systems. The inclusion of virome analysis and genome annotation approaches reflects the growing importance of high-throughput sequencing and bioinformatics in uncovering known and novel viruses in fruit crops. The book also highlights the transformative role of advanced analytical platforms. Proteomics, metabolomics, and integrated multi-omics approaches are discussed as powerful tools to decode host responses, identify biomarkers, and design targeted virus management strategies. These systems-level insights are essential for bridging fundamental research with practical disease control in complex perennial fruit systems. Early detection and prevention form a central theme of this volume. Chapters on rapid on-site diagnostics, tissue culture–based production of healthy planting materials, quarantine regulations, and certification programs emphasize the critical role of preventive plant health measures in minimizing virus spread. These approaches are particularly relevant for perennial fruit crops, where once-established viral infections are difficult, if not impossible, to eliminate. Innovative and future-oriented management strategies are comprehensively covered, including RNA interference and dsRNA-based gene silencing, virus-induced gene silencing for functional genomics, cross-protection using mild viral strains, nanotechnology-based interventions, and post-harvest virus management. The final sections of the book address transgenic approaches and emerging biotechnological solutions for virus resistance in perennial fruit crops, while also considering regulatory, biosafety, and policy perspectives. The editors sincerely acknowledge the invaluable contributions of all authors, whose diverse expertise and collaborative efforts have shaped this volume into a cohesive and forward-looking resource. Their work reflects the evolving nature of fruit virology, where molecular innovation, sustainability, and practical applicability must converge. We hope that *Fruit Crops and Viral Pathogens: Molecular Insights and Sustainable Control* will serve not only as a comprehensive reference but also as a catalyst for new research, interdisciplinary collaboration, and the development of resilient, virus-managed fruit production systems capable of meeting the demands of a changing global landscape.

New Delhi, Delhi, India — Ravinder Kumar
Lucknow, Uttar Pradesh, India — Rahul Kumar Tiwari
Cuttack, Odisha, India — Milan Kumar Lal
New Delhi, Delhi, India — Virendra Kumar Baranwal
Shimla, Himachal Pradesh, India — Brajesh Singh

Contents

Editors and Contributors

About the Editors

Ravinder Kumar is a senior scientist in Plant Pathology at ICAR-Indian Agricultural Research Institute, New Delhi. He earned his M.Sc. and Ph.D. in Plant Pathology from Sardar Vallabhbhai Patel University of Agriculture and Technology, Meerut, and previously served at the ICAR-Central Potato Research Institute, Shimla. His research focuses on host–pathogen interactions, plant virus epidemiology, and disease management, with major contributions in molecular characterization, genome sequencing, and diagnostics of plant viruses. He has developed advanced diagnostic tools, including multiplex RT-PCR, real-time PCR, LAMP, and RPA. Dr. Kumar has published over 150 research articles and edited 20 books with leading publishers, serving on several international journal editorial boards.

Rahul Kumar Tiwari is currently working on melatonin-mediated mitigation of soil and tuber-borne diseases and physiological disorders in crop plants. He has also worked on the signaling mechanism of pathogens, pathogenomics and the management of hemibiotrophic fungi. He has explored the genetic diversity of pathogenic microbes in field and horticultural crops. Additionally, his area of interest is on multiple aspects of pathophysiological alterations in food crops under the influence of diverse biotic and abiotic stresses. He has published more than 130 publications in reputed journals.

Milan Kumar Lal is a plant physiologist specializing in abiotic stress and nutritional aspects of potato and other starchy crops. His research focuses on heat, drought, salinity, and heavy metal stress, as well as plant physiological, biochemical, and molecular responses to biotic stresses caused by fungi, viruses, and bacteria. He also has expertise in nutrition and quality traits of starchy crops, including resistant starch, glycemic index, phytonutrients, bioactive compounds, functional fermented foods and beverages, and processing techniques to enhance nutritional value. His work has resulted in over 180 publications in international peer-reviewed journals.

Virendra Kumar Baranwal is a distinguished plant virologist. He earned his M.Sc. and Ph.D. in Botany (Virology) from Lucknow University and joined the Agricultural Research Service in 1985, and served at CPCRI and NCIPM before joining the Advanced Centre for Plant Virology, IARI, in 1999. He served as Professor (2016–2022) and ICAR–National Professor (2022–2025). He has guided 4 M.Sc. and 20 Ph.D. students and published over 200 peer-reviewed articles. A recipient of the IARI Best Teacher Award, he is a Fellow of NAAS and NASI. His international exposure includes IPM training in Indonesia, DBT–Overseas Associate tenure at The State University of New Jersey, USA, and academic visits to the University of Florida and the University of Queensland. Dr. Baranwal's major contributions include the development of a microarray chip for the detection of over 1100 plant viruses, infectious clones of banana-infecting badnaviruses, studies on virus resistance in banana, HTS–based virus and viroid genomics, and rapid field-based diagnostics of plant viruses.

Brajesh Singh is a Doctor in Botany and currently serves as Director, ICAR–Central Potato Research Institute, Shimla, and Principal Scientist in the Crop Physiology, Biochemistry, and Post-Harvest Technology Division. His research interests focus on horticultural crop storage technologies, post-harvest processing, and nutritional quality assessment, with particular emphasis on improving the value and shelf life of potato crops. He is coordinating programs on genome editing of potato, post-harvest management, and nutritional enhancement, contributing to sustainable productivity and quality improvement. Dr. Singh has an extensive publication record with over 220 peer-reviewed research articles and is the author of four patents, reflecting his strong commitment to innovation, translational research, and advancement of potato science in India.

Contributors

Faheem Ahmad Department of Botany, Aligarh Muslim University, Aligarh, Uttar Pradesh, India

Azmi Nur Karimah Amas Hasanuddin University, Makassar University, Makassar, Indonesia

Zakir Amin Plant Virology Laboratory, Division of Fruit Science, ICAR-Central Institute of Temperate Horticulture, Srinagar, Jammu & Kashmir, India

I. L. F. Andrade Department of Agronomy, Plant Pathology Graduate Program, Federal Rural University of Pernambuco, Recife, Pernambuco, Brazil

Lishini Arucho Amity Institute of Virology & Immunology, Amity University Uttar Pradesh, Noida, Uttar Pradesh, India

Aadil Ayaz Plant Virology Laboratory, Division of Fruit Science, ICAR-Central Institute of Temperate Horticulture, Srinagar, Jammu & Kashmir, India

V. Balasubramanian Molecular Virology Laboratory, ICAR-National Research Center for Banana, Tiruchirappalli, Tamil Nadu, India

Mark Angelo O. Balendres Department of Biology, College of Science, De La Salle University, Manila, Philippines

Plant and Soil Health Research Unit, Center for Natural Science and Environmental Research, College of Science, De La Salle University, Manila, Philippines

Preetam Baruah Department of Plant Pathology, Assam Agricultural University, Jorhat, Assam, India

Laxmipriya Behera ICAR-Central Rice Research Institute, Cuttack, Odisha, India

Tesleem T. Bello Department of Agricultural Science Education, Federal College of Education, Abeokuta, Ogun State, Nigeria

Munmi Borah Department of Plant Pathology, Assam Agricultural University, Jorhat, Assam, India

Swapnil Borah Department of Plant Pathology, Assam Agricultural University, Jorhat, Assam, India

Archita Bordoloi Department of Molecular Biology and Biotechnology, Tezpur University, Sonitpur, Tezpur, Assam, India

R. L. R. Cabral Department of Agronomy, Genetic and Plant Breeding Graduate Program, Federal University of Espírito Santo, Alegre, Espírito Santo, Brazil

Sanjukta Chakraborty Department of Plant Pathology, Assam Agricultural University, Jorhat, Assam, India

Vanita Chandel Amity Institute of Virology & Immunology, Amity University Uttar Pradesh, Noida, Uttar Pradesh, India

Puja Dey Department of Plant Pathology, Assam Agricultural University, Jorhat, Assam, India

Alfredo Diaz-Lara Tecnologico de Monterrey, Queretaro, Mexico

Anutee Dolley Department of Molecular Biology and Biotechnology, Tezpur University, Sonitpur, Tezpur, Assam, India

Y. Durga Molecular Virology Laboratory, ICAR-National Research Center for Banana, Tiruchirappalli, Tamil Nadu, India

Tabia Fayaz Division of Plant Pathology, SKUAST-K, Shalimar, Jammu & Kashmir, India

Patricia M. B. Fernandes Núcleo de Biotecnologia, Universidade Federal do Espírito Santo, Vitória, Espírito Santo, Brazil

Fhilmar Raj Jayaraj Mallika Molecular Virology Laboratory, ICAR-National Research Center for Banana, Tiruchirappalli, Tamil Nadu, India

Darshana Goswami Department of Molecular Biology and Biotechnology, Tezpur University, Sonitpur, Tezpur, Assam, India

Muhammad Yusril Hardiansyah Institute of Biotechnology, College of Bioresources and Agriculture, National Taiwan University, Taipei, Taiwan

Bernadetta Rina Hastilestari Research Center for Genetic Engineering, National Research and Innovation Agency (BRIN), Jakarta, Indonesia

Paola Elizabeth Hernández-Meléndez Facultad de Ingeniería y Ciencias, Universidad Autónoma de Tamaulipas, Centro Universitario Victoria, Ciudad Victoria, Tamaulipas, Mexico

Oscar P. Hurtado-Gonzales USDA-Animal and Plant Health Inspection Service, Riverdale, MD, USA

Rupak Jena ICAR-Central Rice Research Institute, Cuttack, Odisha, India

D. R. S. Jesus Department of Agronomy, Plant Pathology Graduate Program, Federal Rural University of Pernambuco, Recife, Pernambuco, Brazil

Jothishri Chidambaram Molecular Virology Laboratory, ICAR-National Research Center for Banana, Tiruchirappalli, Tamil Nadu, India

Priyam Kafle Department of Plant Pathology, Assam Agricultural University, Jorhat, Assam, India

Pronoy Kalita Department of Plant Pathology, Assam Agricultural University, Jorhat, Assam, India

Uchakankhi Kashyap Department of Molecular Biology and Biotechnology, Tezpur University, Sonitpur, Tezpur, Assam, India

Purnama Isti Khaerani Research Center for Genetic Engineering, National Research and Innovation Agency (BRIN), Jakarta, Indonesia

S. Kruthika ICAR-Central Rice Research Institute, Cuttack, Odisha, India

Awadhesh Kumar ICAR-Central Rice Research Institute, Cuttack, Odisha, India

Ravinder Kumar ICAR-Indian Agricultural Research Institute, New Delhi, India

Milan Kumar Lal ICAR-Central Rice Research Institute, Cuttack, Odisha, India

Darwin M. Landicho Department of Crop Protection, College of Agriculture, Central Luzon State University, Muñoz, Nueva Ecija, Philippines

Bisma Majid Division of Plant Pathology, SKUAST-K, Shalimar, Jammu & Kashmir, India

Subaya Manzoor Plant Virology Laboratory, Division of Fruit Science, ICAR-Central Institute of Temperate Horticulture, Srinagar, Jammu & Kashmir, India

Tabasum Manzoor Plant Virology Laboratory, Division of Fruit Science, ICAR-Central Institute of Temperate Horticulture, Srinagar, Jammu & Kashmir, India

Marlonni Maurastoni Department of Entomology and Plant Pathology, North Carolina State University, Raleigh, NC, USA

Vietgar Membalik Wageningen University and Research, Wageningen, the Netherlands

Bandana Mohapatra ICAR-Central Rice Research Institute, Cuttack, Odisha, India

Aranza Mondragón-Cruz Tecnologico de Monterrey, Queretaro, Mexico

Giovanni Ferreira Montovaneli Núcleo Multidisciplinar de Pesquisa em Biologia, Universidade Federal do Rio de Janeiro, Duque de Caxias, Rio de Janeiro, Brazil

Sajad Un Nabi Plant Virology Laboratory, Division of Fruit Science, ICAR-Central Institute of Temperate Horticulture, Srinagar, Jammu & Kashmir, India

Nagendran Krishnan Molecular Virology Laboratory, ICAR-National Research Center for Banana, Tiruchirappalli, Tamil Nadu, India

Nima D. Namsa Department of Molecular Biology and Biotechnology, Tezpur University, Sonitpur, Tezpur, Assam, India

Adrij Pawan Neog Department of Molecular Biology and Biotechnology, Tezpur University, Sonitpur, Tezpur, Assam, India

P. M. S. Noor Mohamed Ibrahim Department of Botany, Aligarh Muslim University, Aligarh, Uttar Pradesh, India

Fredmoore Orosco Virology and Vaccine Research Program, Industrial Technology Development Institute, Department of Science and Technology, Bicutan, Taguig, Philippines

S&T Fellows Program, Department of Science and Technology, Bicutan, Taguig, Philippines

Department of Biology, College of Science, University of the Philippines Manila, Manila, Philippines

Saswat Swarup Padhi ICAR-Central Rice Research Institute, Cuttack, Odisha, India

Reynaldo P. Peja Jr. Department of Biological Sciences, Faculty of Natural and Mathematical Sciences, Visayas State University, Visca, Baybay, Leyte, Philippines

Bushra Rasool Plant Virology Laboratory, Division of Fruit Science, ICAR-Central Institute of Temperate Horticulture, Srinagar, Jammu & Kashmir, India

Silas Pessini Rodrigues Núcleo Multidisciplinar de Pesquisa em Biologia, Universidade Federal do Rio de Janeiro, Duque de Caxias, Rio de Janeiro, Brazil

Mohor Roy Amity Institute of Virology & Immunology, Amity University Uttar Pradesh, Noida, Uttar Pradesh, India

Souvik Kumar Saha Department of Biotechnology, National Institute of Technology, Durgapur, Paschim Bardhaman, West Bengal, India

Anindita Saikia Department of Molecular Biology and Biotechnology, Tezpur University, Sonitpur, Tezpur, Assam, India

G. A. S. Santos Department of Agronomy, Plant Pathology Graduate Program, Federal Rural University of Pernambuco, Recife, Pernambuco, Brazil

Devayani Sarmah Department of Plant Pathology, Assam Agricultural University, Jorhat, Assam, India

Division of Plant Pathology, ICAR-Central Rice Research Institute, Cuttack, Odisha, India

R. Selvarajan Molecular Virology Laboratory, ICAR-National Research Center for Banana, Tiruchirappalli, Tamil Nadu, India

Manoj K. Singh National Academy of Agriculture Sciences, NASC Complex, Pusa, New Delhi, India

Roshani Singh Amity Institute of Virology & Immunology, Amity University Uttar Pradesh, Noida, Uttar Pradesh, India

Kristian Stevens University of California-Davis, Davis, CA, USA

Rahul Kumar Tiwari ICAR-Indian Sugarcane Research Institute, Lucknow, Uttar Pradesh, India

Juan Manuel Tovar-Pedraza Centro de Investigación en Alimentación y Desarrollo, Sinaloa, Mexico

Greeshma Varghese Department of Plant Pathology, Assam Agricultural University, Jorhat, Assam, India

Ubilfrido Vasquez-Gutierrez Department of Parasitology, Universidad Autónoma Agraria Antonio Narro, Buenavista, Saltillo, Mexico

Marcella S. Vieira Núcleo de Biotecnologia, Universidade Federal do Espírito Santo, Vitória, Espírito Santo, Brazil

J. Vinodhini Molecular Virology Laboratory, ICAR-National Research Center for Banana, Tiruchirappalli, Tamil Nadu, India

Prashant S. Waghrulkar Molecular Virology Laboratory, ICAR-National Research Center for Banana, Tiruchirappalli, Tamil Nadu, India

A. S. Xavier Department of Agronomy, Plant Pathology Graduate Program, Federal Rural University of Pernambuco, Recife, Pernambuco, Brazil

Abbreviations

5mC	5-Methylcytosine
6mA	N6-methyladenine
ACLSV	Apple chlorotic leaf spot virus
ADFv	Apple dimple fruit viroid
AGCaV	Apple green crinkle–associated virus
AgNPs	Silver nanoparticles
AGO	Argonaute protein
AI	Artificial intelligence
ALKBH	AlkB homolog demethylase
ALSV	Apple latent spherical virus
amiRNA	Artificial microRNA
AMP1	Altered meristem program 1
AMV	Apple mosaic virus
ApMV	Apple mosaic virus
ARWV2	Apple rubbery wood virus 2
ASBVd	Avocado sunblotch viroid
ASGV	Apple stem grooving virus
ASPV	Apple stem pitting virus
ASSVd	Apple scar skin viroid
AuNPs	Gold nanoparticles
BAC	Bacterial artificial chromosome
BaYMV	Barley yellow mosaic virus
BBrMV	Banana bract mosaic virus
BBSV	Broad bean stain virus
BBTD	Banana bunchy top disease
BBTV	Banana bunchy top virus
BCLCaV	Blackcurrant leaf chlorosis–associated virus
BCTV	Beet curly top virus
BDBV	Banana die-back virus
BLMoV	Blueberry leaf mottle virus
BlScV	Blueberry scorch virus
BRNV	Black raspberry necrosis virus
BRV	Blackcurrant reversion virus
BSMV	Barley stripe mosaic virus

BSV	Banana streak virus
BVX	Banana virus X
CAC	Codex Alimentarius Commission
Cas	CRISPR-associated protein
CBD	Convention on Biological Diversity
CbLCV	Cabbage leaf curl virus
CCDaV	Citrus chlorotic dwarf–associated virus
CCGaV	Citrus concave gum–associated virus
CeO_2NPs	Cerium oxide nanoparticles
CGMMV	Cucumber green mottle mosaic virus
ChIP-seq	Chromatin immunoprecipitation sequencing
CHS	Chalcone synthase
CLBV	Citrus leaf blotch virus
CLRV	Cherry rasp leaf virus
CMV	Cucumber mosaic virus
CNTs	Carbon nanotubes
CO_2	Carbon dioxide
CP	Coat protein
CPsV	Citrus psorosis virus
CRISPR	Clustered regularly interspaced short palindromic repeats
CRISPR/Cas9	Clustered regularly interspaced short palindromic repeats/Cas9
CTV	Citrus tristeza virus
CVV	Citrus variegation virus
CYSDV	Cucurbit yellow stunting disorder virus
DAMPs	Damage-associated molecular patterns
DCL	Dicer-like protein
DL	Deep learning
DMGs	Differentially methylated genes
DNA	Deoxyribonucleic acid
DP	Diagnostic protocol
dPCR	Digital polymerase chain reaction
dsDNA	Double-stranded DNA
dsRNA	Double-stranded RNA
eIF4E	Eukaryotic translation initiation factor 4E
ELISA	Enzyme-linked immunosorbent assay
EM	Electron microscopy
ENMs	Engineered nanomaterials
EPA	Environmental Protection Agency
EPPO	European and Mediterranean Plant Protection Organization
ET	Ethylene
ETI	Effector-triggered immunity
FAO	Food and Agriculture Organization of the United Nations
FDA	Food and Drug Administration
Fe_3O_4NPs	Iron oxide nanoparticles
GDP	Gross domestic product

GEd	Genome-edited
GFLV	Grapevine fanleaf virus
GFP	Green fluorescent protein
GLRaV	Grapevine leafroll-associated virus
GLRaV-3	Grapevine leafroll-associated virus 3
GM	Genetically modified
GMO	Genetically modified organism
GO-AgNPs	Graphene oxide–silver nanoparticles
GPGV	Grapevine Pinot gris virus
GRSPaV	Grapevine rupestris stem pitting-associated virus
GUS	β-Glucuronidase
GWAS	Genome-wide association study
H3K27me3	Histone H3 lysine 27 trimethylation
H3K4me3	Histone H3 lysine 4 trimethylation
HASTY	Exportin-5 homolog
HC-Pro	Helper component-proteinase
HDR	Homology-directed repair
HIGS	Host-induced gene silencing
HLVd	Hop latent viroid
HR	Hypersensitive response
HTS	High-throughput sequencing
ICTV	International Committee on Taxonomy of Viruses
INSV	Impatiens necrotic spot virus
IoT	Internet of Things
IPM	Integrated pest management
IPPC	International Plant Protection Convention
ISPM	International Standards for Phytosanitary Measures
ISRs	Induced systemic resistance
ISVs	Insect-specific viruses
JA	Jasmonic acid
LDH	Layered double hydroxide
LLM	Large language model
lncRNA	Long non-coding RNA
LUC	Luciferase
m6A	N6-methyladenosine
MAS	Marker-assisted selection
MBCA	Microbiological biocontrol agent
miRNA	MicroRNA
ML	Machine learning
MP	Movement protein
MS	Murashige and Skoog medium
MSCP	Mild strain cross-protection
MYC2	MYC transcription factor 2
NCBI	National Center for Biotechnology Information
ncRNA	Non-coding RNA

NGS	Next-generation sequencing
NHEJ	Non-homologous end joining
NiONPs	Nickel oxide nanoparticles
NPPO	National Plant Protection Organization
NPs	Nanoparticles
OFV	Orchid fleck virus
ONT	Oxford nanopore technology
ORF	Open reading frame
PAM	Protospacer adjacent motif
PAMPs	Pathogen-associated molecular patterns
PCR	Polymerase chain reaction
PDR	Pathogen-derived resistance
PDV	Prune dwarf virus
PepMV	Pepino mosaic virus
PIF	Phytochrome-interacting factor
PMMoV	Pepper mild mottle virus
PMWaV	Pineapple mealybug wilt-associated virus
PNRSV	Prunus necrotic ringspot virus
Pol IV	RNA polymerase IV
Pol V	RNA polymerase V
PPV	Plum pox virus
PR proteins	Pathogenesis-related proteins
PRC2	Polycomb repressive complex 2
PRSV	Papaya ringspot virus
PRSV-P	Papaya ringspot virus (papaya strain)
PSTVd	Potato spindle tuber viroid
PTGS	Post-transcriptional gene silencing
PVX	Potato virus X
PVY	Potato virus Y
qPCR	Quantitative polymerase chain reaction
QTL	Quantitative trait locus
RBDV	Raspberry bushy dwarf virus
RdDM	RNA-directed DNA methylation
RDR	RNA-dependent RNA polymerase
RdRp/RDRp	RNA-dependent RNA polymerase
RH	Relative humidity
RISC	RNA-induced silencing complex
RNAi	RNA interference
RNA-seq	RNA sequencing
RPPO	Regional Plant Protection Organization
RT-PCR	Reverse transcription PCR
RT-qPCR	Reverse transcription quantitative PCR
SA	Salicylic acid
SAR	Systemic acquired resistance
SCFCOI1	Skp1–Cullin–F-box complex COI1

SDN	Site-directed nuclease
sgRNA	Single-guide RNA
SGS3	Suppressor of gene silencing 3
siRNA	Small interfering RNA
SMRT	Single-molecule real-time sequencing
sRNA	Small RNA
ssDNA	Single-stranded DNA
ssRNA	Single-stranded RNA
TE	Transposable element
TFL1	Terminal flower 1
TGS	Transcriptional gene silencing
TiO_2NPs	Titanium dioxide nanoparticles
TMV	Tobacco mosaic virus
ToBRFV	Tomato brown rugose fruit virus
ToLCNDV	Tomato leaf curl New Delhi virus
ToLCV	Tomato leaf curl virus
TRV	Tobacco rattle virus
TSWV	Tomato spotted wilt virus
TYLCSV	Tomato yellow leaf curl Sardinia virus
TYLCV	Tomato yellow leaf curl virus
USD	United States dollar
USDA	United States Department of Agriculture
VAMP	Virus-associated molecular pattern
VIGS	Virus-induced gene silencing
VPg	Viral protein genome-linked
VSR	Viral suppressor of RNA silencing
WHO	World Health Organization
WSMV	Watermelon silver mottle virus
ZFNs	Zinc finger nucleases
ZnONPs	Zinc oxide nanoparticles
ZYMV	Zucchini yellow mosaic virus

Economic and Quality Deterioration in Fruit Crops Due to Viral Diseases

1

Reynaldo P. Peja Jr. and Mark Angelo O. Balendres

Abstract

Although global fruit production grows yearly, it has historically remained insufficient to meet recommended consumption levels. Moreover, viral diseases continue to pose a major barrier to achieving a sufficient fruit supply worldwide. Different fruit crops face distinct viral threats (tobacco mosaic virus, banana bunchy top virus, tomato leaf curl virus, papaya ringspot virus, cucumber mosaic virus, etc.). However, common symptoms include smaller, misshapen, or discolored fruit, altered sugar-acid balance, and reduced shelf life. These undermine productivity and quality, and when outbreaks occur, they inflict multimillion-dollar economic losses that affect everyone, from large-scale farms to smallholder farmers. Even moderate infection levels can reduce yields by 15–40%. Countering these impacts requires early detection and containment, especially since the infection is permanent once a plant harbors the virus. This chapter reviews major fruit-infecting viruses and quantifies their regional economic losses, covering both direct and indirect costs. It also examines how these viral diseases affect key fruit quality traits, such as size, sugar content, acidity, and aroma, thereby undermining marketability.

Keywords

Fruit viruses · Fruit quality · Marketability

R. P. Peja Jr.
Department of Biological Sciences, Faculty of Natural and Mathematical Sciences, Visayas State University, Visca, Baybay, Leyte, Philippines

M. A. O. Balendres (✉)
Department of Biology, College of Science, De La Salle University, Manila, Philippines
e-mail: mark.angelo.balendres@dlsu.edu.ph

R. Kumar et al. (eds.), *Fruit Crops and Viral Pathogens Volume II*, Plant Pathology for Sustainable Agriculture, https://doi.org/10.1007/978-981-95-7261-8_1

1.1 Introduction

Since humans adopted the hunter-gatherer lifestyle during the Paleolithic era, fruits have been essential to human nutrition. They supply the body with vitamins, minerals, fiber, and health-promoting phytochemicals (del Río-Celestino and Font 2020). Fruit-rich diets are associated with lower risks of chronic diseases (Boeing et al. 2012; Arias et al. 2022). The World Health Organization (WHO) recommends consuming at least 400 g of fruits and vegetables per person per day (WHO 2020). World fruit production grew 63% from 2000 to 2022, with a total volume of 933 million tons (FAO 2023). However, availability still lags dietary targets (Jha et al. 2025). Approximately 76% of tropical fruits are primarily grown in Asia and the Americas (Vieira et al. 2024; FAO 2023). These regional imbalances and the general production-supply gap mean that any disruption, like drought, pests, or diseases, can quickly create local shortages and drive up prices, impacting both large-scale operations and smallholder farmers. Plant viruses are among the destructive pathogens of fruit crops. Most of these are single-stranded RNA viruses with positive-strand polarity (+ssRNA). DNA viruses are less common (Roossinck 2003; Umer et al. 2019). These viruses often cause mosaicism, chlorosis, leaf and fruit malformations, sterility, flower or fruit drop, and increased susceptibility to other stresses, cutting yield and diminishing marketable value and quality of downstream products (Amer and Mahmoud 2020; Rodríguez-Verástegui et al. 2022; Jiménez et al. 2024). Infected fruit can be smaller, twisted, or discolored, with altered sugar-acid balance and reduced storage life, as in grapes (Alabi et al. 2016; Rupnik-Cigoj et al. 2018; Rodríguez-Verástegui et al. 2022). Even moderate infections can slash yields by tens of percent, and severe outbreaks can devastate crops. Damage can also accrue from multiple years of infection. In such cases, annual losses in crop production could amount to nearly EUR 40 billion, highlighting the severe threat that viral pathogens pose to agriculture (Lázár and Bisztray 2011). On another important note, indirect costs also accrue, significantly increasing the total impact of viral diseases in plants beyond the primary production losses. These economic burdens include, but are not limited to, costs for disease management, diagnostic testing, certification programs, trade restrictions, and impacts on food security. This chapter provides a comprehensive review of major viral fruit pathogens and their impacts on production and fruit quality. Quantified losses attributable to these pathogens at the regional scale are presented. Alterations of fruit quality, including fruit size, sugar content, acidity, aroma, and shelf life from viral infections, are also examined.

1.2 Viral Pathogens of Major Fruit Crops

As a diverse group of phytopathogens, fruit-infecting viruses cause plant health impairment, yield reduction, and deterioration in fruit quality. Adding a layer of complexity to these is their ability to infect plants through various pathways, including insect vectors, mechanical injury, and propagation practices. Hence, a solid understanding of virus biology and epidemiology is crucial for the efficient and cost-effective control of viruses. This section provides a discussion of the major viral taxa, their hosts, mode of transmission, and global distributions, which inform integrated disease management strategies.

1.3 Taxonomy and Biology of Fruit-Infecting Viruses

There are more than 30 virus families that infect fruit crops, each exhibiting distinct structural features, genome types, and replication strategies. These genomes may be single-stranded DNA (ssDNA), double-stranded DNA (dsDNA), single-stranded RNA (ssRNA), or double-stranded RNA (dsRNA) (Walker et al. 2022; Tatineni and Hein 2023). Depending on the species, the genome may be monopartite, containing all coding sequences within a single molecule, or segmented into two or more molecules, requiring co-transmission for successful infection (Chaitanya 2019; Michalakis and Blanc 2020). Segmentation, while adding complexity to transmission, also allows for the reassignment of genome segments, potentially altering virulence or expanding the host range (Varsani et al. 2018). Although the diversity of fruit-infecting viruses continues to expand, more than 70% are positive-sense RNA viruses (+ssRNA) (Umer et al. 2019). Among these, members of the *Closteroviridae* and *Potyviridae* are the most significant in terms of both agricultural impact and geographic spread (Fuchs et al. 2020; Rodríguez-Verástegui et al. 2022). The *Closteroviridae* are typically long, flexuous +ssRNA viruses with monopartite, bipartite, or tripartite genomes. A hallmark of many species in this family is their restriction to phloem tissues, where they replicate and move exclusively through the plant's vascular system (Qiao et al. 2018). This tissue specificity underlies many chronic closterovirus diseases, which are often characterized by slow development. Citrus tristeza virus (CTV), with its 19.3 kb monopartite genome, is an example, causing chlorosis, vein corking, stem corking, and tree decline, resulting in severe yield and tree losses in orchards (Folimonova 2020). In contrast, *Potyviridae*, which have long, filamentous monopartite +ssRNA, are not confined to the phloem. Instead, they infect mesophyll and epidermal tissues, allowing them to move systematically through plasmodesmata and vascular connections, often leading to faster symptom development (Revers and García 2015). Potyviruses are associated with a diverse range of symptoms, including mosaics, ringspot, and leaf or fruit distortions, as seen with Papaya ringspot virus (PRSV), Turnip mosaic virus (TuMV), and Plum pox virus (PPV) (Revers and García 2015; Gibbs et al. 2020). Other notable virus families that substantially contribute to fruit crop burdens include members of *Bromoviridae* and *Tospoviridae*, both of which have segmented genomes and cause complex symptom syndromes. *Bromoviridae* include the well-known, broad-host-range Cucumber mosaic virus (CMV), which causes chlorotic mottling, stunting, and fruit malformation in tomato, pepper, and cucumber (Seo et al. 2014). Tospoviruses, represented by Tomato spotted wilt virus (TSWV), have an enveloped, three-partite genomic structure with both positive and negative polarity. As illustrated, fruit-infecting viruses exhibit considerable variation in their taxonomy and biological diversity, including genome structure, tissue tropism, transmission strategies, and symptom expression.

1.3.1 Transmission Routes

Fruit-infecting plant viruses spread through multiple routes, the most important being insect and nematode vectors, as well as seed, pollen, and vegetative propagation, such as grafting. Generally, phloem-limited viruses are transmitted by phloem-feeding insects, while viruses infecting non-phloem tissues—such as those targeting epidermal or mesophyll cells—often rely on nonpersistent sap transmission, seed transmission, or mechanical means like grafting. Aphids are among the most widespread and impactful vectors, capable of transmitting both persistent and nonpersistent viruses. Persistent aphid transmission is characteristic of viruses in the phloem, such as certain closteroviruses (Harris 2018). For example, CTV is transmitted semi-persistently by *Aphis gossypii* and *Toxoptera citricida* (Wubshet and Amare 2019). Prolonged feeding of *Myzus persicae* similarly transmits the *luteovirus*, potato leafroll virus, which causes chlorosis, necrosis, and leaf curling in potatoes and tomatoes (Shoala 2025). On the one hand, nonpersistent aphid-transmitted viruses are acquired rapidly on the stylet and do not circulate within the vector, but rather remain stylet-borne, allowing for within-second inoculation (e.g., Plum pox virus (PPV) and Zucchini yellow mosaic virus (ZYMV)) (Moreno et al. 2009; Vučurović et al. 2012). The fact that numerous vectors can carry the same viruses heightens the complexity of their interactions and poses challenges for effective control. Many DNA and RNA viruses are spread to fruit crops by phloem-feeding whiteflies, especially *Bemisia tabaci*. This species transmits Begomoviruses such as tomato yellow leaf curl virus (TYLCV), which infects tomato, melon, pepper, and eggplant (Ning et al. 2015). Whitefly-aided transmission is also essential in spreading Criniviruses such as Cucurbit yellow stunting disorder virus (CYSDV) and Cucurbit chlorotic yellow virus (CCYV) (Orfanidou et al. 2021). Similar to aphids, whiteflies persistently transmit viruses, consistent with the pathogens' phloem-restricted nature (Jiménez et al. 2020). The multipartite genomes of many whitefly-spread viruses also require co-transmission, which is facilitated by encapsidation in the insect's foregut or hemolymph (Kennedy et al. 2023). Tospoviruses are transmitted by thrips (Thysanoptera: Thripidae) in a propagative, persistent manner. For example, Tomato spotted wilt virus (TSWV) must first be acquired by thrips during their larval stage before it infects tomato, pepper, and various fruit and vegetable hosts (Gupta et al. 2018). Phloem-restricted viruses are also efficiently transmitted by other phloem feeders, including sap-sucking scales and mealybugs (Hemiptera: *Pseudococcidae*, *Coccidae*). Examples of such viruses are the Grapevine leafroll-associated virus, vectored by *Planococcus ficus* and *Pseudococcus maritimus* (Krüger et al. 2015; Cabaleiro et al. 2022), and the Little cherry virus 2 (LChV-2), vectored by mealybugs that spread it through grafting (Shires et al. 2022). Several bipartite RNA nepoviruses in the family *Secoviridae* are transmitted by soil-borne nematodes, such as dagger nematodes (*Xiphinema* spp.). For instance, the successful infection by Grapevine fanleaf virus (GFLV) is mediated by *X. index* (Garcia et al. 2019). Meanwhile, other viral fruit pathogens, including tomato ringspot virus and tomato black ring virus, are transmitted by either *Xiphonema* or *Longidorus* species (Bileva et al. 2009). Regardless of vector species, the pathogenic route these viruses take begins with root and phloem invasion, followed by systemic foliar symptoms. Fruit viruses can also be seed-borne, carried internally or externally, ready to infect the emerging seedlings. A prominent example is the Cucumber green mottle mosaic virus (CGMV) infecting cucurbits. Some potyviruses are also seedborne but are transmitted at low rates, such as cucumoviruses (e.g., CMV). Pepper seed transmission of TYLCV has also been reported (Kil et al. 2018). Pollen transmission is another route of virus spread. Prunus necrotic ringspot virus (PNRSV) and Prune dwarf virus (PDV), both Ilarviruses, follow this course in infecting stonefruits, including peach and cherry (Tayal et al. 2024). Virus-carrying pollen grains can fertilize healthy plants, introducing the virus to the ovule that leads to latent infections in the maternal parent. Finally, clonal propagation and grafting facilitate the transmission of viruses in woody fruit crops. CTV is frequently spread via infected budwood or rootstock and aphid vectors (Guarino et al. 2022). Apple viruses such as Apple chlorotic leaf spot virus and Apple stem grooving virus have no known insect vectors and persist primarily through grafting (Eastwell 2017). Once introduced into a nursery or orchard, graft-transmitted viruses are difficult to eliminate without destroying the infected plant.

1.3.2 Host Range and Global Distribution

Fruit-infecting viruses exhibit a broad host range, often spanning multiple genera and botanical families. This increases their potential for transboundary spread and long-term persistence in orchards and plantations. Blueberry scorch virus (BlScV), primarily restricted to North America but also detected in Europe, devastates blueberry production through blossom blight and "scorch" symptoms that prevent fruit set, resulting in season-long losses (Wegener et al. 2006; Raworth et al. 2008; Saad et al. 2021). Viruses with broader host ranges, such as CMV, threaten diverse fruit crops, including cucurbits, solanaceous fruits, and banana, causing leaf mosaics, fruit malformation, and stunting that can cut marketable yield by more than half in susceptible crops (Dheepa and Paranjothi 2010; Jacquemond 2012; Mahjabeen et al. 2012). Apple mosaic virus (ApMV) persists in apple-, pear-, and stone-fruit-growing regions of Europe, North America, and Asia, where chlorotic leaf patches and necrosis reduce photosynthetic capacity and fruit set. However, this incidence has dropped in areas where clean stock programs are implemented (Grimova et al. 2016). High-value tropical crops are also threatened: BBTV (Banana bunchy top virus) remains endemic in Southeast Asia and parts of Africa and the Pacific, causing severe stunting and "choked" growth that prevents fruiting (Qazi 2016), while Banana bract mosaic virus (BBrMV), prevalent in Asia and the Pacific, distorts banana bunches and reduces fruit size and number (Ouattara et al. 2023). Among the most globally damaging viruses, TYLCV impacts solanaceous plants. Cucurbit fruit crops in tropical and subtropical regions, inducing stunting, yellowing, and fruit-setting failure that can result in near-total yield loss in unmanaged outbreaks (Guo et al. 2019). Strawberries (Fragaria × ananassa) are host to a complex of viruses that often occur in mixed infections, including the aphid-transmitted Strawberry mottle virus (SMoV), Strawberry crinkle virus (SCV), Strawberry vein banding virus (SVBV), and Strawberry mild yellow edge virus (SMYEV). These viruses are reported in all major strawberry-growing regions worldwide (Thompson and Jelkmann 2003; Sharma et al. 2018; Fan et al. 2021). Typical individual symptoms of these viruses are only subtle or latent. However, their synergistic interactions produce severe decline syndrome characterized by stunting, leaf discoloration, and dramatic yield reduction (Babini et al. 2003; Martin and Tzanetakis 2013; Luciani et al. 2018). Their frequent co-occurring patterns in strawberry fields emphasize their collective impact and the need for management strategies of such viral complexes to prevent decline (Salazar et al. 2017; Ma et al. 2015). Often, the regional distribution of viruses is reflected in the ecology of vectors, propagation practices, and phytosanitary measures. For instance, Grapevine fanleaf virus (GFLV), which causes reduction of fruit set and sugar content of grapes through vine malformation and reproductive failure, is reported in all major viticulture regions (Digiaro et al. 2017; Martelli 2017). Overall, the combination of broad host ranges, long-distance dissemination via infected planting materials, and significant yield and quality losses underpins the need for integrated management and strict phytosanitary certification programs in global fruit production.

1.4 Economic Impact of Viral Diseases

Viral pathogens have a profound economic impact on fruit crop industries worldwide. Plant viruses collectively cause an estimated $30 billion in crop losses annually, with fruit crops bearing a significant share of that burden (Sastry and Zitter 2014). Unlike many fungal or bacterial diseases, viral infections in fruit trees and vines often cannot be cured, resulting in chronic yield losses, reduced fruit quality, and the need for roguing infected plants (Rodríguez-Verástegui et al. 2022; Diaz-Lara et al. 2023). The economic consequences range from immediate revenue losses due to decreased marketable yields to long-term costs of replanting orchards and maintaining virus-free certification programs (Davis et al. 2021; Fuchs et al. 2021). Viral diseases in fruits threaten farmer livelihoods, local food security, and international trade, necessitating substantial ongoing investments in management and prevention.

1.4.1 Yield Losses and Production Declines

Viruses in fruit crops often cause dramatic yield losses and quality declines. This results in the reduction of both the quantity of fruit harvested and its market value. In many cases, infections result in malformed or unripe fruits that are unmarketable, or premature drop that greatly lowers yields. For example, Plum pox virus (PPV), causing sharka disease in stone fruits, can devastate apricot, plum, and peach production (García et al. 2025). Severe PPV strains have been reported to cause yield losses of 83–100% in susceptible *Prunus* trees, resulting in crop destruction in infected orchards (Cambra et al. 2024). Over the decades, PPV has spread across Europe, Asia, and the Americas, resulting in cumulative global losses exceeding EUR 13 billion since its emergence in the early 1900s (Cambra et al. 2024). In tropical fruit crops, yield impacts can be equally extreme. Banana bunchy top virus (BBTV), transmitted by aphids, causes stunting of banana plants and a "bunching" of leaves that gives the disease its name (Ocimati et al. 2024). Infected banana mats rarely produce marketable bunches. Studies in Africa and Asia document yield losses of 70–100% in the first crop cycle, with subsequent cycles yielding nothing (Tripathi et al. 2016; Jekayinoluwa et al. 2020). BBTV epidemics have historically shrunk banana production areas dramatically. In India, recurring BBTV outbreaks in the late 2000s resulted in annual production losses of approximately $50 million, and replanting with clean suckers can take years to restore productivity (Selvarajan and Balasubramanian 2014; Qazi et al. 2016; Vikrant et al. 2025).

Viruses also severely impact fruit quality, compounding economic losses. Grapevine leafroll-associated virus in wine grapes causes uneven ripening (low sugar, poor color), so that even when yields are not entirely lost, the fruit may fail to meet wine-making standards (Salo et al. 2024; Fust et al. 2025). Leafroll infection typically reduces grape yields by 30–68% and lowers quality, forcing price penalties for growers (Atallah et al. 2012; Ricketts et al. 2015). Severe field outbreaks of tomato spotted wilt virus (TSWV), transmitted by thrips, have resulted in a total loss of commercial value in tomatoes (Sevik and Arli-Sokmen 2012). In papaya, the notorious PRSV can wipe out orchards, causing infections that result in mosaic rings and malformed fruit, often rendering 60–80% of the crop unmarketable, as observed in outbreaks in the Philippines and Brazil (Abreu et al. 2015). Before the development of resistant papaya in Hawaii, PRSV nearly collapsed the papaya industry, with production dropping by over 50% during the 1990s. This economic disaster was later averted by transgenic resistance (Tripathi et al. 2007; Gonsalves 2013).

1.4.2 Direct vs. Indirect Costs

The economic impact of fruit viruses extends beyond yield loss alone. Direct costs include managing diseases and replacing infected trees. In contrast, indirect costs encompass market disruptions, trade restrictions, and long-term research and breeding efforts. A recent comprehensive analysis of sharka of stone fruits caused by PPV over 28 years illustrates this well: of an estimated EUR 2.4 billion total cost (1995–2023), over EUR 2 billion were direct losses from reduced fruit yields, and more than EUR 100 million were spent on eradication and containment programs (Cambra et al. 2024). These direct control measures, from labor for roguing diseased trees to pesticides for vector control, are a significant economic burden on growers. For instance, citrus growers in Brazil spend USD 70–80 million each year on miticides to control the mites that spread Citrus leprosis virus, a necessary expense to prevent potentially 20–100% yield losses in unprotected orchards (Bastianel et al. 2010; de Oliveira et al. 2025). Studies in California show that implementing a roguing and replanting strategy to maintain a vineyard with <5%, Grapevine leafroll disease (GLD) infection still cuts net present value by tens of thousands of dollars per hectare compared to a virus-free scenario (Ricketts et al. 2015; Golino et al. 2017). The GLD significantly diminishes yield by as much as 30–68% (Vega et al. 2011; Jiao et al. 2022). In vineyards with no control measures, annual economic costs can reach up to USD 25,000 per hectare (Atallah et al. 2012). In Taiwan and Thailand, outbreaks of PRSV during the 2000s resulted in substantial declines in papaya cultivation. In Thailand, papaya production and area declined by over 50% between 1997 and 2006, prompting a shift from being a net exporter of papaya to an importer (Sakuanrungsirikul et al. 2005; Napasintuwong and Traxler 2009). In Taiwan, implementing netting as a disease control measure helped delay the geographic displacement of crops. However, many growers still exited papaya farming due to persistently low and unprofitable yields, as well as escalating management costs (Gonsalves et al. 2010).

Certification programs and virus detection are another direct cost of the industry and government. Many fruit industries have established clean planting material programs (e.g., shoot tip culture, virus indexing of nursery stock) to prevent virus spread (Sastry and Zitter 2014). While cost-effective in the long run, these programs require continuous investment. For instance, when PPV was detected in Pennsylvania and Canada, authorities invested over USD 57 million in survey and eradication efforts, including tree removal and testing, to protect stone fruit industries (Cambra et al. 2024). On an ongoing basis, millions are spent on diagnostic testing. In the European Union and Turkey, sharka management included EUR 137 million on ELISA and molecular tests over 10 years to monitor orchards for PPV, an indispensable cost to identify infected trees early (Cambra et al. 2024). Certification of fruit tree nurseries, where mother stock is subjected to regular viral testing, introduces substantial production overhead costs. Such price increases can indirectly constrain disease prevention programs by reducing access to high-quality planting material. Smallholder farmers, positioned at the lowest tier of the value chain, bear the brunt of these elevated input costs. The additional financial pressure erodes their margins, constraining farm-level investment capacity and overall revenue potential. Indirect costs often manifest as market and trade impacts. If a region is known to harbor a particular fruit virus, export markets may close or require expensive safeguards. For example, during the European PPV outbreaks, some countries rejected imports of stone fruits or nursery stock from affected areas, resulting in direct trade losses. One estimate tallied EUR 1.4 million in rejected fruit shipments internationally due to visible symptoms of sharka (Cambra et al. 2024). Although that particular figure is small relative to direct losses, the threat of losing market access exerts a constant economic pressure. To sell fruit or plants, growers may need to meet quarantine or certification requirements, incurring compliance costs (Gergerich et al. 2015). In addition, virus-induced quality reductions can depress market prices. Grape growers with leafroll-infected vines receive less for lower-quality grapes, effectively adding a "virus tax" on their product (Rodríguez-Verástegui et al. 2022). There are also broader ripple effects, such as lost jobs and income in packing, processing, and export sectors when fruit supply declines due to diseases (Gergerich et al. 2015). Furthermore, heirloom varieties suffer from genetic erosion due to the abandonment of farmers, resulting in long-term economic and cultural costs that are difficult to quantify (Aoun 2024). Finally, the need for ongoing research, such as breeding resistant cultivars and developing new diagnostics, is an economic cost. Between 1995 and 2023, the European Union and U.S. agencies, including the U.S. Department of Agriculture, invested approximately EUR 27.7 million and EUR 90 million in PPV research (Cambra et al. 2024). In 2023–2024, a private philanthropic organization also committed approximately $500,000 to support BBTV-resistant screening by screening wild *Musa* species and developing diagnostic tools (Bill and Melinda Gates Foundation 2023). These public and private funds devoted to addressing fruit virus problems illustrate the high stakes of protecting fruit crops.

1.5 Quality Deterioration in Infected Fruit

Viral infections in fruit crops can profoundly degrade both the physical and chemical quality traits of the produce. Infected fruits often exhibit reduced size, abnormal coloration, diminished sugar or acid content, and changes in texture and aroma. These changes occur because plant viruses disrupt fundamental physiological processes, including photosynthesis and carbohydrate metabolism, as well as hormone signaling and gene expression, which are essential for normal fruit development and ripening. The net effect is a decline in organoleptic qualities, such as the fruit's sweetness, flavor, aroma, and texture, as well as marketability, including the fruit's appearance, shelf life, and grading.

1.5.1 Disruption of Carbohydrate Metabolism and Loss of Sweetness

One common impact of viruses is the reduction of sugars in the fruit, which directly lowers sweetness and flavor. Viruses that impair phloem transport or photosynthesis limit the flow of photoassimilates into developing fruits. Grapevine leafroll-associated virus 3 (GLRaV-3) is a striking example. GLRaV-3-infected grapevines show delayed and uneven ripening, yielding berries with significantly lower soluble solids or sugar content, measured in °Brix, and reduced anthocyanin pigmentation (Vega et al. 2011; Wright 2012). Mechanistically, transcriptomic analyses indicate that GLRaV-3 disrupts sugar transport in the source leaves; infected leaves accumulate excess sugars due to repressed export (Song et al. 2022). This means that less sugar is delivered to the grape berries, typically 1–2 °Brix lower than in healthy plants, resulting in a more acidic and unbalanced must (Montero et al. 2016).

Organoleptic consequences are also evident. Grapes from GLRaV-3 vines taste less sweet, and red wine produced from such fruit has a paler color and flatter flavor profile, often requiring amelioration or being relegated to lower-quality blends (Song et al. 2021). Viral interference with carbohydrate allocation is similarly observed in virus-infected cucurbits. In CMV-infected melons, source leaves suffer from a pronounced decline in net photosynthetic rates and unusually high levels of reducing sugars with notably low starch concentrations. These metabolic shifts are synchronized with elevated respiration and reduced photosynthetic output (Shalitin and Wolf 2000). This disruption of source–sink dynamics means that the developing fruits receive an inadequate sugar supply. For instance, Cucurbit yellow stunting disorder virus (CYSDV) infection has been shown to reduce the flesh Brix value of melons so severely that even otherwise marketable melons are rendered unmarketable (Sinclair and Crosby 2002). Farmers find that CMV or potyvirus-infected squash, cucumber, or melon fruits are undersized and bland. Moreover, these viruses often cause striking cosmetic defects, such as green mosaics, yellow spots, or color breaking on the rind, as well as misshapen or knobbly fruit form (Dragich et al. 2014; Akbar et al. 2015). Pepper consumers are likely to reject deformed, off-colored produce, which cannot be sold at market value (Lepage et al. 2025). Powders from infected peppers are also often rejected due to inferior aromatic quality (Lepage et al. 2025). Viruses hijack the fruit's carbohydrate metabolism, leading to poor flavor development and severe economic losses due to quality downgrades.

1.5.2 Hormonal Imbalance, Stunting, and Reduced Fruit Size

Many economically important fruit viruses cause stunting and poor development by disrupting host phytohormone balances and growth processes. Infected plants often exhibit elevated levels of stress hormones, such as abscisic acid, salicylic acid, or ethylene. In contrast, levels of growth-promoting hormones, such as auxins, gibberellins, and cytokinins, are reduced, leading to reduced cell expansion, shorter internodes, and inhibited fruit set (Mishra et al. 2020). For example, the Banana bunchy top virus systemically dwarfs banana plants by disrupting meristem function and suppressing auxin, gibberellin, and cytokinin activity, while increasing abscisic acid levels that further inhibit growth and ripening (Mishra et al. 2020). Early infections typically prevent any bunch formation, while later infections produce only small and distorted bunches of undersized bananas that remain green and hard, consistent with disrupted ethylene signaling. In citrus, transcriptome and hormone profiling of lemon infected by Citrus yellow vein-clearing virus showed extensive transcriptional suppression together with reprogrammed hormone networks that match vein chlorosis, growth restraint, and compromised fruit development (Hu et al. 2024). In stone fruit, PPV perturbed growth and stress hormones in peach with increases in GA_3 and abscisic acid associated with chlorosis. In contrast, graft-induced resistance, achieved by inserting the almond cultivar Garrigues onto susceptible GF305 peach, lowered GA_3 and abscisic acid levels while engaging salicylic acid-linked defenses, accompanied by shifts in trans-zeatin and jasmonate levels. Symptoms abated with recovery of growth (Dehkordi et al. 2018). Collectively, these examples show that virus infection shifts hormonal network toward defense at the expense of growth (Jiang and Zhou 2023), suppressing cell expansion, sink filling, and uniform ripening.

1.5.3 Derailed Fruit Ripening and Quality

Viruses can severely disrupt fruit ripening by altering hormone signaling and metabolic pathways. Ripening contributes to fruit aroma and typically involves coordinated increases in sugars, decreases in acids, and the synthesis of pigments such as carotenoids or anthocyanins, and the formation of esters, aldehydes, and lactones (Kumar et al. 2013; Forlani et al. 2019). Infection can directly affect fruit tissues or indirectly shift hormone balances, leading to uneven or stalled maturation. For example, PPV in stone fruits produces pale chlorotic rings and internal browning where pigment and cell integrity are lost. Infected plums exhibit reduced °Brix and fewer aroma volatiles, resulting in bland, mealy fruit, with premature drop rates of up to 80–100% in susceptible cultivars (García et al. 2025). In grapes, Grapevine leafroll-associated viruses and Grapevine red blotch virus delay ripening and lower sugars by 2–5 °Brix, keep acidity high, and reduce anthocyanins, producing berries with poor flavor and color that winemakers often reject (Martínez-Lüscher et al. 2019; Song et al. 2021; Cieniewicz and Fuchs 2025). Tropical crops show similar patterns. PRSV causes concentric skin rings, reduced size, and low sugar content in papayas, often resulting in bitter or watery flesh (Umer et al. 2022). Infected fruits have impaired pigment formation and altered volatile profiles, leaving them bland and unmarketable (Singh and Shukla 2009). Quantitatively, PRSV can lower °Brix by several points and significantly cut marketable yield, with severe early infections making nearly all fruit unsalable (Thirugnanavel et al. 2015). Although there is a lack of direct quantitative data, viral infection likely leads to lower vitamin levels in fruit, particularly vitamin C, because viruses disrupt normal ripening physiology and increase oxidative degradation. For instance, studies on healthy papaya have shown that ripening alone can reduce vitamin C levels from around 98 mg/kg to 70 mg/kg (Koul et al. 2022; Aroh et al. 2023). Viral infections, such as those caused by PPV and PRSV, compromise pigment development, sugar metabolism, and sink function, all of which are metabolically linked to vitamin C synthesis and retention. Correlative evidence from studies on citrus fruits shows that vitamin C is highly sensitive to oxidative stress and metabolic disturbance during ripening (Miles and Calder 2021). Nonetheless, these reports prove that viral stresses likely exacerbate the natural decline in vitamin C during fruit maturation.

1.6 Summary and Conclusion

Global fruit production is pervasively threatened by virus pathogens. They cause irreversible infections that severely reduce yields and degrade marketable quality, resulting in substantial economic losses. Several case studies highlight the severity of these impacts: Plum pox virus can completely decimate susceptible stone fruit orchards, and Banana bunchy top virus has caused reported losses of 70–100% in affected regions. Penalties for quality deterioration are equally severe. Berries borne from grapevines infected with leafroll or red blotch viruses exhibit delayed ripening and lowered sugar levels by several degrees. Meanwhile, Papaya ringspot virus significantly reduces the sweetness of papaya fruit. Losses from direct costs include clearing of infected trees, applying control vector measures, and implementing certification programs, while indirect costs encompass trade restrictions, price reductions, and long-term impacts on farmer livelihoods and genetic diversity. These combined effects put forward the substantial and multifaceted economic threat posed by viral diseases in fruit crops.

References

Abreu PM, Antunes TF, Magaña-Álvarez A, Pérez-Brito D, Tapia-Tussell R, Ventura JA, Fernandes AA, Fernandes PM (2015) A current overview of the Papaya meleira virus, an unusual plant virus. Viruses 7(4):1853–1870

Akbar A, Ahmad Z, Begum F, Raees N (2015) Varietal reaction of cucumber against cucumber mosaic virus. Am J Plant Sci 6(07):833

Alabi OJ, Casassa LF, Gutha LR, Larsen RC, Henick-Kling T, Harbertson JF, Naidu RA (2016) Impacts of grapevine leafroll disease on fruit yield and grape and wine chemistry in a wine grape (*Vitis vinifera* L.) cultivar. PLoS One 11(2):e0149666

Amer MA, Mahmoud SY (2020) First report of tomato brown rugose fruit virus on tomato in Egypt. New Dis Rep 41(24):2044–0588

Aoun M (2024) Unlocking heirloom diversity: a pathway to bridging global challenges in modern apple cultivation. Front Hort 2:1268970

Arias A, Feijoo G, Moreira MT (2022) Exploring the potential of antioxidants from fruits and vegetables and strategies for their recovery. Innov Food Sci Emerg Technol 77:102974

Aroh AO, Omoniyi KI, Mohammed FR (2023) Determination of Vitamin C and Mineral content of two local species of Carica papaya (samba and hortus gold papaya) at various stages of ripening. Fudma J Sci 7(1):280–283

Atallah SS, Gómez MI, Fuchs MF, Martinson TE (2012) Economic impact of grapevine leafroll disease on *Vitis vinifera* cv. *Cabernet franc* in Finger Lakes vineyards of New York. Am J Enol Vitic 63(1):73–79. https://doi.org/10.5344/ajev.2011.11055

Babini AR, Cieślińska M, Karešová R, Thompson JR, Cardoni M (2003) Occurrence and identification of strawberry viruses in five European countries. In: X International symposium on small fruit virus diseases 656:39–43

Bastianel M, Novelli VM, Kitajima EW, Kubo KS, Bassanezi RB, Machado MA, Freitas-Astúa J (2010) Citrus leprosis: centennial of an unusual mite–virus pathosystem. Plant Dis 94(3):284–292

Bileva T, Choleva B, Hockland S, Ciancio A (2009) Management of virustransmitting nematodes with special emphasis on South-East Europe. In: Integrated management of fruit crops and forest nematodes. Springer Science & Business Media, pp 215–242

Bill & Melinda Gates Foundation (2023) Committed grant: INV-063305. https://www.gatesfoundation.org/about/committed-grants/2023/11/inv-063305

Boeing H, Bechthold A, Bub A, Ellinger S, Haller D, Kroke A, Leschik-Bonnet E, Müller MJ, Oberritter H, Schulze M, Watzl B (2012) Critical review: vegetables and fruit in the prevention of chronic diseases. Eur J Nutr 51(6):637–663

Cabaleiro C, Pesqueira AM, Segura A (2022) *Planococcus ficus* and the spread of grapevine leafroll disease in vineyards: a 30-year-long case study in north-West Spain. Eur J Plant Pathol 163(3):733–747

Cambra M, Madariaga M, Varveri C, Çağlayan K, Morca AF, Chirkov S, Glasa M (2024) Estimated costs of plum pox virus and management of sharka, the disease it causes. Phytopathol Mediterr 63(3):343–365

Chaitanya KV (2019) Structure and organization of virus genomes. In: Genome and genomics: from archaea to eukaryotes. Springer Singapore, Singapore, pp 1–30

Cieniewicz E, Fuchs M (2025) Grapevine red blotch disease: a threat to the grape and wine industries. Annu Rev Virol 12:335

Davis TJ, Gómez MI, Harper SJ, Twomey M (2021) The economic impact of hop stunt viroid and certified clean planting materials. HortScience 56(12):1471–1475

de Oliveira FJ, Farias PRS, Noronha ACDS, Maldonado Junior W, Cardoso LAS (2025) Sequential sampling of the citrus leprosis mite in different cultivation systems in the Eastern Amazon. Exp Appl Acarol 94(4):56

Dehkordi AN, Rubio M, Babaeian N, Albacete A, Martínez-Gómez P (2018) Phytohormone signaling of the resistance to plum pox virus (PPV, sharka disease) induced by almond (*Prunus dulcis* (Miller) Webb) grafting to peach (P. persica L. Batsch). Viruses 10(5):238

del Río-Celestino M, Font R (2020) The health benefits of fruits and vegetables. Foods 9(3):369

Dey KK, Green JC, Melzer M, Borth W, Hu JS (2018) Mealybug wilt of pineapple and associated viruses. Horticulturae 4(4):52

Dheepa R, Paranjothi S (2010) Transmission of cucumber mosaic virus (CMV) infecting banana by aphid and mechanical methods. Emir J Food Agr 22(2):117

Diaz-Lara A, Aguilar-Molina VH, Monjarás-Barrera JI, Vončina D, Erickson TM, Al Rwahnih M (2023) Potential implications and management of grapevine viruses in Mexico: a review. Int J Plant Biol 14(1):177–189

Digiaro M, Elbeaino T, Martelli GP (2017) Grapevine fanleaf virus and other old world nepoviruses. In: Grapevine viruses: molecular biology, diagnostics and management. Springer International Publishing, Cham, pp 47–82

Dragich M, Melzer M, Nelson S (2014) Cucumber mosaic virus in Hawaii. Plant Dis 101:1–10

Eastwell KC (2017) Management of viruses and virus-like agents affecting apple production. In Achieving sustainable cultivation of apples. Burleigh Dodds Science Publishing 407–426

Fan L, He C, Wu M, Gao D, Dong Z, Hou S et al (2021) Incidence, genomic diversity, and evolution of strawberry mottle virus in China. Biocell 45(4):1137

FAO (2023) Agricultural production statistics 2000–2022. FAOSTAT Anal Briefs, 79. Rome

Folimonova SY (2020) Citrus tristeza virus: a large RNA virus with complex biology turned into a valuable tool for crop protection. PLoS Pathog 16(4):e1008416

Forlani S, Masiero S, Mizzotti C (2019) Fruit ripening: the role of hormones, cell wall modifications, and their relationship with pathogens. J Exp Bot 70(11):2993–3006

Fuchs M, Bar-Joseph M, Candresse T, Maree HJ, Martelli GP, Melzer MJ, Menzel W, Minafra A, Report Consortium, I. C. T. V (2020) ICTV virus taxonomy profile: Closteroviridae. J Gen Virol 101(4):364–365

Fuchs M, Almeyda CV, Al Rwahnih M, Atallah SS, Cieniewicz EJ, Farrar K, Foote WR, Golino DA, Gómez MI, Harper SJ, Welliver R (2021) Economic studies reinforce efforts to safeguard specialty crops in the United States. Plant Dis 105(1):14–26

Fust C, Lameront P, Shabanian M, Song Y, Abou Kubaa R, Bester R, Maree HJ, Al Rwahnih M, Meng B (2025) Grapevine leafroll-associated virus 3: a global threat to grapevine and wine industries but a gold mine for scientific discovery. J Exp Bot 76:eraf039

Garcia S, Hily JM, Komar V, Gertz C, Demangeat G, Lemaire O, Vigne E (2019) Detection of multiple variants of grapevine fanleaf virus in single *Xiphinema index* nematodes. Viruses 11(12):1139

García JA, Rodamilans B, Martínez-Turiño S, Valli AA, Simón-Mateo C, Cambra M (2025) Plum pox virus: an overview of the potyvirus behind sharka, a harmful stone fruit disease. Ann Appl Biol 186(1):49–75

Gergerich RC, Welliver RA, Gettys S, Osterbauer NK, Kamenidou S, Martin RR, Golino DA, Eastwell K, Fuchs M, Vidalakis G, Tzanetakis IE (2015) Safeguarding fruit crops in the age of agricultural globalization. Plant Dis 99(2):176–187

Gibbs AJ, Hajizadeh M, Ohshima K, Jones RA (2020) The potyviruses: an evolutionary synthesis is emerging. Viruses 12(2):132

Golino DA, Fuchs M, Sim S, Farrar K, Martelli GP (2017) Improvement of grapevine planting stock through sanitary selection and pathogen elimination. In: Grapevine viruses: molecular biology, diagnostics and management. Springer International Publishing, Cham, pp 561–579

Gonsalves D (2013) Hawaii's transgenic papaya story 1978–2012: a personal account. In: Genetics and genomics of papaya. Springer, pp 115–142

Gonsalves D, Tripathi S, Carr JB, Suzuki JY (2010) Papaya ringspot virus. Plant Health Instr 10:1094

Grimova L, Winkowska L, Konrady M, Ryšánek P (2016) Apple mosaic virus. Phytopathol Mediterr 55:1–19

Guarino S, Mercati F, Fatta Del Bosco S, Motisi A, Abbate L (2022) Rootstocks with different tolerance grade to Citrus Tristeza virus induce dissimilar volatile profile in *Citrus sinensis* and avoidance response in the vector *Aphis gossypii* Glover. Plants 11(24):3426

Guo L, Su Q, Yin J, Yang Z, Xie W, Wang S et al (2019) Amino acid utilization may explain why *Bemisia tabaci* Q and B differ in their performance on plants infected by the tomato yellow leaf curl virus. Front Physiol 10:489

Gupta R, Kwon SY, Kim ST (2018) An insight into the tomato spotted wilt virus (TSWV), tomato and thrips interaction. Plant Biotechnol Rep 12(3):157–163

Harris KF (2018) Aphid transmission of plant viruses. In: Plant viruses. CRC Press, pp 177–204

Hu X, Liang L, Chen X, Deng L, Zou L, Dong M, Wu Q, Qi T (2024) Transcriptomic and phytohormone metabolomics provide insight into the changes in *Citrus limon* infected by citrus yellow vein clearing virus. Horticulturae 10(3):231

Jacquemond M (2012) Cucumber mosaic virus. Adv Virus Res 84:439–504

Jekayinoluwa T, Tripathi L, Tripathi JN, Ntui VO, Obiero G, Muge E, Dale J (2020) RNAi technology for management of banana bunchy top disease. Food Energy Secur 9(4):e247

Jha A, Jayswal DK, Shikha D, Kumar A, Ahmad F (2025) Empowering vital fruit crops with enhanced nutritional contents. Front Plant Sci 16:1519673

Jiang T, Zhou T (2023) Unraveling the mechanisms of virus-induced symptom development in plants. Plants 12(15):2830

Jiao B, Hao X, Liu Z, Liu M, Wang J, Liu L, Liu N, Song R, Zhang J, Fang Y, Xu Y (2022) Engineering CRISPR immune systems conferring GLRaV-3 resistance in grapevine. Hortic Res 9:uhab023

Jiménez J, Moreno A, Fereres A (2020) Transmission of phloem-limited viruses to the host plants by their aphid vectors. In: Progress in botany, vol 82. Springer International Publishing, Cham, pp 357–382

Jiménez J, Sadras VO, Espaillat N, Moreno A, Fereres A (2024) Interplay between drought and plant viruses co-infecting melon plants. Sci Rep 14(1):15833

Kennedy GG, Sharpee W, Jacobson AL, Wambugu M, Mware B, Hanley-Bowdoin L (2023) Genome segment ratios change during whitefly transmission of two bipartite cassava mosaic begomoviruses. Sci Rep 13(1):10059

Kil EJ, Park J, Choi EY, Byun HS, Lee KY, An CG, Lee JH, Lee GS, Choi HS, Kim CS, Lee S (2018) Seed transmission of tomato yellow leaf curl virus in sweet pepper (*Capsicum annuum*). Eur J Plant Pathol 150(3):759–764

Koul B, Pudhuvai B, Sharma C, Kumar A, Sharma V, Yadav D, Jin JO (2022) *Carica papaya* L.: a tropical fruit with benefits beyond the tropics. Diversity 14(8):683

Krüger K, Saccaggi DL, Van der Merwe M, Kasdorf GGF (2015) Transmission of Grapevine Leafroll-associated virus 3 (GLRaV-3): acquisition, inoculation and retention by the mealybugs *Planococcus ficus* and *Pseudococcus longispinus* (Hemiptera: Pseudococcidae). S Afr J Enol Vitic 36(2):223–230

Kumar R, Khurana A, Sharma AK (2013) Role of plant hormones and their interplay in development and ripening of fleshy fruits. J Exp Bot 65(16):4561–4575

Lázár J, Bisztray GD (2011) Virus and virus-like diseases of grapevine in Hungary. Int J Hortic Sci 17(3):25–36

Lepage E, Szadkowski M, Girardot G, Pascal M, Dumeaux P, Papaïx J, Moury B, Rimbaud L (2025) Cucumber mosaic virus degrades pepper fruit production, marketability and organoleptic quality, with isolate-specific effects. Plant Pathol 74(5):1244–1255

Luciani C, Celli MG, Torrico AK, Asinari F, Pozzi E, Peña Malavera A, Kirschbaum DS, Perotto MC, Conci VC (2018) Incidence and prevalence of aphid-borne viruses infecting strawberry in Argentina. Ann Appl Biol 173(1):80–91

Ma X, Cui H, Bernardy M, Tian L, Abbasi P, Wang A (2015) Molecular characterization of a strawberry mild yellow edge virus isolate from Canada. Can J Plant Pathol 37(3):369–375

Mahjabeen, Akhtar KP, Sarwar N, Saleem MY, Asghar M, Iqbal Q, Jamil FF (2012) Effect of cucumber mosaic virus infection on morphology, yield and phenolic contents of tomato. Arch Phytopathol Plant Protect 45(7):766–782

Martelli GP (2017) An overview on grapevine viruses, viroids, and the diseases they cause. In: Grapevine viruses: molecular biology, diagnostics and management. Springer International Publishing, Cham, pp 31–46

Martin RR, Tzanetakis IE (2013) High risk strawberry viruses by region in the United States and Canada; implications for certification, nurseries, and fruit production. Plant Disease 97(10):1358-1362.

Martínez-Lüscher J, Plank CM, Brillante L, Cooper ML, Smith RJ, Al-Rwahnih M, Yu R, Oberholster A, Girardello R, Kurtural SK (2019) Grapevine red blotch virus may reduce carbon translocation leading to impaired grape berry ripening. J Agric Food Chem 67(9):2437–2448

Michalakis Y, Blanc S (2020) The curious strategy of multipartite viruses. Annu Rev Virol 7:203–218

Miles EA, Calder PC (2021) Effects of citrus fruit juices and their bioactive components on inflammation and immunity: A narrative review. Front Immunol, 12:712608

Mishra J, Srivastava R, Trivedi PK, Verma PC (2020) Effect of virus infection on the secondary metabolite production and phytohormone biosynthesis in plants. 3 Biotech 10(12):547

Montero R, Mundy D, Albright A, Grose C, Trought MCT, Cohen D, Chooi KM, MacDiarmid R, Flexas J, Bota J (2016) Effects of Grapevine Leafroll associated virus 3 (GLRaV-3) and duration of infection on fruit composition and wine chemical profile of *Vitis vinifera* L. cv. *Sauvignon blanc*. Food Chem 197:1177–1183

Moreno A, Fereres A, Cambra M (2009) Quantitative estimation of plum pox virus targets acquired and transmitted by a single *Myzus persicae*. Arch Virol 154(9):1391–1399

Napasintuwong O, Traxler G (2009) Ex-ante impact assessment of GM papaya adoption in Thailand. AgBioforum 12(2):209–220

Ning W, Shi X, Liu B, Pan H, Wei W, Zeng Y, Sun X, Xie W, Wang S, Wu Q, Zhang Y (2015) Transmission of tomato yellow leaf curl virus by *Bemisia tabaci* as affected by whitefly sex and biotype. Sci Rep 5(1):10744

Ocimati W, Ogwal G, Tazuba AF, Kubiriba J, Tugume J, Erima R, Okurut W, Mahuku G, Kutunga D, Blomme G (2024) Mapping the vulnerability of banana production landscapes in Uganda to banana bunchy top disease. Front Agron 6:1401478

Orfanidou C, Katsiani A, Papayiannis L, Katis NI, Maliogka VI (2021) Interplay of cucurbit yellow stunting disorder virus with cucurbit chlorotic yellows virus and transmission dynamics by Bemisia tabaci MED. Plant Disease 105(2):416–424

Ouattara B, Sérémé D, Koala M, Koïta K, Kaboré E, Wonni I (2023) Prevalence and Spatial Distribution of Badnavirus in the Banana (Musa spp) Major Growing Areas in Burkina Faso. Am J Plant Sci 14(4):427–447

Qazi J (2016) Banana bunchy top virus and the bunchy top disease. J Gen Plant Pathol 82(1):2–11

Qiao W, Medina V, Kuo YW, Falk BW (2018) A distinct, non-virion plant virus movement protein encoded by a crinivirus essential for systemic infection. MBio 9(6):10–1128

Raworth DA, French CJ, Lowery DT, Bernardy MG, Bouthillier M, Mathur S, Chan CK, Foottit RG, Maw E, Wegener LA, Sweeney M (2008) Temporal trends in the transmission of Blueberry scorch virus in British Columbia, Canada. Can J Plant Pathol 30(2):345–350

Revers F, García JA (2015) Molecular biology of potyviruses. Adv Virus Res 92:101–199

Ricketts KD, Gomez MI, Atallah SS, Fuchs MF, Martinson TE, Battany MC, Bettiga LJ, Cooper ML, Verdegaal PS, Smith RJ (2015) Reducing the economic impact of grapevine leafroll disease in California: identifying optimal disease management strategies. Am J Enol Vitic 66(2):138–147

Rodríguez-Verástegui LL, Ramírez-Zavaleta CY, Capilla-Hernández MF, Gregorio-Jorge J (2022) Viruses infecting trees and herbs that produce edible fleshy fruits with a prominent value in the global market: An evolutionary perspective. Plants 11(2):203

Roossinck MJ (2003) Plant RNA virus evolution. Curr Opin Microbiol 6(4):406–409

Rupnik-Cigoj M, Jež-Krebelj A, Castellarin SD, Trošt K, Sivilotti P, Pompe-Novak M (2018) Grapevine fanleaf virus affects grape (*Vitis vinifera*) berry anthocyanin content via the transcriptional regulation of anthocyanin biosynthetic genes. Funct Plant Biol 45(7):771–782

Saad N, Olmstead JW, Jones JB, Varsani A, Harmon PF (2021) Known and new emerging viruses infecting blueberry. Plants 10(10):2172

Sakuanrungsirikul S, Sarindu N, Prasartsee V, Chaikiatiyos S, Siriyan R, Sriwatanakul M, Lekananon P, Kitprasert C, Boonsong P, Kosiyachinda P, Gonsalves D (2005) Update on the development of virus-resistant papaya: virus-resistant transgenic papaya for people in rural communities of Thailand. Food Nutr Bull 26(4_suppl3):S307–S311

Salazar SM, Conci VC, Luciani CE, Merino MC, Celli MG, Perotto MC, Torrico AK, Pozzi E, Strumia G, Dughetti AC, Asinari F, Conci LR, Fernandez FD, Meneguzzi NG, Kirschbaum DS (2017) Advances in characterization and epidemiology of strawberry viruses and phytoplasmas in Argentina. Acta Hortic 1156:801–810

Salo W, Considine JA, Considine MJ (2024) Influence of mixed and single infection of grapevine leafroll-associated viruses and viral load on berry quality. Tree Physiol 44(5):tpae035

Sastry KS, Zitter TA (2014) Management of virus and viroid diseases of crops in the tropics. In: Plant virus and viroid diseases in the tropics: volume 2: epidemiology and management. Springer Netherlands, Dordrecht, pp 149–480

Selvarajan R, Balasubramanian V (2014) Host–virus interactions in banana-infecting viruses. In: Gaur RK, Hohn T, Sharma P (eds) Plant virus–host interaction: molecular approaches and viral evolution. Elsevier Academic Press, pp 57–78

Seo JK, Lee YJ, Kim MK, Lee SH, Kim KH, Choi HS (2014) A novel set of polyvalent primers that detect members of the genera *Bromovirus* and *Cucumovirus*. J Virol Methods 203:112–115

Sevik MA, Arli-Sokmen M (2012) Estimation of the effect of Tomato spotted wilt virus (TSWV) infection on some yield components of tomato. Phytoparasitica 40(1):87–93

Shalitin D, Wolf S (2000) Cucumber mosaic virus infection affects sugar transport in melon plants. Plant Physiol 123(2):597–604

Sharma A, Handa A, Kapoor S, Watpade S, Gupta B, Verma P (2018) Viruses of strawberry and production of virus free planting material–a critical review. Intl J Environ Sci Technol 7(2):521–545

Shires MK, Wright AA, Harper SJ (2022) Improved detection of Little cherry virus-2 using a hydrolysis probe to manage the Pacific Northwest little cherry disease epidemic. Plant Dis 106(7):1875–1881

Shoala T (2025) Potato leafroll virus. In: Compendium of phytopathogenic microbes in agro-ecology, vol 2 Viruses and viroids. springer nature Switzerland, Cham, pp 265–284

Sinclair JW, Crosby KM (2002) A review of cucurbit yellow stunting disorder virus (CYSDV)-a "new" virus affecting melons in the lower Rio Grande Valley. Subtrop Plant Sci 54:54–58
Singh V, Shukla K (2009) Effect of PRSV infection on pigment content and assimilation of carbohydrate in Carica papaya L. Ann Plant Prot Sci 17(1):152–156
Song Y, Hanner RH, Meng B (2021) Probing into the effects of grapevine leafroll-associated viruses on the physiology, fruit quality and gene expression of grapes. Viruses 13(4):593
Song Y, Hanner RH, Meng B (2022) Transcriptomic analyses of grapevine leafroll-associated virus 3 infection in leaves and berries of 'Cabernet franc'. Viruses 14(8):1831
Tatineni S, Hein GL (2023) Plant viruses of agricultural importance: current and future perspectives of virus disease management strategies. Phytopathology 113(2):117–141
Tayal M, Rodriguez Bonilla F, Powell G, Cieniewicz E (2024) Pollen-borne ilarviruses of peach: biology, ecology, and disease management. J Integr Pest Manag 15(1):23
Thirugnanavel A, Balamohan TN, Karunakaran G, Manoranjitham SK (2015) Effect of papaya ringspot virus on growth, yield and quality of papaya (*Carica papaya*) cultivars. Indian J Agric Sci 85(8):1069–1073
Thompson JR, Jelkmann W (2003) The detection and variation of Strawberry mottle virus. Plant Dis 87(4):385–390
Tripathi L, Tripathi JN, Vroh-Bi I (2007) Bananas and plantains (Musa spp.): Transgenics and biotechnology. Transgenic Plant J 1(1):185–201
Tripathi S, Patil BL, Verma R (2016) Viral diseases of banana and their management. In: Plant viruses: evolution and management. Springer, pp 289–308
Umer M, Liu J, You H, Xu C, Dong K, Luo N, Kong L, Li X, Hong N, Wang G, Xu W (2019) Genomic, morphological and biological traits of the viruses infecting major fruit trees. Viruses 11(6):515
Umer M, Mubeen M, Iftikhar Y, Haider A, Zafar-ul-Hye M, Asghar R, Abbas M, Rehman M, Moya-Elizondo E, He Y (2022) Papaya ring spot virus: an understanding of a severe positive-sense single stranded RNA viral disease and its management. Phyton 91(10):2099
Varsani A, Lefeuvre P, Roumagnac P, Martin D (2018) Notes on recombination and reassortment in multipartite/segmented viruses. Curr Opin Virol 33:156–166
Vega A, Gutiérrez RA, Pena-Neira A, Cramer GR, Arce-Johnson P (2011) Compatible GLRaV-3 viral infections affect berry ripening decreasing sugar accumulation and anthocyanin biosynthesis in Vitis vinifera. Plant Mol Biol 77(3):261–274
Vieira MS, Cabral RLR, Favaratto L, Maciel LS, Xavier ADS, Zerbini FM, Fernandes PM (2024) Tropical fruit virus resistance in the era of next-generation plant breeding. SynBio 2(3):267–284
Vikrant, Chowdhury S, Majumder S (2025) First report and molecular characterisation of banana bunchy top virus infecting Musa× Paradisiaca Cv. Chiniya from West Bengal, India. Vegetos 1–8
Vučurović A, Bulajić A, Stanković I, Ristić D, Berenji J, Jović J, Krstić B (2012) Non-persistently aphid-borne viruses infecting pumpkin and squash in Serbia and partial characterization of Zucchini yellow mosaic virus isolates. Eur J Plant Pathol 133(4):935–947
Walker PJ, Siddell SG, Lefkowitz EJ, Mushegian AR, Adriaenssens EM, Alfenas-Zerbini P, Dempsey DM, Dutilh BE, García ML, Hendrickson RC, Junglen S, Krupovic M, Kuhn JH, Lambert AJ, Łobocka M, Oksanen HM, Orton RJ, Robertson DL, Rubino L et al (2022) Recent changes to virus taxonomy ratified by the International Committee on Taxonomy of Viruses. Arch Virol 167(11):2429–2440. Advance online publication. https://doi.org/10.1007/s00705-022-05516-5
Wegener LA, Punja ZK, Martin RR, Bernardy MG, MacDonald L (2006) Epidemiology and identification of strains of Blueberry scorch virus on highbush blueberry in British Columbia, Canada. Can J Plant Pathol 28(2):250–262
World Health Organization (2020) Healthy diet [Fact Sheet]. Retrieved from https://www.who.int/news-room/fact-sheets/detail/healthy-diet
Wright S (2012) Grapevine Leafroll associated Virus-3 (GLRaV-3) seasonal titer changes and effects on pinot noir fruit in Oregon. Oregon State University Master's Thesis
Wubshet Z, Amare D (2019) Review on citrus triteza virus. Int J Res 5(4):25–36

Viral Threats in Commercially Important Tropical and Subtropical Fruit Crops

2

Paola Elizabeth Hernández-Meléndez
and Ubilfrido Vasquez-Gutierrez

Abstract

The world's population is growing, and with it, the demand for food is increasing exponentially. Tropical and subtropical fruit crops are important for the consumer chain. The emergence of epidemic viruses has turned the production and marketing of papaya (*Carica papaya* L.), avocado (*Persea americana* Mill.), banana (*Musa* spp. L.), and citrus fruits (*Citrus* spp.) upside down, to the point that they are threatened by the presence of important viruses. This chapter presents a classification of viruses that affect fruit trees, including the genera: *Closterovirus*, *Ampelovirus*, *Crinivirus*, *Bymovirus*, *Ipomovirus*, *Macluravirus*, *Rymovirus*, *Badnavirus*, and *Begomovirus*. Viral transmission in trees depends on the epidemiology of the virus, with most being transmitted by arthropods, grafts, and infected rootstocks. The fundamental basis for achieving efficient integrated management begins with the establishment of a structured epidemiological surveillance system, including third-generation early diagnosis tools to prevent the spread of emerging viruses. So far, success has been based on advances in surveillance rather than on virus control itself. The main strategy may be based on controlling the risk posed by viruses, rather than on stopping their emergence in fruit crops.

Keywords

Epidemic viruses · Participatory surveillance · Transmission

P. E. Hernández-Meléndez
Facultad de Ingeniería y Ciencias, Universidad Autónoma de Tamaulipas, Centro Universitario Victoria, Ciudad Victoria, Tamaulipas, Mexico

U. Vasquez-Gutierrez (✉)
Department of Parasitology, Universidad Autónoma Agraria Antonio Narro, Buenavista, Saltillo, Mexico

R. Kumar et al. (eds.), *Fruit Crops and Viral Pathogens Volume II*, Plant Pathology for Sustainable Agriculture, https://doi.org/10.1007/978-981-95-7261-8_2

2.1 Introduction to Viruses in Tropical and Subtropical Fruit Trees

Fruit crops are mostly perennial and economically important, which makes them susceptible to epidemic viruses. Their presence causes a reduction in yield and product quality. *Citrus* production maintains an estimated annual growth rate of 3.34% and a 13.7% increase in exports by 2030 (Valencia Sandoval and Duana Avila 2019). Lime cultivation (*Citrus aurantifolia*) contributes 1.5% of the gross domestic product (GDP), while orange cultivation (*Citrus sinensis*) contributes 1.15%, but both crops are susceptible to citrus tristeza virus (CTV) (Sáenz Pérez et al. 2019). CTV causes systemic infections, affecting the phloem at the graft union (Ruíz-García et al. 2009); it leads to loss of tree vigor, small and deformed fruit, chlorosis, and defoliation (Irracheta and Arrieta 2012). The changes induced by CTV infection depend on nutritional and environmental conditions and on the varieties, as there are tolerant rootstocks such as sweet orange, Cleopatra mandarin, citrumelo CPB 4475, FA5, and F418 (Pedraza et al. 2016). Citrus fruits are susceptible to diseases caused by viroids, such as citrus exocortis viroid (CEVd), which can affect citrus varieties by causing bark cracking and peeling, leaf yellowing, and loss of tree vigor; it does not damage the fruit, but the reduction in the tree's annual yield results in economic losses of up to 60% (Alcántara-Mendoza et al. 2017). Papaya (*Carica papaya* L.) is the third most consumed tropical fruit worldwide, and thanks to its marketing dynamics, it is considered one of the most economically important fruits (Evans and Ballen 2012; SAGARPA 2013; FAO 2022), with an annual per capita consumption of 7.3 kg. Mexico ranks third among global producers with 1,148,546 tons, accounting for 8.2% of the global volume, of which 99.6% is exported to the United States and Canada, predominantly the Maradol variety (SADER-SIAP 2024). However, this important crop is threatened by the papaya ringspot virus (PRSV), which causes concentric rings on the fruit and size reduction, yellowing of young leaves, and oily-looking spots along the stem and petiole (Tripathi et al. 2008), leading to loss of plant vigor. Notably, economic losses exceed 80% of the final product, and there is a reduction in fruit sugar content of approximately 50% (Umer et al. 2022). PRSV is one of the diseases that incur very high control costs (Bermúdez Guzmán et al. 2018). Although other viruses affect the crop, such as the papaya meleira virus (PMeV), which causes the production of a sticky liquid on various plant parts (Abreu et al. 2015), or the papaya mosaic virus (PMV), which induces mosaic patterns, leaf yellowing, multiple deficiency symptoms, and fruit deformation, primarily, both viruses have less common effects on yield and economic impact (Ordaz-Pérez et al. 2017). Avocado (*Persea americana*, Mill.), commonly known as "green gold," is of great economic importance within the agri-food chain, accounting for 4.35% of gross domestic product (GDP) and positioning itself as the leading fruit exporter with almost 50% of global exports and a demand with a 3.4% increase in production and a 5% increase in international trade by 2030 (Cruz-López et al. 2022). Some limitations cause production losses, resulting in virus-induced diseases such as avocado sunblotch (Saucedo-Carabez et al. 2014) caused by the avocado sunblotch viroid (ASBVd), which produces symptoms ranging from yellowing to necrosis and loss of symmetry in fruits (Pérez Álvarez et al. 2015), reduced tree vigor, and defoliation (Paredes Valdivieso 2021). The degree of intensity will depend on nutritional conditions and environmental factors, although some trees have been observed to remain asymptomatic. When the disease presents apparent symptoms, it can cause yield losses of up to 27.3%, a 40.1% reduction in price, and fruit losses exceeding 50% (Beltrán Peña et al. 2014). Pineapple production (*Ananas comosus* (L.) Merr.) increased by 29% over the past decade. Exports have grown, accounting for 0.48% of GDP, with an estimated 1.7% increase in yield and a 5.82% rise in foreign trade over the past 5 years (Rodríguez et al. 2016). Some challenges must be addressed; viruses cause significant damage to this crop, such as pineapple mealybug wilt-associated virus-1 (PMWaV-1), PMWaV-2, and PMWaV-3—as well as two putative viruses, PMWaV-4 and PMWaV-5 (Ochoa Martínez et al. 2016). Symptoms of leaf yellowing, loss of turgor, and plant death appear 2 months after vector infestation (Peña-Prades et al. 2015; Dey et al. 2018); production losses can range from 30% to 40%, and the disease's distribution is under quarantine in Ecuador (Young et al. 2022). Banana cultivation (*Musa paradisiaca* L.) is one of the most important worldwide, ranking fourth and producing over two million tons, with an extensive production chain generating 250,000 direct and indirect jobs (Manzo-Sánchez et al. 2014). However, it is affected by eight viruses (Kumar 2014), including cucumber mosaic cucumovirus (CMV) and banana bunchy top virus (BBTV) (Ndowora et al. 1999), the latter being the most devastating, with reported production losses of up to 95% (Selvarajan and Balasubramanian 2013). Mexico is threatened, with 80,283.26 hectares and an economic impact of over 6965.81 million pesos derived from 450,000 jobs (Drew et al. 1989), although it can also contract diseases such as banana bract mosaic virus (BBrMV), banana mild mosaic virus (BanMMV), abaca mosaic virus (AbaMV), banana die-back virus (BDBV), banana virus X (BVX), and banana streak badnavirus (BSV), which causes banana streak disease (Hu et al. 1993; Javer Higginson, 2007). On the other hand, grape consumption has increased, mainly due to the wine sector, accounting for 0.83% of the national agricultural GDP with estimated production and 2.06% of exports over the next 5 years (Borja-Bravo et al. 2016). It should be noted that the crop faces economic losses exceeding 40% caused by a virus complex, primarily grapevine leafroll-associated virus 3 (GLRaV-3) (Montero et al. 2016), which induces leafroll disease. Symptoms include foliar chlorosis, downward rolling of leaf margins, and reddening of interveinal areas while the rest of the leaf remains green (Čarija et al. 2022).

2.2 Classification of Viruses of Importance in Fruit Trees

Plant viruses can be classified according to the type of genetic material—DNA or RNA—as well as by positive or negative sense (+/−) and single- or double-stranded (ss/ds): ssRNA+, ssRNA−, dsRNA, ssDNA, and dsDNA (Peña Núñez et al. 2003; Sáenz Pérez et al. 2019). The composition of the viral genome will determine replication, gene expression, and interactions with the host. Viruses exhibit structural and functional differences across species, such as the organization of genomic segments, recombination rate, nucleic acid protection mechanisms, and genetic variability (Gutiérrez et al. 2011). These descriptors influence the virus's biology and serve to understand integrated management strategies aimed at reducing viral spread in crops.Understanding the importance of tropical and subtropical fruit crops also involves recognizing the viruses that affect them. Among the wide diversity of viral species described, six main families stand out for their broad distribution and extensive host range in crops of interest (Table 2.1).

- *Closteroviridae*: Plant viruses belonging to this family have a monopartite genome, filamentous virions with a helical structure (Martelli et al. 2002a), and lengths ranging from 250 nm to 2000 nm (Martelli et al. 2012). This family comprises four genera: *Closterovirus*, *Ampelovirus*, *Crinivirus*, and *Velarivirus*, which exhibit distinctive characteristics (Martelli and Boudon-Padieu 2006).
- *Potyviridae*: Most of the genera that make it up have monopartite genomes, except for *Bymovirus*, with flexuous filamentous virions 650–950 nm in length (Díaz et al. 2010); they are divided into five genera: *Bymovirus*, *Ipomovirus*, *Macluravirus*, *Rymovirus*, and *Tritimovirus* (Inoue-Nagata et al. 2022).
- *Badnaviridae*: With viral particles 120–150 nm in size, it has a non-covalently closed circular double-stranded DNA (dsDNA); replication occurs via encoded reverse transcription; three open reading frames (ORFs) are incorporated into the genome (Ishwara Bhat et al. 2023); and viruses of this family are known as endogenous because they can be activated by abiotic stress (Ahlawat et al. 1996). The major crops affected by the presence of viral species belonging to the genus *Badnavirus* include black pepper, banana, cocoa, fig, sugarcane, orange, grapefruit, and others (Agusto et al. 2024). Productivity losses can reach up to 90% (Bhat et al. 2016).
- *Geminiviridae*: This group is characterized by a circular DNA genome within capsids (Morales and Anderson 2001) in quasi-icosahedral particles (Maree et al. 2013); they are known as the smallest viruses, reaching dimensions of 38 nm (Fiallo-Olivé et al. 2021), and may be associated with betasatellites. Fifteen genera with distinct characteristics are grouped (Bell et al. 2018). Due to its broad host range, it can cause economic losses in more than 17 staple and industrial crops (*Begomovirus*), posing a threat to global agriculture (Almeida et al. 2013).
- *Caulimoviridae*: This family of viruses has isometric or bacilliform virions with a circular double-stranded DNA genome (Hull 1995). It includes the first plant virus described with these characteristics. Its viral particles have an approximate diameter of 50 nm, exhibit buoyant density characteristics, and are stable in a pH range of 4–9 (Gong and Han 2018). Currently, 11 genera are recognized, infecting species of the Brassicaceae family, such as papaya, banana, cocoa, and citrus, although other hosts have also been identified (Marín-Montoya and Gutiérrez-Sánchez 2016). Due to their ability to infect agriculturally important crops, these viruses have been reported to cause economic losses of up to 30% and yield losses of up to 75% (Geering et al. 2010).

Table 2.1 Virus of importance for priority crops

Fruit crops	Viral name	Family	Symptoms	Vectors	Host
Tropical	Banana bunchy top virus (BBTV)	*Nanoviridae*	Yellowing, leaf curling, short leaves, and stunting	Aphyds: *Pentalonia nigronervosa* (Hemiptera: Aphididae)	*Musaceae* Varieties of *Musa* sp.
	Papaya ringspot virus (PRSV)	*Potyviridae*	Ring-shaped spots, chlorosis, and leaf distortion	*Myzus persicae*	*Caricaceae* (*Carica papaya*); *Cucurbitaceae* (*Citrullus lanatus*); *Chenopodiaceae* (*Chenopodium quinoa*)
	Citrus tristeza virus (CTV)	*Closteroviridae*	Loss of vigor, fruit deformation, and curling	*Toxoptera citricida*, *Toxoptera aurantii*, *Aphis gossypii*, *Aphis spiraecola*	*Rutaceae* (*Citrus* sp.)
	Cacao Swollen Shoot Virus (CSSV)	*Caulimoviridae*	Chlorotic spots, red bands along the leaf veins, and swelling of stems and roots	*Formicococcus njalensis*, *Planococcus citri*, and *Ferrisia virgata*	*Malvaceae* (*Theobroma cacao*)
Subtropical	Plum pox virus (PPV)	*Potyviridae*	Spots on the exocarp, irregular grooves, and fruit deformities	*Myzus persicae*, *Aphis spiraecola*, *Aphis gossypii*	*Rosaceae* (*P. domestica*, *P. pérsica*, *P. dulcis*, *P. avium*)
	Grapevine leafroll-associated viruses (GLRaVs)	*Closteroviridae*	Interveinal leaf and fruit curling and reddening, and uneven ripening	Species of scale insects (Pseudococcidae) and scales (Coccidae)	Variety of the family *Vitaceae* (*Vitis vinífera*, *V. labrusca*, *V. rotundifolia*)
	Apple stem grooving virus (ASGV)	*Betaflexiviridae*	Necrosis of the xylem and phloem, streaks in the wood, and irregular areas on the tree trunk		Preferably species of *rutáceas* (*Malus domestica*, *Pyrus communis*, *Mespilus germánica*)
	Avocado sunblotch viroid (ASBVd)	*Avsunviroidae* (Viroides)	Scars and reduced fruit size, chlorotic leaves, as well as uneven ripening		Varieties of the species *Persea* sp., Hass, Carmen-Hass

- *Tymoviridae*: It consists of viruses with isometric particles measuring 30 nm in diameter (Ares 1997). Its genome is organized into three open reading frames (ORFs) and the viral capsids are structured from hexamers and pentamers surrounding the viral RNA (Martelli et al. 2002b). The virions have sedimentation properties and remarkable thermal resistance above 80 °C (Gutiérrez Fernández et al. 2002). It comprises three genera that share similar biological and structural characteristics. Viruses in this group cause significant losses in agricultural production, up to 75% in turnip crops (Brassica rapa). It does not affect other economically important hosts such as corn, grapevine, broccoli, cauliflower, and other ornamental species (Komínková et al. 2024).

2.3 Vectors and Routes of Viral Transmission

Diseases caused by viruses and viroids constitute a phytosanitary problem of great importance, as they cause adverse damage to crops with high export demand, resulting in significant economic losses at various stages of the production chain (Mayorga Carrillo et al. 2021). The most efficient strategy for managing them lies in prevention, particularly by controlling insect vectors such as aphids, psyllids, mealybugs, and whiteflies, which are the main agents of dissemination (Jaramillo Zapata et al. 2012). However, there are other important routes of transmission. One of them is vegetative propagation by grafting, where initially asymptomatic plants can act as carriers of viruses and viroids; by using their buds on healthy rootstocks, the disease is transmitted to new plants (Goldschmidt 2014). Some viruses can be transmitted through seeds, a phenomenon known as vertical transmission, in which the embryo is infected and, as long as the seed remains viable, can give rise to diseased seedlings (Pagán 2022). Pollen-mediated transmission has been documented in entomophilous species; insects carry infectious pollen grains to new plants, facilitating viral spread to future generations (Velásquez-Valle et al. 2017). Finally, one of the most common modes of transmission is the use of contaminated tools in agricultural operations; the failure to disinfect them after contact with infected tissues, even from asymptomatic plants, promotes the spread of pathogens to other crops (Garrido and Trujillo 2000).

2.4 Molecular Diagnosis and Surveillance Systems

For the molecular diagnosis of fruit viruses, techniques are employed that enable the specific detection and quantification of pathogenic viruses. Reverse transcription polymerase chain reaction (RT-PCR) is used for the sensitive detection of RNA viruses (Clementi et al. 1993). Real-time PCR (qPCR) allows for the amplification and quantification of viral DNA or RNA (Pabinger et al. 2014). Loop-mediated isothermal amplification (LAMP) enables rapid and efficient DNA amplification under isothermal conditions using multiple specific primers. This technique is valuable for field detection due to its efficiency and reduced need for sophisticated equipment (Wong et al. 2018). Next-generation sequencing (NGS) technologies have revolutionized virogenic surveillance in recent years (Rubio et al. 2020), enabling high-throughput sequencing (HTS) and parallel sequencing of DNA or RNA fragments. With the support of HTS platforms such as Pacific Biosciences (PacBio®) and Oxford Nanopore Technologies (ONT), it is possible to identify viral variants, discover emerging viruses, and generate important information to prevent diseases and promote timely diagnosis. For viruses in tropical and subtropical fruit crops, only limited information is available; most studies have focused on vegetable viruses. Success stories include nine viruses detected in beans: cowpea mild mottle virus (CPMMV), peanut mottle virus (PeMoV), and pelargonium line pattern virus (PLPV) (Chiquito-Almanza et al. 2017). Viral variants such as pepper huasteco yellow vein virus (PHYV) and tomato mosaic virus (ToMV) (Vásquez-Gutiérrez et al. 2024) have been discovered, and new species such as chickpea leafroll virus (CpLRV) have been detected in chickpeas (Kanakala and Kuria 2018). In sugarcane, massive sequencing was performed with 0.5 million reads and over 92% genome coverage for RNA virus detection; the analysis revealed Fiji disease virus (FDV), sugarcane mosaic virus (SCMV), sugarcane streak mosaic virus (SCSMV), sugarcane striate mosaic-associated virus (SCSMaV), sugarcane yellow leaf virus (SCYLV), sugarcane streak Egypt virus (SCSEV), and sugarcane white streak virus (SCWSV) (Malapi-Wight et al. 2021). Genomic databases and bioinformatics platforms have gained prominence, as they enable the comparison of viral sequences, the identification of mutations, and support for epidemiological surveillance programs such as VIROSCOPE, Microbe Finder (MiFi), Phytopipe, Microreact, and others (Del Pozo 2025). In Mexico and other countries, there are government phytosanitary alert platforms that notify about the introduction or detection of high-risk emerging viruses. After an alert is issued, mitigation measures are often limited, so early detection and participatory surveillance represent effective strategies for managing viruses in fruit trees, as has been applied in sweet cherry cultivation for prunus necrotic ringspot virus (PNRSV), apple stem pitting virus (ASPV), and olive leaf yellowing-associated virus 1 (OLYaV) (Morgante et al. 2025). Complementarily, genomic databases and bioinformatics platforms have gained prominence, as they enable the comparison of viral sequences, the identification of mutations, and the support of epidemiological surveillance programs, such as VIROSCOPE, Microbe Finder (MiFi), Phytopipe, Microreact, among others (Del Pozo 2025). In Mexico and other countries, there are government phytosanitary alert platforms that notify of the introduction or detection of emerging, high-risk viruses. After an alert is issued, mitigation measures are often limited, so early detection and participatory surveillance represent effective strategies for managing viruses in fruit trees, as has been applied in sweet cherry cultivation for prunus necrotic ringspot virus (PNRSV), apple stem pitting virus (ASPV), and olive leaf yellowing-associated virus 1 (OLYaV) (Morgante et al. 2025). Finally, it is important to note that one of the main limitations to the timely diagnosis of plant viruses in the field is the high cost of these tests and, in many cases, producers' lack of awareness of their availability and utility, which hinders the adoption of molecular tools in phytosanitary management programs. As a result, participatory surveillance has emerged as an indispensable tool for the early detection of viruses in fruit trees, underscoring the priority of precise vector control and implementing a series of chronological activities to prevent the presence of viral diseases of concern (Marín-Montoya and Gutiérrez-Sánchez 2016): (i) starting with a georeferenced mapping of vulnerable areas; (ii) monitoring to explore specific georeferenced points; and (iii) applying chemical control alternatives, taking into account the operation of phytosanitary epidemiological management areas (AMEFIs) and an epidemic containment area (ACE), focusing on infestation hotspots at production scales. Ongoing training for producers is essential to reduce the disease through courses, participatory workshops, and field visits for public- and private-sector extension agents, producers, and the general community involved in fruit crop production. Supervising the implementation of activities within the epidemiological surveillance program ensures efficient continuity to prevent the proliferation of disease outbreaks and to properly manage vectors and the crop of interest.

2.5 Agronomic, Economic, and Commercial Impacts

Tropical and subtropical crops face a significant challenge due to their high susceptibility to viral infections and the growing global demand for these products. The presence of viruses can cause yield reductions of over 85%, as well as induce symptoms that affect the organoleptic and commercial quality of the product, resulting in substantial economic losses and shortening the shelf life of crops (Font-San-Ambrosio et al. 2023). Eradication programs, preventive practices, and the removal of infected trees—a practice that represents an economic loss for farmers—have all been implemented to address the presence of viruses. In 2024, $853,058 was invested in disease control and surveillance; however, global economic losses due to viruses have reached $30 billion annually, underscoring the importance of strengthening crop health (Zamora-Macorra et al. 2023). One of the challenges is phytosanitary barriers in international trade; the rigorous export requirements, health certifications, and restrictions on agrochemical use imposed by each country are intended to prevent the introduction and spread of viruses, but they also limit producers' competitiveness and market access. This creates a wide gap for the sustainable incorporation of high-value crops into the international market, highlighting the need to strengthen epidemiological surveillance strategies, early molecular diagnostics, and the development of increasingly sustainable management alternatives.

2.6 Integrated Management Strategies and Sustainable Crop Protection

As mentioned in a previous section, the most effective strategy for preventing the impact of viral diseases on fruit crops is participatory surveillance. However, important strategies are still considered within integrated virus management. (a) For the control of arthropod vectors, which transmit numerous viruses, biological control should be prioritized to reduce selection pressure and prevent resistance to chemical products. (b) Certified plant material reduces the risk of introducing viruses into new plantings. Selecting crops or varieties with genetic resistance is a key tool for reducing economic losses. (c) Cultural management practices represent an essential line of defense for reducing the incidence of diseases in trees. (d) In recent years, emerging biotechnologies have strengthened diagnostic and management strategies. Among them, the CRISPR technique (Clustered Regularly Interspaced Short Palindromic Repeats) stands out, enabling genome editing to correct mutations and achieve lasting disease resistance (Martinez-Oliva 2020). RNA interference (RNAi) has become an established mechanism for silencing viral or host susceptibility genes, preventing the establishment of disease. Epigenetic editing modifies gene expression by regulating proteins and manipulating chromatin without altering the DNA sequence, making it a promising tool for disease control in the future.

2.7 Futuristic Challenges and Innovative Opportunities

Plant virologists face a major challenge in the emergence and spread of viruses; the risk of introducing new diseases can have devastating consequences for agricultural production. Climate change is a determining factor, as variations in temperature and relative humidity influence the population dynamics and behavior of the vectors responsible for dissemination. The indiscriminate use of chemical products favors the proliferation of resistant pests and hinders phytosanitary control. One of the main obstacles lies in the high costs associated with cutting-edge molecular techniques, which limit their application in routine monitoring programs. This scenario opens the opportunity to develop more accessible, rapid, and reliable detection methods tailored to the needs of local production systems. Overcoming these limitations requires strengthening cooperation among government institutions, research centers, and the private sector to develop projects that address priority issues such as surveillance, diagnosis, and virus management in fruit trees. It is essential to consolidate biosecurity and phytosanitary monitoring strategies to ensure the safety and quality of Mexican agricultural products, thereby enhancing competitiveness and expansion in international markets (Sochor et al. 2012). “We cannot prevent the emergence of epidemic viruses in fruit trees, but we can monitor the risk they pose to important crops.”

References

Abreu PM, Antunes TF, Magaña-Álvarez A, Pérez-Brito D, Tapia-Tussell R, Ventura JA, Fernandes AA, Fernandes PM (2015) A current overview of the Papaya meleira virus, an unusual plant virus. Viruses 7(4):1853–1870. https://doi.org/10.3390/v7041853

Agusto FB, Leite MCA, Owusu-Ansah F, Domfeh O, Hritonenko N, Chen-Charpentier B (2024) Cacao sustainability: the case of cacao swollen-shoot virus co-infection. PLoS One 19(3):e0294579. https://doi.org/10.1371/journal.pone.0294579

Ahlawat YS, Pant RP, Lockhart BEL, Srivastava M, Chakraborty NK, Varma A (1996) Association of a badnavirus with citrus mosaic disease in India. Plant Dis 80:590–592

Alcántara-Mendoza S, Vergara-Pineda S, García-Rubio O, Cambrón-Sandoval VH, Colmenares-Aragón D, Nava-Díaz C (2017) Caracterización del viroide exocortis de los cítricos en diferentes condiciones de indexado. Rev Mex Fitopatol 35(2):284–303. https://doi.org/10.18781/r.mex.fit.1701-3

Almeida RPP, Daane KM, Bell VA, Blaisdell GK, Cooper ML, Herrbach E, Pietersen G (2013) Ecology and management of grapevine leafroll disease. Front Microbiol 4:94. https://doi.org/10.3389/fmicb.2013.00094

Ares VX (1997) Caracterización de la proteína de 24 kDa de PVX y obtención de plantas transgénicas resistentes a tobamovirus. Tesis doctoral, Universidad de Buenos Aires

Bell VA, Hedderley DI, Pietersen G, Lester PJ (2018) Vineyard-wide control of grapevine leafroll-associated virus 3 requires an integrated response. J Plant Pathol 100:399–408. https://doi.org/10.1007/s42161-018-0085-z

Beltrán Peña H, Soria Ruiz J, Téliz Ortiz D, Ochoa Martínez DL, Nava Díaz C, Ochoa Ascencio S (2014) Detección satelital y molecular del viroide de la mancha de sol del aguacate (Avocado Sunblotch Viroid, ASBVd). Rev Fitotec Mex 37(1):21–29

Bermúdez Guzmán M d J, Guzmán González S, Lara Reyna J, Palmeros Suárez PA, López Muraira IG, Gómez Leyva JF (2018) Presencia de *Papaya ringspot virus* (PRSV) en arvenses asociadas a *Carica papaya* en Colima, México. Rev Mex Fitopatol 36(1):1–15. https://doi.org/10.18781/r.mex.fit.1706-2

Bhat AI, Hohn T, Selvarajan R (2016) Badnaviruses: the current global scenario. Viruses 8(6):177. https://doi.org/10.3390/v8060177

Borja-Bravo M, García-Salazar JA, Reyes-Muro L, Arellano-Arciniega S (2016) Rentabilidad de los sistemas de producción de uva (*Vitis vinifera*) para mesa e industria en Aguascalientes, México. Agric Soc Desarro 13(1):151–168

Čarija M, Černi S, Stupin-Polančec D, Radić T, Gaši E, Hančević K (2022) Grapevine leafroll-associated virus 3 replication in grapevine hosts changes through the dormancy stage. Plants 11(23):3250. https://doi.org/10.3390/plants11233250

Chiquito-Almanza E, Acosta-Gallegos JA, García-Álvarez NC, Garrido-Ramírez ER, Montero-Tavera V, Guevara-Olvera L, Anaya-López JL (2017) Simultaneous detection of both RNA and DNA viruses infecting dry bean. Viruses 9(4):63. https://doi.org/10.3390/v9040063

Clementi M, Menzo S, Bagnarelli P, Manzin A, Valenza A, Varaldo PE (1993) Quantitative PCR and RT-PCR in virology. Genome Res 2(3):191–196. https://doi.org/10.1101/gr.2.3.191

Cruz-López DF, Caamal-Cauich I, Pat-Fernández VG, Reza Salgado J (2022) Competitividad de las exportaciones de aguacate Hass de México. Rev Mex Cienc Agric 13(2):355–362. https://doi.org/10.29312/remexca.v13i2.2885

Del Pozo Á (2025) Bioinformatics and omics data management in genetic diagnosis. An Pediatr. https://doi.org/10.1016/j.anpedi.2025.504013

Dey K, Green J, Melzer M, Borth W, Hu J (2018) Mealybug wilt of pineapple and associated viruses. Horticulturae 4(4):52. https://doi.org/10.3390/horticulturae4040052

Díaz A, Quiñones M, Arana F, Soto M, Hernández A (2010) Potyvirus: características generales y diagnóstico. Rev Protección Veg 25(2):69–79

Drew RA, Moisander JA, Smith MK (1989) The transmission of banana bunchy top virus in micropropagated bananas. Plant Cell Tissue Organ Cult 16:187–193

Evans EA, Ballen FH (2012) Una mirada a la producción, comercio y consumo de papaya a nivel mundial. UF/IFAS Extensión

Fiallo-Olivé E, Lett JM, Martin DP, Roumagnac P, Varsani A, Zerbini FM, Navas-Castillo J (2021) ICTV virus taxonomy profile: geminiviridae 2021. J Gen Virol 102(12):001696. https://doi.org/10.1099/jgv.0.001696102

Font-San-Ambrosio MI, Alfaro Fernández AO, Jordá Gutiérrez C (2023) Virosis en cucurbitáceas en los últimos 35 años. Phytoma España 354:3. ISSN 1131-8988

Food and Agriculture Organization of the United Nations (2022) Perspectives on technological and commercial outlook for pineapple cultivation in Colombia [PDF]. FAO. https://openknowledge.fao.org/server/api/core/bitstreams/

Garrido MJ, Trujillo GE (2000) Pruebas de transmisión mecánica con el virus del bandeado amarillo del sorgo (SYBV) [Tests of mechanical transmission with *Sorghum yellow banding virus* (SYBV)]. Rev Fac Agron (LUZ) 17:269–281

Geering ADW, Scharaschkin T, Teycheney PY (2010) The classification and nomenclature of endogenous viruses of the family *Caulimoviridae*. Arch Virol 155(1):123–131. https://doi.org/10.1007/s00705-009-0488-4

Goldschmidt EE (2014) Plant grafting: new mechanisms, evolutionary implications. Front Plant Sci 5:727. https://doi.org/10.3389/fpls.2014.00727

Gong Z, Han G-Z (2018) Euphyllophyte paleoviruses illuminate hidden diversity and macroevolutionary mode of *Caulimoviridae*. J Virol 92:e02043–e02017. https://doi.org/10.1128/JVI.02043-17

Gutiérrez Fernández J, Piedrola Angulo G, Lucena PH (2002) Características generales de los virus (I): estructura, clasificación y replicación. In: Liébana Ureña J (ed) Microbiología oral, vol 30, 2nd edn, Madrid, pp 201–210. https://doi.org/10.13140/RG.2.1.4134.9600

Gutiérrez P, Bastos-Aristizábal S, Marín M (2011) Modelación estructural de la proteína de la cápside del virus A de la papa (PVA, *Potyvirus*). Actual Biol 33(94):93–102

Hu JS, Xu MQ, Wu ZC, Wang M (1993) Detection of *Banana bunchy top virus* in Hawaii. Plant Dis 77(9):952

Hull R (1995) Genus *Caulimovirus*. In: Murphy FA et al (eds) Virus taxonomy: sixth report of the international committee on taxonomy of viruses. Springer-Verlag, Wien, pp 189–192

Inoue-Nagata AK, Jordan R, Kreuze J, Li F, López-Moya JJ, Mäkinen K, Ohshima K, Wylie SJ (2022) ICTV virus taxonomy profile: potyviridae 2022. J Gen Virol. https://doi.org/10.1099/jgv.0.001738

Irracheta CMM, Arrieta GLC (2012) Detección del virus tristeza de los cítricos mediante anticuerpos contra la proteína recombinante p25 de la cápside bajo un sistema de inmunoimpresión. Rev Mex Fitopatol 30(1):31–42

Ishwara Bhat A, Selvarajan R, Balasubramanian V (2023) Emerging and re-emerging diseases caused by badnaviruses. Pathog 12(2):245. https://doi.org/10.3390/pathogens12020245

Jaramillo Zapata MM, Álvarez JA, Marín Montoya M (2012) Características de los virus asociados a la virosis del tomate de árbol (*Solanum betaceum*) en Colombia. Rev Lasallista Investig 9(1):115–127

Javer Higginson E (2007) Reseña de *Banana streak virus* (BSV): características biológicas, epidemiología e importancia económica. Fitosanidad 11(4):61–69

Kanakala S, Kuria P (2018) *Chickpea chlorotic dwarf virus*: an emerging monopartite dicot-infecting *Mastrevirus*. Viruses 11(1):5. https://doi.org/10.3390/v11010005

Komínková M, Ben Mansour K, Komínek P, Brožová J, Střalková R (2024) Multiple infections with viruses of the family *Tymoviridae* in Czech grapevines. Viruses 16(3):343. https://doi.org/10.3390/v16030343

Kumar LS (2014) Biology, etiology, and control of virus diseases of banana and plantain. Elsevier

Malapi-Wight M, Adhikari B, Zhou J, Hendrickson L, Maroon-Lango CJ, McFarland C, Foster JA, Hurtado-Gonzales OP (2021) HTS-based diagnostics of sugarcane viruses: seasonal variation and its implications for accurate detection. Viruses 13(8):1627. https://doi.org/10.3390/v13081627

Manzo-Sánchez G, Orozco-Santos M, Martínez-Bolaños L, Garrido-Ramírez E, Canto-Canche B (2014) Enfermedades de importancia cuarentenaria y económica del cultivo de banano (*Musa* sp.) en México. Rev Mex Fitopatol 32(2):89–107

Maree HJ, Almeida RPP, Bester R, Chooi KM, Cohen D, Dolja VV, Fuchs MF, Golino DA, Jooste AEC, Martelli GP, Naidu RA, Rowhani A, Saldarelli P, Burger JT (2013) Grapevine leafroll-associated virus 3. Front Microbiol 4:82. https://doi.org/10.3389/fmicb.2013.00082

Marín-Montoya M, Gutiérrez-Sánchez PA (2016) Principios de virología molecular de plantas tropicales. AGROSAVIA. https://doi.org/10.21930/agrosavia.investigation.7402308

Martelli GP, Boudon-Padieu E (2006) Directory of infectious diseases of grapevines and viroses and virus-like diseases of the grapevine: bibliographic report 1998–2004. CIHEAM. http://om.ciheam.org/article.php?IDPDF=800521

Martelli G, Agranovsky A, Bar-Joseph M, Boscia D, Candresse T, Coutts RHA, Dolja VV, Falk BW, Gonsalves D, Jelkmann W, Karasev AV, Minifra A, Namba S, Vetten HJ, Wisler GC, Yoshikawa N (2002a) The family *Closteroviridae* revised. Arch Virol 147:2039–2044. https://doi.org/10.1007/s007050200048

Martelli G, Sabanadzovic S, Abou-Ghanem Sabanadzovic N, Edwards MC, Dreher T (2002b) The family *Tymoviridae*. Arch Virol 147:1837–1846. https://doi.org/10.1007/s007050200045

Martelli GP, Agranovsky AA, Al Rwahnih M, Dolja VV, Dovas CI, Fuchs M, Gugerli P, Hu JS, Jelkmann W, Katis NI, Maliogka VI, Melzer MJ, Menzel W, Minafra A, Rott ME, Rowhani A, Sabanadzovic S, Saldarelli P (2012) Taxonomic revision of the family *Closteroviridae*. J Plant Pathol 94:7–19. https://www.jstor.org/stable/45156004

Martinez-Oliva BG (2020) CRISPR, una herramienta para editar genomas. Gac Med Bol 43(2):179–183. https://doi.org/10.47993/gmb.v43i2.66

Mayorga Carrillo L, Mayorga López AV, Sánchez Leal LC (2021) Evolución del diagnóstico de virus y viroides en cultivos comerciales de Solanáceas en Colombia: una revisión sistemática. Innovación y Ciencia 29(2). https://innovacionyciencia.com/articulos_cientificos/evolucion--del-diagnostico-de-virus-y-viroides-en-cultivos-comerciales-de-solanaceas-en-colombia-una-revision-sistematica

Montero R, Mundy D, Albright A, Grose C, Trought MC, Cohen D, Chooi KM, MacDiarmid R, Flexas J, Bota J (2016) Effects of *grapevine leafroll-associated virus 3* and duration of infection on fruit composition and wine chemical profile of *Vitis vinifera* L. cv. Sauvignon blanc. Food Chem 197(Pt B):1177–1183. https://doi.org/10.1016/j.foodchem.2015.11.086

Morales FJ, Anderson PK (2001) The emergence and dissemination of whitefly-transmitted geminiviruses in Latin America. Arch Virol 146:415–441. https://doi.org/10.1007/s007050170041

Morgante V, Jiménez JC, Ponce C, Urrutia C, Vera F, Farías C, Camps de la Maza R, Caro V, Muñoz M, Pollak B (2025) Viroscope™: a universal solution for plant virus and viroid diagnostics using HTS and cloud-based analysis. Front Microbiol 16. https://doi.org/10.3389/fmicb.2025.1609663

Ndowora T, Dahal G, LaFleur D, Harper G, Hull R, Olszewski NE, Lockhart B (1999) Evidence that badnavirus infection in *Musa* can originate from integrated pararetroviral sequences. Virology 255(2):214–220. https://doi.org/10.1006/viro.1998.9583

Ochoa Martínez DL, Uriza Ávila DE, Rojas Martínez RI, Rodríguez Martínez D (2016) Detection of *pineapple mealybug wilt associated virus* 1 and 3 in Mexico. Rev Mex Fitopatol 34(2):131–141. https://doi.org/10.18781/R.MEX.FIT.1601-1

Ordaz-Pérez D, Gámez-Vázquez J, Hernández-Ruiz J, Espinosa-Trujillo E, Rivas-Valencia P, Castro-Montes I (2017) Resistencia de *Vasconcellea cauliflora* al virus de la mancha anular de la papaya-potyvirus (PRSV-P) y su introgresión en *Carica papaya*. Rev Mex Fitopatol 35(3):571–590. https://doi.org/10.18781/r.mex.fit.1703-4

Pabinger S, Rödiger S, Kriegner A, Vierlinger K, Weinhäusel A (2014) A survey of tools for the analysis of quantitative PCR (qPCR) data. Biomol Detect Quantif 1:23–33

Pagán I (2022) Transmission through seeds: the unknown life of plant viruses. PLoS Pathog 18(8):e1010707. https://doi.org/10.1371/journal.ppat.1010707

Paredes Valdivieso MR (2021) Análisis del riesgo de introducción del viroide ASBVd… Tesis de pregrado, Universidad Técnica de Cotopaxi

Pedraza MR, Ponce SJL, Quintanar AEA, Valdés MD, Villareal RG (2016) Una amenaza agrícola mundial: el virus de la tristeza de los cítricos y su vector el pulgón café *Toxoptera citricida*. Artrópodos y Salud 3(1):30–36
Peña Núñez BDR, Fernández Romero JA, Fiallo Olivé E (2003) Origen y evolución de virus con especial referencia a los virus ARN. Biología 17(2) Editorial Universitaria de la República de Cuba
Peña-Prades M, Rodríguez-Regal M, Matos-García J, Cobas-Elías A, Gámez-Rodríguez H, Peña-Rivera L, Cervera-Duverger G, Barrera-Fontanet M, Barquié-Pérez O (2015) La vigilancia fitosanitaria como herramienta para la toma de decisiones en manos del productor. Ciencia en su PC 2:1–15
Pérez Álvarez S, Ávila Quezada G, Coto Arbelo O (2015) Avocado (*Persea americana* Mill). Cult Trop 36(2):111–123
Rodríguez R, Becquer R, Pino Y, López D, Rodríguez RC, Lorente González GY, Izquierdo RE, González JL (2016) Producción de frutos de piña (*Ananas comosus* (L.) Merr.) MD-2 a partir de vitroplantas. Cult Trop 37:40–48
Rubio S, Pacheco-Orozco RA, Gómez AM, Perdomo S, García-Robles R (2020) Secuenciación de nueva generación (NGS) de ADN: presente y futuro en la práctica clínica. Univ Med 61(2):49–63. https://doi.org/10.11144/javeriana.umed61-2.sngs
Ruíz-García DE, Mora-Aguilera G, Rivas-Valencia P, Góngora-Canul C, Loeza-Kuk E, Ochoa-Martínez D, Ramírez-Valverde G, Gutiérrez-Espinoza MA, Álvarez-Ramos R (2009) Sensibilidad de inmunoimpresión-ELISA y DAS-ELISA en el diagnóstico del virus de la tristeza de los cítricos. Rev Chapingo Ser Hortic 15(1):41–47
SADER. (Secretaría de Agricultura y Desarrollo Rural-Servicio de Información Agroalimentaria y Pesquera) (2024) Panorama Agroalimentario: la ruta de la transformación agroalimentaria 2018–2024. México. 210 p
Sáenz Pérez CA, Hernández EO, Estrada Drouaillet B, Poot Poot WA, Delgado Martínez R, Herrera RR (2019) Principales enfermedades en cítricos. Rev Mex Cienc Agríc 10(7):1653–1665. https://doi.org/10.29312/remexca.v10i7.1827
SAGARPA (Secretaría de Agricultura, Ganadería, Desarrollo Rural, Pesca y Alimentación) (2013) Atlas Agroalimentario, Primera edn. Servicio de Información Agroalimentaria y Pesquera, México. 194 p
Saucedo-Carabez JR, Téliz-Ortiz D, Ochoa-Ascencio S, Ochoa-Martínez D, Vallejo-Pérez MR, Beltrán-Peña H (2014) Effect of *Avocado sunblotch viroid* (ASBVd) on avocado yield in Michoacán, Mexico. Eur J Plant Pathol 138:799–805. https://doi.org/10.1007/s10658-013-0354-9
Selvarajan R, Balasubramanian V (2013) Natural occurrence of *Banana bunchy top virus* in *Ensete superbum* in India. Indian J Virol 24(1):97–98. https://doi.org/10.1007/s13337-012-0123-y
Sochor J, Babula P, Adam B, Krska B, Kizek R (2012) Sharka: the past, the present and the future. Viruses 4(11):2853–2901. https://doi.org/10.3390/v4112853
Tripathi S, Suzuki JY, Ferreira SA, Gonsalves D (2008) Papaya ringspot virus-P: characteristics, pathogenicity, sequence variability and control. Mol Plant Pathol 9(3):269–280. https://doi.org/10.1111/j.1364-3703.2007.00462.x
Umer M, Mubeen M, Iftikhar Y, Ali H, Zafar ul Hye M, Asghar R, Abbas M, Rehman MA, Moya Elizondo EA, He Y (2022) *Papaya ringspot virus*: an understanding of a severe positive-sense single-stranded RNA viral disease and its management. Phyton 91(10):2099–2110. https://doi.org/10.32604/phyton.2022.022013
Valencia Sandoval K, Duana Avila D (2019) Los cítricos en México: análisis de eficiencia técnica. Anál Econ 34(87):269–283
Vásquez-Gutiérrez U, Frías-Treviño GA, López-López H, Delgado-Ortiz JC, Aguirre-Uribe LA, Flores-Olivas A (2024) Evaluation of the pathogenicity of three isolates of *tomato brown rugose fruit virus* in tomato plants (*Solanum lycopersicum* L.) from Coahuila, Mexico. Revista Bio Ciencias 11:e1576. https://doi.org/10.15741/revbio.11.e1576
Velásquez-Valle R, Reveles-Torres LR, Reveles-Hernández M (2017) Identificación de enfermedades causadas por virus en cultivos de Aguascalientes, Durango y Zacatecas. INIFAP

Wong YP, Othman S, Lau YL, Radu S, Chee HY (2018) LAMP: a versatile technique for microorganism detection. J Appl Microbiol 124(3):626–643. https://doi.org/10.1111/jam.13647

Young AJ, Pathania N, Manners A, Pegg KG (2022) Heart rot of Australian pineapples caused by *D. zeae*. Australas Plant Pathol 51(5):525–533. https://doi.org/10.1007/s13313-022-00880-x

Zamora-Macorra EJ, Ávila-Alistac N, Lagunes-Fortiz E, de los Santos-Villalobos, S. (2023) Virus y viroides en jitomate (*Solanum lycopersicum*) y rizobacterias promotoras como alternativa de manejo. Rev Mex Fitopatol 41(4):6. https://doi.org/10.18781/R.MEX.FIT.2023-7

Climate Change-Driven Alterations and New Outbreaks of Viral Diseases in Fruit Crops

3

Munmi Borah, Devayani Sarmah, and Greeshma Varghese

Abstract

Climate change strongly affects the occurrence and spreads of viral diseases in fruit crops, posing serious threats to agricultural productivity and food security. Rising temperatures and altered rainfall patterns and distribution disrupts the fruit tree's physiology and phenology, creating favorable conditions for the emergence of new viral strains and the resurgence of existing ones. Moreover, climate changes also impact the life cycles and geographic distribution of insect vectors accelerating the virus transmission, replication and systemic movement within the host plants. Climate change also possesses challenges in management of the viral diseases, leading to significant yield and quality losses. Integrated strategies combining resistant cultivars, adaptive vector management, quarantine measures, and advanced forecasting tools alongside deeper research into virus-vector-host-climate interactions, will be essential to mitigate future climate-driven viral threats.

Keywords

Climate change · Fruits · Viral disease · Banana bunchy top virus · Plum pox virus · Citrus tristeza virus

M. Borah (✉) · G. Varghese
Department of Plant Pathology, Assam Agricultural University, Jorhat, Assam, India
e-mail: munmi.borah@aau.ac.in

D. Sarmah
Department of Plant Pathology, Assam Agricultural University, Jorhat, Assam, India

Division of Plant Pathology, Indian Agricultural Research Institute, Central Rice Research Institute, Cuttack, Odisha, India

R. Kumar et al. (eds.), *Fruit Crops and Viral Pathogens Volume II*, Plant Pathology for Sustainable Agriculture, https://doi.org/10.1007/978-981-95-7261-8_3

3.1 Introduction

Agriculture holds an important role in human civilization, aiding in global food security, economic development, and promoting environmental sustainability. Being the primary source of food, it helps to reduce hunger and meet nutritional needs. In economic terms, agriculture strengthens rural livelihoods and generates employment opportunities, significantly contributing to the nation's GDP (Gross Domestic Product). It also supplies essential raw materials for the textile, biofuels, and pharmaceutical industries. Fruit crops play an important role in agricultural systems worldwide, providing nutritional, economic, environmental, and social benefits. Rich in essential vitamins, minerals, antioxidants, and fiber, fruits contribute significantly to a balanced diet and overall human health (Slavin and Lloyd 2012). According to guidelines provided by WHO and FAO, 400 grams of fruits and vegetables (minimum), excluding starchy tubers, should be consumed by an individual per day (FAO 2015). With the increase in global population, the food demand is also expected to rise (Sharma et al. 2021; Rai et al. 2018), with strong emphasis on agricultural innovations to achieve the production goals. Despite considerable increases in crop yields over the past 50 years, agricultural productivity continues to be challenged by both biotic and abiotic factors (Choudhary and Senthil-Kumar 2024). Among the biotic stresses, plant diseases are becoming a major issue, causing a yield loss of 20–40% annually (Savary et al. 2019). In addition to yield loss, they also disrupt the surrounding ecosystems and food webs (Ristaino et al. 2021). As a result, plant diseases are now widely recognized as a severe threat to food security, the preservation of biodiversity, and the overall stability of ecosystems. Climate change refers to the long-term changes in the Earth's weather conditions, including temperature, rainfall, wind patterns, and seasonal cycles. This change is caused by natural and human activities, such as burning fossil fuels, cutting down forests etc. (Rivera and Jemielniak 2020). Burning of fossil fuels releases a large amount of greenhouse gases like carbon dioxide into the atmosphere, which raises the Earth's temperature and causes global warming. The temperature increase also accelerates the melting of glaciers and polar ice caps, resulting in rising sea levels. Tornadoes, floods, droughts, and cold or heat waves also pose serious threats to biodiversity and various ecosystems, including agricultural systems (Zia et al. 2017). Plant disease outbreaks frequently occur when weather patterns provide highly favorable conditions for pathogen development, multiplication, and spread (Burdon and Zhan 2020). Climate change influences the pathogen evolution and also modifies the host-pathogen relationship, which can further intensify the disease development (Cohen and Leach 2020). Understanding how climate change affects interactions between pathogens, plants, and their associated microbes is crucial for building resilient agricultural and natural ecosystems. This chapter will discuss the influence of climate change on host, pathogen, and vector relationships in fruit crops, along with their impacts on fruit yields, quality, and economy.

3.2 Components of Climate Change

The Earth is made up of four main spheres: atmosphere, hydrosphere, lithosphere, and biosphere. The atmosphere, i.e., layers of gases surrounding the Earth, is the most dynamic and rapidly changing part and consists of 78.1% nitrogen, 20.9% oxygen, and 0.93% argon. These gases don't interact with incoming sunlight or the Earth's infrared radiation. In contrast, gases such as carbon dioxide (CO_2), methane (CH_4), nitrous oxide (N_2O), and ozone (O_3) that are found in trace amounts absorb and emit infrared energy, which strongly influences climate change. Climate change occurs due to both anthropogenic and natural causes (Fig. 3.1).

3.2.1 Natural Causes

The major natural causes of climate change include the various atmospheric and geological processes. These natural variabilities, arising from phenomena such as the El Niño-Southern Oscillation (ENSO), Pacific Decadal Oscillation (PDO), and Atlantic Multidecadal Oscillation (AMO), play crucial roles in modulating the global warming rate (Wei et al. 2019b).

(a) *Intensity of Solar Radiation*: Climate change is influenced by the variation in the amount of solar radiation reaching the Earth's surface. The light intensity reaching the Earth differs due to the Earth's orbital movement around the Sun and its rotation on a tilted axis, which directly determines the amount of heat energy received at different points on the planet (Khavrus and Shelevytsky 2010).

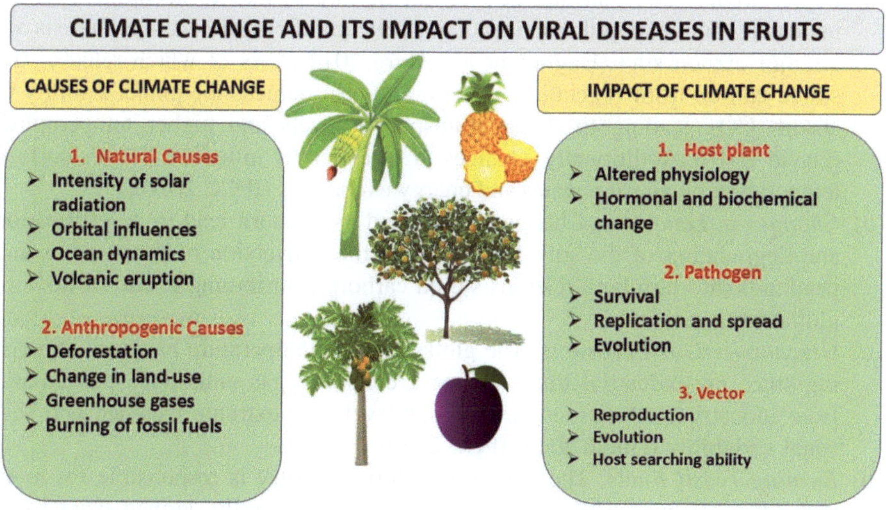

Fig. 3.1 Climate change: causes and impact on fruit viral diseases

(b) *Orbital Influences*: Climatic variations also occur due to factors such as the tilt of Earth's axis, shape of its orbit, and the timing of solstices and equinoxes (Ruddiman 2007).
(c) *Ocean Dynamics*: Wind, wave movement, temperature, and salinity fluctuations all contribute to the continuous, directed flow of seawater known as ocean currents. These circulations are crucial for managing the Earth's climate, as well as maintaining primary productivity and marine ecosystems (Duteil et al. 2014).
(d) *Volcanic Eruptions*: When a volcano erupts, it releases a large quantity of ash and sulfur into the atmosphere. This changes the amount of sunlight reaching the Earth's surface, while cooling the Earth's surface and heating the stratosphere (Robock 2000). The cooling effect can last from months to a few years, depending on eruption size, and is often stronger in the Northern Hemisphere (Kelly and Sear 1984; Sear et al. 1987). Volcanic eruption affects the carbon cycle by slowing soil decomposition and changing ocean dynamics (Frölicher et al. 2011) along with ozone layer depletion.

3.2.2 Anthropogenic Causes

Climate changes that are caused by human activities such as greenhouse gas emissions (CO_2, CH_4, and N_2O), burning of fossil fuels, agriculture, deforestation, urbanization, and industrial activities are called anthropogenic causes. CO_2 is the largest contributor to climate change, while CH_4, and N_2O from farming and fertilizers are more potent but shorter-lived (Ntinyari and Gweyi-Onyango 2021).

(a) *Deforestation*: The constant destruction of forest areas to accommodate the increasing human population is a major concern of climate change. Forests are a major carbon sink (Haghipour and Burg 2014), loss of which releases the stored carbon from vegetation and soil and also reduces the planet's ability to absorb CO_2, leading to rising atmospheric levels and higher temperatures (David 2018). Additionally, changes in forest cover influence Earth's surface temperature by altering water and energy exchanges (IPCC 2020).
(b) *Changes in Land Use*: Changes in the land use pattern lead to desertification and degradation of the soil. For example, the conversion of forest areas and peatlands into farmland releases stored carbon, contributing about 10–15% of global CO_2 emissions.
(c) *Greenhouse Gas Emissions*: The global Earth's temperature has been increasing since pre-industrial times due to greenhouse gas emissions, particularly from industrialized nations, driving sea-level rise, extreme weather, and seasonal variability (Güçlü 2014; IPCC 2020).
(d) *Burning Fossil Fuels*: The agricultural food industry is responsible for more than 30% of the world's energy consumption, primarily derived from fossil fuels, and it contributes approximately 22% to the greenhouse gas emissions caused by human activities. Burning fossil fuels increases CO_2 levels while limiting natural absorption by plants, making rising CO_2 the primary driver of global warming (IPCC 2019).

3.3 Impact of Climate Change on Plant Diseases

The emergence and evolution of viruses are greatly influenced by the interactions between hosts and vectors (Jones 2009; Lefeuvre et al. 2019). Assessing the impact of climate change on plant viruses in fruit crops is a complex task, as it involves multiple interacting factors related to plants, pathogens, and the environment. The arthropod population dynamics also play an important role in the development of viral disease epidemics. Thus, understanding the tripartite relationship between host plants, viruses, and vectors, along with the influencing environmental conditions, is essential to predict the viral epidemics and spread (Canto et al. 2009).

3.3.1 Host

The physiology of fruit trees is severely impacted due to climate change, disrupting their normal growth and development. Extreme weather events such as droughts, floods, waterlogging, etc. induce root hypoxia, impair water relations, stomatal regulation, and photosynthetic activity, leading to hormonal and biochemical imbalances (Habibi et al. 2022). Variations in temperature and rainfall patterns also modify major phenological stages, including flowering, fruiting, and metabolite production. For example, heat stress reduces the photosynthetic efficiency and defense signaling in plants and creates a favorable environment for viral pathogens to infect and establish more easily.

3.3.2 Pathogen

The survival, multiplication, spread, and evolution of fruit viruses are influenced by various climate changes. An increase in temperature shortens the virus replication cycle, resulting in more infection cycles per season, which can interfere with the natural defenses of fruit crops. In several regions, global warming and changes in the rainfall pattern have been linked to increased outbreaks of viral diseases such as papaya ringspot, banana bunchy top, and citrus tristeza. In contrast, excessively high temperatures may suppress viral replication.

3.3.3 Vector

The relationship between virus and vector is very specific. Viruses are mainly transmitted by vectors in persistent, semi-persistent and nonpersistent manner with specificity determined by coat protein (CP) for cucumber mosaic virus, HC-Pro (Helper component protease) for potyvirus etc. Climate change is reshaping these interactions with major implications for agriculture and food security. Rising temperatures and CO_2 levels alter aphid and other sap-sucking insect behavior, distribution, and virus transmission efficiency (Trębicki and Finlay 2018). Change in the physiology of the plant due to stress also influences the host-virus-vector dynamics (Van Munster 2020), making the vectors more efficient carriers (Abbasi 2025). Additionally, some viruses manipulate vectors to promote spread (Ingwell et al. 2012). Climatic factors such as temperature and rainfall affect vector survival, reproduction, and transmission, modifying transmission patterns (Fouque and Reeder 2019). Each vector has specific environmental conditions where it reproduces and transmits the virus efficiently. Climate change, especially high temperature, alters pathogen dynamics by influencing its survival, growth, number of generations etc. For instance, the banana aphid *Pentalonia nigronervosa*, transmitting the banana bunchy top virus, reproduces most efficiently at around 25 °C (Robson et al. 2007).

3.3.4 Environmental Conditions

Different environmental factors affect the viral disease incidence in fruit crops.

3.3.4.1 Temperature

Increase in global temperature significantly influence the replication, transmission, and epidemiology of fruit viruses, thereby limiting agricultural productivity. High temperatures accelerate both local and systemic movement of viruses (Amari et al. 2021), influence viral replication and transmission, resulting in more infection cycles during a single growing season (Tsai et al. 2022). Furthermore, heat stress changes the physiology of fruit tree, affecting their metabolic processes and weakening RNA silencing-based antiviral defenses (Canto et al. 2009). Temperature fluctuations further affect viral evolution by promoting genetic diversity and adaptation within viral populations. Overall, the high temperature poses serious challenges for the management of virus outbreaks, emphasizing the urgent need for improved management strategies and climate-resilient agricultural systems (Tsai et al. 2022).

3.3.4.2 Excess Moisture

With the rise in global temperature, there is a change in moisture availability, rainfall pattern, and rainfall distribution which severely impacts the plant, virus, and vector interactions. Changes in humidity and rainfall patterns alter the habitats of insect vectors, such as changes in geographical distribution, population densities, and migration patterns, indirectly changing the distribution of the viruses (Canto et al. 2009).

3.3.4.3 Drought

Drought changes the physiology and biochemical processes in fruit trees, subsequently altering the occurrence and severity of viral diseases. It intensifies the viral diseases by weakening the plant's defense mechanisms and increasing its vulnerability to biotic stresses. The reduced water availability affects the plant's ability to maintain cellular functions, thus impairing its immune responses and potentially increasing the virulence of pathogens, including viruses (Dikilitas et al. 2016).

Drought alters the calcium signaling, abscisic acid signaling, and protein phosphorylation in the fruit tree, which regulates the tree's early response to the stress (Liu et al. 2023). For instance, Silva et al. (2019) studied the impact of water deficit during CTV (citrus tristeza virus) infection in *Citrus sinensis* (sweet orange) cv. Pera Bianchi, revealing that drought-affected trees showed increased protein and proline accumulation with no effect on CTV population levels, while well-watered plants infected with the severe isolate exhibited higher viral titers.

3.3.4.4 Elevated CO_2

Elevated CO_2 level produces multiple effects on fruit virus interactions, mainly by altering host physiology and dynamics between plants, viruses, and vectors. Rising CO_2 level increases the plant productivity through enhanced photosynthesis but also negatively impact pathology and pest interactions. Elevated CO_2 has been shown to mediate the effects of virus infections on plants and vectors. Studies indicate that CO_2 elevation can influence nitrogen concentrations in plants, which in turn affects the life history and behavior of insect vectors like aphids, as well as the feeding damage they inflict. Notably, aphid populations are reduced under elevated CO_2 on noninfected plants, but this reduction is not observed on virus-infected plants, suggesting that elevated CO_2 could enhance the spread of viruses by maintaining vector populations on infected plants (Trębicki et al. 2016).

3.4 Impact of Climate Change on Major Viral Diseases on Fruit Crops

3.4.1 Banana Bunchy Top Virus (BBTV)

Chakraborty et al. (2021) demonstrated that BBTV acquisition affects the banana aphid, *Pentalonia nigronervosa*. Viruliferous aphids showed shorter lifespans and greater mortality under cold stress, especially at 5 °C. Temperature influenced heat shock protein (Hsp40, Hsp70, Hsp90) expression, with viruliferous aphids showing higher Hsp levels under both heat and cold stress. BBTV titer and transmission efficiency were lowest at 15 °C and highest at 38 °C. Similarly, Dahal et al. (1998) demonstrate prominent symptoms with high virus titer at a reduced temperature of 22 °C than at 28–35 °C in banana plants infected with banana streak badnanvirus (BSV). A climate chamber study evaluated the impact of different temperatures (18.3, 23.9, 29.4, and 35.0 °C) on cucumber, plantain, and banana plants infected with cucumber mosaic virus. Symptom suppression was observed at 35.0 °C, with moderate reduction at 29.4 °C, though some plants became stunted or necrotic. Severe disease symptoms appeared at 23.9 °C in all species, while at 18.3 °C, musaceous plants showed winter-like symptoms and cucumbers developed pronounced mottling (Bird et al. 1971). High climate variability, with extremes in rainfall and temperature, can prevent the establishment and spread of banana bunchy top disease (BBTD). This occurs by reducing aphid populations and lowering virus transmission efficiency (Anhalt and Almeida 2008). BBTD is most common in warm, humid regions at low altitudes. It spreads to higher elevation zones mainly through infected planting materials. However, climate warming might make higher zones more suitable for aphid vectors and BBTV transmission. Cooler high-altitude sites are becoming increasingly at risk due to climate change. The occurrence of BBTD strongly correlates with factors influencing banana production, such as area under Musa cultivation, production zones, yield, and vegetation indices. Geographic assessments show that large areas of Central, West, and East Africa meet favorable conditions for BBTD (Bouwmeester et al. 2023).

3.4.2 Papaya Ringspot Virus (PRSV)

Papaya ringspot virus (PRSV-P) spreads through various aphid species such as *Myzuspersicae*. The damage from this disease is expected to worsen as temperatures rise. Mangrauthia et al. (2009) noted that symptoms were more severe at temperatures between 26 and 31 °C. In a study pertaining PRSV in the Indian subcontinent condition, aphid population dynamics were strongly shaped by seasonal conditions. During the humid, rainy, and windy monsoon months of June–August (mean temp. 27.4–31.8 °C, RH 65–90%, frequent rainfall, wind speeds up to 9 m/s), low aphid activity and PRSV incidence were observed, but favored during the warm, dry, and calm months of October–January (mean temp. 24–30 °C, RH 40–70%, minimal rainfall, wind often <3 m/s, solar radiation >250 W/m^2), when higher aphid populations and PRSV incidence were recorded (Premchand et al. 2025).

3.4.3 Citrus Tristeza Virus (CTV)

Rising temperatures, elevated CO_2 levels, and shifting rainfall patterns also affect the CTV vector brown citrus aphid (*Toxoptera citricida*) population. In the context of climate effects on aphid phenology and survival dynamics in Florida citrus ecosystems, it was found that *T. citricida* thrives on new citrus shoots, with spring offering optimal conditions for growth due to the humid subtropical climate of Florida. Traditionally, cool winters and hot summers limit aphid survival, but warmer winters may reduce dormancy in citrus trees, promote early shoot growth, and support aphid survival between seasons. This could also lead to earlier activity of natural enemies. However, hotter, longer summers may be less favorable for both aphids and their predators, especially in southern Florida. While laboratory studies show upper temperature limits for aphid development, field observations suggest some adaptability to high temperatures (Qureshi 2010). A study showed that certain CTV strains respond negatively to higher temperatures in a strain- and host-specific manner, i.e., strain T36 significantly decreased in titer in *C. macrophylla* and *C. sinensis* but not in *C. aurantium*, while in strain T30, viral titer dropped only in *C. sinensis*, and strain VT decreased across all hosts tested. These temperature-induced changes in virus titer can influence systemic infection and reduce virus spread by vectors, potentially delaying disease onset, especially given the crucial role of specific CTV strain titers in determining disease severity (Cowell et al. 2016).

3.4.4 Plum Pox Virus (PPV)

Climate change, by increasing seasonal temperatures, particularly in regions like Central Europe, shifts the timing and intensity of aphid activity, which are the primary vectors of PPV. This shift means that aphid flights may increasingly coincide with warmer periods when temperatures frequently reach or exceed 30 °C. Under these conditions, PPV isolates that are less negatively affected by higher temperatures in terms of infectivity and accumulation in host plants gain a dissemination advantage. The progress of viral epidemics depends on the retention period of the virus in the vector, which is influenced by temperature. Glasa et al. (2003) studied the influence of temperature on infectivity, symptom expression, and accumulation of five isolates of PPV in *Nicotiana benthamiana* using double antibody sandwich enzyme-linked immunosorbent assay (DAS-ELISA). The results reveal that at 17 °C, all five isolates showed similar infectivity and symptom intensity, but their ability to multiply varied at higher temperatures. PPV replication was inhibited at 32 °C, and both infection rate and antigen levels declined at 30 °C. Among them, the BOR-3 isolate demonstrated greater tolerance to high temperatures, suggesting that adaptation to heat may enhance the epidemiological potential of PPV. Consequently, elevated temperatures can promote the dissemination and epidemiological effects of temperature-adapted PPV strains by prolonging or altering the activity of aphid vectors, which changes virus transmission patterns and may heighten disease pressure in areas that were previously cooler. This dynamic highlights how climate change may modify host-vector-virus interactions, facilitating the spread of PPV and affecting disease management under warmer future scenarios.

3.4.5 Pineapple Mealybug Wilt-Associated Virus (PMWaV)

The importance of pineapple mealybug wilt, caused by the complex called Pineapple mealybug wilt-associated virus (PMWaV-1, PMWaV-2, and PMWaV-3), transmitted and spread by *Dysmicoccus brevipes* and *Dysmicoccus neobrevipes* (Sether et al. 2001), on the other hand, may increase because the vectors could be stimulated by rising temperatures. In a study by Wei et al. (2019a, b) annual mean temperature was identified as the key factor influencing *D. brevipes* distribution, while the minimum temperature of the coldest month and mean temperature of the coldest quarter explained the occurrence of *D. neobrevipes* by the MaxEnt model was used to predict the potential global distribution of these two invasive pineapple mealybugs under future climate change scenarios (2050s and 2070s). The study projected high habitat suitability for both pests in South America, Africa, and South Asia, with distribution shifts varying across climate scenarios.

3.5 Impact on Fruit Crop Yield, Quality, and Economy

Viral infections in fruit crops disrupt key physiological and metabolic processes, causing substantial yield and quality losses with significant economic consequences. Climate change damages chloroplast protein, thereby reducing photosynthesis, plant vigor, and fruit set. It also causes hormonal imbalance and disrupts the cell cycle, resulting in stunted growth, abnormal flower development, and deformed fruits, further reducing marketable produce. Due to the presence of the virus, the resources from fruit development are directed towards defense, severely affecting fruit size, sweetness, and color. Moreover, accumulation of reactive oxygen species (ROS) and damage to plant tissues causes early leaf drop, unhealthy canopies, and lower photosynthetic capacity throughout the production cycle. Symptoms like mottling, necrosis, and leaf curl reduce visual appeal and affect consumer acceptance (Jiang and Zhou 2023).

3.6 Challenges in Disease Diagnosis and Management Under a Changing Climate

The risk of viruses invading new fruit-growing regions is increasing due to climate change, making quarantine and exclusion strategies very important. Moreover, most of the fruit viruses are transmitted by insects like aphids, mealybugs, and whiteflies, whose populations, migratory patterns, and feeding activities are tightly linked to changes in temperature, humidity, and seasonal weather. Increasing global temperatures intensifies the activity of these vectors, decreases the virus incubation period, and increases transmission efficiency, thereby escalating disease pressure in orchards and plantations (Bhattacharjee et al. 2022). Genetic resistance in fruit crop varieties, achieved through conventional breeding or genetic modification, is also challenged by climate change. Changes in the physiology of the tree, such as shifting hormone profiles, altered surface traits, and modification of metabolic pathways under higher CO_2 and temperature, can undermine resistance mechanisms, making previously resistant cultivars more susceptible to viral infection. Additionally, certain resistance genes may be downregulated or function less effectively as growing conditions change, resulting in partial or total failure of disease resistance (Osorio-Marín et al. 2024). Emerging pathogens will likely move into temperate areas, creating additional difficulties for disease control. As a result, developing resistance to pests and diseases remains a top priority for breeding programs and will continue to be essential, alongside the enhancement of other desirable traits, to address the challenges brought on by climate change (Brummer et al. 2011). For chemical management, the main targets are the insect vectors transmitting the majority of the plant viruses. However, pesticide efficiency may decline due to changes in residue dynamics, vector population behavior, or increasing resistance in insect populations. The harmful impacts of chemical pesticides necessitate reduced usage and greater reliance on biological control. Biological control, using natural enemies such as predators, parasitoids, or entomopathogenic fungi, are effected by climate-driven disruptions in ecological balance and vector biology, leading to variable outcomes depending on context and location (Subedi et al. 2023). Biological control agents play a crucial role in managing insect vectors of plant viruses. Predators such as *Coccinella septempunctata* (ladybird beetle) and *Chrysoperla carnea* (green lacewing) feed on aphids and whiteflies, reducing vector populations. Parasitoids like *Aphidius colemani* (targeting aphids) and *Encarsia formosa* (attacking whiteflies) are widely used in integrated pest management. Entomopathogenic fungi, including *Beauveria bassiana* and *Verticillium lecanii*, infect and kill vectors such as aphids, whiteflies, and thrips, offering effective microbial control. Additionally, insect-specific viruses (ISVs) are emerging as novel biocontrol tools to suppress vector fitness and transmission potential (Wyckhuys et al. 2025). To tackle these challenges, developing adaptive, integrated management strategies is crucial. This involves vector management, use of resistant cultivars, strict quarantine procedures, and monitoring of the evolving interactions among viruses, vectors, and hosts under changing environmental conditions.

3.7 Strategies to Mitigate Climate-Driven Viral Threats

Climate change directly impacts the relationships between viruses and hosts in fruit crops, changing the genetic resistance to viral diseases. Changes in the physiology of the fruit tree due to high CO_2 levels influence photosynthesis, papillae production, silicon buildup at infection sites, nutrient distribution, and metabolite profiles (Debela and Tola 2018). While much research has looked at how temperature and rainfall affect pathogen dynamics, like vector populations and virus transmission, there are few direct studies on viral diseases in fruit crops. Under a changing climate, it is crucial to employ integrated approaches such as vector management, climate-based forecasting and early warning systems, along with policy measures and farmer education, to effectively manage viral diseases in fruit crops. The Citrus Advisory System, hosted by AgroClimate, provided growers with climate-based indicators (e.g., rainfall, temperature, growing degree days) to support decision-making in managing diseases such as CTV and citrus greening, by timing monitoring or interventions to reduce vector pressure and disease risk (AgroClimate n.d.). Furthermore, quarantine measures are important to stop the introduction and spread of new fruit viruses into different areas (Chakraborty and Newton 2011).

3.8 Research Gaps and Future Perspectives

Research on fruit trees, viruses, vectors, and climate remains limited, highlighting the need to understand the impact of climate change on vector biology and virus transmission. Novel diagnostic methods like artificial intelligence offers useful way to improve surveillance, early detection and tracking of fruit viruses. As fruit trees are perennial in nature, long-term studies on host physiology, epidemiology, and host-pathogen interaction need to be carried out using various experimental methods, including open-top chambers, free-air CO_2 enrichment, and controlled environment experiments (Ghini et al. 2011). Tools like MaxEnt, SARIMAX, CLIMAX, and process-based simulations forecast outbreaks by integrating environmental, climatic, and biological data, can also be used. Machine learning and artificial intelligence further enhance accuracy by combining climate trends, image analysis, and historical records for real-time disease management, but these tools are not actively used in studies pertaining to viral diseases in fruit crops as of now.

3.9 Conclusion

Climate change is modifying the impact of viral diseases on fruit crops by changing the vector populations, virus behavior, and host-pathogen interactions. Changes in temperature, rainfall, and CO_2 levels etc. alter the host range and efficiency of transmission of the insect vectors. To manage these emerging threats, flexible, integrated strategies need to be adopted, including vector control, resistant plant varieties, and quarantine measures. Long-term research into the physiology, epidemiology, genetics, and evolution of fruit crop viral systems under various climate scenarios is essential. Keeping global fruit production viable in the face of climate-driven viral risks will rely on coordinated research, policy, and management that can anticipate new challenges and respond quickly.

References

Abbasi E (2025) Climate change and vector-borne disease transmission: the role of insect behavioral and physiological adaptations. Integr Org Biol 7(1):obaf011. https://doi.org/10.1093/iob/obaf011

AgroClimate (n.d.) Tools for managing climate risk in agriculture. University of Florida. http://agroclimate.org/

Amari K, Huang C, Heinlein M (2021) Potential impact of global warming on virus propagation in infected plants and agricultural productivity. Front Plant Sci 12:649768. https://doi.org/10.3389/fpls.2021.649768

Anhalt MD, Almeida RPP (2008) Effect of temperature, vector life stage, and plant access period on transmission of banana bunchy top virus to banana. Phytopathology 98(6):743–748. https://doi.org/10.1094/PHYTO-98-6-0743

Bhattacharjee P, Warang O, Das S, Das S (2022) Impact of climate change on fruit crops—a review. Curr World Environ 17(2):319–330

Bird J, Sanchez J, Tió MA, Liu LJ (1971) Effect of various temperatures on symptom expression by different plant species (cucumber, plantain, and banana) affected by cucumber mosaic virus. J Agric Univ P R 55:70–77

Bouwmeester H, Blomme G, Omondi AB, Ocimati W (2023) Banana bunchy top disease in Africa—predicting continent-wide disease risks by combining survey data and expert knowledge. Plant Pathol 72(8):1476–1490. https://doi.org/10.1111/ppa.13764

Brummer EC, Barber WT, Collier SM, Cox TS, Johnson R, Murray SC, Olsen RT, Pratt RC, Thro AM (2011) Plant breeding for harmony between agriculture and the environment. Front Ecol Environ 9(10):561–568. https://doi.org/10.1890/100225

Burdon JJ, Zhan J (2020) Climate change and disease in plant communities. PLoS Biol 18(11):e3000949. https://doi.org/10.1371/journal.pbio.3000949

Canto T, Aranda MA, Fereres A (2009) Climate change effects on physiology and population processes of hosts and vectors that influence the spread of hemipteran-borne plant viruses. Glob Chang Biol 15(8):1884–1894. https://doi.org/10.1111/j.1365-2486.2008.01820.x
Chakraborty S, Newton AC (2011) Climate change, plant diseases and food security: an overview. Plant Pathol 60(1):2–14. https://doi.org/10.1111/j.1365-3059.2010.02411.x
Chakraborty S, Barman M, Samanta S, Roy M, Tarafdar J (2021) Effect of banana bunchy top virus on the heat shock protein genes of Pentalonia nigronervosa during temperature susceptibility and its effect on virus transmission. Agronomy 11(9):Article 9. https://doi.org/10.3390/agronomy11091866
Choudhary A, Senthil-Kumar M (2024) Drought: a context-dependent damper and aggravator of plant diseases. Plant Cell Environ 47(6):2109–2126. https://doi.org/10.1111/pce.14863
Cohen SP, Leach JE (2020) High temperature-induced plant disease susceptibility: more than the sum of its parts. Curr Opin Plant Biol 56:235–241. https://doi.org/10.1016/j.pbi.2020.02.008
Cowell SJ, Harper SJ, Dawson WO (2016) Some like it hot: citrus tristeza virus strains react differently to elevated temperature. Arch Virol 161(12):3567–3570. https://doi.org/10.1007/s00705-016-3083-5
Dahal G, Hughes JDA, Thottappilly G, Lockhart BEL (1998) Effect of temperature on symptom expression and reliability of banana streak badnavirus detection in naturally infected plantain and banana (Musa spp.). Plant Dis 82(1):16–21. https://doi.org/10.1094/PDIS.1998.82.1.16
David E (2018) Forests and water: background analytical study prepared for the thirteenth session of the United Nations Forum on Forests, April 2018
Debela C, Tola M (2018) Effect of elevated CO_2 and temperature on crop–disease interactions under rapid climate change. Int J Environ Resour Res 13(1):1–7. https://doi.org/10.19080/IJESNR.2018.13.555851
Dikilitas M, Ahmad P, Hashem A, Abd Allah EF, Karakas S (2016) Oxidative stress and plant responses to pathogens under drought conditions. In: Water stress and crop plants: a sustainable approach. Wiley, Chichester, pp 102–123. https://doi.org/10.1002/9781119054450.ch8
Duteil O, Böning CW, Oschlies A (2014) Variability in subtropical-tropical cells drives oxygen levels in the tropical Pacific Ocean. Geophys Res Lett 41(24):8926–8934. https://doi.org/10.1002/2014GL061774
FAO (2015) Promotion of fruit and vegetables for health: report of the pacific regional workshop. FAO
Fouque F, Reeder JC (2019) Impact of past and on-going changes on climate and weather on vector-borne diseases transmission: a look at the evidence. Infect Dis Poverty 8(1):51. https://doi.org/10.1186/s40249-019-0565-1
Frölicher TL, Joos F, Raible CC (2011) Sensitivity of atmospheric CO_2 and climate to explosive volcanic eruptions. Biogeosciences 8(8):2317–2339. https://doi.org/10.5194/bg-8-2317-2011
Ghini R, Bettiol W, Hamada E (2011) Diseases in tropical and plantation crops as affected by climate changes: current knowledge and perspectives. Plant Pathol 60(1):122–132. https://doi.org/10.1111/j.1365-3059.2010.02403.x
Glasa M, Labonne G, Quiot J-B (2003) Effect of temperature on plum pox virus infection. Acta Virol 47(1):49–52
Güçlü Y (2014) Sea surface temperature anomalies at the Mediterranean coast of Turkey (period of 1968-2010). Procedia Soc Behav Sci 120:476–486. https://doi.org/10.1016/j.sbspro.2014.02.127
Habibi F, Liu T, Shahid MA, Schaffer B, Sarkhosh A (2022) Physiological, biochemical, and molecular responses of fruit trees to root zone hypoxia. Environ Exp Bot 206:105179. https://doi.org/10.1016/j.envexpbot.2022.105179
Haghipour N, Burg JP (2014) Geomorphological analysis of the drainage system on the growing Makran accretionary wedge. Geomorphology 209:111–132. https://doi.org/10.1016/j.geomorph.2013.11.030
Ingwell LL, Bosque-Pérez NA, Eigenbrode SD (2012) Plant viruses alter insect behavior to enhance their spread. Sci Rep 2(1):578. https://doi.org/10.1038/srep00578

Intergovernmental Panel on Climate Change (IPCC) (2019) Summary for Policymakers. In: Global warming of 1.5°C. An IPCC special report on the impacts of global warming of 1.5°C above preindustrial levels and related global greenhouse gas emission pathways, in the context of strengthening the global response to the threat of climate change, sustainable development, and efforts to eradicate poverty [Masson-Delmotte, V.P. Zhai, H.-O. Pörtner, D. Roberts, J. Skea, P.R. Shukla, A. Pirani, W. Moufouma-Okia, C. Péan, R. Pidcock, S. Connors, J.B.R. Matthews, Y. Chen, X. Zhou, M.I. Gomis, E. Lonnoy, T. Maycock, M. Tignor, and T. Waterfield (Eds.)]. IPCC

Intergovernmental Panel on Climate Change (IPCC) (2020) Climate change and land. An IPCC special report on climate change, desertification, land degradation, sustainable land management, food security, and greenhouse gas fluxes in terrestrial ecosystems. Summary for Policymakers. IPCC

Jiang T, Zhou T (2023) Unraveling the mechanisms of virus-induced symptom development in plants. Plants 12(15):2830. https://doi.org/10.3390/plants12152830

Jones RA (2009) Plant virus emergence and evolution: origins, new encounter scenarios, factors driving emergence, effects of changing world conditions, and prospects for control. Virus Res 141(2):113–130. https://doi.org/10.1016/j.virusres.2008.07.028

Kelly PM, Sear CB (1984) Climatic impact of explosive volcanic eruptions. Nature 311(5988):740–743. https://doi.org/10.1038/311740a0

Khavrus V, Shelevytsky I (2010) Introduction to solar motion geometry on the basis of a simple model. Phys Educ 45(6):641. https://doi.org/10.1088/0031-9120/45/6/010

Lefeuvre P, Martin DP, Elena SF, Shepherd DN, Roumagnac P, Varsani A (2019) Evolution and ecology of plant viruses. Nat Rev Microbiol 17(10):632–644. https://doi.org/10.1038/s41579-019-0232-3

Liu X, Gao T, Mao K, Liu C, Ma F, Li C, Gong X (2023) Fruit crops combating drought: physiological responses and regulatory pathways. Plant Physiol 192(3):1768–1784. https://doi.org/10.1093/plphys/kiad202

Mangrauthia SK, Singh Shakya VP, Jain RK, Praveen S (2009) Ambient temperature perception in papaya for papaya ringspot virus interaction. Virus Genes 38(3):429–434. https://doi.org/10.1007/s11262-009-0336-3

Ntinyari W, Gweyi-Onyango JP (2021) Greenhouse gases emissions in agricultural systems and climate change effects in sub-Saharan Africa. In: African handbook of climate change adaptation. Springer, Cham, pp 1081–1105. https://doi.org/10.1007/978-3-030-45106-6_43

Osorio-Marín J, Fernandez E, Vieli L, Ribera A, Luedeling E, Cobo N (2024) Climate change impacts on temperate fruit and nut production: a systematic review. Front Plant Sci 15:1352169. https://doi.org/10.3389/fpls.2024.1352169

Premchand U, Mesta RK, Basavarajappa MP, Venkataravanappa V, Narasimha Reddy LRC, Shankarappa KS (2025) Epidemiological studies on the incidence of papaya ringspot disease under Indian sub-continent conditions. Sci Rep 15(1):6973. https://doi.org/10.1038/s41598-025-91612-w

Qureshi JA (2010) Implications of climate change for Toxoptera citricida (Kirkaldy), a disease vector of citrus in Florida. In: Kindlmann P, Dixon AFG, Michaud JP (eds) Aphid biodiversity under environmental change: patterns and processes. Springer, Dordrecht, pp 91–106. https://doi.org/10.1007/978-90-481-8601-3_6

Rai PK, Kumar V, Lee S, Raza N, Kim KH, Ok YS, Tsang DC (2018) Nanoparticle-plant interaction: implications in energy, environment, and agriculture. Environ Int 119:1–19. https://doi.org/10.1016/j.envint.2018.06.012

Ristaino JB, Anderson PK, Bebber DP, Brauman KA, Cunniffe NJ, Fedoroff NV et al (2021) The persistent threat of emerging plant disease pandemics to global food security. Proc Natl Acad Sci USA 118(23):e2022239118. https://doi.org/10.1073/pnas.2022239118

Rivera and Jemielniak (2020) Collaborative consumption, sharing economy and sustainability. Emerald Publishing Limited

Robock A (2000) Volcanic eruptions and climate. Rev Geophys 38(2):191–219. https://doi.org/10.1029/1998rg000054

Robson JD, Wright MG, Almeida RP (2007) Biology of Pentalonia nigronervosa (Hemiptera, Aphididae) on banana using different rearing methods. Environ Entomol 36(1):46–52. https://doi.org/10.1603/0046-225X(2007)36[46:BOPNHA]2.0.CO;2

Ruddiman WF (2007) Earth's climate: past and future, 2nd edn. W. H. Freeman

Savary S, Willocquet L, Pethybridge SJ, Esker P, McRoberts N, Nelson A (2019) The global burden of pathogens and pests on major food crops. Nat Ecol Evol 3(3):430–439. https://doi.org/10.1038/s41559-018-0793-y

Sear CB, Goodess CM, Jones PD, Kelly PM (1987) Global surface-temperature responses to major volcanic eruptions. Nature 330(6146):365–367. https://doi.org/10.1038/330365a0

Sether DM, Karasev AV, Okumura C, Arakawa C, Zee F, Kislan MM, Busto JL, Hu JS (2001) Differentiation, distribution, and elimination of two different pineapple mealybug wilt-associated viruses found in pineapple. Plant Dis 85(8):856–864. https://doi.org/10.1094/PDIS.2001.85.8.856

Sharma S, Rana VS, Pawar R, Lakra J, Racchapannavar V (2021) Nanofertilizers for sustainable fruit production: a review. Environ Chem Lett 19(2):1693–1714. https://doi.org/10.1007/s10311-020-01125-3

Silva CC, Molina RO, Back L, Oliveira CM, Frias AA, Zanutto CA, Nunes WM (2019) The effect of drought conditions on sweet orange (Citrus sinensis) plants infected with citrus tristeza virus (CTV). Trop Plant Pathol 44(4):335–342. https://doi.org/10.1007/s40858-019-00291-2

Slavin JL, Lloyd B (2012) Health benefits of fruits and vegetables. Adv Nutr 3(4):506–516. https://doi.org/10.3945/an.112.002154

Subedi B, Poudel A, Aryal S (2023) The impact of climate change on insect pest biology and ecology: implications for pest management strategies, crop production, and food security. J Agric Food Res 14:100733. https://doi.org/10.1016/j.jafr.2023.100733

Trębicki P, Finlay K (2018) Pests and diseases under climate change; its threat to food security. In: Food security and climate change. Wiley, Hoboken, pp 229–249. https://doi.org/10.1002/9781119180661.ch11

Trębicki P, Bosque-Pérez NA, Yen AL, Fitzgerald GJ, Freeman AJ, Powell KS, Dader B, Luck JE, Vandegeer RK (2016) Virus infection mediates the effects of elevated CO2 on plants and vectors. Sci Rep 6(1):22785. https://doi.org/10.1038/srep22785

Tsai WA, Brosnan CA, Mitter N et al (2022) Perspectives on plant virus diseases in a climate change scenario of elevated temperatures. Stress Biol 2:37. https://doi.org/10.1007/s44154-022-00058-x

Van Munster M (2020) Impact of abiotic stresses on plant virus transmission by aphids. Viruses 12(2):216. https://doi.org/10.3390/v12020216

Wei J, Peng L, He Z, Lu Y, Wang F (2019a) Potential distribution of two invasive pineapple pests under climate change. Pest Manag Sci 76(5):1652–1663. https://doi.org/10.1002/ps.5684

Wei M, Qiao F, Yang X, Shu Q, Deng J, Guo Y, Song Z (2019b) Quantifying the importance of interannual, interdecadal and multidecadal climate natural variabilities in the modulation of global warming rates. Clim Dyn 53(11):6715–6727. https://doi.org/10.1007/s00382-019-04955-2

Wyckhuys KAG, Zou Y, Crowder DW, Adriani E, Albaytar AB, Beltran MJB, Ben Fekih I, Camargo-Gil C, Sta. Cruz FC, Cicero L, Colmenarez YC, Cuellar-Palacios CM, Dubois T, Eigenbrode SD, Francis F, Fereres A, Haddi K, Khamis FM, Le Lann C et al (2025) Biological control mitigates spread of vector-borne plant pathogens. Agric Ecosyst Environ 388:109683. https://doi.org/10.1016/j.agee.2025.109683

Zia A, Iqbal MA, Mohsin M, Aslam MK (2017) Climate change and its impact on food security. In: Handbook of climate change management. Springer, pp 1–20

Robson JD, Wright MG, Almeida RP (2007) Biology of Pentalonia nigronervosa (Hemiptera, Aphididae) on banana using different rearing methods. Environ Entomol 36(1):46–52. https://doi.org/10.1603/0046-225X(2007)36[46:BOPNHA]2.0.CO;2
Ruddiman WF (2001) Earth's climate: past and future, 3rd edn. W. H. Freeman
Savary S, Willocquet L, Pethybridge SJ, Esker P, McRoberts N, Nelson A (2019) The global burden of pathogens and pests on major food crops. Nat Ecol Evol 3(3):430–439. https://doi.org/10.1038/s41559-018-0793-y
[illegible]

Viral Disease Complexes in Fruit Crops: Synergism and Mixed Infections

4

R. L. R. Cabral, I. L. F. Andrade, G. A. S. Santos, D. R. S. Jesus, and A. S. Xavier

Abstract

Viral diseases pose a major threat to fruit production worldwide, reducing productivity and fruit quality, and compromising the sustainability of crop systems. Unlike single infections, mixed infections of two or more viruses are recognized for their ability to intensify symptom severity, increase viral load, and exacerbate economic losses. This chapter provides an overview of the mechanisms underlying viral complexes in fruit crops. Special attention is given to the role of coinfections, whether simultaneous or sequential, and the impact of subviral agents, such as satellites and viroids, which can potentiate or attenuate disease expression. Examples from different fruit crops illustrate how viral complexes influence epidemiology and management, highlighting cases where resistance breakdown occurs in mixed infections. The emergence of new viral complexes, often associated with climate change and global trade, highlights the dynamic nature of these pathosystems. Current advances in biotechnology, including RNA interference (RNAi), genome editing (CRISPR-Cas), and the development of transgenic plants, are promising strategies for mitigating the challenges posed by complex infections. In addition to these approaches, integrating traditional methods, such as virus-free planting material, certification programs, and vector management, is also necessary. In conclusion, this chapter emphasizes the need

R. L. R. Cabral
Department of Agronomy, Genetic and Plant Breeding Graduate Program, Federal University of Espírito Santo, Alegre, Espírito Santo, Brazil
e-mail: rafael.cabral@ufes.br

I. L. F. Andrade · G. A. S. Santos · D. R. S. Jesus · A. S. Xavier (✉)
Department of Agronomy, Plant Pathology Graduate Program, Federal Rural University of Pernambuco, Recife, Pernambuco, Brazil
e-mail: ingryd.andrade@ufrpe.br; grenville.antonio@ufrpe.br; deyse.silvino@ufrpe.br; andre.sxavier@ufrpe.br

R. Kumar et al. (eds.), *Fruit Crops and Viral Pathogens Volume II*, Plant Pathology for Sustainable Agriculture, https://doi.org/10.1007/978-981-95-7261-8_4

for multidisciplinary approaches that combine virology, entomology, plant breeding, and bioinformatics. This integration is crucial for developing sustainable management strategies, ensuring long-term disease control and preserving fruit production in a global agricultural landscape.

Keywords

Sustainable agriculture · Plant virology · Epidemiology · Biotechnology · Cross-protection · Crop resilience · Host adaptation · Agroecosystem health

4.1 Introduction

The fruit market is an important sector, not only for its direct food supply to the population, but also for its socioeconomic impact on producing regions. According to recent FAO data, global fruit production has grown consistently in recent decades in 2023, with a total of 2.1 billion tons of fruit being produced, generating a trade turnover of approximately US$466.736 billion. Currently, China is the world's largest producer (269.3 million tons), concentrating various crops such as apples, citrus fruits, melons, pears, and watermelons; India is second in the ranking (113.9 million tons), with emphasis on bananas, mangoes, oranges, and papayas. Brazil ranks third as the world's largest fruit producer (43.2 million tons), with emphasis on oranges, bananas, and tomatoes (FAOSAT 2024). However, global fruit farming faces significant challenges. Climate change, market fluctuations, phytosanitary restrictions, and trade barriers require constant adaptations by producers and supply chains. Problems related to pests and diseases pose a constant threat to fruit productivity and quality, reinforcing the need for integrated management strategies and genetic improvement programs capable of ensuring greater crop resilience (Lindell et al. 2023). In this scenario, plant viruses can cause several symptoms in plants, accounting for significant production losses in agricultural fields. Symptoms of infection include mosaic disease, chlorosis, stunting, and/or leaf necrosis, with impaired growth. Plants may also experience stunted growth and fruit deformation, compromising development and reducing fruit production and quality (Jiang and Zhou 2023). Viral diseases are reported worldwide in fruit crops, particularly in large-scale commercial production, resulting in high rates of agricultural losses. In grapevines, grapevine leafroll caused by Grapevine leafroll-associated virus 3 (GLRaV-3) can compromise 60% of vineyard productivity (Maree et al. 2013). In strawberries, asymptomatic strawberry virus 1 (StrV-1) infection led to a 31.8% loss of productivity (Valentova et al. 2022). In apples, *Apple necrotic mosaic virus* (ApNMV) infection can cause losses of up to 50% of the orchard (Manzoor et al. 2023). In citrus, citrus tristeza disease caused by Citrus tristeza virus (CTV) has destroyed 100 million trees propagated by infected sour orange rootstocks. At another time, destructive tristeza epidemics also occurred in producing countries such as Argentina (1930), Brazil (1937), the United States (1939 and 1951), Israel (1970), Cyprus (1989), Mexico (1995) and Italy (2002), among other countries (Moreno et al. 2008). It is evident that viral infections involve the interaction of three main actors:

the plant, the virus, and the vectors and this triangle is being modulated by the environment. Symptomatic manifestation and, consequently, the economic losses resulting from these diseases can be significantly intensified in the presence of mixed infections (Sánchez-Tovar et al. 2025). According to Xu et al. (2022), mixed infections are characterized by a plant being infected by two or more viral species simultaneously in a single host, a situation that can trigger different types of interactions between pathogens, whether synergistic, antagonistic, or neutral. In this chapter, we will examine in greater depth the concepts underlying mixed infections, the mechanisms governing plant-host interactions, the viral complexes affecting fruit trees, and the current diagnostic approaches. Additionally, we will analyze the diverse mechanisms of plant-pathogen interaction, assess the associated economic impacts, and discuss future perspectives in the field.

4.2 Basic Concepts of Coinfection and Synergism

Plant viruses exhibit generalist behavior, infecting a variety of host plants, both wild and crops. Structurally, they consist of a nucleic acid genome (RNA or DNA) either free or surrounded by capsid proteins, and some may also have a lipid envelope associated with glycoproteins. Due to their genetic nature, they exhibit a high frequency of recombination and mutation, providing opportunities for the emergence of new variants. This characteristic results in the ability to colonize different hosts, expanding the spectrum of infections in crops. Additionally, as obligatory parasites, they tend to seek equilibrium with the host, increasing the chances of different viruses coexisting on the same plant, establishing mixed infections (Singhal et al. 2021). Plant viruses can be transmitted by nematodes, fungi, and arthropods. In addition to disseminating viruses among crop and wild plant species, these vectors also act as transient hosts, facilitating the coexistence of multiple viral agents. This transmission dynamic increases the probability that two or more viruses will infect the same plant, resulting in what is termed a mixed infection. Once inside the host, viral particles may engage in various types of interactions. Beyond neutral interactions, these relationships may be antagonistic or synergistic, depending on the viral combination and host conditions (Alcaide et al. 2020; Andret-Link and Fuchs 2005; Singhal et al. 2021).

4.2.1 Sequential Infection and Simultaneous Infection

Simultaneous infection, also called coinfection, occurs when two or more viruses infect the host plant simultaneously or within a very short interval. In the early stages of infection, the number of viral particles is still low, and there is a large amount of healthy tissue available for colonization. Under these conditions, competition for resources is reduced, which favors the occurrence of coinfection. This type of interaction is known to increase viral fitness, increasing the efficiency of the infectious process (Alcaide et al. 2020; Saldana et al. 2003). On the other hand, sequential infection, also called superinfection, occurs when a plant previously infected by an established virus (or strain) is reinvaded by another virus, with a longer interval between infections, which can range from days to weeks. In this context, the order in which viruses arrive is crucial, as it directly influences the nature of the virus-virus interaction. These interactions can profoundly modify symptoms, disease aggressiveness, and even accumulated viral load, especially when synergistic (Syller and Grupa 2016). Therefore, identifying the differences between infection types is crucial to better understanding how multiple viruses interact and shape the development of plant diseases.

4.2.2 Virus-Virus Interactions

Once established in the host tissues, whether through simultaneous or sequential infection, viruses can interact in different ways, resulting in synergistic, antagonistic, or neutral relationships. In synergism, the presence of one virus favors the replication or spread of another. This phenomenon can manifest itself in several ways: increased viral particle count, altered cell-to-cell movement speed or efficiency, suppression of plant defenses based on RNA silencing by viral proteins, colonization of tissues that would not normally be accessible, and even expanded host range. Consequently, symptom severity tends to be greater and disease aggressiveness is enhanced (Anandalakshmi et al. 1998; Kumar and Dasgupta 2021; Sánchez-Tovar et al. 2025). Antagonism refers to situations in which one organism interferes with another, causing some level of harm to its development by affecting its growth or its ability to produce symptoms. In viruses, when two of them infect a host at the same time, the situation is called coinfection. However, antagonism does not depend only on the presence of both viruses, but also on the timing of infection and the conditions in which it occurs. When they are inoculated together and both can cause disease, it is considered a coinfection. But when only one virus establishes the infection first and the other comes later, the second case is known as a superinfection (Sánchez-Tovar et al. 2025). Finally, in neutral relationships, two viruses infect the same plant without significantly altering its multiplication or symptom expression, when compared to individual infections (Moya-Ruiz et al. 2024). Thus, viral interactions represent a central axis in plant virology, determining the intensity of symptoms, as well as the evolutionary and epidemiological dynamics of plant diseases (Fig. 4.1).

4.2.3 Contribution of Subviral Agents to Viral Disease Complexes

Subviral agents are simpler infectious entities than viruses, generally incapable of encoding essential replication and envelope proteins, making them dependent on a helper virus or host cellular machinery for multiplication (Campbell et al. 2022). Among the main groups of subviral agents are geminivirus-associated satellite DNAs (betasatellites and alphasatellites), satellite RNAs, and viroids (small

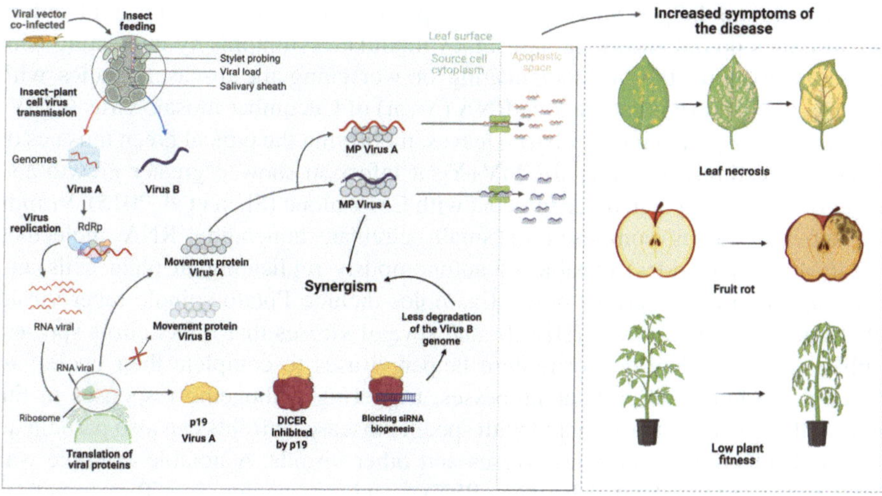

Fig. 4.1 Schematic representation of synergism between generic viruses in plants. The coinfected insect vector simultaneously transmits two viruses to a plant cell (Virus A and Virus B). Within the cell, functional complementation is observed, such as the use of movement proteins from one virus by the other, and suppression of host defense mechanisms, exemplified by the inhibition of the RNAi pathway by suppressor proteins (p19). These processes favor co-replication, increase viral load, and promote more efficient spread between cells. Consequently, symptoms observed in the plant become more severe, including leaf necrosis, fruit rot, and reduced plant fitness

infectious RNAs). These subviral agents modulate pathogenicity and symptom expression. In general, coinfections with satellites or viroids can intensify or attenuate symptoms, modify vector transmission, and pose additional challenges to plant health control (Varma and Singh 2024). Betasatellites, formed by single-stranded circular DNA molecules (~1.3 kb), are primarily associated with monopartite begomoviruses. When present, they often intensify symptom severity, as they depend on the helper virus for movement and encapsidation. They encode a single protein, βC1, which acts as a pathogenicity determinant by interfering with plant defenses, suppressing gene silencing, and modulating resistance pathways. One example is βC1 from Cotton leaf curl virus (CLCuV), capable of interacting with host transcription factors, promoting conditions favorable to viral replication (Varma and Singh 2024; Zhou 2013). Alphasatellites, also composed of circular DNA molecules (~1.4 kb), are also found in begomovirus infections. However, they differ from betasatellites in that they encode a replication protein (Rep), which allows them to replicate autonomously within the plant. Despite this, they still depend on the virus to aid in movement and transmission. In general, these agents reduce viral infection, decreasing viral load and mitigating symptoms (Varma and Singh 2024; Zhou 2013). However, there are exceptions: Mar et al. (2017) described an alphasatellite associated with Euphorbia yellow mosaic (EuYMV) that increased symptom severity and decreased vector transmissibility in coinfected plants. Satellite RNAs (satRNAs) constitute another group of subviral agents composed of small RNAs that depend on helper viruses (usually positive-RNA viruses) for replication. Because they do not encode proteins, they modulate symptoms by interacting with the virus infecting the host, attenuating or worsening the disease. Studies with *Nicotiana* sp. showed that the Y-satRNA (Y-sat) of Cucumber mosaic virus (CMV) induces intense yellowing in tobacco leaves, modifying the typical green mosaics of CMV. Interestingly, this mixed CMV+Y-sat infection showed greater growth and better overall health than those infected with CMV alone (Shen et al. 2015). Viroids are subviral agents composed of small, circular, noncoding RNA molecules (≈246–401 nucleotides) capable of autonomously replicating in plant cells and inducing diseases in various hosts. Examples include Potato spindle tuber viroid (PSTVd), Hop latent viroid (HLVd), and several viruses that infect citrus species. Unlike satellites, they do not require helper viruses to complete their cycle, but directly interfere with cellular processes, triggering pathogenic responses in the host. Although usually associated with specific diseases, viroids can also participate in mixed infections, both with viruses and other viroids. A notable example was described in the coinfection between PSTVd and the begomovirus Tomato yellow leaf curl Sardinia virus (TYLCSV), in which strong antagonism was observed (Torchetti et al. 2016). The viroid induces hypermethylation of the host DNA, drastically reducing infectivity and viral accumulation. In practical terms, PSTVd acts as an "epigenetic disarmament" against TYLCSV, significantly reducing its viral load, even though the virus itself continued to cause damage to the plant (Torchetti et al. 2016). In contrast, research conducted on citrus species naturally infected by multiple viroids has revealed both antagonistic effects—in which one viroid suppresses the other, leading to milder symptoms—and synergistic effects, characterized by mutual worsening of symptom severity. These phenomena illustrate the complexity of subviral interactions, in which different viroids can compete for cellular resources or modulate distinct plant defense pathways, resulting in significant variability in disease manifestation (Wang et al. 2024). In short, subviral agents should not be considered strong components in the formation and maintenance of disease complexes. Their presence profoundly redefines pathogenesis, potentially intensifying or attenuating symptoms, altering transmission patterns, and modifying epidemiological dynamics in the field. Thus, by interacting variably with auxiliary viruses and the host itself, these agents create highly complex scenarios with unpredictable and multifaceted consequences. This reality highlights the need to understand subviral agents from a molecular, ecological, and evolutionary perspective, as they directly influence management strategies, genetic resistance, and phytosanitary control in agricultural systems.

4.3 Molecular Mechanisms of Interaction

4.3.1 RNA Interference

RNA interference (RNAi) is a cellular process that suppresses gene expression after transcription by acting directly on messenger RNA (mRNA). It relies on double-stranded RNA (dsRNA), which, once integrated into an active complex in the cytoplasm, is able to recognize sequences that match the target mRNA, leading to the inhibition of its translation or to its breakdown (França et al. 2010). The discovery of RNA interference (RNAi) began unexpectedly in the early 1990s, in experiments with transgenic petunias. The original idea was to intensify flower color, but the result was the opposite: plants presented white or variegated flowers due to the simultaneous blockade of the endogenous gene and the transgene for chalcone synthase, a key enzyme in the pigmentation pathway (Napoli et al. 1990). This phenomenon, described by R. Jorgensen as "co-suppression," became the first evidence of silencing induced by transgene overexpression (Jorgensen 1995). Later studies showed that mRNA loss was not associated with reduced nuclear transcription but rather with posttranscriptional degradation, leading to the term "post-transcriptional gene silencing" (PTGS) (Francesco et al. 2001). The relevance of this phenomenon was highlighted by the work of Andrew Z. Fire and Craig C. Mello, who won the 2006 Nobel Prize in Medicine or Physiology for their elucidation of the molecular mechanisms involved in RNAi (Fire et al. 1998). Strategically, it is worth highlighting that specific gene silencing is related to two evolutionarily ancient processes: co-suppression in plants and suppression in fungi. It also plays roles in regulatory processes, including transposon control, antiviral defense, gene expression regulation, and chromosome modification (Agrawal et al. 2003). This perspective highlights the importance of RNAi not only as an experimental phenomenon but also as a central mechanism of cellular defense and regulation, paving the way for a discussion of the molecular mechanisms of virus-plant interactions. It is now recognized that this process is widely distributed among eukaryotic organisms (Agrawal et al. 2003) and depends on the action of double-stranded RNA (dsRNA) molecules. These dsRNAs can assume different forms and functions, being classified mainly as microRNAs (miRNAs), small interfering RNAs (siRNAs), and short hairpin RNAs (shRNAs) (França et al. 2010; Meister and Tuschl 2004). MicroRNAs (miRNAs) are small, endogenous double-stranded RNAs of approximately 22 nucleotides whose main function is to act as posttranscriptional regulators, inhibiting the translation of target mRNAs into proteins (França et al. 2010). Genes encoding miRNAs are transcribed by RNA polymerase II into long molecules called primary microRNAs (pri-miRNAs), which are cleaved in the nucleus by a protein complex containing RNase III (Drosha) and the Pasha protein or DGCR8, the latter possessing a dsRNA-binding domain. This processing generates the precursor microRNA (pre-miRNA), approximately 70 base pairs long, with a double-stranded structure and a single loop, forming a hairpin. The pre-miRNA is transported to the cytoplasm by exportin-5 and processed by Dicer, generating the mature miRNA of approximately 22 nucleotides. Dicer, an RNase III essential in RNAi, has an RNA-binding domain, PAZ, which appears to determine the size of the dsRNA produced, in addition to being involved in the incorporation of one of the miRNA strands into the RISC complex (RNA-Induced Silencing Complex), which includes Argonaute (Ago2) and other proteins (Ricarte Filho and Kimura 2006). RISC allows the incorporated miRNA strand to pair with the complementary region of the target mRNA. When there is complete complementarity, mRNA degradation occurs; when there is partial complementarity, translation repression occurs, followed by RNA degradation (Rivas et al. 2005). The Argonaute proteins present in RISC have conserved PAZ and PIWI domains, bind to siRNAs and miRNAs, exhibit endonucleolytic activity directed at the complementary strand of mRNA, and are responsible for selecting the siRNA strand that will be incorporated into the complex (França et al. 2010). *Short hairpin RNAs* (shRNAs) are double-stranded RNAs designed to form structures similar to miRNAs. They can be synthesized externally and introduced directly into the cell or transcribed from vectors containing the shRNA under the control of an RNA polymerase III promoter. In this case, the transcript is processed by Dicer in a manner analogous to miRNAs (Paddison et al. 2002). *Small interfering RNAs* (siRNAs) are synthetic double-stranded RNA molecules of 19–30 base pairs that promote specific gene silencing when they pair with complementary sequences in the target mRNA, leading to its degradation. This strategy allows investigation of the function of proteins encoded by the target genes within the cell (Morris et al. 2004).

RNA interference (RNAi) regulates endogenous genes and transgenes and also constitutes a major antiviral defense pathway in plants. During viral replication, the formation of double-stranded RNA (dsRNA) intermediates triggers the activation of the RNA silencing machinery. These dsRNA molecules are recognized and processed by Dicer-like (DCL) enzymes into small interfering RNAs (siRNAs), which subsequently guide Argonaute (AGO) proteins within the RNA-induced silencing complex (RISC) to degrade complementary viral RNA targets. Consequently, the virus can act as a source of dsRNA intermediates and simultaneously become a target of the RNAi-mediated antiviral response (Agrawal et al. 2003). Plant viruses depend on vectors to spread between hosts and, once inside the plant, exploit the cellular machinery to ensure their replication and movement. Unlike other pathogens, they not only cause localized lesions but also often trigger systemic infections, manifested by symptoms such as chlorosis, malformations, and reduced growth. Most of these viruses have single- or double-stranded RNA genomes, although some families are composed of DNA viruses. Because replication occurs directly within the host cells, RNAi mechanisms play a central role in plant antiviral defense (Muhammad et al. 2019). Plant viruses are subject to intense selective pressures to spread from the initially infected cell to the rest of the organism and to be transmitted to new hosts. To meet these challenges, they have evolved sets of genes whose products interact with cellular components, ensuring the replication of the viral genome, its movement within the plant, and, in many cases, the protection of the genetic material to enable transmission. Among these genes, one group stands out for its high evolutionary conservation, encoding proteins directly involved in viral replication. These genes represent strategic targets in RNA interference (RNAi) studies, since inhibiting their functions compromises critical stages of the viral cycle (Koonin et al. 2022). On the other hand, there are less conserved viral genes, highly variable among different viruses and, in some cases, exclusive to certain groups, with no homologues in other viruses not directly related to replication. This set includes the evolutionarily ancient capsid proteins (CP), which, together with replication proteins, form the so-called "virus hallmark proteins" (Virus Hallmark

Proteins). Also part of this group are movement proteins (MP), encoded by viruses and arising as an evolutionary adaptation to infect plants, as they enable cell-to-cell spread through plasmodesmata. Furthermore, these include viral suppressor of RNA silencing proteins (VSRs), capable of acting against posttranscriptional (PTGS) and transcriptional (TGS) silencing (Atabekova et al. 2023). TGS corresponds to an epigenetic silencing mechanism that promotes the methylation of DNA promoter regions, generating heterochromatin and, consequently, repressing genetic transcription (Waterhouse et al. 2001). VSRs represent an essential viral strategy for neutralizing RNAi-mediated defense mechanisms, significantly reducing the effectiveness of the plant's antiviral responses and allowing successful infection. Many viruses have multiple suppressor genes, each acting at different stages of RNA silencing, which increases infection efficiency. Studies indicate that VSRs have evolved independently in plant and animal viruses, as well as in other nonviral microbial pathogens, highlighting their evolutionary importance for viral success (Atabekova et al. 2023; Burgyán and Havelda 2011). In specific cases, such as Cucumber mosaic virus (CMV) infection, expression of suppressor 2b reduces the accumulation of small viral RNAs (siRNAs) produced by DCL1, DCL2, and DCL3, attenuating the host's antiviral response (Diaz-Pendon et al. 2007). In addition to blocking the RNAi machinery, certain suppressors also play roles in viral replication and movement, depending on host factors to ensure successful infection. Proteins such as CRP, 6K1, and 29K are involved in suppressing RNA silencing during the early phases of viral replication, while TGBp1, HC-Pro, P3, and P3N-PIPO are required for virus movement between cells and over long distances within the plant (Muhammad et al. 2019).

4.3.2 Plant-Pathogen Interaction

Plant-pathogen interactions involve complex hormonal and physiological changes that determine the success of pathogen defense or colonization. To establish themselves in plant tissues, pathogenic microorganisms must bypass or suppress basal defenses, access nutrients and water, and proliferate, often resulting in symptoms such as abnormal growth and the formation of chlorotic or necrotic lesions. Among their virulence strategies, manipulation of plant hormonal signaling is common, allowing them to suppress defenses or alter plant development (Kunkel and Johnson 2021). Plants are immobile organisms, so they cannot run away from pathogens, herbivores, or unfavorable environmental conditions. Thus, when they detect signs of attack from these factors, they can release certain hormones that move across membranes and activate defense responses in different tissues. The classical defense hormones are salicylic acid (SA), jasmonic acid (JA), and ethylene (ET). These hormones can regulate responses to abiotic stresses, while others, such as auxins (AUX), brassinosteroids (BRs), cytokinins (CKs), and abscisic acid (ABA), also play important roles in plant-pathogen interactions (Mishra et al. 2024; Verma et al. 2016). Salicylic acid (SA) is a key hormone in plant immunity, especially in defense against biotrophic and hemibiotrophic pathogens, which trigger its biosynthesis. In addition, it is a phenolic secondary metabolite involved in several aspects of plant development, including germination, growth, photosynthetic balance, flowering, senescence, and stomatal regulation. It also contributes to tolerance against drought, heat, osmotic stress, and heavy metal toxicity. SA signaling can act alone or interact with hormones such as jasmonic acid, ethylene, and abscisic acid, forming a complex crosstalk that depends on the species, tissue, and stimulus involved (Mishra et al. 2024). Abscisic acid (ABA) is a hormone that plays a role in many aspects of plant growth and in how plants handle stress. When the plant is under stressful conditions, it usually produces more ABA in different tissues, and this increase helps trigger several protective responses. One of the most important effects is stomatal regulation, which helps reduce water loss and also limits the entry of pathogens (Lim et al. 2015). These infections show that plant defense involves several interconnected processes, including hormonal signals, metabolites, and changes in gene expression. Understanding what happens in the early stages of infection, when the pathogen can either be blocked or begin to multiply, is essential, especially because most available information comes from dicots, while defense mechanisms in monocots remain less explored (Mandadi and Scholthof 2013). In addition to individual viruses, viral complexes involving helper and satellite viruses demonstrate highly specific and complex molecular interactions. Viral satellites, often composed of single-stranded circular or linear RNA, can perform innovative cellular functions, complementing the helper virus genome to enhance symptoms or, in certain cases, competing with it, attenuating the disease. These interactions are strongly influenced by the host and the helper virus and are essential for evolution, classification, and symptom development (Badar et al. 2020). Some examples are shown below (Table 4.1).

4.4 Agronomic and Epidemiological Impacts of Mixed Viral Infections

Crops are often infected by more than one virus simultaneously, a phenomenon known as mixed infection. These infections lead to different types of interactions between viruses, ranging from synergistic effects, where the presence of one virus favors the replication and/or expression of symptoms of the other, to antagonisms, in which one virus inhibits the infection of another virus (Syller 2012; Moreno and López-Moya 2020). In agriculture, such interactions have important implications for productivity, postharvest fruit quality, the longevity of perennial plants, the epidemiology of viral diseases, and phytosanitary management. Multiple viral infections cause more severe symptoms than single infections. This phenomenon is related to an increase in total viral load, prolonged physiological stress, and, in some cases, the combined manipulation of host defense responses (Xu et al. 2022). A common example of synergism is the interaction between Potato virus X (PVX) and Potato virus Y (PVY) in *Solanaceae*, where coinfected plants exhibit severe necrosis and almost complete loss of production capacity (Syller 2012). In cucurbits, mixed infections involving viruses such as Cucumber mosaic virus (CMV) and

Table 4.1 Viral pathosystems in fruit growing: agents, symptoms, and impact

Cultivate	Viral complex (species involved)	Associated symptoms	Productive impact	Reference
Papaya (*Carica papaya*)	Papaya meleira virus (PMeV1) + Papaya meleira virus 2 (PMeV2)	Latex exudation, light spots on the fruit skin, necrosis on the leaf tips	Drastic reduction in plant lifespan and total loss of marketable fruit	Sa Antunes et al. (2020)
Watermelon (*Citrullus lanatus*)	Squash vein yellowing virus (SqVYV) + Cucurbit leaf crumple virus (CuLCrV) + Papaya ringspot virus type W (PRSV-W) + Cucurbit yellow stunting disorder virus (CYSDV)	Rapid wilting of the vine, decline of the plant close to harvest, leaf curling and wrinkling, yellow spots and mottling, vascular necrosis	Losses of 50–100% in affected crops; losses of US$60–70 million in 2004 alone for Florida producers	Turechek et al. (2010)
Apple (*Malus domestica*)	Apple necrotic mosaic virus (ApNMV) + Apple mosaic virus (ApMV) + Apple stem pitting virus (ASPV) + Apple stem grooving virus (ASGV)	Viruses were found in plants with symptoms of mosaic and mosaic necrosis, but also in asymptomatic plants	Reduction in productivity and fruit quality, resulting in serious economic losses	Nabi et al. (2022)
Peach (*Prunus persica*)	Prunus necrotic ringspot virus (PNRSV) + Peach latent mosaic viroid (PLMVd)	PNRSV: chlorotic rings, mosaic, reduced growth. PLMVd: latent mosaic, leaf deformation, reduced vigor. PNRSV + PLMVd coinfection: exacerbated symptoms, tissue necrosis, greater reduction in vegetative development	Single infections cause moderate losses in vigor and productivity (~20–30%), but coinfection results in marked plant decline, poor fruit formation, and losses exceeding 50% in affected orchards	Herranz et al. (2013)

Zucchini yellow mosaic virus (ZYMV) result in small, deformed fruits and a significant drop in commercial productivity (Moya-Ruiz et al. 2024). Recent work with melons shows that coinfections reduce not only biomass but also the number and quality of marketable fruit, an effect exacerbated by abiotic stresses such as drought (Jiménez et al. 2024). Therefore, from an agronomic perspective, mixed infections are often associated with greater yield losses than those observed with single infections. In addition to reduced productivity, there are consistent records of direct impacts on fruit quality. Coinfections can alter physicochemical properties, such as soluble sugar content, acidity, firmness, and flesh color (Moreno and López-Moya 2020). These changes are relevant to the fresh fruit market and the processing industry, which require specific quality standards. For example, in melons, plants simultaneously infected by Cucurbit aphid-borne yellows virus (CABYV) and Melon necrotic spot virus (MNSV) produce fruits with irregular external color and reduced soluble solids content, affecting flavor and commercial acceptance (Jiménez et al. 2024). In grapevines and citrus, the presence of multiple viruses has been associated with uneven ripening and loss of color uniformity in the fruits, factors that reduce the added value of the product (Naidu et al. 2015). In perennial crops, such as fruit and woody species, the effects of mixed infections can compromise orchard longevity. The accumulation of multiple viruses in a single individual promotes chronic decline, reducing vigor, vegetative growth, and production capacity over the years (Maliogka et al. 2018). This scenario is particularly problematic in citrus, grapevines, and walnut trees, where complex infections involving multiple viruses and viroids can lead to the need for early replanting, increasing maintenance costs and reducing the sustainability of the agricultural system (Naidu et al. 2015). In some cases, premature plant mortality requires the replacement of entire orchards, which represents significant economic losses. Mixed infections also have relevant epidemiological implications. Plant viruses can modify host physiology and morphology, altering volatile emission, leaf color, and nutrient content, which influences the behavior of insect vectors (Stafford et al. 2011). In coinfections, these effects can be cumulative or mutually modify each other. Studies show that vectors can acquire two viruses simultaneously, increasing the likelihood of multiple transmission and, consequently, the maintenance of mixed infections in plant populations (Fereres and Moreno 2009; McLaughlin et al. 2022). In some cases, the presence of one virus can increase the transmission efficiency of another, as observed in interactions involving begomoviruses transmitted by whiteflies (*Bemisia tabaci*) (McLaughlin et al. 2022). On the other hand, there are also examples of antagonism, in which coinfection reduces the efficiency of acquisition or inoculation, depending on the compatibility between virus, vector and host (Syller 2012). From a management perspective, mixed infections pose a significant challenge. Symptoms resulting from coinfection are often atypical, making visual identification difficult and confounding diagnoses based on classic symptom patterns (Kanapiya et al. 2024). Furthermore, variability in viral titer between agents can lead to false negatives in serological and molecular tests targeting a single virus (Massart et al. 2014). In this context, high-throughput methodologies such as multiplex RT-qPCR, next-generation sequencing (NGS), and metagenomic analyses have established themselves as essential tools for the accurate diagnosis of mixed infections (Massart et al. 2014). These approaches allow not only the simultaneous detection of multiple viruses but also the discovery of emerging, previously undescribed viruses. For management, the presence of multiple viruses means that strategies focused on a single pathogen, such as the use of cultivars resistant to a specific virus, may fail in the field if other viruses are present (Moreno and López-Moya 2020). Therefore, the adoption of integrated management programs is recommended, including the use of certified virus-free propagation material, rigorous vector control, and constant monitoring with comprehensive molecular tools. Mixed plant virus infections are a complex phenomenon with profound consequences for agriculture. From an agronomic perspective, they result in productivity losses, reduced product quality, and reduced orchard shelf life. From an epidemiological perspective, they alter transmission dynamics and complicate diagnosis and management. Advances in molecular biology and genomics have made it possible to better understand these interactions and propose more effective management strategies. However, given the variability of interactions and the diversity of viruses involved, this remains one of the greatest challenges in plant virology today.

4.5 Diagnosis of Viral Complexes in Fruit Trees

Epidemiological knowledge and efficient management require accurate and early diagnosis of viral diseases that affect plants, especially fruit trees. By definition, diagnosis is the identification and classification of the cause of a problem, condition, or disease through the analysis of data, signs, and symptoms, supported by tests (Ravelonandro 2021). Complexity increases in scenarios of coinfection and synergism, where different viruses coexist and interact within the same host. In these situations, sensitive and comprehensive methods are essential to distinguish each agent, assess their individual contribution, and identify their combined effects on the plant. Basically, the diagnosis of viral coinfections in plants has accessible but limited, conventional methods. As a result, high-throughput approaches are emerging that offer a more complete and predictive view of the viral landscape. The integration of methodological resources, from RT-qPCR to NGS sequencing, supported by updated databases (Maliogka et al. 2018), proves to be a promising strategy for addressing the challenges posed by mixed infections, thus ensuring greater efficiency in management programs and phytosanitary safety in fruit farming.

4.5.1 Conventional Methods

The first approaches applied to the diagnosis of viruses in fruit trees were based on simple serological and molecular techniques. For example, the ELISA (Enzyme-Linked Immunosorbent Assay) method was widely used due to its ease, low cost, and ability to examine large sample volumes (Boonham et al. 2014). The basic principle involves the selective binding of antibodies to antigens, followed by the addition of an enzyme that produces a color change, enabling detection. However, in mixed infections, the test is inaccurate, as it relies on specific antibodies and may fail to detect divergent isolates (Boonham et al. 2014). Similarly, fruit tissues contain compounds that interfere with the analysis (polyphenols, pectins), limiting virus detection, especially in cases of low viral load. In general, polymerase chain reaction (PCR) analysis offers greater specificity in the diagnosis of plant diseases, enabling the amplification of defined viral sequences. However, this application also has limitations when multiple viruses circulate in the same plant host, especially those that are genetically related or have low concentrations in certain tissues (Singhal et al. 2021). These methods have increased sensitivity and specificity over ELISA, enabling viral sequence detection even at low concentrations. Aligned techniques increase sensitivity and data collection, but increase the risk of contamination. In coinfections, targeted amplifications require carefully designed oligonucleotides to avoid off-targets or cross-amplifications between related species and do not, by themselves, provide a comprehensive view of the virome (Singhal et al. 2021). In short, conventional methods are valuable for targeted screening but insufficient when seeking to characterize viral complexes or discover novel agents.

4.5.2 Reverse Transcriptase Quantitative PCR

The advent of real-time PCR (RT-qPCR) has revolutionized the detection of plant viruses, combining the reverse transcription of RNA to cDNA and real-time quantification. This technique combines sensitivity with high specificity, quantifying viral loads at different stages of infection (Lu et al. 2016). Its premise is quite simple: a fluorescent dye is used that emits a signal only in the presence of the newly amplified cDNA strand, followed by signal recognition by a sensor at each cycle. The amount of viral load is determined by the time of signal capture; that is, the earlier the signal is detected, the higher the viral load. For more complex diagnostics, it is essential to validate the reaction efficiency (90–110%) by constructing standard curves for absolute quantification. Implementation mechanisms include the use of internal controls (exogenous gene to assess extraction/inhibition), negative controls, and standardized collection/storage protocols. When quantification is critical, it is recommended to complement RT-qPCR with other methodologies, such as sequencing analysis (Rubio et al. 2020). In mixed infections, multiplex RT-qPCR panels are often used, in which multiple primers and fluorescent probes facilitate the simultaneous detection of two or more viruses in a single reaction. In addition to revealing the presence of pathogens, the method provides quantitative information on the dynamics of each virus, allowing us to understand competitive or synergistic interactions that affect disease severity (Kader Jailani and Paret 2024). Despite its robustness, the need for known target sequences limits its application to emerging, as yet uncharacterized viruses.

4.5.3 High-Yield Diagnostics

NGS sequencing brought a shift in perspective, providing a more specialized understanding and recognition of the plant virome. Unlike methods based on specific probes, NGS allows the detection of both known and novel viruses, even at low concentrations (Pecman et al. 2017). This advance revealed the true diversity of latent viruses in fruit trees, elucidating cases of synergism where species previously considered secondary gained prominence in the presence of other phytopathogens (Pecman et al. 2017). Strategies used in fruit crops include RNA-seq (total RNA), small RNA-seq (maps plant-generated siRNAs against the virus), dsRNA sequencing, and viral particle enrichment (Massart et al. 2019). Each protocol has its tradeoffs: small RNA-seq is sensitive to RNA viruses and detects silencing interactions; RNA-seq also captures host expression and DNA virions when complemented with DNA-seq. Currently, RNA-seq, viral metagenomics, and purified viral particle sequencing protocols have been successfully applied to citrus (Bester et al. 2021), grapevine (Di Gaspero et al. 2022), and tomato (González-Pérez et al. 2024), among other crops. Despite this, the application has limitations such as high cost, the need for sophisticated laboratory infrastructure, and complex bioinformatics analysis, representing barriers to its widespread adoption in genetic improvement programs.

4.5.4 Genomic Database

Advances in molecular applications and NGS are intrinsically linked to the availability of robust genomic databases. These repositories, such as NCBI Virus (Sayers et al. 2024) and the EPPO Global Database, are freely accessible and facilitate the comparison of field-obtained plant and virus sequences with reference genomes, providing viral sequences, contigs, and annotation tools. Furthermore, they are essential for identifying viral species and strains and for monitoring mutations, recombination, and the potential emergence of more aggressive variants (Rubio et al. 2020). For fruit trees, having regional collections and local reference panels on field isolates and viral variants in rootstocks are essential for interpreting endemic versus emerging variants. In addition to sequences, standardized metadata (host, tissue, site, date, method) are crucial for epidemiological surveillance (Ritsch et al. 2023). Data sharing networks between laboratories and integration with phenotyping databases improve analytical reliability and favor early detection of viral complexes and new threats. In this context, the integration of genomic, epidemiological, and phenotypic information makes collections indispensable tools for applied research and for certification programs, crop quality control, and genetic improvement programs.

4.6 Integrated Plant Viruses Management Strategies

Integrated plant virus management intrinsically depends on the use of virus-free propagation materials, given that many pathogens spread through the planting of contaminated seedlings. Therefore, the selection and use of healthy stock plants form the basis for the production of seedlings with superior sanitary quality. This practice is complemented by seedling and nursery certification programs, ensuring that the materials distributed to farmers undergo rigorous inspection and diagnostic processes (Tatineni and Hein 2023). In addition, vector control measures—such as sucking insects, the main transmitters of viral phytopathogens—and cultural management practices, such as staggered planting, the elimination of volunteer plants, and cleaning the growing environment, become essential to reduce inoculum pressure and minimize the spread of viruses in the field (Jeger 2020). At the same time, the development of resistant or tolerant varieties has proven to be one of the most sustainable tools for long-term management. Breeding programs seek to identify resistance genes for incorporation into commercial cultivars, providing farmers with an efficient alternative to viruses in endemic regions. However, this approach becomes finite due to the overcoming of resistance by emerging strains, supporting the ongoing search for new genetic sources. In recent years, biotechnology-based approaches, such as RNAi and gene editing, have emerged as innovative strategies for controlling viral infections. RNAi was discussed in Sect. 4.3.1 of this chapter. Gene editing offers the possibility of altering susceptibility genes (S genes), making plants less susceptible to infection. These technologies promise significant advances in combating mixed infections and viral complexes (Tatineni and Hein 2023). Another fundamental pillar of integrated management is continuous epidemiological monitoring. Constant surveillance of viral disease occurrence, combined with the use of diagnostic tools (Sect. 4.5), allows for early identification of outbreaks of new infections and guides strategic interventions (Jeger 2020). Furthermore, this analysis generates essential information for understanding the spatial and temporal dynamics of viral diseases, supporting more effective prevention and management policies. In short, successful integrated viral disease management depends on a set of actions that combine multiple complementary strategies. To reduce viral pressure in fields, every step counts, from the implementation of healthy, certified seedlings to investment in biotechnology and epidemiological surveillance. These practices are essential for reducing losses, extending the shelf life of resistant cultivars, and promoting sustainable production in environments constantly challenged by viruses and their vectors.

4.7 Conclusions

Research on viral complexes and mixed infections in plants is fundamental to advancing sustainable and resilient fruit production systems. Understanding the interactions among viruses, plant hosts, and their vectors provides crucial insights into processes ranging from the manifestation of severe disease symptoms to the breakdown of resistance in commercial cultivars. In high-value tropical crops such as papaya and citrus, this knowledge is particularly consequential, as losses caused by single and mixed viral infections directly affect food security, reduce the income of smallholder farmers, and diminish competitiveness in international markets. Addressing these diseases demands an integrated approach encompassing diagnostics, management practices, and genetic improvement. Rapid and precise diagnostic tools enable early detection of viral complexes, integrated management incorporates coordinated preventive and control measures, and genetic improvement offers durable solutions by conferring broad and long-lasting resistance to diverse viral pathogens. Looking ahead, the field is poised for continued scientific and technological advances. Emerging tools such as CRISPR–Cas genome editing, RNA-based antiviral technologies, and transgenic resistance are becoming central pillars in the development of more resilient crop varieties. Nevertheless, the implementation of these strategies must be accompanied by robust public policies, epidemiological surveillance, and specialized technical training to ensure their safe, responsible, and effective deployment. Thus, research on viral complexes extends far beyond the boundaries of plant virology; it constitutes a strategic foundation for global agricultural sustainability. By integrating scientific innovation with practical crop management, the challenges posed by mixed viral infections can be transformed into opportunities to advance production systems that are more efficient, secure, and environmentally balanced.

References

Agrawal N, Dasaradhi PV, Mohmmed A, Malhotra P, Bhatnagar RK, Mukherjee SK (2003) RNA interference: biology, mechanism, and applications. Microbiol Mol Biol Rev 67(4):657–685

Alcaide C, Rabadán MP, Moreno-Perez MG, Gomez P (2020) Implications of mixed viral infections on plant disease ecology and evolution. Adv Virus Res 106:145–169

Anandalakshmi R, Pruss GJ, Ge X, Marathe R, Mallory AC, Smith TH, Vance VB (1998) A viral suppressor of gene silencing in plants. Proc Natl Acad Sci 95(22):13079–13084

Andret-Link P, Fuchs M (2005) Transmission specificity of plant viruses by vectors. J Plant Pathol 87:153–165

Atabekova AK, Selitsky SR, Ryabov EV, Morozov SY (2023) Role of plant virus movement proteins in suppression of host RNAi defense. Int J Mol Sci 24(10):9049

Badar U, Venkataraman S, AbouHaidar M, Hefferon K (2020) Molecular interactions of plant viral satellites. Virus Genes 57(1):1–22

Bester R, Cook G, Breytenbach JH, Steyn C, De Bruyn R, Maree HJ (2021) Towards the validation of high-throughput sequencing (HTS) for routine plant virus diagnostics: measurement of variation linked to HTS detection of citrus viruses and viroids. Virol J 18(1):61

Boonham N, Kreuze J, Winter S, van der Vlugt R, Bergervoet J, Tomlinson J, Mumford R (2014) Methods in virus diagnostics: from ELISA to next generation sequencing. Virus Res 186:20–31

Burgyán J, Havelda Z (2011) Viral suppressors of RNA silencing. Trends Plant Sci 16:265–272

Campbell AJ, Anderson JR, Wilusz J (2022) A plant-infecting subviral RNA associated with poleroviruses produces a subgenomic RNA which resists exonuclease XRN1 in vitro. Virology 566:1–8

Di Gaspero G, Radovic S, De Luca E, Spadotto A, Magris G, Falginella L et al (2022) Evaluation of sensitivity and specificity in RNA-Seq-based detection of grapevine viral pathogens. J Virol Methods 300:114383

Diaz-Pendon JA, Li F, Li WX, Ding SW (2007) Suppression of antiviral silencing by cucumber mosaic virus 2b protein in Arabidopsis is associated with drastically reduced accumulation of three classes of viral small interfering RNAs. Plant Cell 19(6):2053–2063

Fereres A, Moreno A (2009) Behavioural aspects influencing plant virus transmission by homopteran insects. Virus Res 141(2):158–168

Fire A, Xu S, Montgomery MK, Kostas SA, Driver SE, Mello CC (1998) Potent and specific genetic interference by double-stranded RNA in *Caenorhabditis elegans*. Nature 391:806–811

Food and Agriculture Organization of the United Nations (2024) FAO statistical yearbook 2024: world food and agriculture. FAO, Rome. Available: https://www.fao.org/statistics/yearbook. Accessed in 10 Sept 2025

França NR, Mesquita Júnior D, Lima AB, Pucci FV, Andrade LE, Silva NP (2010) RNA interference: a new alternative for rheumatic diseases therapy. Rev Bras Reumatol 50(6):695–702

Francesco DS, Hanspeter S, Alejandro L, Cornia T, Estelle B, Frederick M (2001) Sense and antisense mediated gene silencing in tobacco is inhibited by the same viral suppressors and is associated with accumulation of small RNAs. Proc Natl Acad Sci USA 96:6506–6510

González-Pérez E, Chiquito-Almanza E, Villalobos-Reyes S, Canul-Ku J, Anaya-López JL (2024) Diagnóstico e caracterização de vírus de plantas utilizando HTS para apoiar o manejo de vírus e melhoramento de tomate. Vírus 16(6):888

Herranz MC, Niehl A, Rosales M, Fiore N, Zamorano A, Granell A, Pallas V (2013) A remarkable synergistic effect at the transcriptomic level in peach fruits doubly infected by prunus necrotic ringspot virus and peach latent mosaic viroid. Virol J 10(1):164

Jeger MJ (2020) The epidemiology of plant virus disease: towards a new synthesis. Plants 9(12):1768

Jiang T, Zhou T (2023) Unraveling the mechanisms of virus-induced symptom development in plants. Plants 12(15):2830

Jiménez J, Sadras VO, Espaillat N, Moreno A, Fereres A (2024) Interplay between drought and plant viruses co-infecting melon plants. Sci Rep 14(1):15833

Jorgensen RA (1995) Cosuppression, flower color patterns, and metastable gene expression states. Science 268(5211):686–691

Kader Jailani AA, Paret ML (2024) A multiplex RT-qPCR assay for simultaneous detection of cucurbit viruses from individual whitefly and plant samples. Plant Dis 108(12):3508–3517

Kanapiya A, Amanbayeva U, Tulegenova Z, Abash A, Zhangazin S, Dyussembayev K, Mukiyanova G (2024) Recent advances and challenges in plant viral diagnostics. Front Plant Sci 15:1451790

Koonin EV, Dolja VV, Krupovic M (2022) The logic of virus evolution. Cell Host Microbe 30:917–929

Kumar G, Dasgupta I (2021) Variability, functions and interactions of plant virus movement proteins: what do we know so far? Microorganisms 9(4):695

Kunkel BN, Johnson JMB (2021) Auxin plays multiple roles during plant–pathogen interactions. Cold Spring Harb Perspect Biol 13(9):a040022

Lim CW, Baek W, Jung J, Kim J-H, Lee SC (2015) Function of ABA in stomatal defense against biotic and drought stresses. Int J Mol Sci 16(7):15251–15270. https://doi.org/10.3390/ijms160715251

Lindell CA, Irish-Brown A, Rothwell NL, Wallis AE (2023) Pest and disease risk and management in high-density perennial crops: current knowledge and areas of future research. Crop Prot 165:106150

Lu R, Liu Z, Fang P, Chen G, Sun F, Fan Y et al (2016) Reference genes: essential criteria for assessment of the real-time PCR based virus detection in plants virology. Hosts Viruses 3(1):6

Maliogka VI, Minafra A, Saldarelli P, Ruiz-García AB, Glasa M, Katis N, Olmos A (2018) Recent advances on detection and characterization of fruit tree viruses using high-throughput sequencing technologies. Viruses 10(8):436

Mandadi KK, Scholthof KBG (2013) Plant immune responses against viruses: how does a virus cause disease? Plant Cell 25(5):1489–1505

Manzoor S, Nabi SU, Baranwal VK, Verma MK, Parveen S, Rather TR et al (2023) Overview on century progress in research on mosaic disease of apple (Malus domestica Borkh) incited by apple mosaic virus/apple necrotic mosaic virus. Virology 587:109846

Mar TB, Mendes IR, Lau D, Fiallo-Olivé E, Navas-Castillo J, Alves MS, Murilo Zerbini F (2017) Interaction between the New World begomovirus Euphorbia yellow mosaic virus and its associated alphasatellite: effects on infection and transmission by the whitefly Bemisia tabaci. J Gen Virol 98(6):1552–1562

Maree HJ, Almeida RP, Bester R, Chooi KM, Cohen D, Dolja VV et al (2013) Grapevine leafroll-associated virus 3. Front Microbiol 4:82

Massart S, Olmos A, Jijakli H, Candresse T (2014) Current impact and future directions of high throughput sequencing in plant virus diagnostics. Virus Res 188:90–96

Massart S, Chiumenti M, De Jonghe K, Glover R, Haegeman A, Koloniuk I et al (2019) Virus detection by high-throughput sequencing of small RNAs: large-scale performance testing of sequence analysis strategies. Phytopathology 109(3):488–497

McLaughlin AA, Hanley-Bowdoin L, Kennedy GG, Jacobson AL (2022) Vector acquisition and co-inoculation of two plant viruses influences transmission, infection, and replication in new hosts. Sci Rep 12(1):20355

Meister G, Tuschl T (2004) Mechanisms of gene silencing by double-stranded RNA. Nature 431:343–349

Mishra S, Roychowdhury R, Ray S, Hada A, Kumar A, Sarker U, Aftab T, Das R (2024) Salicylic acid (SA)-mediated plant immunity against biotic stresses: an insight on molecular components and signaling mechanism. Plant Stress 11:100427. https://doi.org/10.1016/j.stress.2024.100427

Moreno AB, López-Moya JJ (2020) When viruses play team sports: mixed infections in plants. Phytopathology 110(1):29–48

Moreno P, Ambrós S, Albiach-Martí MR, Guerri J, Pena L (2008) Citrus tristeza virus: a pathogen that changed the course of the citrus industry. Mol Plant Pathol 9(2):251–268

Morris KV, Chan SW, Jacobsen SE, Looney DJ (2004) Small interfering RNA-induced transcriptional gene silencing in human cells. Science 305(5688):1289–1292

Moya-Ruiz CD, Ferriol I, Gómez P (2024) The temporal order of mixed viral infections matters: common events that are neglected in plant viral diseases. Viruses 16(12):1954

Muhammad T, Zhang F, Zhang Y, Liang Y (2019) RNA interference: a natural immune system of plants to counteract biotic stressors. Cells 8(1):38

Nabi SU, Madhu GS, Rao GP, Baranwal VK (2022) Development of multiplex RT-PCR assay for simultaneous detection of four viruses infecting apple (*Malus domestica*). Lett Appl Microbiol 74(4):586–592

Naidu RA, Maree HJ, Burger JT (2015) Grapevine leafroll disease and associated viruses: a unique pathosystem. Annu Rev Phytopathol 53(1):613–634

Napoli C, Lemieux C, Jorgensen R (1990) Introduction of a chimeric chalcone synthase gene into *Petunia* results in reversible co-suppression of homologous genes in trans. Plant Cell 2:279–289

Paddison PJ, Caudy AA, Bernstein E, Hannon GJ, Conklin DS (2002) Short hairpin RNAs (shRNAs) induce sequence-specific silencing in mammalian cells. Genes Dev 16:948–958

Pecman A, Kutnjak D, Gutiérrez-Aguirre I, Adams I, Fox A, Boonham N, Ravnikar M (2017) Next generation sequencing for detection and discovery of plant viruses and viroids: comparison of two approaches. Front Microbiol 8:1998

Ravelonandro M (2021) Reliable methodologies and impactful tools to control fruit tree viruses. Crops 1(1):32–41

Ricarte Filho JCM, Kimura ET (2006) MicroRNAs: novel class of gene regulators involved in endocrine function and cancer. Arq Bras Endocrinol Metabol 50(6):1102–1107

Ritsch M, Cassman NA, Saghaei S, Marz M (2023) Navigating the landscape: a comprehensive review of current virus databases. Viruses 15(9):1834

Rivas FV, Tolia NH, Song JJ, Aragon JP, Liu J, Hannon GJ et al (2005) Purified Argonaute2 and a siRNA form recombinant human RISC. Nat Struct Mol Biol 12(4):340–349

Rubio L, Galipienso L, Ferriol I (2020) Detection of plant viruses and disease management: relevance of genetic diversity and evolution. Front Plant Sci 11:1092

Sa Antunes TF, Maurastoni M, Madroñero LJ, Fuentes G, Santamaría JM, Ventura JA et al (2020) Battle of three: The curious case of papaya sticky disease. Plant Dis 104(11):2754–2763

Saldana J, Elena SF, Solé RV (2003) Coinfection and superinfection in RNA virus populations: a selection–mutation model. Math Biosci 183(2):135–160

Sánchez-Tovar MR, Rivera-Bustamante RF, Saavedra-Trejo DL, Guevara-González RG, Torres-Pacheco I (2025) Mixed plant viral infections: complementation, interference and their effects, a review. Agronomy 15(3):620

Sayers EW, Beck J, Bolton EE, Brister JR, Chan J, Connor R et al (2024) Database resources of the National Center for Biotechnology Information in 2025. Nucleic Acids Res 53(D1):D20

Shen WX, Au PCK, Shi BJ, Smith NA, Dennis ES, Guo HS et al (2015) Satellite RNAs interfere with the function of viral RNA silencing suppressors. Front Plant Sci 6:281

Singhal P, Nabi SU, Yadav MK, Dubey A (2021) Mixed infection of plant viruses: diagnostics, interactions and impact on host. J Plant Dis Prot 128(2):353–368

Stafford CA, Walker GP, Ullman DE (2011) Infection with a plant virus modifies vector feeding behavior. Proc Natl Acad Sci 108(23):9350–9355

Syller J (2012) Facilitative and antagonistic interactions between plant viruses in mixed infections. Mol Plant Pathol 13(2):204–216

Syller J, Grupa A (2016) Antagonistic within-host interactions between plant viruses: molecular basis and impact on viral and host fitness. Mol Plant Pathol 17(5):769–782

Tatineni S, Hein GL (2023) Plant viruses of agricultural importance: current and future perspectives of virus disease management strategies. Phytopathology 113(2):117–141

Torchetti EM, Pegoraro M, Navarro B, Catoni M, Di Serio F, Noris E (2016) A nuclear-replicating viroid antagonizes infectivity and accumulation of a geminivirus by upregulating methylation-related genes and inducing hypermethylation of viral DNA. Sci Rep 6(1):35101

Turechek WW, Kousik CS, Adkins S (2010) Distribution of four viruses in single and mixed infections within infected watermelon plants in Florida. Phytopathology 100(11):1194–1203

Valentova L, Rejlova M, Franova J, Cmejla R (2022) Symptomless infection by strawberry virus 1 (StrV-1) leads to losses in strawberry yields. Plant Pathol 71(5):1220–1228

Varma A, Singh MK (2024) The role of satellites in the evolution of begomoviruses. Viruses 16(6):970

Verma V, Ravindran P, Kumar PP (2016) Plant hormone-mediated regulation of stress responses. BMC Plant Biol 16:86. https://doi.org/10.1186/s12870-016-0771-y

Wang Y, Shi Y, Li H, Chang J (2024) Understanding citrus viroid interactions: experience and prospects. Viruses 16(4):577

Waterhouse PM, Wang MB, Lough TJN (2001) Gene silencing as an adaptive defence against viruses. Nature 411:834–842

Xu Y, Ghanim M, Liu Y (2022) Mixed infections of plant viruses in nature and the impact on agriculture. Front Microbiol 13:922607

Zhou X (2013) Advances in understanding begomovirus satellites. Annu Rev Phytopathol 51(1):357–381

Mechanisms of Viral Pathogenesis in Fruit Crops

5

Subaya Manzoor, Tabasum Manzoor, Bushra Rasool, Zakir Amin, Sajad Un Nabi, and Aadil Ayaz

Abstract

Since the advent of human civilization, fruits have been recognized as an essential component of the daily diet. They play a critical role in maintaining human health and physiological development, owing to their rich content of vitamins, minerals, and other bioactive compounds. Also, fruit crops are the cash crops that proportionally affect global nutrition and economy. The reduction in productivity is mainly due to diseases caused by a variety of pathogens and among them, viruses occupy a significant place as they limit productivity, quality, and market value. Understanding the molecular mechanisms of viral pathogenesis is essential for developing sustainable control strategies. This chapter highlights the intricate interactions between host and pathogen interactions, focusing on key processes such as viral entry, replication, gene expression, expression pattern, movement, and systemic spread. Particular emphasis is placed on multifunctional viral proteins, coat proteins (CPs), movement proteins (MPs), and RNA-dependent RNA polymerases (RdRps) that not only support replication and transport but also act as major determinants of virulence and system expression dynamics. The manipulation of host defense mechanisms, including RNA silencing and effector-triggered immunity (ETI), and the hypersensitive response (HR), is discussed in the context of viral adaptation and counter-defense strategies. The illustration of specific genomic mutations which modulate the pathogenic kinetics, host range, and symptom development is gracefully depicted by Apple mosaic virus, Cucumber mosaic virus, Plum pox virus, and Tomato mosaic virus. The integration of omics approaches has further strengthened our understanding and enabled us to decipher the obscure phenomenon of host–virus interactions and resistance mechanisms. Overall, this chapter provides a comprehensive

S. Manzoor · T. Manzoor · B. Rasool · Z. Amin · S. U. Nabi · A. Ayaz (✉)
Plant Virology Laboratory, Division of Fruit Science, ICAR-Central Institute of Temperate Horticulture, Srinagar, Jammu & Kashmir, India

R. Kumar et al. (eds.), *Fruit Crops and Viral Pathogens Volume II*, Plant Pathology for Sustainable Agriculture, https://doi.org/10.1007/978-981-95-7261-8_5

overview of viral pathogenesis in fruit crops and emphasizes the importance of molecular insights in designing durable, eco-friendly strategies for managing viral diseases.

Keywords

Fruit crops · Viral pathogenesis · Host–pathogen interactions · Viral replication · Hypersensitive response · Resistance mechanisms

5.1 Introduction

Fruit crops represent important agricultural commodities with substantial nutritional, economic, and sociocultural importance. Fruits such as apples, pears, grapes, citrus, peaches, and cherries support millions of livelihoods by contributing to fresh consumption, processed food industries, and export economies. Their cultivation enhances food security and provides vital sources of vitamins, minerals, and antioxidants crucial for human health (Ray et al. 2020). Fruit crops face severe biotic stresses, among which viral diseases represent a particularly challenging threat. Fruit viruses are highly infectious and can inflict severe damage on host trees. Their impact is most evident in reduced yield and overall productivity, though they are also responsible for a wide range of other complications that demand substantial financial resources to manage. The most apparent consequences include significant decreases in both crop production and productivity. These viral infections not only harm farmers and producers but also affect consumers by impairing plant growth and reproduction, leading to sterility, diminished vigor, and weaker resistance to environmental stresses. In severe cases, they can cause complete crop failure or loss of commercial value. Moreover, the presence of such viruses often triggers strict quarantine regulations and necessitates the eradication of affected plants. This, in turn, escalates the expenses associated with disease management, including the implementation of control strategies and the execution of reliable detection programs (Wani et al. 2024). Viruses, being the obligate parasites, use host cell machinery for propagation (Rodríguez-Verástegui et al. 2022). Productive infection is achieved only when the virus encounters compatible host factors that facilitate its replication and successfully circumvent the host's defense mechanisms. Consequently, the majority of plant–virus interactions fail to result in successful infections. Moreover, even in cases where viral replication occurs, it does not necessarily lead to the manifestation of pathogenic symptoms. Pathogenesis takes place when virus infection disturbs host physiology, causing developmental abnormalities and other phenotypic manifestations that are considered disease symptoms (García and Pallás 2015). The replication of plant viruses entails a highly intricate network of molecular interactions between viral components and host cellular factors, which collectively dictate the trajectory and outcome of infection. Consequently, the infection outcome is determined by the dynamic interplay between proviral host factors that enhance viral fitness and antiviral mechanisms that suppress viral proliferation

(Garcia-Ruiz 2019). Pathogenesis in plant–virus interactions can arise not only from the direct effects of viral replication within the host but also from the host's defense responses activated against the invading pathogen. In certain cases, a single viral factor may serve as the principal determinant of pathogenicity (Syller and Grupa 2016). The deleterious effects associated with such factors can be intrinsically linked to their essential roles in viral multiplication—for instance, RNA silencing suppressors that simultaneously facilitate infection and induce host cellular disturbances (Moissiard and Voinnet 2004). However, in other instances, the functions of viral determinants contributing to replication and those responsible for symptom development are genetically and mechanistically distinct. Only specific viral isolates harboring particular allelic variants of these pathogenicity factors are capable of inducing characteristic disease phenotypes. Notable examples include elicitors that trigger hypersensitive response (HR)-associated symptoms (García and Pallás 2015). Although HR is frequently accompanied by localized necrosis, these two processes can occur independently, reflecting the complexity of virus–host molecular interactions underlying symptom expression. Moreover, plant viral pathogenesis often results from multifactorial interactions involving multiple viral gene products and host components, or even entire viral regulatory pathways. Importantly, while specific variants of individual viral factors are often associated with enhanced pathogenicity, the expression of disease symptoms may also depend on the synergistic contribution of additional viral elements (Komatsu et al. 2010). This chapter seeks to provide a comprehensive yet concise synthesis of current understanding regarding the molecular basis of plant viral pathogenesis, with an emphasis on the roles of viral proteins and nucleic acids in mediating infection, symptom induction, and host response modulation (Rodríguez-Verástegui et al. 2022). Ultimately, these insights lay the groundwork for advancing disease management strategies and protecting global fruit production from viral threats.

5.2 Overview of Fruit Crop Viruses

Global fruit production is dominated by pome, banana, citrus grapevine, mango, and stone fruit trees, which together account for nearly 90% of the world's woody fruit tree cultivation. It is reported that total, 163 viruses have been documented in woody fruit trees, distributed across 45 genera and 23 families (Fig. 5.1). Grapevine harbors the highest diversity, with 80 reported viruses, while other fruit trees each host fewer than 35, and only a single virus has been identified in mango (Umer et al. 2019). The majority of these viruses (over 70%) are positive-sense single-stranded RNA (+ssRNA), while the remainder belong to the −ssRNA, dsRNA, ssRNA-RT, ssDNA, or dsDNA-RT groups, each comprising less than 8% (ICTV 2025). Morphologically, most viruses are icosahedral or isometric (79 out of 163) with diameters ranging from 16 to 80 nm, predominantly 25–30 nm. Cross-infection is frequent between pome and stone fruit species but rarely occurs among banana, citrus, and grapevine. Transmission largely occurs through vegetative propagation, grafting, and natural root grafts within orchards, with mealybugs, soft scales, aphids,

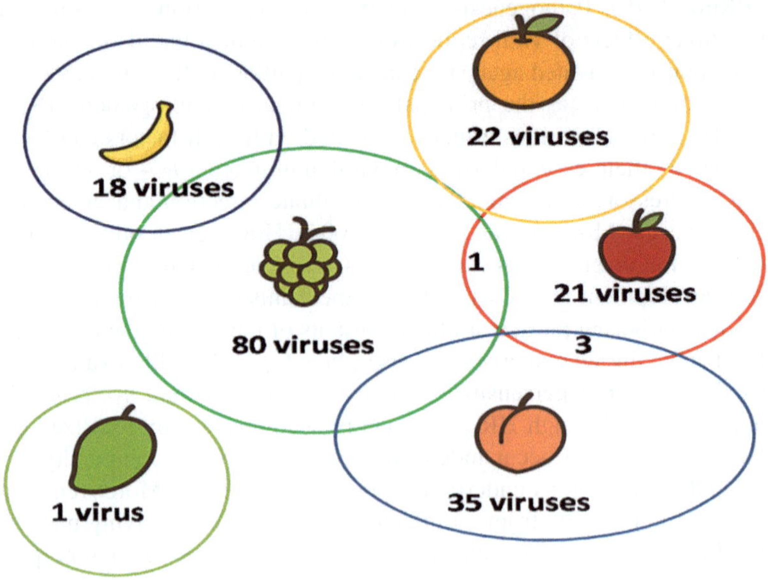

Fig. 5.1 Viruses infecting fruit crops

mites, and thrips acting as vectors. These viral infections impose serious agronomic impacts by reducing tree vigor, longevity, yield, and fruit quality (Bujarski 2019; Rubino et al. 2025) (Fig. 5.1).

5.3 Significance of Understanding Viral Pathogenesis

Comprehending the mechanisms of viral pathogenesis in fruit crops is of paramount importance, as these processes exert substantial influence on agricultural productivity, fruit quality, and the overall economic sustainability of horticultural systems. Viruses cause a range of physiological disorders in fruit plants by altering cellular processes such as photosynthesis, hormone regulation, and nutrient transport, ultimately leading to stunted growth, reduced fruit size, poor flavor, and premature fruit drop (Manzoor et al. 2024). These symptoms translate into significant yield losses and diminished marketability, thereby affecting growers' profitability and food supply chains. Moreover, viruses can establish latent infections that remain asymptomatic until stressed conditions arise, complicating timely detection and management (Nabi et al. 2020). At a molecular level, elucidating the mechanisms by which viruses manipulate host cellular machinery and evade immune defenses provides insights necessary for developing resistant cultivars and targeted antiviral strategies. Additionally, understanding transmission dynamics allows for the formulation of effective control measures, including vector management and the use of certified virus-free planting material (Nabi et al. 2024). Ultimately, knowledge of viral pathogenesis enables integrated disease management approaches that mitigate the spread and impact of viral diseases, ensuring the long-term sustainability and resilience of fruit production systems worldwide.

5.4 How Viruses Invade and Persist in Fruit Crops

The viral entrance, transmission, and establishment in fruit crops occur through multiple intertwined biological and mechanical pathways critical for successful host infection and systemic disease development. Aphids, whiteflies, mealybugs, and eriophyid mites are major biological transmitters, acquiring virus particles during feeding on infected plants and inoculating them into healthy hosts while feeding intermittently as banana streak virus (BSV) transmission occurs by mealybugs, Citrus tristeza virus (CTV), transmitted by aphids (*Toxoptera citricida*) and Papaya leaf curl virus (PaLCuV) is spread to hosts by whiteflies (Vieira et al. 2024). These vectors facilitate both circulative and non-circulative transmission modes depending on virus-vector specificity. Soil-borne nematodes, such as [Xiphinema; cherry rasp leaf virus (James et al. 2001)] and Longidorus species, transmit several nepoviruses by feeding on root tissues, while fungal vectors, though less common, can also assist in virus dissemination. Mechanical transmission arises when contaminated horticultural tools, pruning wounds, or human activities introduce virus particles into plant tissues. In perennial fruit crops, vegetative propagation through grafting and budding is a predominant mode, ensuring virus transfer directly into the vascular system via infected scion or rootstock material. Although seed transfer has also reported in apple, six viruses infecting apple trees, namely, apple chlorotic leaf spot virus (ACLSV), apple green crinkle-associated virus (AGCaV), apple rubbery wood virus 2 (ARWV2), apple stem grooving virus (ASGV), apple stem pitting virus (ASPV), and citrus concave gum-associated virus (CCGaV), occur in seeds extracted from apple fruits produced by infected maternal trees (Wunsch et al. 2024). Viral entry is initiated when virions gain access through wounds or vector feeding sites, subsequently uncoating to release viral genomes inside cells. These genomes then hijack host cell machinery for replication and expression, while viral movement proteins mediate cell-to-cell movement through plasmodesmata, facilitating initial infection foci establishment. Systemic spread follows via vascular tissues, primarily the phloem, leading to whole-plant colonization. This complex series of events, ranging from vector transmission or mechanical inoculation to cellular entry, uncoating, and early replication, underpins the establishment of viral infections and symptom development in fruit crops, highlighting important targets for intervention in disease management strategies.

5.5 Viral Gene Expression Pathways

Viral replication and gene expression encompass the fundamental processes by which plant viruses reproduce their genetic material and produce proteins necessary for infection and systemic spread within host plants. Replication strategies vary significantly depending on the nature of the viral genome, positive-sense RNA, negative-sense RNA, or DNA, each engaging distinct molecular mechanisms and host interactions (Walsh et al. 2013). Positive-sense RNA viruses, which constitute the majority of plant viruses, possess genomes that function directly as messenger RNA upon entry into the host cell cytoplasm. These viral RNAs are immediately translated by host ribosomes to produce replication-associated proteins, including RNA-dependent RNA polymerase (RdRp). The RdRp then synthesizes a complementary negative-sense RNA strand serving as a template for generating numerous positive-sense progeny genomes. Like family *Closteroviridae*, grapevine leafroll disease is the most damaging viral disease afflicting global grape and wine production (Fust et al. 2025). Additionally, subgenomic RNAs are transcribed to allow expression of downstream viral genes. Family representatives such as *Bromoviridae* (apple mosaic virus/apple necrotic mosaic virus) (Nabi et al. 2020; Manzoor et al. 2025) and *Potyviridae* employ this replication mechanism, capitalizing on the host's translational machinery while evading RNA silencing defenses through viral suppressors. The potyviral genome acts as mRNA and codes for a single polyprotein resembling those of the animal-infecting picornaviruses: they possess a small viral protein covalently bound to their 5′ ends (VPg) (Revers and García 2015), and they are polyadenlylated at their 3′ ends (Adams et al. 2005). Negative-sense RNA viruses carry genomes complementary to the required mRNA and thus cannot be translated directly. Upon entry, these viruses bring prepackaged viral RdRp enzymes that transcribe the negative-sense genome into complementary positive-sense mRNAs. These transcripts are used for both translation and replication. Negative-sense RNA plant viruses are less common but include genera-like Tospoviruses, which replicate in the cytoplasm and utilize intricate ambisense strategies to control gene expression (TeVelthuis et al. 2021). Like citrus virus A (CiVA) and citrus concave gum-associated virus (CCGaV) (Liu et al. 2020). DNA viruses, particularly those with single-stranded DNA genomes (geminiviruses), enter the nucleus where the viral DNA is converted to double-stranded DNA forms that serve as transcription templates. These viruses rely heavily on host DNA replication and transcription enzymes, recruiting cellular polymerases to produce viral mRNAs and replicate DNA genomes, like temperate fruit decay-associated virus (TFDaV) (Al Rwahnih et al. 2009; Manzoor et al. 2025). Viral gene expression involves coordinated transcription, translation, and sometimes polyprotein processing or posttranslational modifications. Many plant viruses encode multifunctional proteins, such as movement proteins facilitating cell-to-cell transport MPs carry out multitasking, which includes gene silencing suppression, viral replication and modulation of host protein turnover machinery, coat proteins for virion assembly, and suppressors of host RNA silencing for immune evasion. Protein synthesis is fundamentally dependent on host ribosomes and factors, although viruses may induce alterations in host translation to favor viral mRNA (Kumar and Dasgupta 2021). Host factors play indispensable roles in supporting viral replication and gene expression. These include host RNA-dependent RNA polymerases that may complement viral enzymes, helicases that modulate RNA structures, chaperones assisting folding of viral proteins, and membrane-associated complexes facilitating replication complex assembly. Viruses manipulate host translation initiation factors (e.g., eIF4E) to enhance viral mRNA access to ribosomes. The interplay with host factors is finely tuned and critical for successful viral propagation and symptom development in fruit crops (Nagy and Pogany 2012). Understanding these diverse replication strategies and host dependencies guides the development of antiviral resistance mechanisms, including targeting viral replication enzymes, disrupting host–virus interactions, and engineering RNA interference pathways to degrade viral RNAs.

5.6 Intracellular and Intercellular Movement

Intracellular and intercellular movement of plant viruses are critical phases in the establishment and systemic spread of infection within fruit crops, underpinning symptom development and disease progression. Following replication, newly synthesized viral genomes and proteins must be effectively transported within and between host cells to facilitate colonization beyond the initial infection site (Park et al. 2014). Intracellular trafficking involves the directed transport of viral components through the host cytoplasm to reach plasmodesmata, the plant intercellular channels connecting adjacent cells. Viral movement proteins (MPs) play an essential role in this process by associating with viral genomes and modifying plasmodesmata to permit the passage of viral ribonucleoprotein complexes or entire virions. These MPs can interact with host cytoskeletal elements, such as microtubules and actin filaments, and recruit host membrane trafficking pathways, enabling efficient intracellular delivery of viral cargo to the cell periphery (Atabekova et al. 2023). In systemic infection, viruses enter the plant's vascular system, particularly the phloem, which is responsible for long-distance nutrient transport. Viral particles or ribonucleoprotein complexes are translocated via the phloem sieve elements, reaching sink tissues such as growing meristems, roots, and reproductive organs. Movement proteins and coat proteins are implicated in facilitating phloem loading and unloading processes, with interactions with host phloem-associated proteins aiding viral trafficking. Efficient long-distance movement ensures the rapid colonization of multiple plant organs and is a critical determinant of disease severity and spread within orchards (Atabekova et al. 2023).

5.7 Host–Virus Interactions

Host–virus interactions in fruit crops represent a complex and dynamic battleground where viruses manipulate host cellular processes to facilitate their replication, movement, and evasion of defenses, while the host attempts to restrict infection and mitigate damage. These interactions profoundly alter host metabolism, hormonal signaling, gene expression, and cellular architecture, collectively shaping disease outcomes and symptom expression. One of the key viral strategies is the modulation of host metabolism to create a conducive environment for viral replication and particle assembly. Viruses can redirect primary metabolic pathways such as carbohydrate metabolism, amino acid biosynthesis, and energy production. For instance, infections often result in increased glycolysis and energy flux to supply the energetic demands of viral replication. Concurrently, biosynthesis of nucleotides and amino acids may be enhanced to support genome replication and protein synthesis. Secondary metabolite pathways, including phenylpropanoids and flavonoids, are also altered, sometimes contributing to symptom development and defense responses (Hyodo and Okuno 2020). Viruses manipulate host hormonal pathways crucial for plant growth and defense regulation to facilitate infection and symptom induction. Auxin pathways are often targeted, leading to abnormalities in cell division and differentiation, which manifest as stunting, leaf curling, or tumor-like growths. Salicylic acid (SA), jasmonic acid (JA), and ethylene pathways, central to plant immune signaling, are modulated to suppress effective antiviral responses or induce systemic acquired resistance depending on the virus–host context. SA signaling is typically associated with resistance to biotrophic pathogens, including viruses, but some viruses produce suppressors that interfere with SA-mediated defense activation, allowing enhanced replication and spread. Similarly, JA and ethylene pathways may be hijacked to modulate defense gene expression or hormonal crosstalk, impacting symptom severity and disease progression (Zhao and Li 2021). Manipulation of the host transcriptional machinery is another critical aspect of viral pathogenesis. Plant viruses can alter transcription factor activity and epigenetic modifications to selectively upregulate or downregulate host genes involved in immunity, cell cycle regulation, and metabolic processes. Viral proteins may interact with host RNA polymerases or transcription factors directly or indirectly, reprogramming the host transcriptome to favor viral persistence. High-throughput transcriptomic studies reveal widespread changes in mRNA abundance during infection, including suppression of photosynthesis-related genes and induction of stress and defense genes (Ziegler-Graff 2020). Subcellular reorganization, including virus-induced organelle modifications, is a hallmark of infection and integral to viral replication complex formation. Viruses frequently remodel membranes derived from the endoplasmic reticulum, chloroplasts, peroxisomes, or mitochondria to create specialized compartments that concentrate viral replication machinery and protect viral RNA/DNA from degradation. For example, replication complexes of many positive-sense RNA viruses are anchored to remodeled ER membranes, forming vesicular or tubular structures. Chloroplast morphology and function are often disturbed, contributing to symptoms such as chlorosis and mosaic. These virus-induced organelle alterations play critical roles in facilitating efficient replication and shielding viral genomes from host immune responses (Laliberté and Sanfaçon 2010). Understanding these intricate host–virus interactions at molecular and cellular levels offers pathways to develop innovative disease control strategies,

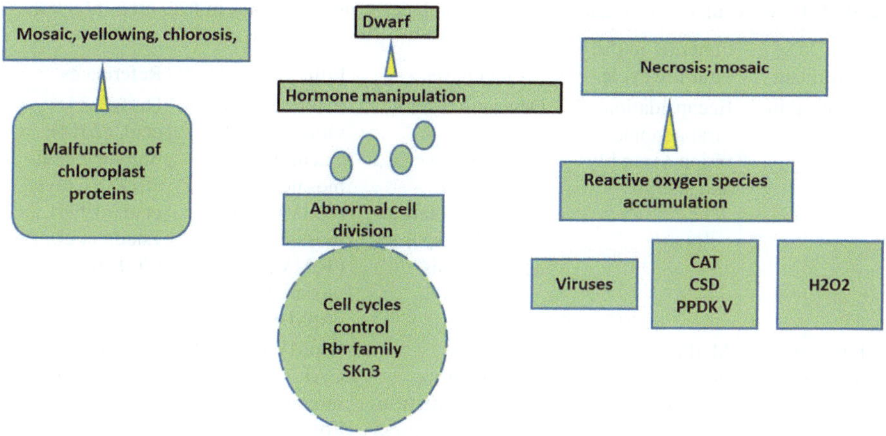

Fig. 5.2 Cellular functions manipulated by fruit viruses

including metabolic engineering, hormone-based treatments, and interference with viral replication complex formation (Fig. 5.2).

5.8 Molecular Determinants of Viral Pathogenesis

Molecular determinants of viral pathogenesis are specific viral components that directly influence infection efficiency, symptom development, and host range in fruit crops. These determinants include various viral proteins and noncoding genome regions that function as key pathogenicity factors by interfacing with host cellular machinery. The coat protein (CP) plays a dual role in both protecting viral genomes and mediating host interactions (Table 5.1). Beyond encapsidation, CPs determine viral particle assembly, stability, and vector transmission efficiency. Importantly, CPs interact with host factors to influence viral movement and symptom induction, and can modulate host immune recognition. Mutations in CP genes often alter symptom severity or viral host range, reflecting their role as critical pathogenicity determinants (Agranovsky 2021). The N-terminal region of the coat protein (CP) in pepino mosaic virus (PepMV) has recently been identified as a key determinant of pathogenicity (Duff-Farrier et al. 2014). In several plant viruses, even a single amino acid substitution in the CP can dramatically influence host responses and disease outcomes. For example, in tobacco mosaic virus (TMV), a single amino acid alteration enables the virus to trigger a hypersensitive response (HR) in *Nicotiana sylvestris* (Knorr and Dawson 1988), while in cucumber mosaic virus (CMV), similar mutations can modify the symptom expression and host range in infected tobacco plants (Shintaku et al. 1992). Structural studies of TMV CP have shown that amino acid changes that disrupt the quaternary structure (without affecting the tertiary structure) can lead to host recognition and HR induction (Culver et al. 1994). In the pepo strain of CMV, virulence is enhanced by specific CP

Table 5.1 Molecular determinants of viral pathogenesis and their roles in fruit crop infections

Viral determinant	Function in the viral life cycle	Role in pathogenesis	Fruit viruses	References
Coat protein (CP)	Encapsidation of viral genome; virion assembly and stability	Determines host specificity, symptom severity, and defense evasion; mutations alter HR response and chlorosis	Pepino mosaic virus (PepMV), Cucumber mosaic virus (CMV), Tomato mosaic virus (ToMV), Apple mosaic virus (ApMV)	Duff-Farrier et al. (2014); Shintaku et al. (1992); Culver et al. (1994); Zhang et al. (2006a, b)
Movement protein (MP)	Mediates intercellular and systemic transport via plasmodesmata	Alters host physiology; interacts with host factors to suppress defense; triggers HR in R-gene hosts	PepMV (TGBp1/ CAT1 interaction), ToMV (Tm-2/ Tm-22 resistance), Barley stripe mosaic virus (BSMV)	Mathioudakis et al. (2013); Weber et al. (2004); Amari et al. (2012)
RNA-dependent RNA polymerase (RdRp) / Replicase	Catalyzes viral RNA replication	Mutations enhance or attenuate symptom severity; they can act as an elicitor of ETI and HR	Plum pox virus (PPV), Tobacco mosaic virus (TMV), Cucumber mosaic virus (CMV)	García and Pallás (2015); Syller and Grupa (2016)
RNA silencing suppressors (RSS/VSRs)	Counteract host RNAi-mediated antiviral defense	Disrupt small RNA pathways; misregulate hormone and defense genes; induce developmental abnormalities	Cucumber mosaic virus (2b), Potyviruses (HC-Pro, P1), Beet necrotic yellow vein virus (p25, p31)	Wang et al. (2012); Jay et al. (2011)
Noncoding regions (UTRs/ NCRs)	Regulate replication, translation, and genome stability	Mutations alter RNA secondary structure, affecting pathogenicity and host adaptation	Apple stem grooving virus (ASGV), Citrus tristeza virus (CTV)	Kumar and Chakraborty (2021)
Host–Virus protein interactions	Facilitate replication, trafficking, and immune suppression	Determine compatibility or resistance; define symptom patterns	CMV (CP–chloroplast interaction), ToMV (MP–IP-L interaction), PepMV (TGBp1–CAT1)	Pallas and García (2011); Zhang et al. (2006a, b)

mutations that cause suppression of chloroplast- and photosynthesis-related genes, resulting in chloroplast deformities and the appearance of chlorosis in infected plants (Balasubramaniam et al. 2014). Similarly, the tomato mosaic virus (ToMV) CP interacts with a tobacco protein known as IP-L, which is localized in the thylakoid membranes (Zhang et al. 2006a, b). This interaction is believed to disrupt chloroplast stability and function, ultimately leading to chlorotic symptoms in infected leaves. Movement proteins (MPs) facilitate cell-to-cell transport of viral genomes through plasmodesmata. MPs interact with viral nucleic acids and host cytoskeletal elements to enable virus trafficking within and between cells. MPs also modulate plasmodesmal permeability and can interfere with host defense responses. The efficiency and specificity of MPs strongly affect virus spread rates and symptom expression, marking them as prime molecular determinants of pathogenesis (Kan and Citovsky 2025). Movement proteins (MPs) can exploit host cellular pathways and factors to assist in viral genome transport, thereby altering plant physiology and directly influencing symptom development. Numerous host proteins that interact with MPs either promoting or restricting virus movement, have been identified (Pallas and García 2011). Many of these interactions are closely associated with symptom expression. A notable example is the interaction between pepino mosaic virus (PepMV) triple gene block protein 1 (TGBp1/p26) and tomato catalase 1 (CAT1), an enzyme responsible for detoxifying hydrogen peroxide (H_2O_2) (Mathioudakis et al. 2013). It has been suggested that during PepMV infection, the p26–CAT1 interaction enhances H_2O_2 scavenging, thereby suppressing plant defense responses and facilitating viral infection. MPs can also interact with resistance (R) gene products to trigger a hypersensitive response (HR). One of the most well-studied cases is the tomato mosaic virus (ToMV) 30 kDa MP, which interacts with the tomato resistance genes Tm-2 and Tm-22. Weber et al. (2004) demonstrated that the domains and structural elements responsible for inducing necrosis differ for each resistance gene. In the model grass *Brachypodium distachyon*, the ND strain of barley stripe mosaic virus (BSMV) fails to infect the Bd3-1 inbred line carrying the resistance gene Bsr1, whereas the NW strain can successfully infect this genotype. Moreover, plant viral MPs can modulate host susceptibility by functioning as viral enhancers of RNA silencing (VERs), promoting the intercellular spread of silencing signals (Amari et al. 2012). This activity can lead to a wave of small RNA-mediated gene expression changes ahead of the infection front, thereby increasing host vulnerability to the invading virus. RNA silencing serves as a central mechanism not only for antiviral defense but also for the regulation of plant growth and development. Viral RNA silencing suppressors (RSSs), while crucial for viral survival, often disrupt host gene regulatory networks, leading to developmental abnormalities and disease symptoms (Wang et al. 2012). By interfering with small RNA pathways such as sequestration of viral and endogenous small RNAs, inactivation of miRNAs (e.g., miR167), or direct targeting of key silencing components like AGO1-RSSs can misregulate genes controlling auxin signaling, defense responses, and hormone pathways. These disruptions frequently result in morphological defects, necrosis, and altered stress responses (Zhang et al. 2006a, b). Moreover, some RSSs can trigger effector-triggered immunity (ETI), producing hypersensitive responses or systemic cell death, further contributing to symptom manifestation. Thus, while viral RSSs enhance viral fitness by evading RNA silencing, their interference with the host's finely tuned regulatory systems underscores a delicate balance between viral pathogenicity and host defense, ultimately shaping the outcome of plant–virus interactions. RNA silencing suppressors (VSRs) are pivotal in counteracting host RNA interference (RNAi), a major antiviral defense. VSRs such as HC-Pro, p19, and 2b proteins bind small RNAs or interfere with RNAi machinery, allowing viral RNAs to evade degradation and persist (Jay et al. 2011). The presence, strength, and specificity of VSR activity directly correlate with viral virulence and symptom onset (Atabekova et al. 2023). Noncoding regions (NCRs) of viral genomes, including untranslated regions (UTRs) and intergenic sequences, are crucial for replication, translation regulation, and genome stability. NCRs mediate interactions with host and viral proteins, influencing RNA secondary structures that regulate pathogenicity. Mutations or rearrangements within NCRs can drastically alter viral replication efficiency and host adaptation (Kumar and Chakraborty 2021). Virus–host protein–protein interactions form the molecular interface where viral components hijack host processes or trigger defense pathways. High-throughput interactome studies reveal that viral proteins target key host factors involved in immunity, transcription, translation, and intracellular trafficking. These interactions define the infection outcome, with specific virus–host protein pairs dictating host susceptibility or resistance. Disrupting critical virus–host protein interactions is a promising antiviral strategy under active investigation (Rahman et al. 2021). In summary, viral pathogenicity is a multifactorial process arising from the collective action of numerous viral components rather than a single determinant. Studies have revealed that pathogenicity factors are distributed throughout the viral genome, involving both proteins and nucleic acids. Several viral proteins, including potyviral P1, P3-6K1, CI, and 6K2, as well as p25 and p31 of Beet necrotic yellow vein virus, and the triplet p33, p18, and p13 of Citrus tristeza virus, contribute to symptom development and host specificity, although the exact molecular mechanisms remain poorly understood. Viral RNA molecules also play a crucial role in pathogenicity, as seen in the formation of defective interfering (DI) RNAs that modulate disease severity, and in viral or satellite-derived siRNAs that target host genes, such as those involved in chlorophyll biosynthesis, resulting in characteristic yellowing symptoms. Furthermore, both the 5′ and 3′ untranslated regions (UTRs) of viral genomes, as well as silent nucleotide mutations within coding sequences, can influence symptom expression. Notably, in Apple stem grooving virus (ASGV), even a single silent mutation within an open reading frame has been shown to attenuate disease symptoms, highlighting the sensitivity of virus–host interactions to subtle genomic changes. Collectively, these findings emphasize that viral pathogenicity is governed by complex interactions between viral genetic elements and host defense systems, which together shape the nature and severity of symptoms in virus-infected plants. Hypersensitive response (HR) is a well-characterized plant defense mechanism against viral infection, involving localized programmed cell death (PCD) that restricts pathogen spread. In *Nicotiana benthamiana*, systemic necrosis induced by Plantagoasiatica mosaic virus (PlAMV) infection exhibits hallmark features of HR, including PCD, biochemical alterations, and specific gene expression profiles. The induction of necrosis during PlAMV infection depends on key regulatory components such as SGT1, RAR1, and the downstream mitogen-activated protein kinase (MAPK) cascade involving MAPKKKα and MEK2. Interestingly, while silencing of SGT1 and RAR1 enhances PlAMV accumulation, disruption of the MAPKKKα–MEK2 cascade does not, suggesting that viral multiplication is partly restrained during systemic necrosis through mechanisms that require SGT1 and RAR1 but are independent of MAPK signaling. Similarly, in the *Rx*-mediated HR to Potato virus X (PVX), both SGT1 and MAPKKKα are necessary for PCD induction, yet only SGT1 is involved in limiting viral replication. These observations indicate that HR and systemic necrosis involve two distinct yet coordinated processes: programmed cell death and suppression of viral multiplication, the latter being mediated through pathways independent of canonical MAPK signalling (Komatsu et al. 2010). Emerging insights from omics approaches have revolutionized the understanding of host–virus interactions in fruit crops by providing comprehensive, high-resolution data on molecular changes during infection. Transcriptomic analyses reveal global shifts in host gene expression upon viral infection, identifying key defense genes, metabolic pathways, and hormonal networks altered by virus presence (Samarskaya et al. 2025). These studies enable dissection of temporal and spatial response patterns, facilitating identification of biomarkers and targets for resistance breeding. Proteomics complements transcriptomics by profiling changes in protein abundance, posttranslational modifications, and protein–protein interactions, thus illuminating how viruses remodel host cellular machinery and modify defense proteins. Metabolomics has uncovered virus-induced reprogramming of primary and secondary metabolite pools, including sugars, amino acids, and antioxidants, which contribute to symptom development and defense modulation. Small RNA profiling elucidates the dynamics of host RNA interference (RNAi) pathways during viral infection, revealing viral suppressors' effects and host small RNA populations critical in antiviral immunity (Samarskaya et al. 2025) (Fig. 5.3).

5.9 Future Perspectives and Research Gaps

The effective management of viral diseases in fruit crops relies on the implementation of multifaceted, integrated strategies that encompass epidemiological understanding, vector control, host resistance, and advanced diagnostic and biotechnological interventions. The foremost step is the use of virus-free, certified planting material, ensuring that nurseries adhere to stringent certification protocols to prevent the introduction of viruses into orchards. Breeding and deployment of resistant or tolerant cultivars, empowered by biotechnological advances such as RNA interference (RNAi) and CRISPR/Cas genome editing, offer durable genetic resistance against prevalent viruses. Cross-protection, involving the deliberate inoculation with mild virus strains to shield plants from severe strains, provides a biological control option in certain crops, notably citrus. Vector management is critical

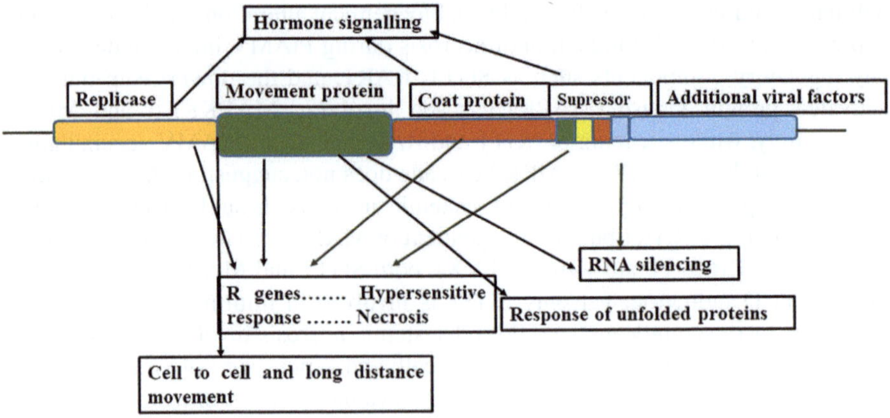

Fig. 5.3 Viral factors involved in plant pathogenesis

and involves integrated pest management (IPM) strategies combining chemical controls, biological agents, cultural practices (e.g., timely pruning, sanitation), and habitat management to reduce populations of aphids, mealybugs, nematodes, and other vectors. Moreover, sensitive and rapid diagnostic tools facilitate early detection and monitoring, enabling prompt remedial actions. Quarantine and regulatory measures to limit virus dissemination, coupled with grower education and coordinated surveillance, complete a holistic approach that minimizes viral impact, safeguards fruit quality, and supports sustainable horticultural production in the face of evolving viral threats.

5.10 Conclusion

Viral diseases of fruit crops constitute a complex and persistent threat to global horticultural productivity, adversely affecting yield, fruit quality, and overall economic returns. The pathogenic potential of plant viruses stems from their obligate intracellular parasitism, genomic diversity, and intricate interactions with both host plants and biological vectors. These viruses exploit host cellular machinery to accomplish replication, intracellular and intercellular movement, and evasion of host immune responses, culminating in a continuum of disease phenotypes ranging from asymptomatic or latent infections to severe tissue degeneration and systemic dysfunction. A comprehensive understanding of viral taxonomy, genome organization, and transmission mechanisms is fundamental for effective disease surveillance, risk assessment, and management. At the molecular level, elucidation of processes such as viral genome replication, transcriptional and translational regulation, and intracellular trafficking has provided critical insights into the mechanisms underlying viral pathogenesis. Furthermore, virus-induced suppression of host defense responses, manipulation of phytohormonal signaling, and reprogramming of primary and secondary metabolism collectively contribute to symptom manifestation and disease severity. Recent advances in multi-omics technologies, including genomics, transcriptomics, proteomics, and metabolomics, have significantly enhanced our understanding of virus–host and vector interactions. Concurrently, novel biotechnological platforms such as RNA interference (RNAi) and CRISPR/Cas-mediated genome editing are emerging as powerful tools for engineering virus-resistant cultivars and developing next-generation precision diagnostics. The implementation of integrated viral disease management strategies, encompassing the use of virus-free planting material, vector population suppression, and deployment of genetically resistant cultivars, remains central to sustainable horticultural production systems. Bridging the existing gaps in our knowledge of virus–host and vector dynamics and translating molecular insights into field-applicable solutions are imperative for safeguarding fruit crop productivity, ensuring food security, and sustaining economic stability in the global horticultural sector.

References

Adams MJ, Antoniw JF, Beaudoin F (2005) Overview and analysis of the polyprotein cleavage sites in the family Potyviridae. Mol Plant Pathol 6:471–487. https://doi.org/10.1111/j.1364-3703.2005.00296.x

Agranovsky A (2021) Enhancing capsid proteins' capacity in plant virus-vector interactions and virus transmission. Cells 10(1):90

Al Rwahnih M, Daubert S, Golino D, Rowhani A (2009) Deep sequencing analysis of RNAs from a grapevine showing Syrah decline symptoms reveals a multiple virus infection that includes a novel virus. Virol 387(2):395–401

Amari K, Vazquez F, Heinlein M (2012) Manipulation of plant host susceptibility: an emerging role for viral movement proteins? Front Plant Sci 3:10

Atabekova AK, Solovieva AD, Chergintsev DA, Solovyev AG, Morozov SY (2023) Role of plant virus movement proteins in suppression of host RNAi defense. Int J Mol Sci 24(10):9049

Balasubramaniam M, Kim BS, Hutchens-Williams HM, LoeschFries LS (2014) The photosystem II oxygen-evolving complex protein PsbP interacts with the coat protein of Alfalfa mosaic virus and inhibits virus replication. Mol Plant-Microbe Interact 27:1107–1118

Bujarski J (2019) ICTV virus taxonomy profile: Bromoviridae. J Gen Virol 100:1206. https://www.ictv.global/report/bromoviridae

Culver JN, Stubbs G, Dawson WO (1994) Structure–function relationship between tobacco mosaic virus coat protein and hypersensitivity in *Nicotiana sylvestris*. J Mol Biol 242:130–138

Duff-Farrier CR, Bailey AM, Boonham N, Foster GD (2014) A pathogenicity determinant maps to the N-terminal coat protein region of the *Pepino mosaic virus* genome. Mol Plant Pathol 15:107–116

Fust C, Lameront P, Shabanian M, Song Y, AbouKubaa R, Bester R et al (2025) Grapevine leafroll-associated virus 3: a global threat to grapevine and wine industries but a gold mine for scientific discovery. J Exp Bot 76:eraf039

García JA, Pallás V (2015) Viral factors involved in plant pathogenesis. Curr Opin Virol 11:21–30

Garcia-Ruiz H (2019) Host factors against plant viruses. Mol Plant Pathol 20(11):1588–1601

Hyodo K, Okuno T (2020) Hijacking of host cellular components as proviral factors by plant-infecting viruses. In: Advances in virus research, vol 107. Academic Press, pp 37–86

ICTV (2025) About virus taxonomic classification. International Committee on Taxonomy of Viruses. https://ictv.global/about/taxonomy

James D, Howell WE, Mink GI (2001) Molecular evidence of the relationship between a virus associated with flat apple disease and cherry rasp leaf virus as determined by RT-PCR. Plant Dis 85:47–52. https://doi.org/10.1094/PDIS.2001.85.1.47

Jay F, Wang Y, Yu A, Taconnat L, Pelletier S, Colot V, Renou J-P, Voinnet O (2011) Misregulation of *AUXIN RESPONSE FACTOR 8* underlies developmental abnormalities caused by viral silencing suppressors in *Arabidopsis*. PLoS Pathog 7:e1002035

Kan Y, Citovsky V (2025) The roles of movement and coat proteins in the transport of tobamoviruses between plant cells. Front Plant Sci 16:1580554

Knorr DA, Dawson WO (1988) A point mutation in the tobacco mosaic virus capsid protein gene induces hypersensitivity in Nicotiana sylvestris. Proc Natl Acad Sci USA 85:170–174

Komatsu K, Hashimoto M, Ozeki J, Yamaji Y, Maejima K, Senshu H et al (2010) Viral-induced systemic necrosis in plants involves both programmed cell death and the inhibition of viral multiplication, which are regulated by independent pathways. Mol Plant-Microbe Interact 23(3):283–293

Kumar K, Chakraborty S (2021) Roles of long non-coding RNAs in plant virus interactions. J Plant Biochem Biotechnol 30(4):684–697

Kumar G, Dasgupta I (2021) Variability, functions and interactions of plant virus movement proteins: what do we know so far?. Microorg, 9(4):695

Laliberté JF, Sanfaçon H (2010) Cellular remodeling during plant virus infection. Annu Rev Phytopathol 48(1):69–91

Liu H, Wang G, Yang Z, Wang Y, Zhang Z, Li L et al (2020) Identification and characterization of a pear chlorotic leaf spot-associated virus, a novel Emaravirus associated with a severe disease of pear trees in China. Plant Dis 104:2786–2798. https://doi.org/10.1094/PDIS-01-20-0040-RE

Manzoor S, Nabi SU, Rather TR, Verma MK, Alrobaish SA, Raja WH et al (2024) Comparative biochemical and physiological responses to the virus-induced mosaic disease in apple (Malus domestica). J Plant Dis Prot 131(6):1945–1958

Manzoor S, Nabi SU, Ayaz A, Rasool B, Sharma SK, Chesti MH et al (2025) Unveiling viral threats to temperate pome fruits: characterization, transmission, and sustainable management strategies. Front Microbiol 16:1560720

Mathioudakis MM, Veiga RS, Canto T, Medina V, Mossialos D, Makris AM, Livieratos I (2013) *Pepino mosaic virus* triple gene block protein 1 (TGBp1) interacts with and increases tomato catalase 1 activity to enhance virus accumulation. Mol Plant Pathol 14:589–601

Moissiard G, Voinnet O (2004) Viral suppression of RNA silencing in plants. Mol Plant Pathol 5(1):71–82

Nabi SU, Baranwal VK, Yadav MK, Rao GP (2020) Association of Apple necrotic mosaic virus (ApNMV) with mosaic disease in commercially grown cultivars of apple (*Malus domestica* Borkh) in India. 3 Biotech 10:1–9

Nabi SU, Baranwal VK, Mir JI, Raja WH, Yasmin S, Shah MA et al (2024) Transmission efficiency protocol of two Ilarviruses (ApMV and ApNMV) associated with mosaic disease of apple (Malus domestica) through cleft grafting and T-budding. Indian Phytopathol 77(2):433–439

Nagy PD, Pogany J (2012) The dependence of viral RNA replication on co-opted host factors. Nat Rev Microbiol 10(2):137–149

Pallas V, García JA (2011) How do plant viruses induce disease? Interactions and interference with host components. J Gen Virol 92:2691–2705

Park MR, Jeong RD, Kim KH (2014) Understanding the intracellular trafficking and intercellular transport of potexviruses in their host plants. Front Plant Sci 5:60

Rahman A, Sinha KV, Sopory SK, Sanan-Mishra N (2021) Influence of virus–host interactions on plant response to abiotic stress. Plant Cell Rep 40(11):2225–2245

Ray PK, Singh KM, Kumar A, Singh RN, Singh RR (2020) Chapter-1 Horticulture for food security and livelihood improvement. In: advances in agricultural and life sciences, 3, p1

Revers F, García JA (2015) Chapter three-molecular biology of potyviruses. Adv Virus Res 92:101–199. https://doi.org/10.1016/bs.aivir.2014.11.006

Rodríguez-Verástegui LL et al (2022) Viruses infecting trees and herbs that produce edible fruits: taxonomy and host range. Front Plant Sci 13:8778216

Rubino L et al (2025) Summary of taxonomy changes ratified by the International Committee on Taxonomy of Viruses (ICTV). Arch Virol 170(7):1903–1913. https://pubmed.ncbi.nlm.nih.gov/40711908/

Samarskaya V, Spechenkova N, Kalinina NO, Love AJ, Taliansky M (2025) The emerging role of omics-based approaches in plant virology. Viruses 17(7):986

Shintaku MH, Zhang L, Palukaitis P (1992) A single amino acid substitution in the coat protein of *Cucumber mosaic virus* induces chlorosis in tobacco. Plant Cell 4:751–757

Syller J, Grupa A (2016) Antagonistic within-host interactions between plant viruses: molecular basis and impact on viral and host fitness. Mol Plant Pathol 17(5):769–782

TeVelthuis AJ, Grimes JM, Fodor E (2021) Structural insights into RNA polymerases of negative-sense RNA viruses. Nat Rev Microbiol 19(5):303–318

Umer M, Liu J, You H, Xu C, Dong K, Luo N et al (2019) Genomic, morphological and biological traits of the viruses infecting major fruit trees. Viruses 11(6):515

Vieira MS, Cabral RLR, Favaratto L, Maciel LS, Xavier ADS, Zerbini FM, Fernandes PM (2024) Tropical fruit virus resistance in the era of next-generation plant breeding. SynBio 2(3):267–284

Walsh D, Mathews MB, Mohr I (2013) Tinkering with translation: protein synthesis in virus-infected cells. Cold Spring Harb Perspect Biol 5:a012351. https://doi.org/10.1101/cshperspect.a012351

Wang MB, Masuta C, Smith NA, Shimura H (2012) RNA silencing and plant viral diseases. Mol Plant-Microbe Interact 25:1275–1285

Wani S, Ashaq M, Dar GH, Ahmad M, Padder BA, Shah MD (2024) Major viruses infecting temperate fruit crops and their impact on the fruit industry. In: Industrial applications of soil microbes, vol 4. Bentham Science Publishers, pp 300–323

Weber H, Ohnesorge S, Silber MV, Pfitzner AJP (2004) The *Tomato mosaic virus* 30 kDa movement protein interacts differentially with the resistance genes *Tm-2* and *Tm-22*. Arch Virol 149:1499–1514

Wunsch A, Hoff B, Sazo MM, van Zoeren J, Lamour KH, Hurtado-Gonzales OP, Fuchs M (2024) Viruses of apple are seedborne but likely not vertically transmitted. Viruses 16(1):95

Zhang CZ, Liu YY, Sun XC, Qian WD, Zhang DD, Qiu BS (2006a) Characterization of a specific interaction between IP-L, a tobacco protein localized in the thylakoid membranes, and Arabidopsis AGO1 cleavage activity to counter plant defense. Genes Dev 20:3255–3268

Zhang X, Yuan Y-R, Pei Y, Lin S-S, Tuschl T, Patel DJ, Chua N-H (2006b) Cucumber mosaic virus-encoded 2b suppressor inhibits Arabidopsis Argonaute1 cleavage activity to counter plant defense. Genes Dev 20:3255–3268

Zhao S, Li Y (2021) Current understanding of the interplays between host hormones and plant viral infections. PLoS Pathog 17:e1009242

Ziegler-Graff V (2020) Molecular insights into host and vector manipulation by plant viruses. Viruses 12(3):263

Biochemical and Molecular Insights on Virus-Vector Interactions in Fruit Crops

6

Laxmipriya Behera, Saswat Swarup Padhi, Bandana Mohapatra, S. Kruthika, Rupak Jena, Ravinder Kumar, Rahul Kumar Tiwari, Awadhesh Kumar, and Milan Kumar Lal

Abstract

Viral infections, mostly spread by insect vectors, cause large productivity and quality losses (30–70%) in fruit crops such as papaya, bananas, citrus, grapes, and cucurbits. The interactions between viruses and vectors are extremely specialized: viruses alter host volatiles, defense signaling, and metabolism to improve vector attractiveness and performance, while insect vectors use salivary effectors and stylet-mediated feeding to acquire, maintain, and spread viruses. Chlorosis, mosaic, deformation, and decreased fruit set are the results of infections that affect photosynthesis, pigment biosynthesis, chloroplast structure, and hormone balance at the biochemical level. Viral proteins target transcription factors, RNA silencing pathways, and ubiquitin-mediated degradation to reduce host immunity, according to molecular research. Nonpersistent, semi-persistent, and similarly persistent modes of transmission are all influenced by changes in host physiology and vector behaviors brought on by the virus. Recent advancements in genome sequencing, transcriptomics, and RNA-based technologies are uncovering these intricate tripartite relationships. These developments are also driving innovative management techniques, such as RNA interference, topical dsRNA treatments, and resistant cultivars. In order to safeguard fruit crops and guarantee global food security, sustainable ways must be designed with a mechanistic understanding of virus–vector–plant relationships.

L. Behera · S. S. Padhi · B. Mohapatra · S. Kruthika · R. Jena · A. Kumar · M. K. Lal (✉)
ICAR-Central Rice Research Institute, Cuttack, Odisha, India

R. Kumar
ICAR-Indian Agricultural Research Institute, New Delhi, India

R. K. Tiwari
ICAR-Indian Sugarcane Research Institute, Lucknow, Uttar Pradesh, India

R. Kumar et al. (eds.), *Fruit Crops and Viral Pathogens Volume II*, Plant Pathology for Sustainable Agriculture, https://doi.org/10.1007/978-981-95-7261-8_6

Keywords

Viral protein · Genomics · Pathophysiology · Antioxidant mechanism

6.1 Introduction

Fruit crops have a significant impact on environmental sustainability, economic growth, and human well-being (Meena et al. 2022). They have numerous impacts on various areas, each making a distinct contribution to society and the environment (Mume et al. 2024). Consuming fruit and fruit-derived products has been linked to protective effects against several chronic illnesses (Boeing et al. 2012). The growing emphasis on healthier eating practices, as evidenced by global dietary trends, has also contributed to this increase in demand for fruit (Medici et al. 2020). Fruits are vital to human nutrition, full of vitamins A, C, and E, minerals potassium and magnesium, and antioxidants that help fight against lifestyle diseases and malnutrition (Kostecka-Gugała et al. 2020). Including fruits in daily diets promotes healthy eating habits, increases dietary diversity, and supports the development of both the body and mind. These diets are essential for obtaining fiber, vitamins, and minerals. For the agricultural and commercial sectors, the growth of fruit production and markets into new regions presents both new potential and difficulties. Fruit production is essential to rural livelihoods, nutrition, and food security (Cosme et al. 2022). However, a constant danger is their susceptibility to viral infection, which is frequently spread by insect vectors, including mealybugs, aphids, and whiteflies. Since viral infections are incurable once they have taken hold, prevention and early detection are crucial, unlike bacterial or fungal diseases (Mittapelly and Rajarapu 2020).

Fruit crop viral infection is affecting agricultural productivity worldwide, especially in tropical and subtropical areas (Kaur et al. 2016). In addition to lowering fruit quality and productivity, these diseases have a negative financial impact on farmers, businesses, and national economies (Medici et al. 2020). With a focus on direct and indirect losses, management expenses, and wider socioeconomic implications, this book examines the complex economic effects of viral infections in fruit crops (Cembali et al. 2003). Yield decreases of 30–70% can be caused by viral illnesses like papaya ringspot virus (PRSV), banana bunchy top virus (BBTV), and citrus tristeza virus (CTV) (Whitfield et al. 2018). Several symptoms, including mosaics, dead patterns, stunted development, and yellowed foliage, are displayed by the virus-infected host plant (Jiang and Zhou 2023). Fruits that are deformed, undersized, or discolored are frequently produced by infected plants, which can result in market rejection or lower prices (Zakaria 2023). These losses directly impact the availability of food and farm income. Plant pathogenic viruses can spread mechanically by direct seed contact or through a vector-borne virus. Vector-borne viruses are transmitted by fungi, nematodes, insects, and mites (Ray and Casteel 2022). The epidemiology and transmission of plant viruses are significantly influenced by insect vectors (Catto et al. 2022). Plants can identify a wide range of attack-related elicitors, such as three types of molecular patterns: HAMPs, pathogen-associated molecular patterns (PAMPs), and damage-associated molecular patterns (DAMPs). Immunity activated by patterns is the outcome of elicitor recognition, which takes place at the cell surface by receptor-like proteins and receptor kinases (Jangra et al. 2024). Plants use RNA silencing as a primary immunological mechanism. Plants generate tiny interfering molecules in response to double-stranded RNA derived from virus genomes (siRNAs) (Tang et al. 2003). Plants use siRNAs to silence viral DNAs and RNAs through transcriptional and posttranscriptional gene silencing (Alimanović 2021). The host's immune responses are silenced by some siRNAs produced from viral genomes that are identical to plant targets. Plants also use small RNAs (sRNAs) to silence nonviral pathogen genes, and certain fungi and oomycetes produce sRNAs that silence host defense genes (Prasad et al. 2019).

6.2 Plant Viral Vectors Interaction

Mealybugs, aphids, leafhoppers, and whiteflies are among the insects in the order Hemiptera that spread the most plant viruses (*Thysanoptera* order). Certain viral species have important vectors. More than 70% of plant viruses identified to date spread among hosts through insect vectors. While thrips and beetles are particularly effective vectors because they eat using stylets, which are needle-like mouthparts that allow them to consume a variety of plant cell types as well as individual cells, as shown in Table 6.1. Hemipterans inject saliva into the plant during the first stages of host interaction, feeding, and viral transmission. Both effectors and elicitors are present in this saliva. Effectors can assist insects in accessing the chemical makeup and nutritional profile of plants, suppress plant defenses, and conceal insect detection. Additionally, certain hemipterans produce gelling saliva while feeding, which envelops the stylet in a sheath that shields it from the apoplast and gives it a path to follow. Whiteflies and aphids are by far the most significant hemipteran vectors, as they together spread more than 500 virus species. Whiteflies mainly spread begomoviruses (*Geminiviridae*), criniviruses (*Closteroviridae*), torradoviruses (*Secoviridae*), and ipomoviruses, while there have been a few isolated reports of whiteflies spreading viruses from other families. It has been demonstrated that aphids can spread viruses belonging to the *Potyviridae, Bromoviridae, Secoviridae, Caulimoviridae, Closteroviridae, Nanoviridae,* and *Reovirodae* families (Rosario et al. 2015). Aphids and whiteflies both use stylets for feeding and release elicitors and effectors, but their feeding biology differs greatly. Whiteflies go through plant cells via their stylet. Sometime, puncturing a cell before establishing phloem feeding sites in a cellular context. Equally, aphids pierce and taste almost every cell along the route to the phloem using their stylets. Adult whiteflies with wings and the crawlers that hatch from the eggs are both mobile; however, the whitefly crawlers become immobile as they molt into the nymphal stages. This indicates that whitefly nymphs frequently spend extended periods of time (up to 21–30 days) feeding from a single feeding spot. Mutualisms, or mutually beneficial interactions between species, are common in nature (Benckiser et al. 2017). Plants supply resources for

Table 6.1 The content of the table shows that the interactions are mostly mediated by certain viral and insect proteins, and the interactions determine whether the transmission is nonpersistent, semi-persistent, or persistent

Virus Family/Genus	Mode of transmission	Type of biochemical & molecular insight	Vector	Examples in fruit crops	References
Potyviridae	Nonpersistent	Aphids	Virus attaches to the vector's stylet (mouthparts) via a helper component (HC-Pro) and the viral coat protein (CP). The HC-Pro acts as a molecular bridge	Plum pox virus (PPV) in stone fruits (e.g., plum, peach)	Coleman et al. (2015), Ng and Perry (2004)
Cucumoviridae	Nonpersistent	Aphids	The virus is carried on the tip of the stylet, mediated by the coat protein (CP). The transmission is fast, and the virus is lost during molting	Cucumber mosaic virus (CMV) affects various fruits like citrus and berries	Ng and Perry (2004)
Geminiviridae	Persistent circulative	Whiteflies, leafhoppers	The virus enters the body through the midgut, travels through the hemolymph, and is subsequently released by the salivary glands. In order to the insect's salivary glands and gut must contain particular receptor-like proteins that interact with the viral coat protein (CP)	Grapevine leafroll-associated viruses (GLRaVs) transmitted by mealybugs	Coleman et al. (2015) and Hommay et al. (2021)

Closteroviridae	Semi-persistent	Mealybugs, aphids	The virus is retained in the foregut of the vector. The viral p94 protein and other accessory proteins anchor the virus particles to the foregut lining, allowing for a longer retention period than nonpersistent viruses	Grapevine leafroll-associated viruses (GLRaVs) are transmitted by mealybugs and soft scales	Krebelj et al. (2015)
Luteoviridae	Persistent circulative (non-propagative)	Aphids	The virus does not reproduce inside the vector, but it does move through it. Transmission is highly specific, involving a read-through domain (RTD) on the viral protein that facilitates passage through the vector's tissues	Barley yellow dwarf virus (BYDV), which can affect fruit crops	del Río-Celestino and Font Villa (2020), Meena et al. (2022)
Nepovirus	Persistent (non-propagative)	Nematodes	The nematode's pharynx and esophagus have cuticle linings that the virus particles adhere to. An important factor in this particular bining is the viral coat protein	Grapevine fanleaf virus (GFLV) in grapes	Krebelj et al. (2015)

(continued)

Table 6.1 (continued)

Virus Family/Genus	Mode of transmission	Type of biochemical & molecular insight	Vector	Examples in fruit crops	References
Tospovirus	Persistent propagative	Thrips	The virus multiplies inside the body of the vector. The glycoprotein (G-N/G-C) on the viral envelope binds to specific receptors in the thrips' midgut, and the virus then moves to the salivary glands where it replicates before transmission	Tomato spotted wilt virus (TSWV), which can affect a wide range of fruit crops	Fereres et al. (2016)
Protein-protein interaction	Nonpersistent	Aphids	The HC-Pro of a potyvirus interacts with the coat protein to form a complex that binds to aphid stylet receptors, ensuring efficient transmission	Plum pox virus (PPV)	Gadhave et al. (2020)
Virus-receptor interaction	Persistent circulative	Whiteflies	The coat protein of geminiviruses binds to specific proteins in the whitefly gut and salivary glands, acting as receptors for virus entry and exit	Tomato yellow leaf curl virus (TYLCV) affecting tropical and sub-tropical fruit crops	Luan et al. (2013)

Retention site	Semi-persistent	Mealybugs	The virus is retained in the foregut of the vector, a specific anatomical location that prevents its circulation but allows for temporary transmission	Grapevine leafroll-associated virus 3 (GLRaV-3)	Maree et al. (2013)
Retention site	Nonpersistent	Aphids	The virus is held on the stylet tip, a very superficial retention site, which explains the quick acquisition and inoculation times	Papaya ringspot virus (PRSV) in papaya	Gonsalves et al. (2008) and Sa Antunes et al. (2020)
Viral protein-mediated movement	Persistent	Leafhoppers	A read-through protein on the viral coat is essential for the virus to pass through the midgut and enter the hemolymph of the vector	Grapevine fleck virus (GFkV)	Sabanadzovic et al. (2000)
Tripartite interaction	Nonpersistent	Aphids	Viruses can alter the host plant's metabolism, affecting the emission of volatile organic compounds that either attract or repel vectors, influencing transmission rates	Papaya ringspot virus (PRSV)	Kalleshwaraswamy and Kumar (2008)

(continued)

Table 6.1 (continued)

Virus Family/Genus	Mode of transmission	Type of biochemical & molecular insight	Vector	Examples in fruit crops	References
Virus-induced host change	Persistent	Whiteflies	A virus can induce changes in the host's nutritional content, making it more suitable for vector feeding and reproduction, thereby increasing the vector population and viral spread	Tomato yellow leaf curl virus (TYLCV)	Pakkianathan et al. (2015)
Viral genome modification	Nonpersistent	Aphids	The helper component (HC-Pro) of a potyvirus can also act as an RNA silencing suppressor, counteracting the plant's defense system and aiding in virus replication and systemic spread	Plum pox virus (PPV)	Kamenova et al. (2002)

Gaining an understanding of these mechanisms is essential to creating fruit crop viral disease control plans that work pollinators, while mycorrhizal fungi benefit plants from a variety of resources. Mutualism occurs when one species modifies its morphology, physiology, or behavior to benefit the other, such as viral transfer between hosts. The partner gives nutritional resources to the host, including increased nutrient content and decreased viral defenses. Some mutualist interactions involve obligate partners who rely on each other for survival, while others are not as strictly tied (Kiers et al. 2011).

6.3 Plant Virus Symptoms: Physiological, Cellular, and Biochemical Alterations

Virus infection can be classified into two types: systemic and local virus infections. A systemic infection is progressive, beginning at the site of infection using the host vasculature to spread throughout the entire plant, whereas a local infection is defined as the virus infection occurring within and close to the site of infection (Waigmann et al. 2004). Infected plants can display several symptoms, including the formation of colorful patterns (mosaics). Symptoms may include chlorosis, necrosis, or chlorotic patches on leaves, twisting or curling of branches, and discoloration or streaking on flower petals. The ion travels throughout the plant via the host vasculature (Balamurugan 2024). Viruses can alter the size, shape, and quality of fruit, causing obvious symptoms such as ringspots, pits, mottling, or line patterns that reduce its appearance and organoleptic characteristics (Umer et al. 2019). Leaf chlorosis is a common symptom of viral infection that results from a change in the pigments produced by the chloroplasts. Consequently, numerous alterations have been noted in these organelles, including a reduction or clustering of chloroplasts (Almási et al. 2001). Among other things, these include swelling or an unusual shape (globular or amoeboid), the development of vesicles at the periphery, envelope rupture, variations in the chloroplast content, stroma disappearance or thylakoid dilatation, and disordered grana. Chlorophyllase was found to be responsible for the development of chlorotic patches, ringspots, and mosaics in cucumbers when viruses affected chlorophyll content (Ushiyama and Matthews 1970). A 50% decrease in photosynthetic activity occurs during viral infection due to decreased production of chloroplast proteins involved in the electron transport chain and the Calvin-Benson cycle. Reduced photosynthesis causes problems for the growth, development, and/or reproduction of diseased plants. An excellent illustration of this is the banana bunchy top virus (BBTV; *Babuvirus, Nanoviridae*), which affects banana plants and results in induced dwarfism in the leaves and stems, as well as little to no fruit output (Rahoutei et al. 2000). The result of elevated chlorophyllase enzymatic activity after CMV (cucumber mosaic virus) infection. The leaves of pumpkins (*Cucurbita pepo* cv Eskandarani) infected with the zucchini yellow mosaic virus (ZYMV; *Potyvirus, Potyviridae*) have also shown this. The breaking or variegation of flower petals displaying dots, lines, or tissue segments of varying colors, is another sign of pigment loss (Radwan et al. 2007). For instance, yellowing of the leaves is a defining sign that apple mosaic virus (ApMV; *Ilavirus, Bromoviridae*) in apple plants, while irregular light/dark green patterns are characteristic of tomato yellow leaf curl virus (TYLCV; *Begomovirus, Geminiviridae*) or CTV in tomato and citrus trees, respectively. By blocking chloroplast formation, viral agents such as the coat protein (CP) have been found to cause symptoms. For instance, when Ferredoxin I binds to the tobacco mosaic virus (TMV; Tobamovirus, Virgaviridae) CP, it results in mosaic and chlorosis in leaves, while mosaic and chlorosis are encouraged when the potato virus Y (PVY; *Potyvirus, Potyviridae*) CP interacts with RuBisCo enzyme subunits. When viruses affect chloroplasts collectively, photosynthetic activity often declines.

6.4 Effectors of Viruses that Target Transcription Factors in Plants

Plant defense signaling pathways can be targeted at the transcriptional level by viral effectors interacting with transcription factors, as seen in Fig. 6.1. Certain Begomoviruses (family *Geminiviridae*) have connections to betasatellites and are spread by whiteflies (Hanley-Bowdoin et al. 2013). These ssDNA molecules encode a single protein and rely on the begomo virus for replication, encapsidation, motility, and insect transmission. The betasatellite bC1 affects the pathogenicity of Tomato yellow leaf curl China virus (*TYLCCNV*) and enhances whitefly (*B. tabaci*) eating and reproduction by binding to transcription factors that govern plant defense mechanisms. bC1 binds to MYC2, a transcription factor that regulates jasmonic acid-dependent pathways (Zou et al. 2020). Binding to MYC2 hinders dimerization and transcription of genes involved in the synthesis of glucosinolates, an essential

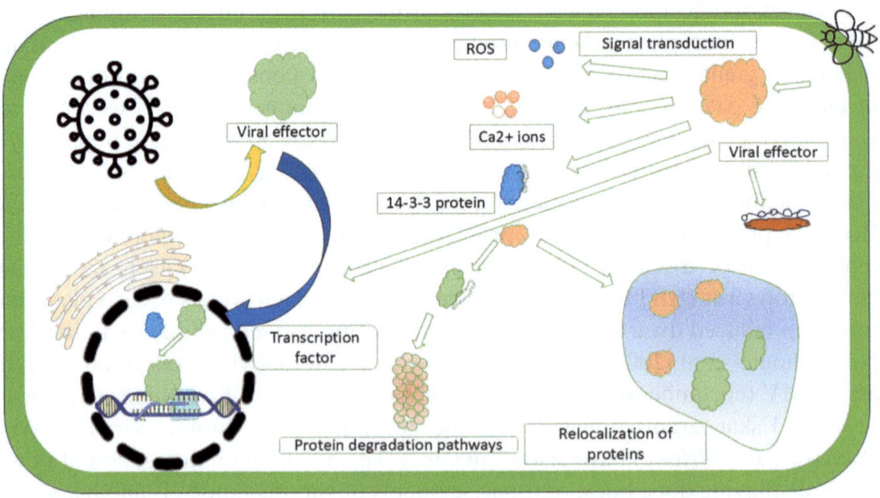

Fig. 6.1 Plant targets of the viral and vector effectors and their downstream effects on plant signaling pathways. The green circular structures stand for vector effectors, while yellow round structures stand for viral effectors

anti-insect defense molecule. bC1 binds to and activates ASYMMETRICLEAVES1, a negative regulator of plant AS1, which decreases the expression of PLANT DEFENSIN 1.2, a JA- and ethylene (ET)-dependent defense gene implicated in plant defense against whiteflies (Dombrecht et al. 2007). bC1 improves whitefly performance on infected host plants by inhibiting and boosting the activity of several plant transcription factors. Plants infected with TYLCCNV and bC1 exhibit higher whitefly attraction and lower terpene production compared to plants without the betasatellites. bC1 interacts with MYC2 and transcription factors, leading to increased whitefly attraction and enhanced volatile production (Zhou 2013). Factors that interact with phytochromes, such as PIFs, interact with MYC2 to favorably control light-regulated plant defenses. They also bind directly to the terpene promoter's synthase genes, which promote their transcription. The begomovirus-encoded bC1 suppresses PIF and MYC2 transcriptional activity by disrupting its dimerization through direct contacts. bC1 binds to several transcription factors, contributing to improved whitefly performance and attractiveness (Zhou 2013).

6.5 Virus Effectors that Target the Protein Degradation Pathways

According to Fig. 6.2, plant viruses enhance both their own performance and the attractiveness and functionality of their insect vectors by targeting proteolytic degradation pathways. Whiteflies benefit from TYLCV-infected plants through altered plant protein turnover, and TYLCV benefits from whitefly feeding, as was

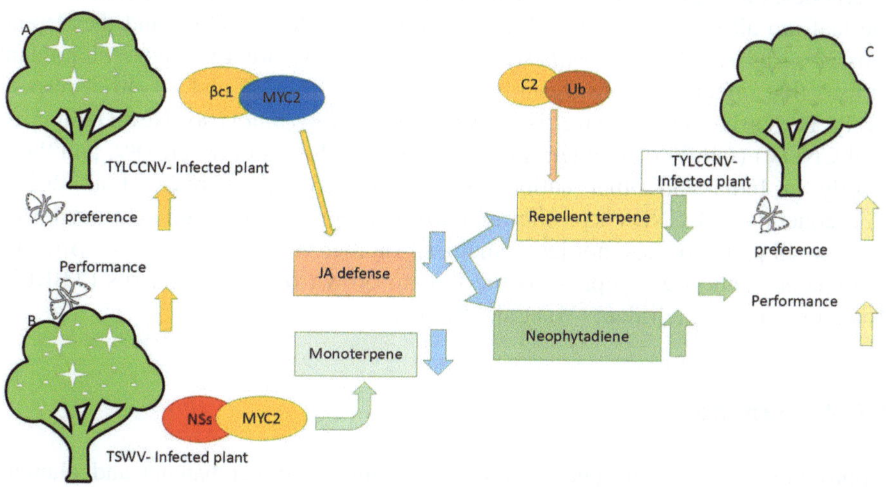

Fig. 6.2 Interactions between plants, viruses, and vectors in transmission that are nonpersistent, semi-persistent, and persistent. Aphids can spread nonpersistent plant viruses like CMV; whiteflies can spread semi-persistent plant viruses like ToCV; and thrips and whiteflies can spread persistent plant viruses like TYLCV and TSWV

previously mentioned (Benckiser et al. 2017). The E3 ubiquitin ligase SCFCOi1 increases JA signaling and defense responses when JA is identified by recruiting JAZs for ubiquitination and degradation through the 26S proteasome pathway (Wang et al. 2016). By interacting with and inhibiting the ubiquitin precursor protein RPS27A, the TYLCV protein C2 stops JAZ1 from being ubiquitinated or broken down. In plants that express the C2 protein of TYLCV, this results in the suppression of MYC2 and JA-responsive terpene synthase genes, which enhances whitefly performance (Li et al. 2019).

6.6 Vectors Associated with Viral Transmission in Fruit Crops

6.6.1 Tomato

The tomato, or *S. lycopersicum*, is a native American herbaceous plant. As a vegetable, tomatoes are the most consumed fruit worldwide, and they are used to make a variety of goods, including paste, soup, juices, and tomato concentrates (Ilić et al. 2014). Viruses from 12 families and 15 genera live in this plant species. TYLCV, tomato bushy stunt virus (TBSV; *Tombosvirus*), *Begomovirus, Geminiviridae*; virus that causes tomato leaf curl (ToLCV). Due to the presence of *Tombusviridae* and beet curly top virus (BCTV; *Curtovirus, Geminiviridae*), production yields have been drastically decreased from 18% to 100% (Massumi et al. 2009). The whitefly bemisia tabaci spreads TYLCV and its recombinants, which attack tomato plants grown in greenhouses and in the field (Rodríguez-Verástegui et al. 2022). Short internodes, thin leaves with yellowish margins, and a spoon-shaped upward curve of the leaf are all symptoms of the disease (Zou et al. 2020). TMV and the Pepino mosaic virus (PepMV; *Alphaflexiviridae; Potexvirus*) are two more significant viruses that infect tomatoes, resulting in losses of 19–33% and 20–40%, respectively. The curl of tomato leaves is caused by another *begomovirus*, New Delhi virus (ToLCNDV) (family *Geminiviridae*), which causes losses between 18% and 99%, mostly in India and other nations like Spain and Italy (Fortes et al. 2025). Consequently, ToLCNDV's proliferation to other regions may result in severe losses in up to 43 plant species, not just tomatoes. Another virus that may be detrimental to tomatoes and sweet peppers is the tomato brown rugose fruit virus (TBRFV; Tobamovirus, Virgaviridae) (Vo et al. 2022).

6.6.2 Banana

Approximately 20 viral species across five families infect banana and plantain crops; the most significant of these viruses in terms of economic impact are Several banana streak virus species, including BBTV, *Caulimoviridae, Badnavirus,* and BBrMV (Kuan et al. 2022). The leaf has a few dark green stripes or spots on the lower portion of the leaf blade, which are the most distinctive signs of

BBTV. Furthermore, brittle-textured, smaller leaves are seen, presenting curled upwards and with chlorotic borders. Younger plants are known to exhibit stunting and typically do not bear fruit; when they do, the fruits are twisted. *BBTV* is regarded as an economically significant disease since it can infect both banana and plantain plants, lowering yields by as much as 100% (Kumar et al. 2015). This virus is spread through the use of contaminated plants for propagation or by the aphid *Pentalonia nigronervosa* (Manohar Jebakumar et al. 2018). The *BBTV*-caused illness was initially discovered in the Fiji Islands before spreading to other nations due to commerce and human migration (Ngatat et al. 2024).

6.6.3 Cucurbitaceae

Watermelon (*C. lanatus*), cucumbers (*C. sativus*), squash (*Cucurbita* sp.), and melon (*C. melo*) are the cucurbits that are grown and eaten the most globally (Rubatzky and Yamaguchi 1997). These EFFs are all members of the second-most economically significant family, the *Cucurbitaceae* (Tian et al. 2023). There are at least 39 viruses from eight families that damage cucurbit crops in underdeveloped nations; however, the degree of infection varies depending on the climate and agricultural management circumstances that allow the virus's vectors to spread (Rabadán et al. 2021). Potyviruses result in significant financial losses in underdeveloped nations, and they are spread by several aphid species in a nonpersistent way. Watermelon mosaic virus (WMV; *Potyvirus, Potyviridae*), cucumber green mottle mosaic virus (CGMMV; *Tobamovirus, Virgaviridae*), and *CMV* are some of the viruses that infect cucurbits (De Moya-Ruiz et al. 2023). There are at least 39 viruses from eight families that damage cucurbit crops in underdeveloped nations; however, the degree of infection varies depending on the climate and agricultural management circumstances that allow the virus's vectors to spread (Radouane et al. 2021). WMV has responsibility for the declines in watermelon, cucumber, and squash yield and quality. Yield losses in watermelon ranged from 18% to 73%, in cucumbers from less than 2%, and in squash from 9% to 49%. Symptoms can vary depending on the host; typically, WMV results in bending of the leaf blades and the veins have a dark green mosaic (Cheng et al. 2018). Fruit discoloration and deformation, mosaic, leaf distortion, and plant growth retardation are all caused by this virus. The symptoms on the leaves of watermelon and melon include dark necrotic lesions, mosaic, and mottling (Parrella 2025). Should the symptoms be severe, the plant may wilt. Watermelon and other melon fruits contain varying degrees of distortion. There is internal sponginess, yellowing, or rotting, as well as superficial netting and mottling in the situation of melon (Rubatzky and Yamaguchi 1997).

6.6.4 Apples

Apples (*M. domestica*) are one of the most traded fruits worldwide and are a member of the Rosaceae family. There have been reports of 20 viruses infecting apples, including *ApMV*, apple chlorotic leaf spot, and apple stem grooving virus (*ASGV*; *Capillovirus, Betaflexiviridae*). Virus (ASSVd; *Apscaviroid, Pospiviroidae*) and apple scar skin virus (ACLSV; *Trichovirus, Betaflexiviridae*) (Nabi et al. 2022). Yellow dots along the major veins of the leaves are the first signs of ApMV infection. However, if the illness worsens, leaves may fall off too soon, delaying the growth of the apple tree. If these diseased trees bear fruit, it will be colorless and sour in flavor. The virus spreads more readily in colder climes because symptoms are more noticeable in the spring (Pedrelli et al. 2025). The primary distribution regions for *ApMV* include Europe, Oceania, America, and a few nations in Asia and Africa. With losses of up to 50%, "Golden Delicious," "Granny Smith," and "Jonathan" are the most vulnerable apple cultivars. Additionally, in ApMV, ASGV, ASPV, and ASSVd can cause mixed infections in apple trees, making the situation more challenging for disease control measures (Akbaş and Ilhan 2005).

6.6.5 Grapes

Grapes, scientifically known as *V. vinifera*, yield delectable fruits that can be enjoyed fresh or dried, such as raisins (Belkina et al. 2025). Grapes are also the primary element in wines, jellies, vinegar, oils, or juices. *Vitis vinifera* hosts around 90 viruses, but only a few are economically significant, including grapevine fanleaf virus (*GFLV*; *Nepovirus*, *Secoviridae*) (Fall et al. 2020). Numerous *Closteroviridae* species are viruses that cause grapevine leafroll. According to Naidu et al. (2025), GFLV is transmitted by the ectoparasite Xiphinera index, which feeds on plant roots and can persist in the soil for long periods of time. According to Krebelj et al. (2015), GFLV symptoms usually manifest during the growing season and include yellow mottling of the leaves, fan-like deformation, double nodes, short and malformed internodes, poor fruit taste, and a reduced shelf life. Grafting and mealybugs are two ways that grapevine leafroll-associated viruses might proliferate (Tsai et al. 2010). These viruses can cause yield losses of up to 40%, with symptoms being more apparent near the end of the growth season and young tissues often asymptomatic. Symptoms of adult plants appear on mature leaves, starting at the base of the canes and moving up to the youngest leaves (Hommay et al. 2021).

6.6.6 Citrus

Citrus fruits are highly popular due to their unique flavor, nutritional value, and functionality, both fresh and processed. Citrus fruits include oranges, lemons, grapefruit, tangerines, and mandarins (Liu et al. 2012). Citrus fruit production is hampered by CTV, a family of Citrus tristeza, which infects plants in the Rutaceae family, specifically *Citrus*. CTV is transmitted through plants spread contaminated with viruses or aphids, such as *Toxoptera citricida*, *Toxoptera aurantii*, and *Aphis spiraecola* (Hosni and Mohamed 2009). *CTV* is a global virus with varying symptoms depending on host genotype (Dawson et al. 2013). When lemon (*C.*

macrophylla), sweet orange (*C. sinensis*), grapefruit (*Citrus x paradisi*), or mandarin (*C. reticulata*) grafts are combined with sour orange (*C. aurantifolia*), the virus can cause yellowing of leaves, root rot, and plant death within two or three years (Baker et al. 2014). Citrus trees that are resistant to *CTV*, such as trifoliate orange (*Poncirus trifoliate)*, rangpur lime (*C. limonia*), Cleopatra mandarin (*C. reshni*), and rough lemon (*C. jambhiri*), are utilized as rootstocks to create resistant hybrids (Khalilzadeh et al. 2025). Despite attempts, almost 50% of the world's cultivated plants are lost owing to the virus. Research shows that *Halminton orange* (*C. sinensis*) yields in Brazil have decreased by up to 87%. Kagzi lime (*C. aurantifolia*) plant's output can decline by up to 12%, according to reports (Tatineni and Hein 2023). Finally, the yield of oranges and other cultivars loses 40–98% of the citrus fruits: Frost Lisbon lemons, tangelos, Frost Washington Navel oranges, tangerines, and Valencia oranges (Strano et al. 2022).

6.7 Modes of Transmission

6.7.1 Nonpersistent Transmission

Plant viruses change the smells of their host plants, attracting insect vectors for nonpersistent transmission that necessitates probing and quick spread. Changes in plant volatile profiles correlate with this vector behavior (Shi et al. 2021). The cucumber mosaic virus (CMV) increased the attraction of *Aphis gossypii* by causing volatile changes in its squash host (Mauck et al. 2010). By altering plant defense responses, nonpersistent plant viruses can change the feeding habits, preferences, and fitness of insect vectors. As an example, consider tobacco plants. Aphid feeding behavior changed, and virus transmission increased because of CMV infection, which also raised Reactive Oxygen Species (ROS) levels and defense signals in tissues (Table 6.1) (Clemente-Orta et al. 2024).

6.7.2 Semi-persistent Transmission

ToCV (family: *Closteroviridae*, genus: Crinivirus) is semi-persistently transmitted by several cryptic species of the *B. tabaci* complex, as well as *Tri-. Aleurodes abutiloneus* and *T. vaporariorum ToCV* transmission in four weeds identified *Solanum nigrum* as the primary viral source, with transmission efficiency correlated with the host preference of *B. tabaci* MED (Mediterranean), suggesting that the vector may prefer a specific host for viral acquisition rather than injection (Fereres et al. 2016). Cucurbit chlorotic yellows virus (CCYV) (family: *Closteroviridae*, genus: Crinivirus) is transmitted semi-persistently by *B. tabaci. B. tabaci* MED increased the frequency of insects (Yan et al. 2023). MED is a more effective vector for CCYV transmission than MEAM1 due to feeding-site translocation and shorter feeding bouts, leading to higher viral inoculation rates (Sarıkaya et al. 2023).

6.7.3 Persistent

MED is a more effective vector for CCYV plant defensive responses associated with JA and SA in virus-insect vector-plant during persistent viral infection (Li et al. 2016). Defenses are decreased, allowing vectors to perform better. Infection with the tomato yellow leaf curl China virus (TYLCCNV) suppressed JA-associated defense responses in tobacco plants, enhancing the performance of the B (MEAM1) whitefly, *B. tabaci*, on virus-infected tobacco plants (Li et al. 2025). Plant defense responses to JA and SA in virus-insect vector-plant interactions during persistent transmission, as shown in Fig. 6.2. JA after viral infection, defense is reduced, allowing vectors to perform better (Abe et al. 2011). Infection with the tomato yellow leaf curl China virus (TYLCCNV) (family: *Geminiviridae*, genus: Bomovirus) suppressed JA-associated defense responses in tobacco plants, enhancing the performance of the B (MEAM1) whitefly, *B. tabaci*, on infected tobacco plants (Luan et al. 2013).

6.8 Conclusions

Determining the reasons behind virus-induced symptoms in plants requires an understanding of the intricate mechanisms through which viruses interact with host factors and manipulate cellular processes. Recent studies have shed light on these complex mechanisms, demonstrating how viral infections impair hormone regulation, cell cycle control, ROS accumulation, and chloroplast protein function. Host responses, such as metabolic diseases and modifications in gene expression, also influence the development of symptoms. The impact of environmental conditions on viral infection and symptom severity further highlights the complex nature of virus-plant interactions. By combining these discoveries, this thorough analysis provides a comprehensive understanding of the mechanisms behind virus-induced plant symptoms, paving the way for the development of efficient crop protection and disease management plans. Novel interference options to lessen the impact of viral infections on crops can be investigated with a better understanding of how viruses interfere with plant cellular functions, expanding on our knowledge of the alterations in gene expression and metabolic abnormalities brought on by viral infections. These RNA-based approaches provide potential strategies for creating crop plants resistant to viruses. These methods can effectively prevent viral infections in plants by introducing specific RNA molecules that either target viral genes or inhibit viral replication. Furthermore, non-transgenic techniques have been developed, including topical dsRNA, hairpin RNA, and artificial microRNA application. In conclusion, we will better understand the mechanisms behind virus-induced symptoms in plants and develop more efficient methods for crop protection and disease management by conducting an in-depth study on virus-vector plant interactions. This will have a significant impact on food security and the sustainable development of agriculture.

References

Abe H, Tomitaka Y, Shimoda T, Seo S, Sakurai T, Kugimiya S, Tsuda S, Kobayashi M (2011) Antagonistic plant defense system regulated by phytohormones assists interactions among vector insect, thrips and a tospovirus. Plant Cell Physiol. 53(1):204–212. https://doi.org/10.1093/pcp/pcr173

Akbaş B, Ilhan D (2005) Widespread distribution of Apple mosaic virus on Apple in Turkey. Plant Disease 89(9):1010–1010. https://doi.org/10.1094/PD-89-1010C

Alimanović A (2021) Pathogenic viruses and viral vectors in plants: a comprehensive review. Bioeng Stud 2(1):11–18. https://doi.org/10.37868/bes.v2i1.id179

Almási A, Harsanyi A, Gáborjányi R (2001) Photosynthetic alterations of virus infected plants. Acta Phytopathol Entomol Hung 36(1–2):15–29. https://doi.org/10.1556/APhyt.36.2001.1-2.3

Baker R, Bragard C, Candresse T, Gilioli G, Grégoire J-C, Holb I, Jeger MJ, Karadjova OE, Magnusson C, Makowski D, Manceau C, Navajas M, Rafoss T, Rossi V, Schans J, Schrader G, Urek G, Vloutoglou I, Winter S, Van Der Werf W (2014) Citrus tristeza virus. EFSA J 12(12):3923. https://doi.org/10.2903/j.efsa.2014.3923

Balamurugan A, Ashajyothi M, Velmurugan S, Charishma K, Sakthivel K, Muthamilan M, Kumar A (2024) Bacillus amyloliquefaciens: a versatile antimicrobial Firmicute for the suppression of wilt caused by Ralstonia pseudosolanacearum in tomato. Biocontrol Sci Technol 34(4):336–354. https://doi.org/10.1080/09583157.2024.2342475

Belkina D, Stepanov I, Makarkina M, Porotikova E, Lifanov I, Kozhevnikov E, Gorislavets S, Vinogradova S (2025) In-depth population genetic study of Vitis vinifera ssp. sylvestris from the Black Sea region and its virome. Front Plant Sci 16.1536862. https://doi.org/10.3389/FPLS.2025.1536862/PDF

Boeing H, Bechthold A, Bub A, Ellinger S, Haller D, Kroke A, Leschik-Bonnet E, Müller MJ, Oberritter H, Schulze M, Stehle P, Watzl B (2012) Critical review: vegetables and fruit in the prevention of chronic diseases. Eur J Nutr 51(6):637–663. https://doi.org/10.1007/s00394-012-0380-y

Catto MA, Mugerwa H, Myers BK, Pandey S, Dutta B, Srinivasan R (2022) A review on transcriptional responses of interactions between insect vectors and plant viruses. Cells 11(4). https://doi.org/10.3390/cells11040693.

Cembali T, Folwell RJ, Wandschneider P, Eastwell KC, Howell WE (2003) Economic implications of a virus prevention program in deciduous tree fruits in the US. Crop Prot 22(10):1149–1156. https://doi.org/10.1016/S0261-2194(03)00156-X

Cheng F, Wu J, Cai X, Liang J, Freeling M, Wang X (2018) Gene retention, fractionation and subgenome differences in polyploid plants. Nat Plants 4(5):258–268. https://doi.org/10.1038/s41477-018-0136-7

Clemente-Orta G, Cabello Á, Garzo E, Moreno A, Fereres A (2024) Aphidius colemani behavior changes depending on volatile organic compounds emitted by plants infected with viruses with different modes of transmission. Insects 15(2):92. https://doi.org/10.3390/insects15020092

Coleman AD, Wouters RHM, Mugford ST, Hogenhout SA (2015) Persistence and transgenerational effect of plant-mediated RNAi in aphids. J Exp Bot 66(2):541–548. https://doi.org/10.1093/jxb/eru450

Cosme F, Pinto T, Aires A, Morais MC, Bacelar E, Anjos R, Ferreira-Cardoso J, Oliveira I, Vilela A, Gonçalves B (2022) Red fruits composition and their health benefits—a review. Foods 11(5). https://doi.org/10.3390/foods11050644. MDPI

Dawson WO, Garnsey SM, Tatineni S, Folimonova SY, Harper SJ, Gowda S (2013) Citrus tristeza virus-host interactions. Front Microbiol 4:88. https://doi.org/10.3389/FMICB.2013.00088/PDF

De Moya-Ruiz C, Gómez P, Juárez M (2023) Occurrence, distribution, and management of aphid-transmitted viruses in cucurbits in Spain. Pathogens 12(3). https://doi.org/10.3390/PATHOGENS12030422

del Río-Celestino M, Font R (2020) The health benefits of fruits and vegetables. Foods 9(3):369. https://doi.org/10.3390/foods9030369

Dombrecht B, Xue GP, Sprague SJ, Kirkegaard JA, Ross JJ, Reid JB, Kazan K (2007) MYC2 differentially modulates diverse jasmonate-dependent functions in Arabidopsis. The Plant Cell 19(7):2225–2245. https://doi.org/10.1105/tpc.106.048017

Fall ML, Xu D, Lemoyne P, Ben Moussa IE, Beaulieu C, Carisse O (2020) A diverse virome of leafroll-infected grapevine unveiled by dsRNA sequencing. Viruses 12(10). https://doi.org/10.3390/V12101142

Fereres A, Peñaflor MFGV, Favaro CF, Azevedo KEX, Landi CH, Maluta NKP, Bento JMS, Lopes JRS (2016) Tomato infection by whitefly-transmitted circulative and non-circulative viruses induce contrasting changes in plant volatiles and vector behaviour. Viruses 8(8). https://doi.org/10.3390/V8080225

Fortes IM, Díaz-Martínez L, Moriones E, Grande-Pérez A (2025) Virus–host interactions and genetic exchange in mixed infections of tomato yellow leaf curl virus (TYLCV), tomato leaf curl New Delhi virus (ToLCNDV), and tomato chlorosis virus (ToCV). Agronomy 15(5):1006. https://doi.org/10.3390/AGRONOMY15051006/S1

Gadhave KR, Gautam S, Rasmussen DA, Srinivasan R (2020) Aphid transmission of potyvirus: the largest plant-infecting RNA virus genus. Viruses 12(7):773. https://doi.org/10.3390/V12070773

Gonsalves D, Suzuki JY, Tripathi S, Ferreira SA (2008) Papaya ringspot virus. Encyclopedia Virol 1–5:1–8. https://doi.org/10.1016/B978-012374410-4.00731-7

Hanley-Bowdoin L, Bejarano ER, Robertson D, Mansoor S (2013) Geminiviruses: masters at redirecting and reprogramming plant processes. Nat Rev Microbiol 11(11):777–788. https://doi.org/10.1038/nrmicro3117

Hommay G, Alliaume A, Reinbold C, Herrbach E (2021) Transmission of Grapevine leafroll-associated virus-1 (Ampelovirus) and Grapevine virus A (Vitivirus) by the Cottony Grape Scale, Pulvinaria vitis (Hemiptera: Coccidae). Viruses 13(10). https://doi.org/10.3390/V13102081

Ilić ZS, Kapoulas N, Šunić L (2014) Tomato fruit quality from organic and conventional production. In: Organic agriculture towards sustainability. InTech.284 https://doi.org/10.5772/58239

Jangra S, Chinnaiah S, Patil SR, Shukla B, Devendran R, Kumar M (2024) Deciphering the role of virus receptors in plant–virus–vector interactions. Receptor 3(2):255–279. https://doi.org/10.3390/receptors3020013

Jiang T, Zhou T (2023) Unraveling the mechanisms of virus-induced symptom development in plants. Plants 12(15):2830. https://doi.org/10.3390/plants12152830.

Kalleshwaraswamy CM, Kumar NKK (2008) Transmission efficiency of papaya ringspot virus by three aphid species. Phytopathology 98(5):541. https://doi.org/10.1094/PHYTO-98-5-0541

Kamenova I, Lohuis D, Peters D (2002) Loss of aphid transmissibility of plum pox virus isolates. Biotechnol Biotechnol Equip 16(1):48–54. https://doi.org/10.1080/13102818.2002.10819155

Kaur N, Hasegawa DK, Ling KS, Wintermantel WM (2016) Application of genomics for understanding plant virus-insect vector interactions and insect vector control. Phytopathology 106(10):1213–1222. https://doi.org/10.1094/PHYTO-02-16-0111-FI

Khalilzadeh M, Aldrich DJ, Maree HJ, Levy A (2025) Complex interplay: The interactions between citrus tristeza virus and its host. Virology 110388. https://doi.org/10.1016/j.virol.2024.110388.

Kiers ET, Duhamel M, Beesetty Y, Mensah JA, Franken O, Verbruggen E, Fellbaum CR, Kowalchuk GA, Hart MM, Bago A, Palmer TM, West SA, Vandenkoornhuyse P, Jansa J, Bücking H (2011) Reciprocal rewards stabilize cooperation in the mycorrhizal symbiosis. Science 333(6044):880–882. https://doi.org/10.1126/SCIENCE.1208473

Kostecka-Gugała A, Kruczek M, Ledwożyw-Smoleń I, Kaszycki P (2020) Antioxidants and health-beneficial nutrients in fruits of eighteen cucurbita cultivars: analysis of diversity and dietary implications. Molecules 25(8). https://doi.org/10.3390/molecules25081792

Krebelj AJ, Čepin U, Ravnikar M, Novak MP (2015) Spatio-temporal distribution of Grapevine fanleaf virus (GFLV) in grapevine. Eur J Plant Pathol 142(1):159–171. https://doi.org/10.1007/S10658-015-0600-4/FIGURES/7

Kuan CP, Tsai CH, Tseng CS, Yang TC (2022) Development of a bead-based assay for detection of three banana-infecting viruses. PeerJ 10:e13409. https://doi.org/10.7717/peerj.13409
Kumar PL, Selvarajan R, Iskra-Caruana ML, Chabannes M, Hanna R (2015) Biology, etiology, and control of virus diseases of banana and plantain. Adv Virus Res 91(1):229–269. https://doi.org/10.1016/bs.aivir.2014.10.006.
Li J, Liang X, Wang X, Shi Y, Gu Q, Kuo YW, Yan F (2016) Direct evidence for the semipersistent transmission of Cucurbit chlorotic yellows virus by a whitefly vector. Sci Rep 6(1):36604. https://doi.org/10.1038/srep36604
Li Y, Ljhofbji Z, Cui S, Jagadamma S, Zhang Q (2019) Residue retention and minimum tillage improve physical environment of the soil in croplands: a global meta-analysis. Soil Till Res 194:104292. https://doi.org/10.1016/j.still.2019.06.009
Li P, Zhang L, Yan XT, Zheng C, Zhang XY, Chen JP, Wang XW (2025) Suppression of TGA2-mediated Salicylic Acid Defence by Tomato Yellow Leaf Curl Virus C2 via Disruption of TCP7-like Transcription Factor Activity in Tobacco. Plant, Cell & Environ 48(6):4039–4050. https://doi.org/10.1111/pce.15411
Liu Y, Heying E, Tanumihardjo SA (2012) History, global distribution, and nutritional importance of citrus fruits. Compr Rev Food Sci Food Saf 11(6):530–545. https://doi.org/10.1111/J.1541-4337.2012.00201.
Luan JB, Wang YL, Wang J, Wang XW, Liu SS (2013) Detoxification activity and energy cost is attenuated in whiteflies feeding on Tomato yellow leaf curl China virus-infected tobacco plants. Insect Mol Biol 22(5):597–607. https://doi.org/10.1111/IMB.12048
Manohar Jebakumar R, Velusamy Balasubramanian, Selvarajan R (2018) Virus titre determines the efficiency of Pentalonia nigronervosa (Aphididae: Hemiptera) to transmit banana bunchy top virus. Virusdisease 29. https://doi.org/10.1007/s13337-018-0493-x
Maree HJ, Almeida RP, Bester R, Chooi KM, Cohen D, Dolja VV, Burger JT (2013) Grapevine leafroll-associated virus 3. Front Microbiol 4:82. https://doi.org/10.3389/fmicb.2013.00082.
Massumi H, Shaabanian M, Pour AH, Heydarnejad J, Rahimian H (2009) Incidence of viruses infecting tomato and their natural hosts in the southeast and central regions of Iran. Plant Dis 93(1):67–72. https://doi.org/10.1007/s13313-012-0164-0
Mauck KE, De Moraes CM, Mescher MC (2010) Deceptive chemical signals induced by a plant virus attract insect vectors to inferior hosts. Proc Natl Acad Sci 107(8):3600–3605. https://doi.org/10.1073/pnas.0907191107
Medici M, Canavari M, Toselli M (2020) Interpreting environmental impacts resulting from fruit cultivation in a business innovation perspective. Sustainability (Switzerland) 12(23):1–14. https://doi.org/10.3390/su12239793
Meena VS, Gora JS, Singh A, Ram C, Meena NK, Pratibha A, Rouphael Y, Basile B, Kumar P (2022) Underutilized fruit crops of Indian arid and semi-arid regions: importance, conservation and utilization strategies. Horticulturae 8(2). https://doi.org/10.3390/horticulturae8020171
Mittapelly P, Rajarapu SP (2020) Applications of proteomic tools to study insect vector–plant virus interactions. Life 10(8):1–16. https://doi.org/10.3390/life10080143.
Mohamed AHI (2009) Use of molecular and biochemical methods to determine citrus tristeza virus (CTV) viral components and resistance in candidate rootstocks to replace sour orange. University of Florida ProQuest Dissertations & Theses:3367560.
Mume AA, Turyasingura B, Abdi E, Umer Y, Amanzi LN, Uwimbabazi A, Ndeke C, Bosco NJ, Chavula P (2024) Impact of climate change on the environment: a synthesis study. Asian J Res Agric For 10(2):86–96. https://doi.org/10.9734/ajraf/2024/v10i2288
Nabi SU, Baranwal VK, Rao GP, Mansoor S, Vladulescu C, Raja WH, Jan BL, Alansi S (2022) High-throughput RNA sequencing of mosaic infected and non-infected apple (Malus × domestica Borkh.) cultivars: from detection to the reconstruction of whole genome of viruses and viroid. Plants 11(5):675. https://doi.org/10.3390/PLANTS11050675/S1
Naidu AY, Singh S, Sreelash K, Srivastava HS, Nikam BR (2025) Contributions of meteorological and vegetation factors to surface and root zone soil moisture variability across India's agroecological regions. Remote Sens Appl: Soc Environ 101803. https://doi.org/10.1016/j.rsase.2025.101803

Ng JCK, Perry KL (2004) Transmission of plant viruses by aphid vectors. Mol Plant Pathol 5(5):505–511. https://doi.org/10.1111/J.1364-3703.2004.00240.X

Ngatat S, Hanna R, Fotio ARPD, Lienou JA, Nanga SN, Kuate AF, Osundahunsi B, Fiaboe KKM, Ndemba B, Dossa GS, Alabi T, Kumar PL (2024) Distribution and diversity of emergent Banana bunchy top virus infecting banana and plantain in Cameroon, Central Africa "BBTV mitigation: Community Management in Nigeria and Screening Wild Banana Progenitors for Resistance, Grant/Award Number: OPP1130226"; CGIAR Initiative on "Plant Health and Rapid Response to Protect Food Security and Livelihoods"; Programme de la Relance de la Filière Plantain-PRFP (contract MINADER-IITA); The Program for Sustainable Management of Natural Resources (PSMNR-SW). J Phytopathol 172(2):2016–2033. https://doi.org/10.1111/jph.13279

Pakkianathan BC, Kontsedalov S, Lebedev G, Mahadav A, Zeidan M, Czosnek H, Ghanim M (2015) Replication of tomato yellow leaf curl virus in its whitefly vector, Bemisia tabaci. J Virol 89(19):9791–9803. https://doi.org/10.1128/JVI.00779-15

Parrella JA, Rice K, Le Gall M (2025) Understanding apple growers' perspectives on the sterile insect technique for codling moth (Lepidoptera: Tortricidae) management in Virginia: implications for adoption and outreach. J Econ Entomol toaf333. https://doi.org/10.1093/jee/toaf333

Pedrelli A, Nali C, Panattoni A, Pellegrini E, Cotrozzi L (2025) Apple chlorotic leaf spot virus infection affects the physiology and biochemistry of leaves and the quality of fruits of Tuscany (Italy) autochthonous apple varieties. Plant Pathol 74(1):158–170. https://doi.org/10.1111/PPA.14004

Prasad A, Sharma N, Muthamilarasan M, Rana S, Prasad M (2019) Recent advances in small RNA mediated plant-virus interactions. Crit Rev Biotechnol 39(4):587–601. https://doi.org/10.1080/07388551

Rabadan MP, Juarez M, De Moya-Ruiz C, Gomez P (2021) Aphid-borne viruses infecting cultivated watermelon and squash in Spain: characterization of a variant of cucurbit aphid-borne yellows virus (CABYV). Plant Pathol 70(6):1476–1485. https://doi.org/10.1111/ppa.13390

Radouane N, Ezrari S, Belabess Z, Tahiri A, Tahzima R, Massart S, Jijakli H, Benjelloun M, Lahlali R (2021) Viruses of cucurbit crops: current status in the Mediterranean Region. Phytopathol Mediterr 60(3):493–519. https://doi.org/10.36253/phyto-12340

Radwan DEM, Fayez KA, Younis Mahmoud S, Hamad A, Lu G (2007) Physiological and metabolic changes of Cucurbita pepo leaves in response to zucchini yellow mosaic virus (ZYMV) infection and salicylic acid treatments. Plant Physiol Biochem 45(6–7):480–489. https://doi.org/10.1016/J.PLAPHY.2007.03.002

Rahoutei J, García-Luque I, Barón M (2000) Inhibition of photosynthesis by viral infection: effect on PSII structure and function. Physiol Plant 110(2):286–292. https://doi.org/10.1034/J.1399-3054.2000.110220.X

Ray S, Casteel CL (2022) Effector-mediated plant–virus–vector interactions. Plant Cell 34(5):1514–1531. https://doi.org/10.1093/plcell/koac058.

Rodriguez-Verastegui LL, Ramirez-Zavaleta CY, Capilla-Hernandez MF, Gregorio-Jorge J (2022) Viruses infecting trees and herbs that produce edible fleshy fruits with a prominent value in the global market: An evolutionary perspective. Plants 11(2):203. https://doi.org/10.3390/plants11020203

Rosario K, Seah YM, Marr C, Varsani A, Kraberger S, Stainton D, Moriones E, Polston JE, Duffy S, Breitbart M (2015) Vector-enabled metagenomic (VEM) surveys using whiteflies (Aleyrodidae) reveal novel begomovirus species in the new and old worlds. Viruses 7(10):5553–5570. https://doi.org/10.3390/v7102895

Rubatzky VE, Yamaguchi M (1997) Cucumber, melons, watermelons, squash, and other cucurbits. In: World vegetables, pp 577–639. https://doi.org/10.1007/978-1-4615-6015-9_24

Sa Antunes TF, Maurastoni M, Madroñero LJ, Fuentes G, Santamaria JM, Ventura JA, Abreu EF, Fernandes AAR, Fernandes PMB (2020) Battle of three: the curious case of papaya sticky disease. Plant Dis 104(11):2754–2763. https://doi.org/10.1094/PDIS-12-19-2622-FE

Sabanadzovic S, Abou-Ghanem N, Castellano MA, Digiaro M, Martelli GP (2000) Grapevine fleck virus-like viruses in Vitis. Arch Virol 145(3):553–565. https://doi.org/10.1007/S007050050046/METRICS

Sarıkaya P, Yıldız K, Fırat G, Fidan H (2023) Research on drawing attention to overlooked viruses in plant viruses causing yellowing-type symptoms in cucurbitaceae. Int J Innov Approaches Agric Res 7(1):74–84. https://doi.org/10.29329/ijiaar.2023.536.5

Shi X, Zhang Z, Zhang C, Zhou X, Zhang D, Liu Y (2021) The molecular mechanism of efficient transmission of plant viruses in variable virus–vector–plant interactions. Hortic Plant J 7(6):501–508. https://doi.org/10.1016/j.hpj.2021.04.006

Strano MC, Altieri G, Allegra M, Di Renzo GC, Paterna G, Matera A, Genovese F (2022) Postharvest technologies of fresh citrus fruit: advances and recent developments for the loss reduction during handling and storage. Horticulturae 8(7). https://doi.org/10.3390/HORTICULTURAE8070612

Tang G, Reinhart BJ, Bartel DP, Zamore PD (2003) A biochemical framework for RNA silencing in plants. Genes Dev 17(1):49–63. https://doi.org/10.1101/gad.1048103

Tatineni S, Hein GL (2023) Plant viruses of agricultural importance: current and future perspectives of virus disease management strategies. Phytopathology 113(2):117–141. https://doi.org/10.1094/PHYTO-05-22-0167-RVW

Tan XL, Chen JL, Benelli G, Desneux N, Yang XQ, Liu TX, Ge F (2017) Pre-infestation of tomato plants by aphids modulates transmission-acquisition relationship among whiteflies, tomato yellow leaf curl virus (TYLCV) and plants. Front Plant Sci 8:1597. https://doi.org/10.3389/fpls.2017.01597

Tian S, Zhang Z, Qin G, Xu Y (2023) Parthenocarpy in cucurbitaceae: advances for economic and environmental sustainability. Plants 12(19). https://doi.org/10.3390/PLANTS12193462

Tsai CW, Rowhani A, Golino DA, Daane KM, Almeida RP (2010) Mealybug transmission of Grapevine leafroll viruses: An analysis of virus-vector specificity. Phytopathology 100(8):830–834. https://doi.org/10.1094/PHYTO-100-8-0830. PMID: 20626287

Umer M, Liu J, You H, Xu C, Dong K, Luo N, Kong L, Li X, Hong N, Wang G, Fan X, Kotta-Loizou I, Xu W (2019) Genomic, morphological and biological traits of the viruses infecting major fruit trees. Viruses 11(6):515. https://doi.org/10.3390/V11060515

Ushiyama R, Matthews REF (1970) The significance of chloroplast abnormalities associated with infection by turnip yellow mosaic virus. Virology 42(2):293–303. https://doi.org/10.1016/0042-6822(70)90273-4

Vo TTB, Troiano E, Lal A, Hoang PT, Kil EJ, Lee S, Parrella G (2022) ToLCNDV-ES infection in tomato is enhanced by TYLCV: Evidence from field survey and agroinoculation. Front Microbiol 13:954460. https://doi.org/10.3389/fmicb.2022.954460

Wang LL, Wang XR, Wei XM, Huang H, Wu JX, Chen XX, Wang XW (2016) The autophagy pathway participates in resistance to tomato yellow leaf curl virus infection in whiteflies. Autophagy 12(9):1560–1574. https://doi.org/10.1080/15548627.2016.1192749

Waigmann E, Ueki S, Trutnyeva K, Citovsky V (2004) The Ins and Outs of Nondestructive Cell-to-Cell and Systemic Movement of Plant Viruses. Crit Rev Plant Sci 23(3):195–250. https://doi.org/10.1080/07352680490452807

Whitfield AE, Huot OB, Martin KM, Kondo H, Dietzgen RG (2018) Plant rhabdoviruses—their origins and vector interactions. Curr Opin Virol 33:198–207. https://doi.org/10.1016/j.coviro.2018.11.002.

Yan M, He H, Zhang Z, Zhang B, Zhu C, Yan W, Zhao C, Li J, Yan F (2023) Molecular basis of mutual benefits between cucurbit chlorotic yellows virus (CCYV) transmission and imidacloprid resistance in Bemisia tabaci. J Pest Sci 96(2):489–497. https://doi.org/10.1007/s10340-022-01553-w

Zakaria L (2023) Fusarium species associated with diseases of major tropical fruit crops. Hortic 9(3):322. https://doi.org/10.3390/horticulturae9030322.

Zhou X (2013) Advances in understanding begomovirus satellites. Annu Rev Phytopathol 51:357–381. https://doi.org/10.1146/ANNUREV-PHYTO-082712-102234

Zou C, Shu YN, Yang JJ, Pan LL, Zhao J, Chen N, Wang XW (2020) Begomovirus-associated betasatellite virulence factor βC1 attenuates tobacco defense to whiteflies via interacting with plant SKP1. Front Plant Sci 11:574557. https://doi.org/10.3389/fpls.2020.574557

Viral Suppressors of RNA Silencing in Fruit Crops

7

Jothishri Chidambaram, J. Vinodhini, Y. Durga, Filmar Raj Jayaraj Mallika, Nagendran Krishnan, and R. Selvarajan

Abstract

RNA silencing is an evolutionarily conserved, homology-dependent gene inactivation mechanism that plays a vital role in plant immune responses. To overcome this defense, viruses have evolved specific viral suppressor proteins (VSRs) that counteract host silencing pathways. These VSRs play a crucial role in the molecular arms race between host and pathogen. VSRs disrupt host antiviral RNA silencing by targeting essential components of the silencing machinery or by mimicking host cellular proteins. They interfere with diverse processes, including the biogenesis of small interfering RNAs (siRNA), sequestration of host RNA-induced silencing complex (RISC) and suppression of host-mediated DNA methylation. Well-characterized examples include cucumber mosaic virus (CMV) 2b, potato virus S (PSV) 2b, and plum pox virus (PPV) HC-Pro, which serve as models for understanding silencing suppression. A variety of assays, such as transient expression, reversal of silencing, and stable expression systems, have been instrumental in identifying and characterizing VSR activity. Insights from mechanistic studies of VSRs deepen our understanding of complex plant-virus interactions and suggest strategies for durable viral disease management. Exploiting RNA silencing pathways and targeting host–VSR interfaces represent promising approaches for breeding virus-resistant fruit crop varieties. In this review, we highlight the antiviral silencing strategies employed by fruit crops and explore the molecular functions of their associated VSRs.

J. Chidambaram · J. Vinodhini · Y. Durga · F. R. Jayaraj Mallika · N. Krishnan (✉) · R. Selvarajan
Molecular Virology Laboratory, ICAR-National Research Center for Banana, Tiruchirappalli, Tamil Nadu, India

R. Kumar et al. (eds.), *Fruit Crops and Viral Pathogens Volume II*, Plant Pathology for Sustainable Agriculture, https://doi.org/10.1007/978-981-95-7261-8_7

Keywords

RNA silencing · VSR · Plant-virus interaction · Antiviral defense · Antiviral management

7.1 Introduction

RNA silencing is a conserved, sequence-specific gene regulation mechanism that forms the core of antiviral defense in plants (Hannon 2002; Voinnet 2005a; Ding 2010). Triggered by double-stranded RNA (dsRNA) generated during viral replication, the pathway produces small RNAs of 20–24 nucleotides that guide effector complexes to inactivate viral transcripts. Central components of this machinery include Dicer-like proteins (DCLs), which cleave dsRNA into small interfering RNAs (siRNAs) or microRNAs (miRNAs); Argonaute proteins (AGOs), which incorporate these small RNAs into the RNA-induced silencing complex (RISC) to direct cleavage or translational repression of targets; RNA-dependent RNA polymerases (RDRs), which amplify silencing signals; and Suppressor of Gene Silencing (SGS) proteins, which stabilize dsRNA substrates (Ipsaro and Joshua-Tor 2015; Zhao et al. 2020). Together, these proteins orchestrate a robust immune response that restricts viral infection. However, viruses have evolved viral suppressors of RNA silencing (VSRs) that interfere with different steps of this pathway, such as inhibiting siRNA production, sequestering small RNAs, blocking RISC assembly, or preventing systemic signaling (Pumplin and Voinnet 2013). Notable examples include the 2b protein of cucumber mosaic virus, P19 of Tombusvirus, and HC-Pro of plum pox virus (Nicola-Negri et al. 2005; Lewsey et al. 2010; Jay et al. 2023). The impact of such mechanisms is particularly severe in perennial fruit crops such as citrus, grapevine, plum, banana, and apple, where systemic viral infections accumulate over years, leading to yield loss, reduced fruit quality, and tree decline (Moreno et al. 2008; Niu et al. 2009; Yaegashi et al. 2007a, b; Tatineni and Hein 2023; Wani et al. 2024). Understanding how fruit-infecting viruses deploy VSRs is therefore essential for unraveling viral pathogenesis and developing durable resistance strategies.

7.2 RNA Silencing Pathways in Plants

Plant RNA silencing comprises two major arms: transcriptional gene silencing (TGS) and posttranscriptional gene silencing (PTGS). TGS operates in the nucleus through RNA-directed DNA methylation (RdDM), which modifies viral or host promoters, thereby suppressing transcription (Baulcombe 2004; Castel and Martienssen 2013). PTGS, in contrast, acts in the cytoplasm, where DCL proteins process viral dsRNA into siRNAs that are loaded into AGO-containing RISC complexes to degrade or repress viral transcripts (Deleris et al. 2006; Ding 2010). Collectively, these mechanisms are referred to as RNA interference (RNAi), a central component of plant immunity against viruses (Hannon 2002; Voinnet 2005a).

Beyond siRNAs, miRNAs also contribute to antiviral defense by regulating host gene networks that influence susceptibility and immunity (Rogers and Chen 2013). Amplification of silencing signals by RDRs and cofactors such as SGS3 produces secondary siRNAs, which strengthen the response and enable systemic signalling, allowing silencing to spread through plasmodesmata and the phloem to uninfected tissues (Obbard et al. 2009; Voinnet 2005b). This multilayered network ensures both local and systemic protection, though it is frequently targeted by viral suppressors that disrupt one or more of these processes (Csorba et al. 2015).

7.3 Viral Suppressors of RNA Silencing (VSRs)

To overcome host defenses, plant viruses encode viral suppressors of RNA silencing (VSRs), multifunctional proteins that interfere with different stages of the RNA silencing pathway (Fig. 7.1). A common strategy is the interference with Dicer-like proteins (DCLs), preventing the cleavage of viral double-stranded RNA into small interfering RNAs (siRNAs). Another widespread mechanism is siRNA sequestration, in which viral proteins bind siRNAs to prevent their incorporation into Argonaute-containing RNA-induced silencing complexes (RISC). Some VSRs directly target Argonaute proteins through binding or inhibition, thereby blocking slicer activity or destabilizing RISC. Others interfere with the systemic signaling of silencing, preventing the movement of mobile silencing signals through plasmodesmata and the phloem, which allows viruses to spread despite local silencing. Additionally, certain VSRs suppress DNA methylation by disrupting RNA-directed DNA methylation (RdDM), thereby maintaining viral promoter activity and facilitating replication. Well-studied model examples include the 2b protein of cucumber mosaic virus (CMV), which binds siRNAs and inhibits AGO1 activity (Lewsey et al. 2010); the P19 protein of tombusvirus, which acts as a molecular decoy by sequestering siRNA duplexes (Sagan 2009); and the HC-Pro of plum pox virus (PPV), which not only sequesters siRNAs but also interferes with their biogenesis and RDR6-dependent amplification (Csorba et al. 2015). Together, these mechanisms illustrate the diverse strategies viruses employ to disable host RNA silencing, underscoring the central role of VSRs in plant–virus interactions and viral pathogenesis (Fig. 7.1).

7.4 Experimental Approaches to Identify VSRs

The identification of viral suppressors of RNA silencing (VSRs) relies on experimental approaches that can capture their activity at different levels of the silencing pathway. Three major strategies, such as transient assays, reversal assays, and stable expression assays, have been widely employed, each providing unique insights into how viral proteins interact with host defense mechanisms. When applied to perennial fruit crops, these methods not only reveal mechanistic details of silencing suppression but also highlight the evolutionary adaptations of viruses in long-lived hosts.

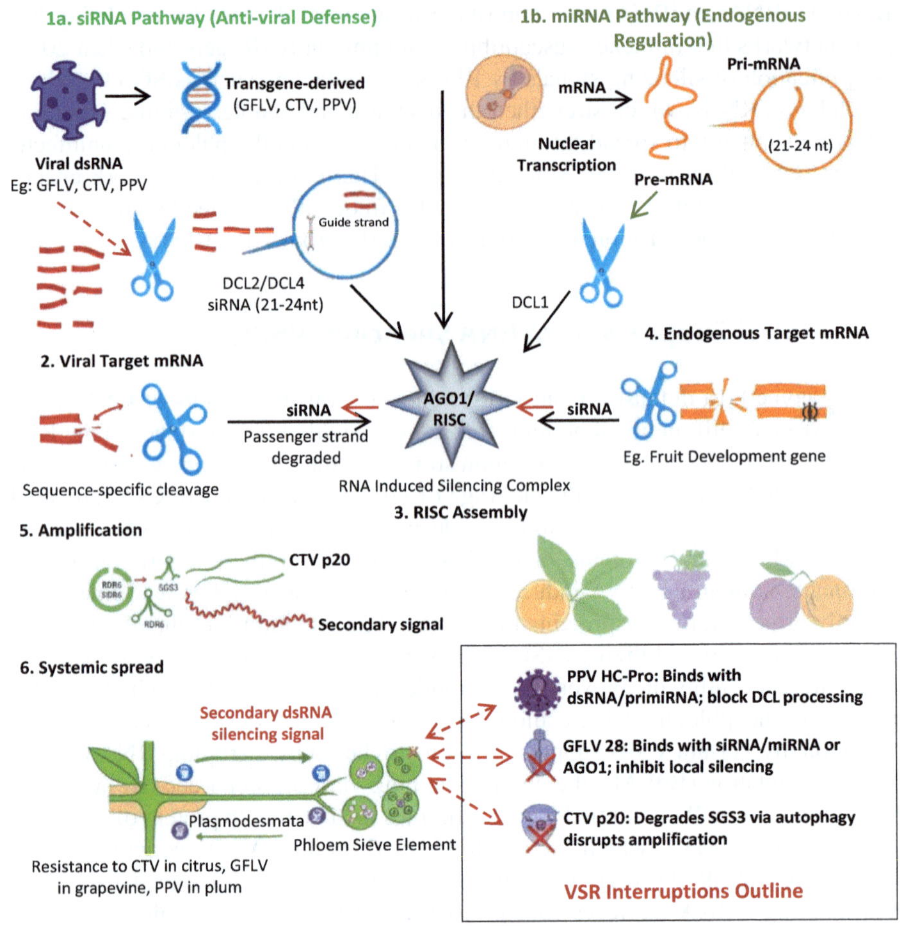

Fig. 7.1 RNA silencing pathways and viral suppressor interactions in fruit crops. Schematic representation of RNA silencing pathways in plants, illustrating both the siRNA branch (left) and the miRNA branch (right). In the siRNA pathway, viral double-stranded RNA (dsRNA) generated during replication (e.g., by citrus tristeza virus [CTV], grapevine fanleaf virus [GFLV], or plum pox virus [PPV]) is processed by Dicer-like enzymes (DCL2/DCL4) into siRNAs. These are incorporated into the RNA-induced silencing complex (RISC) containing Argonautes (AGO1/AGO2/AGO4) to direct cleavage of viral RNA. Amplification by RNA-dependent RNA polymerase (RDR6) and suppressor of gene silencing 3 (SGS3) strengthens silencing, which then spreads systemically through plasmodesmata and the phloem. In the miRNA pathway, primary miRNAs (pri-miRNAs) are processed by DCL1 into mature miRNAs that load into AGO1-containing RISC complexes to regulate endogenous gene expression via mRNA cleavage or translational repression. Viral suppressors of RNA silencing (VSRs) disrupt these pathways at multiple points: CTV p23 binds siRNAs/miRNAs to block RISC loading, CTV p20 degrades SGS3 through autophagy, and the CTV coat protein (CP) interferes with systemic spread; GFLV 2b suppresses local silencing by binding siRNAs/AGO1; and PPV HC-Pro sequesters dsRNA or pri-miRNAs to inhibit DCL processing. Fruit crop contexts (citrus, grapevine, plum) are highlighted, with VSR-mediated disruptions indicated in red

7.4.1 Transient Assays

Transient assays are the most widely used and rapid approach for detecting VSR activity. They typically involve the co-expression of a reporter gene such as green fluorescent protein (GFP), luciferase (LUC), or β-glucuronidase (GUS) with candidate viral proteins using *Agrobacterium*-mediated agroinfiltration. In this system, RNA silencing is triggered by the introduction of double-stranded RNA or inverted repeat constructs that normally lead to degradation of reporter transcripts through posttranscriptional gene silencing (PTGS). The presence of a functional VSR interferes with this process and restores reporter expression, which can be visualized as enhanced fluorescence or enzymatic activity. Such assays have become standard in *Nicotiana benthamiana*, where agroinfiltration enables high-throughput screening of viral proteins. In fruit crops, transient assays have been adapted despite challenges in transformation efficiency. For example, optimized agroinfiltration protocols in grapevine (*Vitis vinifera*) have allowed the rapid identification of suppressors encoded by grapevine leafroll-associated virus (GLRaV) with transformation efficiencies exceeding 90% (Gambino et al. 2005). Similar approaches have been used to characterize the V2 protein of citrus chlorotic dwarf-associated virus (CCDaV), which functions as a weak suppressor of local silencing in *N. benthamiana* (Yan et al. 2024), and the coat protein of apple green crinkle-associated virus (AGCaV), which displayed moderate suppression of silencing in transient systems (James et al. 2013). Although these assays are ideal for rapid, high-throughput testing and for dissecting local suppression mechanisms, their limitation lies in the fact that they primarily capture local rather than systemic suppression, which is particularly relevant in perennial crops.

7.4.2 Reversal Crop Assay

Reversal assays provide an important complementary approach by testing whether a VSR can reverse pre-established silencing, thereby mimicking the natural counter-defense observed during viral infection. In this system, reporter-silenced plants, such as GFP-silenced *N. benthamiana* lines, are challenged with viral constructs or transgenes encoding potential suppressors. If the viral protein has strong suppressor activity, the previously silenced reporter gene is re-expressed, providing evidence of silencing reversal. This approach is particularly informative for identifying VSRs that act at systemic levels. Classical studies demonstrated reversal by the HC-Pro of potyviruses, which successfully restored GFP expression in silenced plants (Tatineni et al. 2024). In perennial fruit crops, reversal assays have been used to dissect the silencing suppression activity of Citrus tristeza virus (CTV), which encodes three VSRs (p20, p23, and p25) that act synergistically to restore GUS expression in silenced tobacco lines and enable systemic infection in citrus (Lu et al. 2004; Ruiz-Ruiz et al. 2013). Grafting-based reversal assays have also proven valuable, as silenced scions grafted onto VSR-expressing rootstocks regain reporter expression, providing evidence for intercellular and systemic silencing suppression (Majhi et al.

2023). In contemporary research, reversal by a persistent RNA virus in plants, showing weak VSR mutations fail to reverse silencing, with implications for fruit crop viruses like those in banana (Vieira et al. 2024). Although highly informative, reversal assays are more time-consuming than transient assays and often require specialized silenced reporter lines, which makes them less accessible for routine screening.

7.4.3 Stable Expression Assays

Stable expression assays extend the analysis of VSR function by generating transgenic plants that constitutively express candidate suppressors under strong promoters such as CaMV 35S. These stable lines provide long-term insights into the effects of VSR activity on silencing pathways, viral susceptibility, and host development. For instance, transgenic *N. benthamiana* lines expressing the P0 protein of poleroviruses showed Argonaute degradation and increased susceptibility to luteoviruses, confirming its role as a silencing suppressor relevant to phloem-limited grapevine viruses (Pazhouhandeh 2007). In a similar context, stable Arabidopsis lines expressing geminiviral V2 proteins revealed their ability to block small RNA loading and interfere with RNA-directed DNA methylation, findings that shed light on suppression mechanisms in berry crops (Yang et al. 2018). In vitro-stable hybrid assay in tobacco confirmed *Closterovirus* p6 suppression via RISC inhibition, bridging transient and stable readouts for citrus applications (Hussain et al. 2021). More recently, CRISPR-based approaches have been integrated with stable expression studies in banana, where lines co-expressing the CMV 2b protein restored silenced eIF4E transgenes, which were displaying resistance against multiple viruses (Tripathi et al. 2021). Stable expression assays are particularly valuable in woody perennials, where they enable the study of systemic silencing over long developmental periods. However, they remain labor-intensive, time-consuming, and technically challenging, especially given the low transformation efficiency of many fruit crops.

7.4.4 Comparative Insights in Perennial Fruit Crops

Perennial fruit crops such as citrus, grapevine, and banana present unique challenges for VSR characterization due to their long lifespan, systemic accumulation of viral infections, and inherently low transformation efficiency. In such systems, transient assays offer a practical means of rapidly identifying candidate suppressors and are increasingly being optimized for woody hosts. Reversal assays, though laborious, are especially relevant for perennials, as they allow direct assessment of systemic silencing suppression, which is critical for understanding long-term virus–host interactions. Stable expression assays, while the most demanding, provide the most reliable insights into the durability and systemic nature of VSR activity and are indispensable for studying the coevolution of viruses and hosts in perennial systems. A comparative application of all three approaches—using transient assays for rapid discovery, reversal assays for systemic validation, and stable assays for long-term functional analysis—provides a comprehensive framework for elucidating the role of VSRs in fruit crops.

7.5 Counter-Defense Strategies of RNA Viruses in Fruit Crops

Fruit-infecting RNA viruses deploy a diverse arsenal of viral suppressors of RNA silencing (VSRs) that target different nodes of the host RNAi machinery. Because fruit crops are often perennial and accumulate systemic infections over years, successful viruses combine local silencing evasion with strategies that disable or bypass systemic signaling; the paragraphs below describe the major examples and the mechanistic themes that recur across hosts. The major examples and mechanistic themes are described below, and representative RNA viruses and their VSRs are summarized in Table 7.1.

7.5.1 Apple Viruses

RNA viruses that infect apple use different strategies to suppress the plant's RNA silencing defense. Apple stem grooving virus (ASGV), a member of the *Betaflexiviridae* family, produces a P1 protein that blocks both local and systemic silencing. In pear shoots infected with ASGV, most virus-derived small RNAs (vsiRNAs) were 21 or 22 nucleotides long, with uridine as the most common base at the 5′ end. This indicates that these vsiRNAs are mainly loaded into AGO1 complexes. Interestingly, the strength of RNA silencing and virus accumulation was influenced by temperature. At higher temperatures (37 °C), plants accumulated more vsiRNAs and less viral RNA compared to normal conditions (24 °C). This reduced virus accumulation was linked to the activation of several host defense genes, including *PpDCL2*, *PpDCL4*, *PpAGO1*, *PpAGO2*, *PpAGO4*, and *PpRDR1* (Liu et al. 2016). In contrast, apple mosaic virus (ApMV) encodes a viral suppressor of RNA silencing (VSR) that is capable of blocking systemic silencing, while apple chlorotic leaf spot virus (ACLSV) produces a P50 protein that functions as a much weaker systemic suppressor (Yaegashi et al. 2007a, b; Lukhovitskaya et al. 2024).

7.5.2 Banana Viruses

Banana is infected by several RNA viruses that use specialized proteins to suppress the plant's RNA silencing defense. Banana bract mosaic virus (BBrMV), a potyvirus with a genome similar to tobacco etch virus (TEV), encodes the HC-Pro (Helper Component-Proteinase) protein. HC-Pro is one of the best-studied viral suppressors of RNA silencing and is known to strongly block posttranscriptional gene silencing (PTGS). Its activity takes place downstream of transgene methylation and the

Table 7.1 RNA viruses infecting fruit crops and their viral suppressors of RNA silencing (VSRs)

Host species	Virus name	Virus family/genus	Genome type	Reported VSR protein(s)	Probable suppression mechanism	Reference
Apple	Apple stem grooving virus (ASGV)	*Betaflexiviridae*	ssRNA(+)	Replicase helicase motif	Local and systemic suppression	Xie et al. (2024)
	Apple chlorotic leaf spot virus (ACLSV)	*Betaflexiviridae*	ssRNA(+)	P50	Weak systemic suppression	Yaegashi et al. (2007a, b)
	Apple mosaic virus (ApMV)	*Ilarvirus*	ssRNA(+)	CP-RT (presumed)	Weak silencing suppression	Lukhovitskaya et al. (2024)
	Apple stem pitting virus (ASPV)	*Betaflexiviridae*	ssRNA(+)	Coat protein	siRNA sequestration	Ma et al. (2019)
	Apple latent spherical virus (ALSV)	*Secoviridae*	ssRNA(+)	Vp20	Systemic suppression	Yaegashi et al. (2007a, b)
Apricot	Apricot latent ringspot virus (ALRSV)	*Secoviridae*	ssRNA(+)	Unknown	VSR activity not characterized	Khalili et al. (2024)
Banana	Banana bract mosaic virus (BBrMV)	*Potyviridae*	ssRNA(+)	HC-Pro	Local & systemic suppression	Chelliah et al. (2015)
	Banana mild mosaic virus (BanMMV)	*Betaflexiviridae*	ssRNA(+)	Unknown	VSR activity not characterized	Junique (2024)
	Banana virus X	*Potexvirus*	ssRNA(+)	TGB (presumed)	Not studied	Teycheney et al. (2005)

Blackberry	Blackberry yellow vein-associated virus (BYVaV)	*Closteroviridae*	ssRNA(+)	p22 (presumed)	Systemic suppression	Rubio et al. (2013)
	Blackberry virus E (BVE)	*Unclassified*	ssRNA(+)	Unknown	VSR activity not characterized	Prajapati et al. (2023)
	Blackberry chlorotic ringspot virus (BCRV)	*Ilarvirus*	ssRNA(+)	Unknown	VSR activity not characterized	Tzanetakis et al. (2007)
Blackcurrant	Blackcurrant reversion virus (BRV)	*Nepovirus*	ssRNA(+)	Unknown	VSR activity not characterized	Susi (2004)
	Blackcurrant leaf chlorosis-associated virus (BCLCaV)	*Idaeovirus*	ssRNA(+)	1b (presumed)	Strong local suppression	James et al. (2018)
Blueberry	Blueberry scorch virus (BlScV)	*Amalgaviridae*	dsRNA	Unknown	VSR activity not characterized	Moretti et al. (2011)
	Blueberry shock virus (BlShV)	*Ilarvirus*	ssRNA(+)	Unknown	VSR activity not characterized	Thomas-Sharma et al. (2018)
	Blueberry necrotic ring blotch virus (BNRBV)	*Kitaviridae*	ssRNA(+)	Unknown	VSR activity not characterized	Saad et al. (2021)
Cherry	Cherry virus F (CVF)	*Ilarvirus*	ssRNA(+)	MP	Inhibits systemic silencing	Lotos et al. (2024)
	Cherry virus A (CVA)	*Capillovirus*	ssRNA(+)	Unknown	VSR activity not characterized	Chirkov et al. (2023)
	Little cherry virus 1	*Velarivirus*	ssRNA(+)	P21	Inhibits systemic silencing	Fákó et al. (2024)

(continued)

Table 7.1 (continued)

Host species	Virus name	Virus family/genus	Genome type	Reported VSR protein(s)	Probable suppression mechanism	Reference
Citrus	Citrus tristeza virus (CTV)	*Closteroviridae*	ssRNA(+)	p23, p20, CP	Alters siRNA accumulation & DNA methylation	Fagoaga et al. (2006)
	Citrus leaf blotch virus (CLBV)	*Betaflexiviridae*	ssRNA(+)	MP	Weak local suppression	Renovell et al. (2012)
	Citrus psorosis virus (CPsV)	*Ophioviridae*	ssRNA(−)	24K–25K MP	Binds dsRNA; affects miRNA pathway	Luna et al. (2017)
	Orchid fleck virus (OFV)	*Rhabdoviridae*	ssRNA(−)	MP	Reduces siRNA accumulation; inhibits systemic silencing	Oliveira Leastro et al. (2022)
	Citrus leprosis virus N (CiLV-N)	*Rhabdoviridae*	ssRNA(−)	Unknown	VSR activity not characterized	Leastro et al. (2020)
	Citrus leprosis virus C (CiLV-C)	*Kitaviridae/Cilevirus*	ssRNA(+)	p29, p61, p15	Reduces siRNA accumulation	Leastro et al. (2020)
Dragon fruit	Pitaya virus X (PiVX)	*Potexvirus*	ssRNA(+)	Coat protein	Homodimer formation; siRNA sequestration	Zhao et al. (2023)
Fig	Fig mosaic virus (FMV)	*Emaravirus*	ssRNA(−)	Unknown	VSR activity not characterized	Elbeaino et al. (2009)
	Fig virus A & B	*Closteroviridae*	ssRNA(+)	p23	siRNA sequestration	Park et al. (2021)

Grapevine	Grapevine fanleaf virus (GFLV)	*Secoviridae*	ssRNA(+)	1A, 1BHel	siRNA suppression; AGO2/DCL2/RDR6 modulation	Choi et al. (2023)
	Grapevine leafroll-associated virus 2 (GLRaV-2)	*Closteroviridae*	ssRNA(+)	p24	Function not yet studied	Li et al. (2018)
	Grapevine leafroll-associated virus 3 (GLRaV-3)	*Closteroviridae*	ssRNA(+)	p19.7	siRNA sequestration	Gouveia et al. (2012)
	Grapevine virus A (GVA)	*Betaflexiviridae*	ssRNA(+)	ORF 5 (p10)	siRNA sequestration	Zhou et al. (2006)
	Grapevine virus B (GVB)	*Betaflexiviridae*	ssRNA(+)	Unknown	VSR activity not characterized	Chitarra et al. (2018)
	Grapevine rupestris stem pitting-associated virus (GRSPaV)	*Betaflexiviridae*	ssRNA(+)	Unknown	VSR activity not characterized	Selmi et al. (2020)
	Grapevine pinot gris virus (GPGV)	*Trichovirus*	ssRNA(+)	Coat protein	Interferes with PTGS pathways	Xu Tarquini et al. (2021)
Kiwi	Actinidia virus 1 (AV1)	*Tymovirus*	ssRNA(+)	Unknown	VSR activity not characterized	Blouin et al. (2013)
	Potato virus X (PVX)	*Potexvirus*	ssRNA(+)	P25	Inhibits AGO proteins	Chiu et al. (2010)
Melon	Melon necrotic spot virus (MNSV)	*Tombusviridae*	ssRNA(+)	Coat protein	Prevents systemic small RNA transport	Serra-Soriano et al. (2017)

(continued)

Table 7.1 (continued)

Host species	Virus name	Virus family/genus	Genome type	Reported VSR protein(s)	Probable suppression mechanism	Reference
Papaya	Papaya ringspot virus (PRSV)	*Potyviridae*	ssRNA(+)	HC-Pro	siRNA binding; AGO inhibition	Huang et al. (2019)
	Tomato spotted wilt virus (TSWV)	*Bunyaviridae*	dsRNA	NSs	Inhibits dicer, suppresses siRNAs, interferes with miRNA pathway	Schnettler et al. (2010)
	Papaya meleira virus (PMeV)	Unclassified (totivirus-like)	dsRNA	Unknown	VSR activity not characterized	Sa Antunes et al. (2016)
	Papaya mosaic virus (PapMV)	*Potexvirus*	ssRNA(+)	Unknown	VSR activity not characterized	Chávez-Calvillo et al. (2016)
	Papaya leaf distortion mosaic virus (PLDMV)	*Potyvirus*	ssRNA(+)	HC-Pro, CP	siRNA sequestration	Maoka and Hataya (2005)
Passion fruit	Passion fruit mottle virus	*Potyviridae*	ssRNA(+)	HC-Pro	Reduces siRNAs	Do et al. (2023)
	Passion fruit woodiness virus (PWV)	*Potyviridae*	ssRNA(+)	HC-Pro	siRNA binding; AGO inhibition	Alfenas et al. (2005)
Plum	Plum pox virus (PPV)	*Potyviridae*	ssRNA(+)	P1, HC-Pro	siRNA sequestration; AGO inhibition	Di Nicola-Negri et al. (2005)
	Plum bark necrosis stem pitting-associated virus (PBNSPaV)	*Ampelovirus*	ssRNA(+)	Unknown	VSR activity not characterized	Marais et al. (2014)

Peach	Prunus necrotic ringspot virus (PNRSV)	*Ilarvirus*	ssRNA(+)	Coat protein	Facilitates systemic silencing	He et al. (2023)
	Peach-associated luteovirus (PaLV)	*Tombusviridae*	ssRNA(+)	P4	Weak local suppressor	Fákó et al. (2024)
Pear	Pear chlorotic leafspot virus (PCLSaV)	*Emaravirus*	ssRNA(−)	P5	Enhances movement by interacting with MP (p4)	Ren et al. (2024)
Pineapple	Pineapple mealybug wilt-associated virus 1/2 (PMWaV-1/2)	*Ampelovirus*	ssRNA(+)	p20, CP	Local & systemic suppression	Dey et al. (2015)
Raspberry	Raspberry bushy dwarf virus (RBDV)	*Idaeovirus*	ssRNA(+)	1b	Local silencing suppression	Isogai et al. (2019)
	Raspberry leaf blotch virus (RLBV)	*Emaravirus*	ssRNA(−)	p6/p7 (presumed)	Systemic suppression	Lu et al. (2015)
	Raspberry leaf mottle virus (RLMV)	*Closteroviridae*	ssRNA(+)	Unknown	VSR activity not characterized	Cieślińska (2021)
	Raspberry latent virus (RpLV)	*Reoviridae*	dsRNA	Unknown	VSR activity not characterized	Quito-Avila et al. (2011)
	Raspberry ringspot virus (RpRSV)	*Nepovirus*	ssRNA(+)	Unknown	VSR activity not characterized	Wetzel and Krczal (2007)
	Raspberry vein chlorosis virus (RVCV)	*Rhabdoviridae*	ssRNA(−)	Unknown	VSR activity not characterized	McGavin et al. (2011)
	Tomato ringspot virus (ToRSV)	*Nepovirus*	ssRNA(+)	Coat protein	Binds and destabilizes AGO	Karran and Sanfaçon (2014)
	Raspberry enamovirus 1	*Enamovirus*	ssRNA(+)	Unknown	VSR activity not characterized	Koloniuk et al. (2023)
	Black raspberry necrosis virus	*Sadwavirus*	ssRNA(+)	Unknown	VSR activity not characterized	Susi et al. (2018)

(continued)

Table 7.1 (continued)

Host species	Virus name	Virus family/genus	Genome type	Reported VSR protein(s)	Probable suppression mechanism	Reference
Black raspberry	Black raspberry necrosis virus	*Sadwavirus*	ssRNA(+)	Unknown	VSR activity not characterized	Jayaraj Mallika (2018)
Strawberry	Strawberry mottle virus (SMoV)	*Secoviridae*	ssRNA(+)	Pro2Glu, P28	siRNA sequestration; systemic suppression	Fan et al. (2022)
	Strawberry crinkle virus (SCV)	*Rhabdoviridae*	ssRNA(−)	p3 (presumed)	Movement & replication interference	Hajizadeh and Zandan (2021)
	Strawberry mild yellow edge virus (SMYEV)	*Potexvirus*	ssRNA(+)	Unknown	VSR activity not characterized	Ma et al. (2015)
	Strawberry latent ringspot virus (SLRSV)	*Nepovirus*	ssRNA(+)	Unknown	VSR activity not characterized	El-Morsy et al. (2017)
	Strawberry polerovirus 1 (SPV-1)	*Luteoviridae*	ssRNA(+)	p0	AGO degradation	Holste (2020)

mobile silencing signal, where it specifically prevents the build-up of siRNAs. A conserved RNA-binding domain is present in BBrMV HC-Pro, containing characteristic motifs such as H(Y)HAKRFF, GY, PNC(G), and AIG, which are shared with other potyviruses. Beyond its role as a suppressor, HC-Pro is multifunctional, playing essential roles in aphid transmission, virus replication, and both cell-to-cell and systemic movement within the host (Chelliah et al. 2015). Another important viral suppressor in banana is the 2b protein of Cucumber mosaic virus (CMV). This protein mainly blocks RNA silencing by binding to siRNAs. It has an N-terminal domain and two nuclear localization signals that are important for its function (Dong et al. 2016). The 2b protein can also interact with AGO1 and AGO4, but it needs to form dimers to properly bind siRNAs (Xu et al. 2013). Mutations in two serine positions reduced disease symptoms without affecting siRNA binding, showing that suppression and symptom expression are controlled separately (Lewsey et al. 2010). The 2b protein also interferes with the host defense pathway by lowering RDR6 activity, increasing miR168 levels to reduce AGO1, and blocking the production of secondary siRNAs (Csorba et al. 2015). The effectiveness of both HC-Pro and 2b, as well as the host defense proteins, may be influenced by tropical conditions such as high temperature and humidity.

7.5.3 Berry Crop Viruses

Berry crops such as raspberry, strawberry, blueberry and black currant are infected by several RNA viruses that use viral suppressors to weaken the plant RNA silencing pathway. Raspberry bushy dwarf virus (RBDV) encodes a strong local silencing suppressor, 1b, which also elevates viral genomic RNA in pollen grains, thereby increasing the probability that pollen tubes carry sufficient virus to penetrate stigmas and seed systemic spread via cell-to-cell movement and the phloem, thereby strengthening horizontal pollen transmission. This mechanistic link is supported by reduced gRNA levels when 1b is disrupted and by increased accumulation when 1b is expressed in trans, accentuating its central role in silencing suppression (Isogai et al. 2019). Deep sequencing analysis of virus-derived siRNAs from RBDV infected plants revealed distinct targeting patterns, with vsiRNAs mapping preferentially to 5′ regions of RNA1 and RNA2, while 3′ untranslated regions showed reduced coverage (Jayaraj Mallika 2018; Susi et al. 2018). The genome of black raspberry necrosis virus (BRNV) is a bipartite virus with a notable AlkB oxygenase domain and a putative VPg on RNA-1. Although no RNA-silencing suppressor has yet been identified in BRNV, deep siRNA profiling from wild and cultivated raspberries revealed siRNA accumulation across the highly similar 3′ UTRs of both RNAs (with clear 22-nt hotspots), while key coding regions such as RdRp and the small coat protein show markedly reduced siRNA coverage. The preferential DCL targeting of terminal regions and depletion over certain coding sequences suggests selective exposure or protection of genomic regions which may have implications on systemic silencing signals and possible VSR activity (Jayaraj Mallika 2018; Susi et al. 2018). Raspberry leaf mottle virus (RLMV) and other viruses infecting raspberries could employ VSR mechanisms typical of their family, contributing to the raspberry mosaic disease complex (McGavin and Macfarlane 2010). The synergistic interactions between RBDV, BRNV, and RLMV result in severe symptoms that exceed those caused by individual infections and this viral synergism could be facilitated by VSR activity and virus genome dynamics of the accumulating virus (Quito-Avila and Martin 2012).

7.5.4 Citrus Viruses

Citrus viruses have some of the most well-studied silencing suppressors. Citrus tristeza virus (CTV) produces three proteins, namely p23, p20, and p25 (coat protein), that work together to block silencing (Lu et al. 2004). The p23 protein interferes with both PTGS and DNA methylation-based silencing, p20 helps block systemic silencing signals, and p25 supports movement of the virus and alters vsiRNA profiles (Soler Calvo 2013; Lu et al. 2004; Ruiz-Ruiz et al. 2013). Citrus psorosis virus (CPsV) encodes a 24 K movement protein that binds double-stranded RNA and pri-miRNAs, stopping DCL1 from processing them and disturbing the host's miRNA balance (Reyes et al. 2016; Luna et al. 2017). Orchid fleck virus (OFV), which infects citrus relatives and orchids, uses its movement protein to reduce siRNA accumulation and block systemic silencing (Kondo et al. 2003). This illustrates the variety of VSR mechanisms found in viruses that infect citrus crops.

7.5.5 Grapevine Viruses

Grapevines harbor a diverse virome, which is reflected in the variety of VSR mechanisms employed by viruses infecting the crop. Grapevine fanleaf virus (GFLV) displays intricate interactions with the host's silencing machinery by using a combination of proteins that target different parts of the silencing pathway. The virus comprises silencing suppressors such as 1A, 1BHel, and 1ABHel proteins that regulate the activity of AGO2, DCL2, and RDR6, (Choi et al. 2023). Grapevine leafroll-associated virus 3 (GLRaV-3) utilizes a p19-like protein, whereas grapevine virus A (GVA) uses a protein encoded by ORF5 for siRNA sequestration (Zhou et al. 2006; Gouveia et al. 2012). The weak silencing suppressor encoded by the grapevine rupestris stem pitting-associated virus (GRSPaV) is involved in the regulation of symptoms (Selmi et al. 2020). Studies on virus-induced gene silencing showed that a *Trichovirus* called grapevine pinot gris virus (GPGV) has the capacity to inhibit and silence expression (Xu Tarquini et al. 2021).

7.5.6 Negative-Sense RNA Viruses

Negative-sense RNA viruses infecting fruit crops employ suppressors that must accommodate their unique replication strategy. The three ssRNA viruses known in citrus use different approaches to overcome RNA silencing. For instance, the movement protein (MP) of orchid fleck virus (OFV) selectively inhibits the systemic spread of silencing signals but does not block local silencing. This selective activity may involve interference with RNA-dependent RNA polymerase 6 (RDR6) or its cofactor suppressor of gene silencing 3 (SGS3), which are required for the amplification of secondary siRNAs (Leastro et al. 2020). In contrast, citrus psorosis virus (CPsV) encodes a 24K protein that binds directly to precursor microRNAs (pri-miRNAs) inside the nucleus. By blocking their processing by Dicer-like 1 (DCL1), the protein prevents the formation of mature miRNAs, which reduces host defense regulation and promotes infection (Reyes et al. 2016). The segmented nature of many ssRNA virus genomes further supports the evolution of specialized suppressors encoded on different genome segments. The most widely studied example is the NSs protein of tomato spotted wilt virus (TSWV), which was the first silencing suppressor identified in this group (Takeda et al. 2002). NSs bind strongly to small double-stranded RNAs, including siRNA and miRNA duplexes, thereby preventing their incorporation into the RNA-induced silencing complex (RISC). It also interferes with Dicer-mediated cleavage of long dsRNA, reducing the production of siRNAs and weakening antiviral silencing. By binding miRNA duplexes, NSs further disrupt host microRNA pathways and lower the efficiency of gene regulation, making it a powerful suppressor of RNA silencing (Schnettler et al. 2010; Hedil et al. 2015). Despite these advances, direct evidence of silencing suppressors is still missing for several other fruit-infecting ssRNA viruses, such as citrus virus A (CiVA) (*Phenuiviridae*, *Coguvirus*), apple rubbery wood virus 1 and 2 (ARWV-1, ARWV-2) (*Phenuiviridae, Rubodvirus*), fig mosaic virus (FMV) (*Fimoviridae*, *Emaravirus*), and grapevine muscat rose virus (GMRV) (*Phenuiviridae*, *Rubodvirus*). However, studies on related plant viruses provide supporting evidence. For example, the Pns10 protein of rice dwarf virus (*Phytoreovirus*) has been shown to act as a silencing suppressor (Ren et al. 2010), and NS3 from tenuiviruses also functions as a VSR (Yang et al. 2011). These findings indicate that fruit-infecting ssRNA viruses may also harbor yet-undiscovered suppressors that operate through similar mechanisms.

7.5.7 Double-Stranded RNA Viruses and Suppression

Double-stranded RNA viruses face unique challenges during RNA silencing suppression due to their genome organization and replication strategy. These viruses must prevent recognition of their dsRNA genomes while maintaining effective silencing suppression of host responses. Raspberry latent virus, a member of the *Reoviridae* family, likely encodes suppressors similar to those found in Rice ragged stunt virus, which target multiple steps of the silencing pathway (Quito-Avila et al. 2011). It is presumed that the compartmentalized replication of dsRNA viruses within viral factories provides some protection from host silencing pathways, but the role and mechanism of systemic suppression remain elusive. The VSRs of these viruses may be involved in targeting multiple components of the host defense system simultaneously. Although comprehensive characterization is still lacking, viral suppressors of RNA silencing have been encoded by RNA viruses with negative-sense single-stranded RNA (-ssRNA) and double-stranded RNA (dsRNA) genomes, which are not much studied in fruit crops.

7.6 Counter-Defense Strategies of DNA Viruses in Fruit Crops

Plants employ multiple layers of innate immunity against viral pathogens, with RNA silencing serving as a central antiviral defense. In DNA viruses, silencing is triggered when double-stranded RNA (dsRNA) molecules are recognized during infection (Hohn 2015). This pathway functions at both the posttranscriptional and transcriptional levels. At the posttranscriptional level, small RNAs (21–24 nucleotides) guide the degradation of viral RNAs, while microRNAs (miRNAs) regulate endogenous genes, and transcriptional gene silencing (TGS) operates through DNA methylation to suppress viral transcription (Napoli et al. 1990; Hamilton and Baulcombe 1999; Baulcombe 2004). To overcome these host defenses, DNA viruses encode viral suppressors of RNA silencing (VSRs) that target different stages of silencing, including small RNA biogenesis, AGO-mediated cleavage, and methylation-dependent TGS (Burgyán and Havelda 2011).

7.6.1 Emergence of VSRs in DNA Viruses

Several DNA virus families have been reported in fruit crops, including *Caulimoviridae, Geminiviridae, Badnaviridae*, and *Nanoviridae*. Each of these groups encodes proteins with silencing suppression functions. For example, citrus chlorotic dwarf-associated virus (*Caulimoviridae*) carries V2 proteins that act as silencing suppressors. Geminiviruses such as grapevine red blotch virus (GRBV), apple geminivirus (AGV), and mulberry mosaic dwarf-associated virus (MMDaV) encode multiple VSRs, including V2/AV2, TrAP, Rep, and C4/AC4. Badnaviruses such as banana streak virus (BSV) and citrus yellow mosaic badnavirus (CiYMV) produce ORF proteins with suppressor activity. Nanoviruses, most notably banana bunchy top virus (BBTV), encode Clink and movement proteins (MPs) with silencing suppression roles (Venkataraman and Selvarajan 2019; Vadlamudi et al. 2021; Ye et al. 2024). These functions are outlined in Table 7.2, which summarizes the reported roles of VSR/RSS proteins in DNA viruses.

The columns help in understanding the clear snapshot role of viral suppressors of RNA silencing in fruit crops. The table comprises the list of DNA viruses in a specific fruit host, providing an at-a-glance view of which virus encodes it, which host it infects, when it was first reported, and the molecular role it plays (Table 7.3).

Table 7.2 Functional roles of VSR/RSS-encoded proteins in DNA viruses

S. no.	Virus group	Protein (gene)	Reported function	Reference
1.	*Begomovirus*	Coat protein (CP, V1/AV1)	Structural protein; essential for virion assembly and vector transmission	Briddon et al. (1990), Poornima Priyadarshini et al. (2011)
2.	*Begomovirus*	V2/AV2	Viral suppressor of RNA silencing (VSR)	Amin et al. (2011)
3.	*Begomovirus*	C4/AC4	Viral suppressor of RNA silencing	Glick et al. (2008)
4.	*Begomovirus*	Rep (C1/AC1)	Initiates virion-sense DNA replication	Hanley-Bowdoin et al. (2013)
5.	*Begomovirus*	TrAP (C2/AC2)	Suppresses RNA silencing; modulates host defense responses	Raja et al. (2008), Nawaz-ul-Rehman and Fauquet (2009)
6.	*Begomovirus*	Ren (C3/AC3)	Required for efficient viral replication	Settlage et al. (2005)
7.	*Babuvirus*	Rep (DNA-R)	Initiates rolling-circle replication	Timchenko et al. (2006)
8.	*Babuvirus*	Clink (DNA-C; babuvirus-specific)	Manipulates host cell cycle; exhibits VSR activity	Venkataraman and Selvarajan (2019)
9.	*Babuvirus*	MP (DNA-M)	Facilitates cell-to-cell movement	Wanitchakorn et al. (2000)
10.	*Babuvirus*	NSP (DNA-N)	Mediates nuclear–cytoplasmic shuttling of viral DNA	Bashir et al. (2022)
11.	*Babuvirus*	CP (DNA-S)	Involved in virion assembly and vector transmission	Ji et al. (2019)
12.	*Nanovirus*	U1/U2/U4	Contribute to movement, symptom development, and replication support	Yan et al. (2024), Niu et al. (2009), Grigoras et al. (2009), Grigoras et al. (2018), Thomas et al. (2021)

7.6.2 RNA Interference Activation and Silencing in Babuvirus and Badnavirus

DNA viruses of the *nanoviridae*, like BBTV and *caulimoviridae*, like BSV, have VSRs to counteract the defense mechanism in banana. Upon detection of viral DNA, which is being transcribed into the nucleus. In case of BBTV, the clink protein (DNA-C) interacts with retinoblastoma-related proteins (RBR), forming SCF-SKP1 complexes in order to control cell cycle regulation (Aronson et al. 2000). The SKP1 complex also promotes S-phase and viral replication. It suppresses RNA silencing directly or indirectly via inhibiting AGO/RISC-mediated cleavage of viral RNAs (Lageix et al. 2007). MP (DNA-M) rather than facilitating cell-to-cell

Table 7.3 List of DNA viruses infecting fruit crops and their viral suppressors of RNA silencing (VSR)

S. no.	Fruit crop	Virus family	Virus (genus/species)	VSR proteins	Experimental evidence	Notes	Reference
1.	Apple (*Malus domestica*)	*Geminiviridae*	Apple geminivirus (*Capulovirus*)	V2	AGV proteins tested; VSR activity confirmed in functional assays	First record of AGV infecting crabapple	Sun et al. (2023)
2.	Citrus (*C. sinensis*)	*Geminiviridae*	*Citlodavirus*/citrus chlorotic dwarf-associated virus (CCDaV)	V2 (RSS/VSR)	GFP fluorescence assays showed V2 suppression of RNA silencing	First report of RNA-silencing suppressor in citlodavirus	Ye et al. (2024)
3.	Citrus (*C. sinensis*)	*Caulimoviridae*	*Badnavirus*/citrus yellow mosaic virus (CiYMV)	ORF I	ORF I showed RNA-silencing suppression; ORF II, IV, V, VI, and CP showed none	First report of VSR in CiYMV	Vadlamudi et al. (2021)
4.	Banana (*Musa* spp.)	*Nanoviridae*	*Babuvirus*/Banana bunchy top virus (BBTV)	Clink, MP	Both proteins demonstrated VSR activity	First report of BBTV Clink and MP as silencing suppressor	Herrera-Valencia et al. (2006), Aronson et al. (2000)
5.	Grapevine (*Vitis vinifera*)	*Geminiviridae*	*Grablovirus*/Grapevine red blotch virus (GRBV)	C2, V2	Multiple lines of evidence confirmed strong silencing suppression	C2 and V2 may act independently or share host targets; both likely suppress distinct host defense pathways	Weligodage et al. (2023)

6.	Mulberry (*Morus* spp.)	*Geminiviridae*	*Mulcrilevirus*/ Mulberry mosaic dwarf-associated virus (MMDaV)	V2, RepA, Rep	V2 suppressed RNA silencing in GFP assays; RepA/Rep affected PVX pathogenicity	First report of VSRs in a woody plant-infecting geminivirus; provides insight into MMD mechanism	Yang et al. (2018)
7.	Papaya (*Carica papaya*)	*Geminiviridae*	*Begomovirus*/ papaya leaf curl virus complex (PaLCuV)	V2/AV2, C2/ AC2 (TrAP), C4/AC4	Canonical begomovirus VSRs (V2, C2, C4) reported in PaLCuV; supported by reviews and functional studies	Papaya-specific assays are limited, but VSR profile is consistent with other begomoviruses	Veluthambi and Sunitha (2021), Saxena et al. (2013)
8.	Pomegranate (*Punica granatum*)	*Caulimoviridae*	*Badnavirus* (badnavirus-like sequences)	Not yet studied	No direct experimental evidence available	VSRs remain uncharacterized	Ruiz-García et al. (2023)
9.	Fig (*Ficus carica*)	*Caulimoviridae*	Fig badnavirus 1	Not yet studied	VSR activity was not reported in primary studies	First detection of fig badnavirus; VSRs remain uncharacterized	Laney et al. (2012)

movement. It is believed to sequester viral siRNA, which reduces its availability to load onto RISC and dampens the silencing activity (Guyot et al. 2024). In the case of BSV, TAV proteins and ORF III-encoded proteins disrupt DCL activity. It inhibits RDR6-mediated amplification and suppresses both the transcriptional and PTGS pathways. Whereas in integrated endogenous BSV possesses intact promoter elements that activate under stress, leading to production of viral transcripts. It evades typical silencing control due to altered promoter regulation (Richert-Pöggeler et al. 2021). These mechanisms are depicted schematically in Fig. 7.2.

7.6.3 Endogenous Pararetroviruses (EPRVs) and RNA Silencing

Endogenous pararetroviruses (EPRVs) are integrated sequences within plant genomes that are often fragmented and transcriptionally impaired (Richert-Pöggeler et al. 2021). They can coexist with exogenous viruses and viroids, and like other viruses, undergo an RNA phase during replication (De Tomás and Vicient 2022). Activation of EPRVs is relatively rare and is usually triggered by stress factors such as temperature shifts, hybridization, plant aging, or tissue culture (Ndowora et al. 1999; Lockhart et al. 2000; Richert-Pöggeler et al. 2003; Hansen et al. 2005; Kuriyama et al. 2020). Similar to retrotransposons, these stresses can induce transcriptional reactivation (Grandbastien 1998). Complete viral transcripts may arise through recombination of genomic segments or tandemly arranged EPRV copies

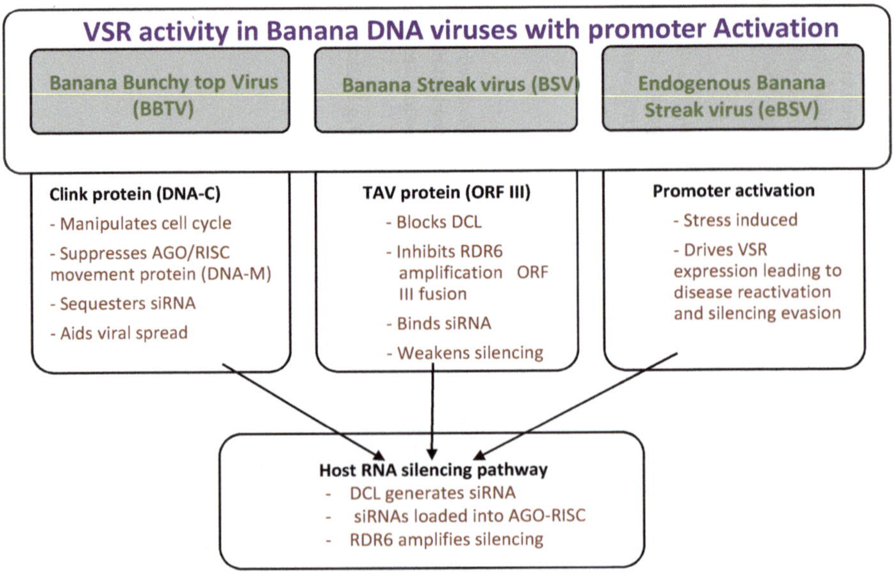

Fig. 7.2 Schematic representation of VSR activity in banana DNA viruses, along with eBSV promoter activation under stress (biotic/abiotic)

(Richert-Pöggeler et al. 2003). The resulting RNA serves as a template for protein synthesis, DNA synthesis, and also acts as a substrate for RNA interference (Hohn and Rothnie 2013; Pooggin 2013; Prasad et al. 2019). Kuriyama et al. (2020) demonstrated the interaction of integrated petunia vein clearing virus (PVCV) with host silencing pathways (Fig. 7.3). In young petunia, infection results in white-veined petals due to transcriptional silencing of PVCV promoters, while posttranscriptional gene silencing (PTGS) suppresses the host CHS-A transcript. In older plants, transient episomal replication occurs as transcriptional gene silencing (TGS) weakens. Viral suppressors of RNA silencing (VSRs) produced upon activation counteract PTGS, enhancing replication and altering pigmentation. Ultimately, the accumulation of viral transcripts stimulates the production of both 21–22 nt and 24 nt siRNAs, reinstating PTGS and TGS to repress viral activity from both chromosomal and episomal DNA sources. Representative examples of EPRV regulation and their interaction with RNA silencing (Richert-Pöggeler et al. 2021) are summarized in Table 7.4.

7.6.4 RNA Interference Activation and Silencing in Geminivirus

Apple geminivirus (AGV) is a single-stranded DNA (ssDNA) virus infecting the apple. Although its genome is DNA-based, replication occurs via a double-stranded intermediate in the plant nucleus, rendering it susceptible to both posttranscriptional gene silencing (PTGS) and transcriptional gene silencing (TGS). To counteract these defenses, AGV relies on the multifunctional V2 protein, which acts as a major viral suppressor of RNA silencing (VSR). V2 suppresses PTGS by binding 21–24 nt siRNAs and blocking their incorporation into argonaute (AGO) effector complexes, thereby preventing the amplification of secondary siRNAs required for systemic silencing and cytoplasmic RNA cleavage (Zhan et al. 2018). In the nucleus, V2 interacts with AGO4 in Cajal bodies, disrupting the RNA-directed DNA methylation (RdDM) pathway, reducing methylation of the viral genome, and maintaining transcriptional activity of viral promoters (Wang et al. 2020). In the cytoplasm, V2 associates with AGO1, reducing slicer activity against viral RNAs, and interferes with RDR6 and SGS3, blocking dsRNA template production and the generation of secondary siRNAs and mobile silencing signals. A schematic model of these suppression mechanisms is presented in Fig. 7.4. Geminiviral suppressors also impair DCL2/3/4 activity, thereby lowering the accumulation of functional small RNAs (Rishishwar and Dasgupta 2019). In addition, V2 cooperates with proteins such as AC2/C2 to inhibit the recruitment of DRM2 DNA methyltransferase and associated histone modifiers, further reducing methylation of viral minichromosomes (Veluthambi and Sunitha 2021). Thus, V2 functions as a nuclear–cytoplasmic shuttle, simultaneously blocking RdDM in the nucleus and siRNA pathways in the cytoplasm, ensuring protection of viral transcripts and the DNA genome, and promoting AGV replication and persistence.

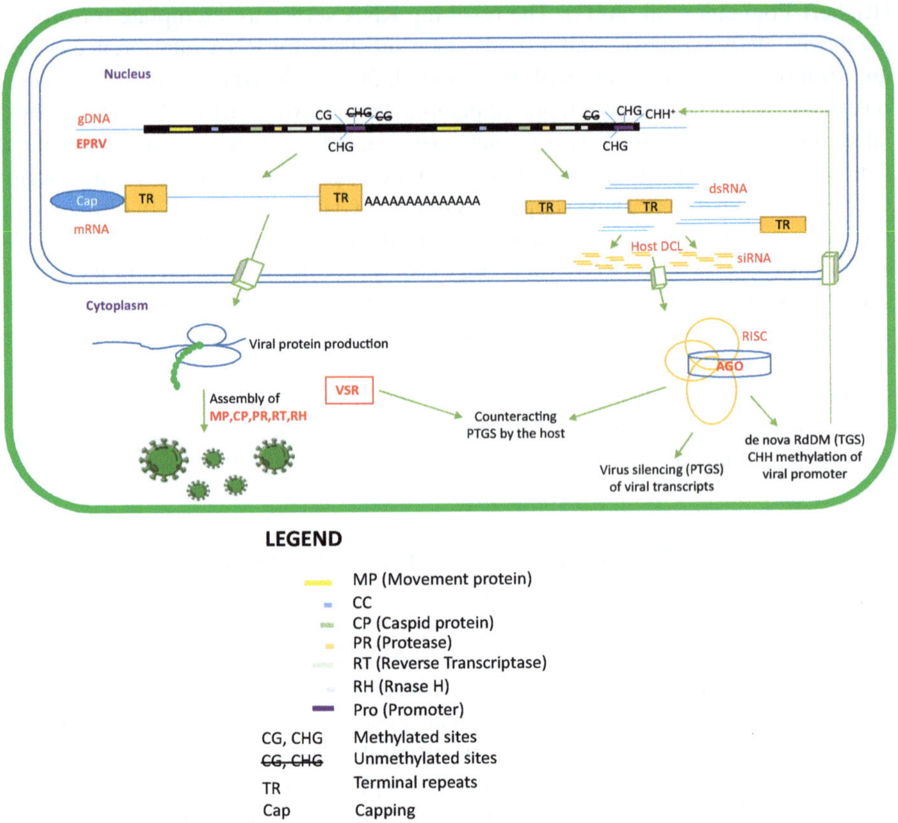

Fig. 7.3 EPRVs which are integrated into the plant genome are transcriptionally silent here by promoter methylation. In most cases, the promoters are heavily methylated at CG and CHG sites. This enforces TGS through RdDM pathway. Epigenetic repression weakens under stress or aging conditions, where demethylation occurs in the promoter. Upon activation, incomplete viral RNA silencing and dsRNA trigger the host RNA silencing pathway. DCL proteins process these dsRNA into siRNA (mainly TGS and PTGS). These siRNAs are loaded into AGO complexes to guide sequence-specific silencing of viral and host targets. Expression of the viral proteins is the crucial step. VSRs are central to the virus-host arms race. VSRs block the siRNA, sequester siRNA into AGO proteins by impairing both PTGS and TGS, and some VSRs interact with RdDM machinery, hindering de novo methylation of viral promoters and favoring viral transcriptional reactivation. A feedback loop of silencing mechanism was developed in *P. hybrida* as a result of PVCV activation, amplifying CHH methylation of EPRV promoters. Viral VSRs, tip the scales by reducing this silencing reaction, enabling the virus to continue protein synthesis and episomal reproduction. This illustrates the evolutionary "arms race" between viral suppressor proteins and host RNA silencing processes

7.7 Conclusion and Future Directions

Understanding viral suppressors of RNA silencing (VSRs) in fruit crops opens new opportunities for innovative disease management. Biotechnological tools such as RNA interference (RNAi), CRISPR/Cas systems, and synthetic siRNAs provide

Table 7.4 Epigenetic and RNA silencing-based regulation of endogenous and episomal plant pararetroviruses

S. no.	Virus/EPRV system	Mechanistic regulation	Reference
1.	Petunia vein clearing virus (PVCV)	Regulation involves H3K9 histone methylation and cytosine DNA methylation	Noreen et al. (2007), Kuriyama et al. (2020)
2.	Endogenous tobacco vein-clearing virus-like sequences in *Solanum* spp.	Silencing mediated by CHH-context DNA methylation, reinforcing RdDM-dependent transcriptional repression	Staginnus et al. (2007)
3.	Fritillaria endogenous pararetroviruses (FriEPRVs)	Suppression of expression via RNA-directed DNA methylation (RdDM)	Becher et al. (2014)
4.	Beet florendovirus EPRVs	Targeted by 18–30 nt small RNAs, mediating both transcriptional (TGS) and posttranscriptional gene silencing (PTGS)	Schmidt et al. (2021)
5.	Citrus endogenous pararetroviruses (CitEPRVs)	Stress conditions activate expression, accompanied by elevated siRNA accumulation in symptomatic tissues	Barrero et al. (2017), Matsumura et al. (2017)
6.	Episomal badnaviruses in *Musa acuminata*	Associated with abundant 22-nt siRNAs, supporting PTGS as the predominant antiviral defense mechanism	Rajeswaran et al. (2014)

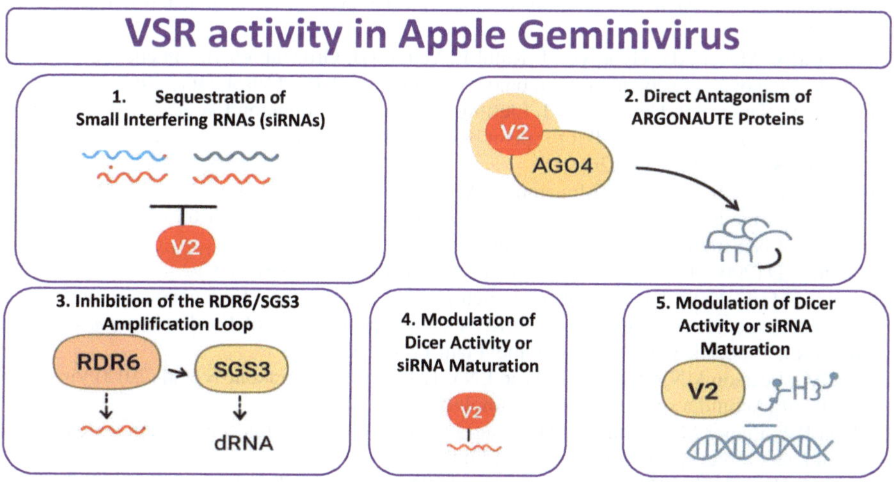

Fig. 7.4 Schematic representation of VSR activity in AGV where the V2 protein is the significant viral suppressor for RNA silencing, which acts on various fronts, including siRNA sequestration, ARGONAUTE inhibition, suppression of RDR6/SGS3 amplification, interference with Dicer activity, and blockage of RdDM-mediated DNA methylation

powerful platforms to selectively target viral genomes or silence essential viral factors. Parallel efforts in breeding for VSR-resilient crops, either through natural resistance genes or genome engineering, offer long-term solutions that can be integrated into crop improvement programs. Another promising avenue is the strategic targeting of host–VSR interfaces, such as AGO–VSR interactions or viral disruption of the RdDM pathway, to block suppressor activity and restore host silencing defenses. Together, these approaches highlight the prospects for durable virus management in fruit crops, moving beyond conventional practices toward molecularly informed resistance strategies. The integration of mechanistic insights into RNA silencing with applied crop protection holds great potential for building resilience against viral diseases. However, VSR activity often shows strong host- and virus-specific variation, underscoring the need for continued research in fruit crop systems. Advancing knowledge of silencing pathways and their viral countermeasures will guide the development of precise, durable, and sustainable resistance. Ultimately, combining molecular virology, biotechnology, and breeding will be essential to safeguard fruit production against the evolving threat of plant viruses.

References

Alfenas PF, Braz ASK, Torres LB, Santana EN, Nascimento AVSD, Carvalho MGD et al (2005) Transgenic passionfruit expressing RNA derived from Cowpea aphid-borne mosaic virus is resistant to passionfruit woodiness disease. Fitopatol Bras 30:33–38

Amin I, Patil BL, Briddon RW, Mansoor S, Fauquet CM (2011) Comparison of phenotypes produced in response to transient expression of genes encoded by four distinct begomoviruses in *Nicotiana benthamiana* and their correlation with the levels of developmental mirnas. Virol J 8(1):238

Aronson MN, Meyer AD, Györgyey J, Katul L, Vetten HJ, Gronenborn B, Timchenko T (2000) Clink, a nanovirus-encoded protein, binds both prb and SKP1. J Virol 74(7):2967–2972

Barrero RA, Napier KR, Cunnington J, Liefting L, Keenan S, Frampton RA et al (2017) An internet-based bioinformatics toolkit for plant biosecurity diagnosis and surveillance of viruses and viroids. BMC Bioinf 18(1):26

Bashir S, Farrakh S, Yasmin T, Muhammad A, Bashir T, Manghwar H et al (2022) Quantitation of multipartite banana bunchy top virus genomic components and their transcripts in infected tissues of banana (*Musa acuminata*). Agronomy 12(12):2990

Baulcombe D (2004) RNA silencing in plants. Nature 431(7006):356–363

Becher H, Ma L, Kelly LJ, Kovarik A, Leitch IJ, Leitch AR (2014) Endogenous pararetrovirus sequences associated with 24 nt small RNA s at the centromeres of *Fritillaria imperialis* L.(L iliaceae), a species with a giant genome. Plant J 80(5):823–833

Blouin AG, Pearson MN, Chavan RR, Woo ENY, Lebas BSM, Veerakone S et al (2013) Viruses of kiwifruit (Actinidia species). J Plant Pathol 95(2):221–235

Briddon RW, Pinner MS, Stanley J, Markham PG (1990) Geminivirus coat protein gene replacement alters insect specificity. Virology 177(1):85–94

Burgyán J, Havelda Z (2011) Viral suppressors of RNA silencing. Trends Plant Sci 16(5):265–272

Castel SE, Martienssen RA (2013) RNA interference in the nucleus: roles for small rnas in transcription, epigenetics and beyond. Nat Rev Genet 14(2):100–112

Chávez-Calvillo G, Contreras-Paredes CA, Mora-Macias J, Noa-Carrazana JC, Serrano-Rubio AA, Dinkova TD et al (2016) Antagonism or synergism between Papaya ringspot virus and papaya mosaic virus in *Carica papaya* is determined by their order of infection. Virology 489:179–191

Chelliah A, Velusamy B, Ramasamy S (2015) Sequence motif comparison and homology modeling of helper component proteinase (HC-pro) of Banana bract mosaic virus. Plant Pathol J 14(3):123

Chirkov S, Sheveleva A, Tsygankova S, Slobodova N, Sharko F, Petrova K, Mitrofanova I (2023) First report and complete genome characterization of cherry virus A and little cherry virus 1 from Russia. Plants (Basel, Switz) 12(18):3295

Chitarra W, Cuozzo D, Ferrandino A, Secchi F, Palmano S, Perrone I, Boccacci P, Pagliarani C, Gribaudo I, Mannini F, Gambino G (2018) Dissecting interplays between *Vitis vinifera* L. and grapevine virus B (GVB) under field conditions. Mol Plant Pathol 19(12):2651–2666

Chiu MH, Chen IH, Baulcombe DC, Tsai CH (2010) The silencing suppressor P25 of potato virus X interacts with Argonaute1 and mediates its degradation through the proteasome pathway. Mol Plant Pathol 11(5):641–649

Choi J, Pakbaz S, Yepes LM, Cieniewicz EJ, Schmitt-Keichinger C, Labarile R et al (2023) Grapevine fanleaf virus RNA1-encoded proteins 1A and 1bhel suppress RNA silencing. Mol Plant-Microbe Interact 36(9):558–571

Cieślińska M (2021) First report of raspberry leaf mottle virus infecting raspberry in Poland. Plant Dis 105(3):714–714

Csorba T, Kontra L, Burgyán J (2015) Viral silencing suppressors: tools forged to fine-tune host-pathogen coexistence. Virology 479(480):85–103

De Tomás C, Vicient CM (2022) Genome-wide identification of reverse transcriptase domains of recently inserted endogenous plant pararetrovirus (Caulimoviridae). Front Plant Sci 13:1011565

Deleris A, Gallego-Bartolome J, Bao J, Kasschau KD, Carrington JC, Voinnet O (2006) Hierarchical action and inhibition of plant dicer-like proteins in antiviral defense. Science 313(5783):68–71

Dey KK, Borth WB, Melzer MJ, Wang ML, Hu JS (2015) Analysis of pineapple mealybug wilt associated virus-1 and-2 for potential RNA silencing suppressors and pathogenicity factors. Viruses 7(3):969–995

Di Nicola-Negri E, Brunetti A, Tavazza M, Ilardi V (2005) Hairpin RNA-mediated silencing of Plum pox virus P1 and HC-Pro genes for efficient and predictable resistance to the virus. Transgenic Res 14(6):989–994

Ding SW (2010) RNA-based antiviral immunity. Nat Rev Immunol 10(9):632–644

Do DH, Nguyen TBN, Ha VC, Raja JA, Yeh SD (2023) Generation of attenuated Passiflora mottle virus through modification of the helper component protease for cross-protection. Phytopathology 113(8):1605–1614

Dong K, Wang Y, Zhang Z, Chai LX, Tong X, Xu J, Li D, Wang XB (2016) Two amino acids near the N-terminus of Cucumber mosaic virus 2b play critical roles in the suppression of RNA silencing and viral infectivity. Mol Plant Pathol 17(2):173–183

Elbeaino T, Digiaro M, Alabdullah A, De Stradis A, Minafra A, Mielke N et al (2009) A multipartite single-stranded negative-sense RNA virus is the putative agent of fig mosaic disease. J Gen Virol 90(5):1281–1288

El-Morsy SI, El-Sheikh MA, Abd El-Razik RA, Youssef SA, Shalaby AA (2017) Molecular identification of Strawberry latent ring spot virus (SLRSV) in Egypt. J Basic Environ Sci 4(1):24–33

Fagoaga C, López C, de Mendoza AH et al (2006) Post-transcriptional gene silencing of the p23 silencing suppressor of Citrus tristeza virus confers resistance to the virus in transgenic Mexican lime. Plant Mol Biol 60:153–165

Fákó V, Jaksa-Czotter N, Várallyay É (2024) In the search of viral suppressors of Prunus infecting viruses. G Agric 28(Suppl. 1):25–30

Fan L, He C, Gao D, Xu T, Xing F, Yan J et al (2022) Identification of silencing suppressor protein encoded by strawberry mottle virus. Front Plant Sci 13:786489

Gambino G, Gribaudo I, Leopold S, Schartl A, Laimer M (2005) Molecular characterization of grapevine plants transformed with GFLV resistance genes: I. Plant Cell Rep 24(11):655–662

Glick E, Zrachya A, Levy Y, Mett A, Gidoni D, Belausov E et al (2008) Interaction with host SGS3 is required for suppression of RNA silencing by tomato yellow leaf curl virus V2 protein. Proc Natl Acad Sci 105(1):157–161

Gouveia P, Dandlen S, Costa Â, Marques N, Nolasco G (2012) Identification of an RNA silencing suppressor encoded by grapevine leafroll-associated virus 3. Eur J Plant Pathol 133(1):237–245

Grandbastien MA (1998) Activation of plant retrotransposons under stress conditions. Trends Plant Sci 3(5):181–187

Grigoras I, Timchenko T, Katul L, Grande-Perez A, Vetten HJ, Gronenborn B (2009) Reconstitution of authentic nanovirus from multiple cloned dnas. J Virol 83(20):10778–10787

Grigoras I, Vetten HJ, Commandeur U, Ziebell H, Gronenborn B, Timchenko T (2018) Nanovirus DNA-N encodes a protein mandatory for aphid transmission. Virology 522:281–291

Guyot V, Trieu TD, Insisiengmay O, Zhang T, Iskra-Caruana ML, Pooggin MM (2024) A new genus of alphasatellites associated with banana bunchy top virus in Southeast Asia. Virus Evol 10(1):vead076

Hajizadeh M, Zandan NG (2021) Iranian Strawberry crinkle cytorhabdovirus variation assessed using its movement protein (P3) gene. Mol Biol Rep 48:7035–7040

Hamilton AJ, Baulcombe DC (1999) A species of small antisense RNA in posttranscriptional gene silencing in plants. Science 286(5441):950–952

Hanley-Bowdoin L, Bejarano ER, Robertson D, Mansoor S (2013) Geminiviruses: masters at redirecting and reprogramming plant processes. Nat Rev Microbiol 11(11):777–788

Hannon GJ (2002) RNA interference. Nature 418(6894):244–251

Hansen CN, Harper G, Heslop-Harrison JS (2005) Characterisation of pararetrovirus-like sequences in the genome of potato (*Solanum tuberosum*). Cytogenet Genome Res 110(1–4):559–565

He C, Xing F, Zhao X, Li S, Zhan B, Liu Z et al (2023) The coat protein of the ilarvirus prunus necrotic ringspot virus mediates long-distance movement. J Gen Virol 104(2):001829

Hedil M, Sterken MG, de Ronde D, Lohuis D, Kormelink R (2015) Analysis of tospovirus nss proteins in suppression of systemic silencing. PLoS One 10(8):e0134517

Herrera-Valencia VA, Dugdale B, Harding RM, Dale JL (2006) An iterated sequence in the genome of Banana bunchy top virus is essential for efficient replication. J Gen Virol 87(11):3409–3412

Hohn T (2015) RNA based viral silencing suppression in plant pararetroviruses. Front Plant Sci 6:398

Hohn T, Rothnie H (2013) Plant pararetroviruses: replication and expression. Curr Opin Virol 3(6):621–628

Holste N (2020) Polerovirus genomic variation and mechanisms of silencing suppression by P0 protein. School of Biological Sciences: Dissertations, Theses, and Student Research

Huang XD, Fang L, Gu QS, Tian YP, Geng C, Li XD (2019) Cross protection against the watermelon strain of Papaya ringspot virus through modification of viral RNA silencing suppressor. Virus Res 265:166–171

Hussain MD, Farooq T, Chen X, Tariqjaveed M, Jiang T, Liu S, Zhou T (2021) Viral suppressors from members of the family Closteroviridae combating antiviral RNA silencing: a tale of a sophisticated arms race in host-pathogen interactions. Phytopathol Res 3(1):27

Ipsaro JJ, Joshua-Tor L (2015) From guide to target: molecular insights into eukaryotic RNA-interference machinery. Nat Struct Mol Biol 22(1):20–28. https://doi.org/10.1038/nsmb.2931

Isogai M, Matsudaira T, Ito M, Yoshikawa N (2019) The 1b gene of raspberry bushy dwarf virus is a virulence component that facilitates systemic virus infection in plants. Virology 526:222–230

James D, Varga A, Jesperson GD, Navratil M, Safarova D, Constable F, Horner M, Eastwell K, Jelkmann W (2013) Identification and complete genome analysis of a virus variant or putative new foveavirus associated with apple green crinkle disease. Arch Virol 158(9):1877–1887

James D, Phelan J, Sanderson D (2018) Blackcurrant leaf chlorosis associated virus: evidence of the presence of circular RNA during infections. Viruses 10(5):260

Jay F, Brioudes F, Voinnet O (2023) A contemporary reassessment of the enhanced transient expression system based on the tombusviral silencing suppressor protein P19. Plant J 113(1):186–204

Jayaraj Mallika FR (2018) Srna deep sequencing aided plant RNA virus detection in cultivated raspberries and molecular characterization of raspberry bushy dwarf virus and black raspberry necrosis virus from Finland. Master's thesis, University of Helsinki. http://hdl.handle.net/10138/236079

Ji XL, Yu NT, Qu L, Li BB, Liu ZX (2019) Banana bunchy top virus (BBTV) nuclear shuttle protein interacts and re-distributes BBTV coat protein in *Nicotiana benthamiana*. 3 Biotech 9(4):121. https://doi.org/10.1007/s13205-019-1656-1. Epub 2019 Mar 5. PMID: 30863700; PMCID: PMC6401052

Junique E (2024) Plant-virus interactions: exploring the impact of banana mild mosaic virus infection on the growth of in vitro banana plants. Master thesis of the Gembloux Agro-Bio Tech (GxABT)

Karran RA, Sanfaçon H (2014) Tomato ringspot virus coat protein binds to ARGONAUTE 1 and suppresses the translation repression of a reporter gene. Mol Plant-Microbe Interact 27(9):933–943

Khalili M, Candresse T, Faure C et al (2024) Complete genome sequences of three Prunus-infecting nepoviruses: apricot latent ringspot virus, myrobalan latent ringspot virus, and a novel virus from smooth stone peach (Prunus mira Koehne). Arch Virol 169:168

Koloniuk I, Fránová J, Přibylová J, Sarkisova T, Špak J, Tan JL, Zemek R, Čmejla R, Rejlová M, Valentová L, Sedlák J, Holub J, Skalík J, Blystad DR, Sapkota B, Hamborg Z (2023) Molecular characterization of a novel Enamovirus infecting raspberry. Viruses 15(12):2281

Kondo H, Maeda T, Tamada T (2003) Orchid fleck virus: Brevipalpus californicus mite transmission, biological properties and genome structure. Exp Appl Acarol 30(1):215–223

Kuriyama K, Tabara M, Moriyama H, Kanazawa A, Koiwa H, Takahashi H, Fukuhara T (2020) Disturbance of floral colour pattern by activation of an endogenous pararetrovirus, petunia vein clearing virus, in aged petunia plants. Plant J 103(2):497–511

Lageix S, Catrice O, Deragon JM, Gronenborn B, Pélissier T, Ramírez BC (2007) The nanovirus-encoded Clink protein affects plant cell cycle regulation through interaction with the retinoblastoma-related protein. J Virol 81(8):4177–4185

Laney AG, Hassan M, Tzanetakis IE (2012) An integrated badnavirus is prevalent in fig germplasm. Phytopathology 102(12):1182–1189

Leastro MO, Castro DYO, Freitas-Astúa J, Kitajima EW, Pallás V, Sánchez-Navarro JÁ (2020) Citrus leprosis virus C encodes three proteins with gene silencing suppression activity. Front Microbiol 11:1231

Lewsey MG, González I, Kalinina NO, Palukaitis P, Canto T, Carr JP (2010) Symptom induction and RNA silencing suppression by the cucumber mosaic virus 2b protein. Plant Signal Behav 5(6):705–708

Li M, Zhang J, Feng M, Wang X, Luo C, Wang Q, Cheng Y (2018) Characterization of silencing suppressor p24 of Grapevine leafroll-associated virus 2. Mol Plant Pathol 19(2):355–368

Liu J, Zhang X, Yang Y et al (2016) Characterization of virus-derived small interfering rnas in Apple stem grooving virus-infected in vitro-cultured *Pyrus pyrifolia* shoot tips in response to high temperature treatment. Virol J 13:166

Lockhart BE, Menke J, Dahal G, Olszewski NE (2000) Characterization and genomic analysis of tobacco vein clearing virus, a plant pararetrovirus that is transmitted vertically and related to sequences integrated in the host genome. J Gen Virol 81(6):1579–1585

Lotos L, Katsiani A, Katis NI, Maliogka VI (2024) Evaluation of the RNA silencing suppression activity of three cherry virus F-encoded proteins. Plants (Basel, Switz) 13(2):264

Lu R, Folimonov A, Shintaku M, Li Y, Verdecchio WB, Castaño A et al (2004) Three distinct suppressors of RNA silencing encoded by a 20-kb viral RNA genome. Proc Natl Acad Sci 101(44):15742–15747

Lu Y, McGavin W, Cock PJA, Schnettler E, Yan F, Chen J, macfarlane S (2015) Newly identified rnas of raspberry leaf blotch virus encoding a related group of proteins. J Gen Virol 96(11):3432–3439

Lukhovitskaya N, Brown K, Hua L, Pate AE, Carr JP, Firth AE (2024) A novel ilarvirus protein CP-RT is expressed via stop codon readthrough and suppresses RDR6-dependent RNA silencing. PLoS Pathog 20(5):e1012034

Luna GR, Reyes CA, Peña EJ, Ocolotobiche E, Baeza C, Borniego MB et al (2017) Identification and characterization of two RNA silencing suppressors encoded by ophioviruses. Virus Res 235:96–105

Ma X, Cui H, Bernardy M, Tian L, Abbasi P, Wang A (2015) Molecular characterization of a strawberry mild yellow edge virus isolate from Canada. Can J Plant Pathol 37(3):369–375

Ma X, Hong N, Moffett P et al (2019) Functional analysis of apple stem pitting virus coat protein variants. Virol J 16:20

Majhi PK, Bhoi TK, Sahoo KC, Mishra N, Tudu S, Das S, Ray M, Singh S, Samal I, Behera PP, Saini V (2023) Understanding the genetics and genomics of vegetable grafting to ensure yield stability. In: Smart plant breeding for vegetable crops in post-genomics era. Springer Nature Singapore, Singapore, pp 69–98

Maoka T, Hataya T (2005) The complete nucleotide sequence and biotype variability of papaya leaf distortion mosaic virus. Phytopathology 95(2):128–135

Marais A, Faure C, Couture C, Bergey B, Gentit P, Candresse T (2014) Characterization by deep sequencing of divergent plum bark necrosis stem pitting-associated virus (pbnspav) isolates and development of a broad-spectrum pbnspav detection assay. Phytopathology 104(6):660–666

Matsumura EE, Coletta-Filho HD, Nouri S, Falk BW, Nerva L, Oliveira TS et al (2017) Deep sequencing analysis of rnas from citrus plants grown in a citrus sudden death-affected area reveals diverse known and putative novel viruses. Viruses 9(4):92

McGavin WJ, Macfarlane SA (2010) Sequence similarities between raspberry leaf mottle virus, raspberry leaf spot virus and the closterovirus raspberry mottle virus. Ann Appl Biol 156(3):439–448

McGavin WJ, Cock PJA, Macfarlane SA (2011) Partial sequence and RT-PCR diagnostic test for the plant Rhabdovirus raspberry vein chlorosis virus. Plant Pathol 60(3):462–467

Moreno P, Ambrós S, Albiach-Martí MR, Guerri J, Peña L (2008) Citrus tristeza virus: a pathogen that changed the course of the citrus industry. Mol Plant Pathol 9(2):251–268

Moretti M, Ciuffo M, Gotta P et al (2011) Molecular characterization of two distinct strains of blueberry scorch virus (blscv) in northern Italy. Arch Virol 156:1295–1297

Napoli C, Lemieux C, Jorgensen R (1990) Introduction of a chimeric chalcone synthase gene into petunia results in reversible co-suppression of homologous genes in trans. Plant Cell 2(4):279–289

Nawaz-ul-Rehman MS, Fauquet CM (2009) Evolution of geminiviruses and their satellites. FEBS Lett 583(12):1825–1832

Ndowora T, Dahal G, lafleur D, Harper G, Hull R, Olszewski NE, Lockhart B (1999) Evidence that badnavirus infection inmusacan originate from integrated pararetroviral sequences. Virology 255(2):214–220

Niu S, Wang B, Guo X, Yu J, Wang X, Xu K et al (2009) Identification of two RNA silencing suppressors from banana bunchy top virus. Arch Virol 154(11):1775–1783

Noreen F, Akbergenov R, Hohn T, Richert-Pöggeler KR (2007) Distinct expression of endogenous Petunia vein clearing virus and the DNA transposon dtph1 in two Petunia hybrida lines is correlated with differences in histone modification and sirna production. Plant J 50(2):219–229

Obbard DJ, Gordon KH, Buck AH, Jiggins FM (2009) The evolution of rnai as a defence against viruses and transposable elements. Philos Trans R Soc B 364(1513):99–115

Oliveira Leastro M, Pallás V, Sánchez-Navarro JÁ (2022) Orchid fleck dichorhavirus movement protein shows RNA silencing suppressor activity. J Gen Virol 103(11):001805

Park D, Goh CJ, Hahn Y (2021) Two novel closteroviruses, fig virus A and fig virus B, identified by the analysis of the high-throughput RNA-sequencing data of fig (Ficus carica) latex. Acta Virol 65(1):42

Pazhouhandeh M (2007) The mechanism of action of polerovirus P0 in RNA silencing suppression. University of Louis Pasteur, Strasbourg

Pooggin MM (2013) How can plant DNA viruses evade sirna-directed DNA methylation and silencing? Int J Mol Sci 14(8):15233–15259

Poornima Priyadarshini CG, Ambika MV, Tippeswamy R, Savithri HS (2011) Functional characterization of coat protein and V2 involved in cell to cell movement of cotton leaf curl Kokhran virus-Dabawali. PLoS One 6(11):e26929

Prajapati MR, Singh J, Kumar P, Baranwal VK (2023) First report of blackberry virus E (BVE) infecting garlic (*Allium sativum* L.) in India. J Plant Pathol 105(1):353–354

Prasad A, Sharma N, Muthamilarasan M, Rana S, Prasad M (2019) Recent advances in small RNA mediated plant-virus interactions. Crit Rev Biotechnol 39(4):587–601

Pumplin N, Voinnet O (2013) RNA silencing suppression by plant pathogens: defence, counter-defence and counter-counter-defence. Nat Rev Microbiol 11(11):745–760. https://doi.org/10.1038/nrmicro3120

Quito-Avila DF, Martin RR (2012) Real-time RT-PCR for detection of Raspberry bushy dwarf virus, Raspberry leaf mottle virus and characterizing synergistic interactions in mixed infections. J Virol Methods 179(1):38–44

Quito-Avila DF, Jelkmann W, Tzanetakis IE, Keller K, Martin RR (2011) Complete sequence and genetic characterization of raspberry latent virus, a novel member of the family Reoviridae. Virus Res 155(2):397–405

Raja P, Sanville BC, Buchmann RC, Bisaro DM (2008) Viral genome methylation as an epigenetic defense against geminiviruses. J Virol 82(18):8997–9007

Rajeswaran R, Golyaev V, Seguin J, Zvereva AS, Farinelli L, Pooggin MM (2014) Interactions of Rice tungro bacilliform pararetrovirus and its protein P4 with plant RNA-silencing machinery. Mol Plant-Microbe Interact 27(12):1370–1378

Ren B, Guo Y, Gao F, Zhou P, Wu F, Meng Z et al (2010) Multiple functions of Rice dwarf phytoreovirus Pns10 in suppressing systemic RNA silencing. J Virol 84(24):12914–12923

Ren Q, Zhang Z, Zhang Y, Zhang Y, Gao Y, Zhang H et al (2024) Protein P5 of pear chlorotic leaf spot-associated virus is a pathogenic factor that suppresses RNA silencing and enhances virus movement. Mol Plant Pathol 25(10):e70015

Renovell A, Vives MC, Ruiz-Ruiz S, Navarro L, Moreno P, Guerri J (2012) The Citrus leaf blotch virus movement protein acts as silencing suppressor. Virus Genes 44(1):131–140

Reyes CA, Ocolotobiche EE, Marmisollé FE, Robles Luna G, Borniego MB, Bazzini AA et al (2016) Citrus psorosis virus 24 K protein interacts with citrus miRNA precursors, affects their processing and subsequent mirna accumulation and target expression. Mol Plant Pathol 17(3):317–329

Richert-Pöggeler KR, Noreen F, Schwarzacher T, Harper G, Hohn T (2003) Induction of infectious petunia vein clearing (pararetro) virus from endogenous provirus in petunia. EMBO J 22:4836

Richert-Pöggeler KR, Vijverberg K, Alisawi O, Chofong GN, Heslop-Harrison JS, Schwarzacher T (2021) Participation of multifunctional RNA in replication, recombination and regulation of endogenous plant pararetroviruses (eprvs). Front Plant Sci 12:689307

Rishishwar R, Dasgupta I (2019) Suppressors of RNA silencing encoded by geminiviruses and associated DNA satellites. Virus 30(1):58–65

Rogers K, Chen X (2013) Biogenesis, turnover, and mode of action of plant microRNAs. Plant Cell 25(7):2383–2399

Rubio L, Guerri J, Moreno P (2013) Genetic variability and evolutionary dynamics of viruses of the family Closteroviridae. Front Microbiol 4:151

Ruiz-García AB, Olmos A, Marais A, Faure C, Candresse T (2023) Natural infection of pomegranate (*Punica granatum*) by apple dimple fruit viroid. Cells 12:49

Ruiz-Ruiz S, Soler N, Sánchez-Navarro J, Fagoaga C, Lopez C, Navarro L et al (2013) Citrus tristeza virus p23: determinants for nucleolar localization and their influence on suppression of RNA silencing and pathogenesis. Mol Plant-Microbe Interact 26(3):306–318

Sa Antunes TF, Amaral RJV, Ventura JA, Godinho MT, Amaral JG, Souza FO et al (2016) The dsrna virus papaya meleira virus and an ssrna virus are associated with papaya sticky disease. PLoS One 11(5):e0155240

Saad N, Olmstead JW, Jones JB, Varsani A, Harmon PF (2021) Known and new emerging viruses infecting blueberry. Plants 10(10):2172

Sagan SM (2009) Development of protein-and RNA-based tools for studying small rnas using Tombusvirus p19 and native target RNA microarrays. Doctoral dissertation, University of Ottawa

Saxena S, Kesharwani RK, Singh V, Singh S (2013) Designing of putative sirna against geminiviral suppressors of rnai to develop geminivirus-resistant papaya crop. Int J Bioinforma Res Appl 9(1):3–12

Schmidt N, Seibt KM, Weber B, Schwarzacher T, Schmidt T, Heitkam T (2021) Broken, silent, and in hiding: tamed endogenous pararetroviruses escape elimination from the genome of sugar beet (*Beta vulgaris*). Ann Bot 128(3):281–299

Schnettler E, Hemmes H, Huismann R, Goldbach R, Prins M, Kormelink R (2010) Diverging affinity of tospovirus RNA silencing suppressor proteins, nss, for various RNA duplex molecules. J Virol 84(21):11542–11554

Selmi I, Pacifico D, Lehad A, Stigliano E, Crucitti D, Carimi F, Mahfoudhi N (2020) Genetic diversity of grapevine rupestris stem pitting-associated virus isolates from Tunisian grapevine germplasm. Plant Pathol 69(6):1051–1059

Serra-Soriano M, Antonio Navarro J, Pallás V (2017) Dissecting the multifunctional role of the N-terminal domain of the Melon necrotic spot virus coat protein in RNA packaging, viral movement and interference with antiviral plant defence. Mol Plant Pathol 18(6):837–849

Settlage SB, See RG, Hanley-Bowdoin L (2005) Geminivirus C3 protein: replication enhancement and protein interactions. J Virol 79(15):9885–9895

Soler Calvo N (2013) Transgenic resistance against citrus tristeza virus (CTV) and analysis of the viral p23 protein as pathogenicity determinant in citrus. Doctoral dissertation, Universitat Politècnica de València

Staginnus C, Gregor W, Mette MF, Teo CH, Borroto-Fernández EG, Machado MLDC et al (2007) Endogenous pararetroviral sequences in tomato (*Solanum lycopersicum*) and related species. BMC Plant Biol 7(1):24

Sun PP, Zhang L, Xu XZ, Zhu M, Zhang B, Li ZN (2023) Molecular characterization of three Apple geminivirus isolates in crabapples detected in Inner Mongolia, China. Plants 12(1):195

Susi P (2004) Black currant reversion virus, a mite-transmitted nepovirus. Mol Plant Pathol 5(3):167–173

Susi H, Rajamäki ML, Artola K, Jayaraj-Mallika FR, Valkonen JPT (2018) Molecular detection and characterisation of black raspberry necrosis virus and raspberry bushy dwarf virus isolates in wild raspberries. Ann Appl Biol 173:97–111. https://doi.org/10.1111/aab.12438

Takeda A, Sugiyama K, Nagano H, Mori M, Kaido M, Mise K et al (2002) Identification of a novel RNA silencing suppressor, nss protein of Tomato spotted wilt virus. FEBS Lett 532(1–2):75–79

Tatineni S, Hein GL (2023) Plant viruses of agricultural importance: current and future perspectives of virus disease management strategies. Phytopathology 113(2):117–141

Tatineni S, Alexander J, Kovacs F (2024) The HC-Pro cistron of Triticum mosaic virus is dispensable for systemic infection in wheat but is required for symptom phenotype and efficient genome amplification. Virus Res 339:199277

Teycheney PY, Marais A, Svanella-Dumas L, Dulucq MJ, Candresse T (2005) Molecular characterization of banana virus X (BVX), a novel member of the Flexiviridae family. Arch Virol 150(9):1715–1727

Thomas JE, Gronenborn B, Harding RM, Mandal B, Grigoras I, Randles JW, Sano Y, Timchenko T, Vetten HJ, Yeh HH, Ziebell H (2021) ICTV virus taxonomy profile: Nanoviridae. J Gen Virol 102(3):001544

Thomas-Sharma S, Wells-Hansen L, Page R, Kartanos V, Saalau-Rojas E, Lockhart BE, mcmanus PS (2018) Characterization of Blueberry shock virus, an emerging Ilarvirus in cranberry. Plant Dis 102(1):91–97

Timchenko T, Katul L, Aronson M, Vega-Arreguin JC, Ramirez BC, Vetten HJ, Gronenborn B (2006) Infectivity of nanovirus dnas: induction of disease by cloned genome components of Faba bean necrotic yellows virus. J Gen Virol 87(6):1735–1743

Tripathi L, Ntui VO, Tripathi JN, Kumar PL (2021) Application of CRISPR/Cas for diagnosis and management of viral diseases of banana. Front Microbiol 11:609784

Tzanetakis IE, Postman JD, Martin RR (2007) First report of Blackberry chlorotic ringspot virus in Rubus sp. in the United States. Plant Dis 91(4):463–463

Vadlamudi T, Kaldis A, Divi VSG, Patil BL, Voloudakis AE (2021) The citrus yellow mosaic badnavirus ORFI functions as a RNA-silencing suppressor. Virus Genes 57(5):469–473

Veluthambi K, Sunitha S (2021) Targets and mechanisms of geminivirus silencing suppressor protein AC2. Front Microbiol 12:645419
Venkataraman S, Selvarajan R (2019) Recent advances in understanding the replication initiator protein of the ssdna plant viruses of the family Nanoviridae. VirusDis 30(1):22–31
Vieira MS, Cabral RLR, Favaratto L, Maciel LS, Xavier ADS, Zerbini FM, Fernandes PM (2024) Tropical fruit virus resistance in the era of next-generation plant breeding. SynBio 2(3):267–284
Voinnet O (2005a) Induction and suppression of RNA silencing: insights from viral infections. Nat Rev Genet 6(3):206–220
Voinnet O (2005b) Non-cell autonomous RNA silencing. FEBS Lett 579(26):5858–5871. https://doi.org/10.1016/j.febslet.2005.09.040
Wang L, Ding Y, He L, Zhang G, Zhu JK, Lozano-Duran R (2020) A virus-encoded protein suppresses methylation of the viral genome through its interaction with AGO4 in the Cajal body. elife 9:e55542
Wani S, Ashaq M, Dar GH, Ahmad M, Padder BA, Shah MD (2024) Major viruses infecting temperate fruit crops and their impact on the fruit industry. In: Industrial applications of soil microbes, vol 4. Bentham Science Publishers, pp 300–323
Wanitchakorn R, Hafner GJ, Harding RM, Dale JL (2000) Functional analysis of proteins encoded by banana bunchy top virus DNA-4 to-6. J Gen Virol 81(1):299–306
Weligodage HDS, Jin G, Kaur M, Rock CD, Sunitha S (2023) Grapevine red blotch virus C2 and V2 are suppressors of post-transcriptional gene silencing. Heliyon 9(3):e14528
Wetzel T, Krczal G (2007) Molecular biology of raspberry ringspot nepovirus. Plant Viruses 1(1):45–51
Xie YS, Zeng Q, Huang WT, Wang JY, Li HW, Yu SZ, Liu C, Zhang XQ, Feng CL, Zhang WH, Li TZ, Cheng YQ (2024) A novel RAV transcription factor from pear interacts with viral RNA-silencing suppressors to inhibit viral infection. Plant J 120(3):1079–1093
Xu Tarquini G, Pagliari L, Ermacora P, Musetti R, Firrao G (2021) Trigger and suppression of antiviral defenses by grapevine Pinot gris virus (GPGV): novel insights into virus-host interaction. Mol Plant-Microbe Interact 34(9):1010–1023
Xu A, Zhao Z, Chen W, Zhang H, Liao Q, Chen J, Carr JP, Du Z (2013) Self-interaction of the cucumber mosaic virus 2b protein plays a vital role in the suppression of RNA silencing and the induction of viral symptoms. Mol Plant Pathol 14(8):803–812
Yaegashi H, Takahashi T, Isogai M, Kobori T, Ohki S, Yoshikawa N (2007a) Apple chlorotic leaf spot virus 50 kda movement protein acts as a suppressor of systemic silencing without interfering with local silencing in Nicotiana benthamiana. J Gen Virol 88(Pt 1):316–324
Yaegashi H, Yamatsuta T, Takahashi T et al (2007b) Characterization of virus-induced gene silencing in tobacco plants infected with apple latent spherical virus. Arch Virol 152:1839–1849
Yan D, Han K, Lu Y, Peng J, Rao S, Wu G et al (2024) The nanovirus U2 protein suppresses RNA silencing via three conserved cysteine residues. Mol Plant Pathol 25(1):e13394
Yang X, Tan SH, Teh YJ, Yuan YA (2011) Structural implications into dsrna binding and RNA silencing suppression by NS3 protein of Rice Hoja Blanca Tenuivirus. RNA 17(5):903–911
Yang X, Ren Y, Sun S, Wang D, Zhang F, Li D, Li S, Zhou X (2018) Identification of the potential virulence factors and RNA silencing suppressors of mulberry mosaic dwarf-associated geminivirus. Viruses 10(9):472
Ye X, Ding D, Chen Y, Liu C, Li Z, Lou B, Zhou Y (2024) Identification of RNA silencing suppressor encoded by citrus chlorotic dwarf-associated virus. Front Microbiol 15:1328289
Zhan B, Zhao W, Li S, Yang X, Zhou X (2018) Functional scanning of apple geminivirus proteins as symptom determinants and suppressors of posttranscriptional gene silencing. Viruses 10(9):488
Zhao Y, Yang X, Zhou G, Zhang T (2020) Engineering plant virus resistance: from RNA silencing to genome editing strategies. Plant Biotechnol J 18(2):328–337
Zhao Y, Yang S, Jiang L, Yang Q, Luo L, Jiang J, Malichan S, Zhao J, Xie X (2023) Pitaya virus X coat protein acts as an RNA silencing suppressor and can be used as a specific target for detection using RT-LAMP. Plant Dis 107(11):3378–3382
Zhou ZS, Dell'Orco M, Saldarelli P, Turturo C, Minafra A, Martelli GP (2006) Identification of an RNA-silencing suppressor in the genome of Grapevine virus A. J Gen Virol 87(8):2387–2395

Molecular Characterization and Genome Annotation of Emerging Fruit Viruses

8

Alfredo Diaz-Lara, Aranza Mondragón-Cruz, Kristian Stevens, Oscar P. Hurtado-Gonzales, and Juan Manuel Tovar-Pedraza

Abstract

Due to the growing threat of emerging viral pathogens to global fruit production, understanding their molecular biology is essential for effective disease management. This chapter analyzes contemporary methodologies for molecular characterization and genome annotation of novel and re-emerging fruit viruses, including advanced sequencing technologies, such as short-read and long-read sequencing, which have improved the discovery and genomic profiling (genetic variants, mutations, genome structure, and expression patterns) of these pathogens. Plus, several molecular techniques for virus identification, genetic diversity, and viral population analysis, such as CRISPR-based assays and amplicon-based metagenomics, are covered. Principles and practical applications of *de novo* genome assembly and annotation pipelines are emphasized, incorporating artificial intelligence. Additionally, this manuscript includes approaches for the identification of open reading frames, predicting protein functions, mapping regulatory elements, and distinguishing evolutionary relationships through phylogenetic analyses. Furthermore, the chapter expands on complementary strategies used to validate genomic findings and refine structural and functional predictions by integrating genomic-transcriptomic evidence and

A. Diaz-Lara (✉) · A. Mondragón-Cruz
Tecnologico de Monterrey, Queretaro, Mexico
e-mail: diazlara@tec.mx

K. Stevens
University of California-Davis, Davis, CA, USA

O. P. Hurtado-Gonzales
USDA-Animal and Plant Health Inspection Service, Riverdale, MD, USA

J. M. Tovar-Pedraza
Centro de Investigación en Alimentacióny Desarrollo, Sinaloa, Mexico

R. Kumar et al. (eds.), *Fruit Crops and Viral Pathogens Volume II*, Plant Pathology for Sustainable Agriculture, https://doi.org/10.1007/978-981-95-7261-8_8

combining computational tools with targeted experimental verification. Finally, this work also covers challenges researchers face during the annotation of complex viral genomes, showing endogenous viruses and diverse replication strategies examples, highlighting common sources of bioinformatic error, and proposing solutions to reduce misclassification or incomplete genome reconstruction. Summarizing, the overall objective was to provide a robust framework for high-quality genomic data generation, significant for developing accurate diagnostic tools, reporting on plant resistance improvement programs (breeding), and guiding sustainable control strategies to protect fruit crop health and support resilient agricultural systems capable of addressing emerging viral threats.

Keywords

Emerging viruses · High-throughput sequencing · Plant virome · Virus diagnostics · Fruit crops · Genome annotation · De novo assembly

8.1 Introduction

Fruit production has a key role worldwide, not only in agriculture but also in food security, facing inevitable challenges. Crop losses occur for distinct reasons; however, those that have had the greatest impact and are the most difficult to control are plant diseases. Around the world, plants are susceptible to diverse diseases, both parasitic and nonparasitic. Viruses are a principal cause of infectious plant diseases, which are obligate parasites, meaning they are dependent entirely on living plant cells for their replication to occur (Ertunc 2020). A large number of viruses can infect fruit plants, with many showing high incidence rates (Hou et al. 2020). However, known viruses are not the only challenge agriculture faces; the presence of other factors, such as climate change, global commerce, and an increase in food demand, intensifies the emergence of novel fruit viruses. The rapid appearance of these pathogens in healthy cultivars stands as a significant challenge for producers, who are not fully prepared with a contingency plan. The risk of emerging fruit viruses is increasing worldwide. Recent studies highlight the substantial damage these pathogens cause to numerous crops. In southern Minas Gerais, Brazil, rising potexvirus infections pose a major threat to dragon fruit production, with incidence levels reaching 80%–100% and spreading rapidly through infected seedlings and cultural practices that facilitate mechanical transmission (Silveira et al. 2024). Likewise, the detection of a newly identified virus has clarified the cause of previously uncharacterized deterioration symptoms in Japanese pear orchards, underscoring the impact of emerging viral agents on fruit crop health and the need for improved monitoring and sanitary measures (Bertin et al. 2025). The impact of these emerging viruses poses an important problem for global fruit production crop yields, leading to significant economic losses and compromising food security. For example, citrus tristeza virus (CTV) has been responsible for the loss of over

50 million trees worldwide, while the banana bunchy top virus (BBTV) causes annual losses of 50 million USD in India alone (Rao and Reddy 2020). In grapevine production, infections caused by viruses such as grapevine leafroll-associated virus 3 (GLRaV-3) and grapevine fanleaf virus (GFLV) can lead to severe yield losses ranging from 50% to 98%, resulting in substantial economic damage estimated at 25,000–40,000 USD per hectare (Rao and Reddy 2020). These cases highlight the serious global impact of viruses and reveal the need to continue researching and developing advanced diagnostic tools to help manage their spread. Characterizing a new virus is complex due to traditional PCR-based diagnostics, which are often ineffective at detecting novel pathogens because of the essential requirement of a homologous genomic sequence. While metagenome sequencing, such as high-throughput sequencing (HTS), shows promise in overcoming this requirement, data generation and analysis are more complex and expensive. Additionally, many novel viruses cannot propagate in standard plant cell cultures, hindering their isolation and compliance with Koch's postulates for pathogenicity confirmation. This leads to serious damage due to the frequent presence of latent or persistent infections, making it difficult to distinguish between a harmless virus and the agent causing disease. The delayed detection of these emerging threats is a direct consequence of ecological complexity and inadequate monitoring in agricultural regions. This allows infections to cause significant economic losses before they are identified, for example, viral diseases in crops around the world cost farmers tens of billions of dollars. This chapter aims to deliver a comprehensive summary of advanced molecular methods and techniques for the discovery, characterization, and genomic annotation of novel or re-emerging viruses. It will specifically focus on synthesizing methodologies, evaluating innovations involving artificial intelligence (AI), and addressing the challenges that researchers currently face in virus annotation.

8.2 Virus Discovery Using Second- and Third-Generation Sequencing Technologies

8.2.1 High-Throughput Sequencing (HTS), Next Generation Sequencing (NGS), Short-Read Sequencing

HTS refers to the set of second-generation sequencing technologies that allow for the rapid, parallel sequencing of large amounts of genetic material. Short-read sequencing is the most common type of HTS, characterized by the generation of sequence fragments ranging from 50 to 300 base pairs in length. For sequence fidelity, these technologies employ a clonal amplification step, which is a limiting factor in the generation of longer sequences—hence the shorter read lengths. There are various platforms, such as MiSeq, NextSeq and NovaSeq by Illumina, Inc.; and Ion Torrent PGM, Proton, S5, and BGISeq-500 (Mehetre et al. 2021), which are all commercially available with Illumina's platforms as the most widely used, with existing workflows that include library preparation (including automation), sequencing, and multiple data analysis pipelines. HTS quickly became essential for plant virology since it has enabled the rapid and accurate sequencing of plant virus and viroid genomes (Rithesh et al. 2024). Its applications are immense, but one of the most relevant is the accurate detection and identification of new pathogens and strains by obtaining complete or partial genomes, as well as the analysis of their genetic makeup (Nizamani et al. 2023). For example, in fruit trees like avocado and citrus, HTS has been instrumental in the identification of re-emerging pathogens such as avocado sunblotch viroid (ASBVd) (Adams et al. 2018) or CTV (Villamor et al. 2019). In addition, this technology is used to understand plant defense mechanisms, which provides a valuable complement to virome analysis (Kanapiya et al. 2024). A broad range of nucleic acid sources can serve as input material for HTS-based virus detection in plants, including total DNA, total RNA, rRNA-depleted RNA, small interfering RNA, double-stranded RNA, and virion-associated nucleic acids (VANA). Among these, total RNA and small interfering RNA have been the most extensively examined (Anaya-López et al. 2025). Even if HTS is a powerful technology, it faces challenges for its optimal implementation in plant disease management. Away from technical limitations, such as the choice of the appropriate platform, complex sample preparation (library prep), and extensive data management (Nizamani et al. 2023), practical considerations are equally important. In laboratories, the use of HTS must be clearly defined, as well as the performance of risk analyses and the definition of appropriate controls, including controls for exotic species (Massart et al. 2022). It is also critical to standardize both protocols and bioinformatics processes, ensure access to reference sequence databases, and confirm HTS results with complementary methods (Maina et al. 2024). Furthermore, developing bioinformatics capacity and seeking collaboration between research institutions and regulatory agencies is key to integrating HTS into routine phytosanitary and epidemiological workflows (Pacheco-Dorantes et al. 2025). The work of Kutnjak et al. (2021) and Massart et al. (2022) demonstrates the various approaches used to analyze HTS data for virus detection in plants, providing insight into addressing several challenges when implementing HTS as a diagnostic tool. Among the remaining challenges are the legal and social implications of using this technology, as well as its difficult accessibility, given by the significant investment of time and money required (Anaya-López et al. 2025).

8.2.2 Long-Read Single Molecule Sequencing: SMRT and ONT

Long-read sequencing is classified as a third-generation technology. It allows for the omission of the clonal amplification step and the sequencing of individual molecules, generating sequences larger than 10 kb with fast execution times, requiring a minimal amount of genetic material, and enabling real-time analysis (Rithesh et al. 2024). It is a highly valuable technology for plant genotyping. Pacific Biosciences (PacBio) and Oxford Nanopore Technology (ONT) are the leading platforms for this type of technology (Mehetre et al. 2021). While single-molecule sequencing has been associated with higher error rates, over the years, both PacBio and ONT have been able to improve this using multiple passes and sensors. PacBio uses SMRT (Single-Molecule Real-Time) technology, while ONT is based on nanopore technology. The portability and low cost, such as that of the ONT MinION device, make long-read sequencing a powerful tool for real-time field diagnostics, unlike short-read sequencing (Mehetre et al. 2021). Table 8.1 shows the comparison between sequencing platforms and their main application in plant virology, focusing on fruit crops.

SMRT and ONT are revolutionizing the detection and identification of emerging or re-emerging viruses in fruit plants, as they have a vast capacity to generate ultra-long reads, allowing for rapid and straightforward sequencing of *de novo* viral genomes, even from samples with mixed infections (Hamim et al. 2022). Contemporary investigations have demonstrated the effectiveness of these technologies in the complete identification and characterization of viral genomes, such as Sri Lankan cassava mosaic virus (SLCMV), an emerging pathogen in Southeast Asia (Leiva et al. 2020) and to identify a new bipartite begomovirus in cowpea (Naito et al. 2019). Despite all the challenges, both long- and short-read sequencing allow for the detailed study of novel and emerging pathogens since these yields vital data for devising effective control measures and significantly improves our knowledge of pathogen biology and host-plant interaction (Nizamani et al. 2023). Each technology has its strengths. Short-read technology can more deeply sample a metagenome, leading to higher sensitivity in a diagnostic setting. Longer reads may improve the sensitivity of finding novel viruses with more distant homologies, requiring more contiguous sequence to characterize.

8.2.3 AI-Powered Basecalling

In recent years, AI has gained significant popularity in all areas of study, and plant virology is no exception. Machine learning and, more recently, neural network-based deep learning are subsets of AI that have been particularly good at solving classification problems—like determining the correct nucleotide given a set of observations. Various tools, pipelines, and methods have been developed that leverage AI and more specifically, neural network-based deep learning algorithms to improve the analysis by short and long-read sequencing. One example is the tool DeepQC, which employs neural networks to compress the quality values of FASTQ files without loss of information. Another example is Racer, which uses neural networks to correct errors in RNA-seq reads (Bandaru et al. 2024; Ilie and Molnar 2013). Basecalling is the fundamental process that converts sequencer signals into nucleotide sequences (FASTQ) on third-generation platforms (ONT and PacBio). Traditionally, this process used ML algorithms based on Hidden Markov Models (HMMs) (Boufounos et al. 2004); more recently, deep neural networks (deep learning) were integrated, which improved accuracy and speed. On ONT, the platform's basecaller is integrated with MinKNOW, which has been augmented by Dorado, an open-source, easy-to-use, high-performance analysis engine for ONT readings. It uses Graphics Processing Units (GPUs), initially designed for rendering graphics and video games (UMA Technology 2025). This accelerated the process, as

Table 8.1 Comparison of sequencing platforms used to study plant viruses

Technology	Read length	Accuracy (Q-score)	Main advantage for viromes	Main limitation	Relative cost per Gb	Ideal application in fruit plants
Sanger sequencing (first-generation sequencing)	~500–1000 bp	Very high (Q40+)	Variants validation, "gold standard"	Very low throughput	Very high	Specific sequences verification
Illumina	50–300 bp	High (Q30+)	High throughput, low cost per base	Short reads difficult *de novo* assembly of complex genomes	Low	Metagenomics (siRNA-seq, total RNA-seq), classification
Pacific Biosciences	10–25 kb	High (Q30+)	Accurate and long reads	Requires high DNA input, moderate costs	Medium	Complete viral genomes, haplotype resolution, quasispecies
Oxford Nanopore	10 to >100 kb	Moderate-high (Q20–Q30+)	Ultra-long reads, portability, and real-time sequencing	Higher raw error rate (correction through IA-basecalling)	Medium	In-field fast detection, complete genomes, epigenetics

basecalling is now in real time (Davis 2024). The addition of this enhancement has become the platform's default. Likewise, PacBio collaborated with Google to develop DeepConsensus, a deep learning tool capable of correcting errors in sequences using its high fidelity (HiFi) technology, achieving 99.9% accuracy and time reduction (Pacific Biosciences of California, Inc. 2022a, b). Basecalling by computer algorithm has been a classic AI problem from the days of Sanger sequencing, when the goal was to replace the arduous task of reading a sequence from a gel with a computer program. As AI algorithms have advanced, so too has their utility and scope in basecalling. Today, they are the foundation for generating high-quality genomic data for emerging viruses from all sequencing platforms.

8.2.4 Raw Data Preprocessing

Preprocessing the resulting data after sequencing is crucial, as this determines the quality of the subsequent bioinformatics analysis. To do this, a three-step process ought to be followed: quality control, cleaning and filtering, and computational enrichment. For the first step, FastQC is the standard tool for assessing the quality of sequencing reads. The program generates an HTML report with graphs of different analyses performed, allowing for the identification of common sequencing problems (quality degradation at read ends, adapter contamination, GC content bias, overrepresentation of k-mers, and the presence of duplicates) (Andrews 2010). This is a vital step in decision-making regarding the filtering of high-quality data optimal for analysis. After quality control, there are three main tools for cleaning and filtering: Trimmomatic or Trim-galore (short reads), Porechop (long reads), and Cutadapt (both short and long reads). Thus, the tool used depends on the selected sequencing platform. For short-read data with Illumina, Trimmomatic is the most popular and works efficiently to trim adapters, remove low-quality ends, and filter out suboptimal reads (Bolger et al. 2020). For long reads, Porechop is a specialized tool designed to detect and remove barcode adapters, even those with low identity (Wick et al. 2017). Cutadapt provides a more general and highly configurable alternative for managing both types of reads (Martin 2011). The third step is critical to avoid low viral abundance in fruit plant samples caused by host RNA dominance. Reads are digitally filtered using the host reference genome, employing tools such as Bowtie2 (Langmead and Salzberg 2012). Non-host reads show a significant abundance of viral content. User-friendly software, such as QIAGEN's CLC Genomics Workbench, already includes workflows for this type of process (https://digitalinsights.qiagen.com/products-overview/discovery-insights-portfolio/qiagen-clc-genomics/). Also, Host Contamination Removal Tool (HoCoRT) is a modern and specialized tool that allows the user to choose the best combination of speed and accuracy depending on the data type (short or long) and the level of contamination (Rumbavicius et al. 2023).

8.3 Molecular and Computational Tools for Detection and Genetic Diversity

8.3.1 Targeted Detection

Molecular techniques for virus detection include hybridization and amplification of nucleic acids, and more recently, nuclease activation using clustered regularly interspaced short palindromic repeats (CRISPR) systems. Amplification techniques include polymerase chain reaction (PCR) and isothermal amplification (Rubio et al. 2020). This chapter focuses on three major approaches: PCR and its recent variants, which remain the most widely used techniques; isothermal amplification, valued for its speed and portability; and CRISPR systems, which have gained growing importance in the biotechnology industry in recent years.

8.3.1.1 Polymerase Chain Reaction (PCR) and Its Variants

Conventional PCR is a fundamental, reliable, and accurate molecular technique for amplifying target DNA. It is employed to identify plant viruses possessing DNA genomes. However, most recognized plant viruses possess RNA genomes, so reverse transcription PCR (RT-PCR) is used, converting the RNA molecule into complementary DNA (cDNA) for subsequent amplification by conventional PCR. On the other hand, quantitative real-time PCR (qPCR) allows for the real-time quantification and detection of target DNA using fluorescent signals. This variant of the technique offers greater speed, sensitivity, specificity, and a lower risk of contamination. However, the cost of equipment and reagents is higher compared to conventional PCR or RT-PCR (Roy et al. 2024). A variant exists that combines both technologies, resulting in real-time quantitative reverse transcription PCR (RT-qPCR), which is recognized as a sensitive and accurate method for determining the amount of RNA in a sample. A one-step RT-qPCR assay was developed for the detection of emerging viruses in apple trees, such as the citrus concave gum-associated virus (CCGaV), showing satisfactory results (Rubio et al. 2020). Digital PCR (dPCR) is a technology that allows for the absolute quantification of nucleic acids without relying on standard curves, which distinguishes it from qPCR (Kang et al. 2025). In stone fruits such as plums and almonds, dPCR and its RNA variant (RT-dPCR) have proven to be important tools for the precise quantification of the effectiveness of breeding programs against emerging viruses (Mehraj et al. 2025; Schoenfeld et al. 2023). RT-dPCR was used to quantify apple stem groove virus (ASGV) in apple seedlings in vitro; it was reported to have a sensitivity approximately 10 times greater than RT-qPCR and is useful for "ASGV-free" certification (Kim et al. 2022). Thus, dPCR proves to be ideal for confirming the presence of a *de novo* virus in an asymptomatic plant or one with a low viral load (Luigi et al. 2023); however, the cost of implementing it in routine diagnoses is still limited. Although there are many more variants of PCR, such as immunocapture PCR (IC-PCR), multiplex PCR, nested PCR, cooperative PCR (Co-PCR), several recent studies have shown that RT-qPCR and RT-dPCR are more suitable for the specific detection of emerging viruses, including low-titer viruses (Roy et al. 2024).

8.3.1.2 Isothermal Amplification

Isothermal amplification has become a valuable alternative because it uses amplification mechanisms that do not require sophisticated lab equipment like PCR; instead, it uses strand-displacing DNA polymerases and simple equipment such as heating blocks (Venbrux et al. 2023). In contrast, this technology does not require highly purified genetic material (Venbrux et al. 2023). A variety of isothermal amplification methods have been developed, with loop-mediated isothermal amplification (LAMP) and recombinase polymerase amplification (RPA) being the best known for detecting plant pathogens (Kanapiya et al. 2024; Shu et al. 2025). LAMP is a technique that stands out for its simplicity, high specificity, and sensitivity, allowing for rapid amplification of sequences (Kanapiya et al. 2024). It uses multiple primers (internal, external, and loop) that bind to different sections of the nucleic acid, in synergy with the polymerase's translocation activity, generating hairpin-shaped DNA structures. This facilitates continuous amplification at a constant temperature in less than 30 min. Its advantages include inhibitor tolerance in plant samples and detection via colorimetry, fluorimetry, or lateral flow spectroscopy. However, its main limitation is the complex design of the necessary primers, which can result in zero specificity if not implemented correctly (Mehetre et al. 2021). Since most plant viruses have RNA genomes, a variant called reverse transcription-LAMP (RT-LAMP) was developed by Zhang et al. (2020), which incorporates the use of reverse transcriptase into the reaction to convert the RNA into cDNA, which can then be amplified. This adaptation has been crucial for the emerging viruses' detection without losing the advantages of the original technique. LAMP and RT-LAMP have become important techniques for detecting emerging viruses in fruit trees. In kiwifruit crops, they were used to identify the emerging Actinidia virus A (AcVA), achieving results in 20 min without the need for RNA extraction (Wang et al. 2022). Another example was reported for stone fruits such as plums and almonds, in which a multiplex RT-LAMP assay was developed that simultaneously and with high sensitivity detected the emerging plum pox virus (PPV) and Prunus necrotic ringspot virus (PNRSV) (Zhang et al. 2023). Similarly, in the work of Selvaraj et al. (2019), a rapid and sensitive immunocapture-RT-LAMP (IC-RT-LAMP) assay was created to detect important CTV isolates. By combining immunocapture with RT-LAMP, eliminating the need for RNA extraction, allowing for direct detection of the virus in citrus. The new assay demonstrated high sensitivity, detecting 11.3 copies of viral RNA in just 10 min. Evaluated on real samples, its results matched 100% with the standard RT-qPCR method, demonstrating its effectiveness. This new technique offers an efficient and cost-effective way to monitor and control dangerous virus strains.

8.3.1.3 CRISPR-Based Assay

The use of CRISPR is considered revolutionary for the diagnosis of plant pathogens due to its high specificity, sensitivity, and portability. It stands out compared to traditional methods, such as PCR, since CRISPR allows detection without the need for complex equipment. Platforms like SHERLOCK and DETECTR integrate isothermal amplification for field use (Kanapiya et al. 2024). However, its main challenges include the need to optimize primer design, off-target effects of some subtypes, reagent sensitivity under varying environmental conditions, and a limited capacity for multiple samples (Karimi et al. 2025). In fruit crops, CRISPR has been adapted to detect emerging viruses with a significant economic impact. This technology allows for the identification of pathogens in raw samples without the need to extract nucleic acids, using portable platforms such as lateral flow strips or fluorometers (Karimi et al. 2025). A recent study applied SHERLOCK to identify tomato brown rugose fruit virus (ToBRFV) and potato virus Y (PVY) in less than 1 h and with minimal detection limits (10 copies/μL) (Karimi et al. 2025). A complementary approach, based on the work of Jiao et al. (2021), demonstrates a CRISPR/Cas12a-based field assay for the simultaneous detection of several RNA viruses and viroids in apples. Combining RT-PCR with CRISPR and reading via the aggregation of gold nanoparticles visible to the human eye, this assay achieves a detection density of 250 viral copies per reaction for apple stem pitting virus (ASPV) and apple stem grooving virus (ASGV) and 2500 copies for other viruses and viroids, completing the process in less than 1 h and with 100% concordance with RT-PCR in field samples (Tanny et al. 2023). These studies illustrate the versatility of CRISPR-based diagnostics, ranging from highly sensitive, multiplex SHERLOCK/DETECTR platforms to simplified visual detection systems that facilitate on-site pathogen identification without specialized infrastructure.

8.3.2 Computational Pipelines for Viral Identification

Virus identification in sequencing data requires the use of computational pipelines that combine various approaches to maximize sensitivity and specificity. These are crucial tools for detecting known and novel viruses in complex samples, such as fruit plant viromes. There are widely used homology-based tools, such as BLAST or DIAMOND, but they are only for the identification of known viruses, which will not be discussed in this chapter. In Fig. 8.1, a computational strategy workflow for emerging virus detection depending on the data type can be seen.

8.3.2.1 Automated Pipelines

For users with less expertise in bioinformatics, automated pipelines have been developed to simplify the workflow. Virfind stands out as an accessible web-based tool that performs rapid screening through a workflow that combines mapping, assembly, and filtering, making it friendly to inexperienced users (Ho and Tzanetakis 2014). On the other hand, the VSD Toolkit offers an optimized approach to analyzing small RNA-seq data (crucial for the analysis of RNA silencing mechanisms in the host plant) (Barrero et al. 2017). Complementarily, Viroscope adds unique biological validation with the use of Viral Genome Assembly Coverage (VGAC), an algorithm that requires an assembled viral genome with sufficient coverage addressing essential viral genes (replicases, coat proteins, etc.) to be considered an actively replicating virus (Valenzuela et al. 2022). The recently developed PhytoPipe includes read classification, assembly-based annotation, and reference-based

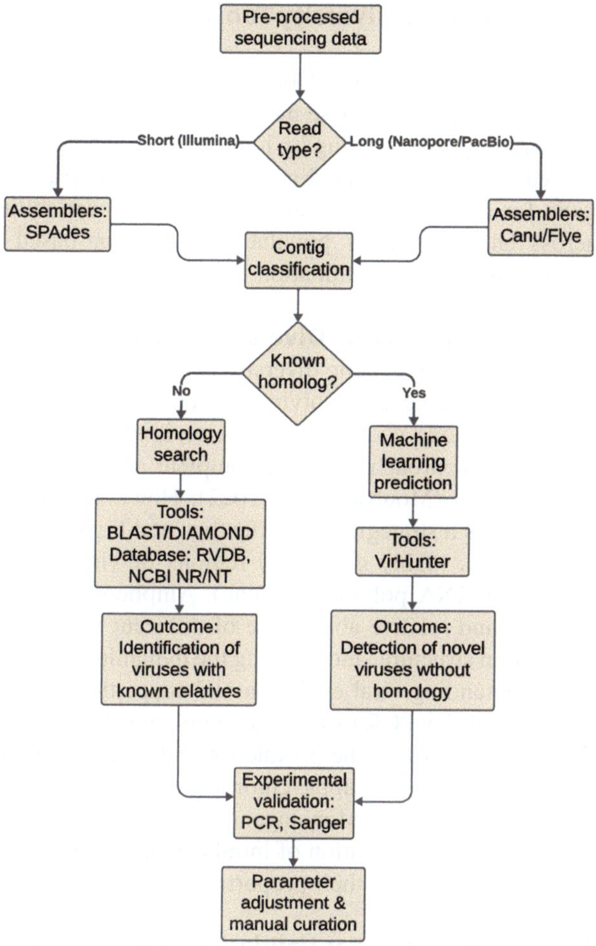

Fig. 8.1 Current computational strategies for viral identification based on data type

mapping to create a comprehensive, easy-to-digest report (Hu et al. 2023). Finally, Viral Genome Assembler (VIGA) is an all-in-one tool designed to identify and assemble eukaryotic virus genomes from HTS data; according to Fu et al. (2023), it can assemble more complete viral genomes than other competitors.

8.3.2.2 Pipelines Based on Machine Learning

Machine learning is used to detect highly divergent or novel viruses that lack homologs in databases, as they learn intrinsic patterns from viral sequences. VirHunter is a deep learning method that uses convolutional neural networks (CNNs) trained on sequences from different hosts to identify viral genomic patterns (Sukhorukov et al. 2022). This tool does not rely on similarity to known sequences, making it ideal for detecting novel viruses compared to other traditional tools.

8.3.3 Genetic Diversity Tools

There are several advanced tools and methodologies for analyzing the genetic diversity of plant viruses. These include quasispecies HTS, single-cell RNA sequencing (scRNA-seq), haplotyping by long-read sequencing, phylodynamic analysis, capillary electrophoresis, and oligonucleotide microarrays. However, there are two main tools for quantifying and analyzing genetic variability within a viral species or across an entire virus population: Amplicon-based metagenomics and Sanger sequencing for strain validation.

8.3.3.1 Virome-Wide Diversity (Amplicon-Based Metagenomics)

To determine the genetic diversity of viruses affecting fruit plants, a metagenomic approach capable of resolving the complex viral communities often present in mixed infections, such as quasispecies, is essential. Shotgun metagenomics, such as the VANA approach, offers a complete and unbiased view of viral nucleic acid amplification (Moubset et al. 2022). However, amplicon-based metagenomics is emerging as a strategic, cost-effective, and high-throughput tool because it specifically targets highly conserved regions of viral genomes (coat protein or RNA-dependent RNA polymerase gene). Amplicon sequencing allows quantifying the richness and relative abundance of different viral variants within a host, profiling population structure, and tracking the dynamics of minority variants, which are vital to understanding viral evolution and adaptation (Katsiani et al. 2020). In a recent study, multiplex PCR using small amplicons (120–135 bp) targeting various viruses and viroids revealed the prevalence and coinfection of previously uncharacterized variants of ASPV and apple chlorotic leaf spot virus (ACLSV) in citrus (Costa et al. 2022). Amplicon-based metagenomics has proven to be an excellent technique for the efficient quantification of intrahost and intrafield diversity; it not only tells you the variants, but also their proportions.

8.3.3.2 Viral Varieties Validation (Sanger Sequencing)

After sequencing, validating viral variants is essential to confirm their presence, spread, and biological importance. This validation goes beyond just verifying the sequence and uses various complementary techniques; it is important to consider the repeatability. Sanger sequencing remains the top standard for validating a specific variant's sequence with near-absolute accuracy (Anaya-López et al. 2025). For example, Xiao et al. (2019) used a combination of RT-PCR and Sanger sequencing to confirm a new variant of GLRaV-3. However, its low throughput makes it impractical for analyzing the full diversity of a quasispecies.

8.4 De Novo Viral Genome Assembly

8.4.1 Assembly Tools and Pipelines

Most *de novo* genome assembly tools can be used for fruit plant viruses. Nonetheless, the efficacy of these methods can differ according to the particular features of the genome under assembly, and, for viral genomes, the unique attributes of the fruit plant samples and their respective hosts. Among the principal challenges associated with assembling fruit viruses is their low abundance due to latent or asymptomatic infections, generating extremely low sequencing coverage, since these types of viruses are more common in this group of plants (Valenzuela et al. 2022). Additionally, mixed infections by viruses and viroids generate overlapping reads, making the assembly of individual genomes difficult (Maina et al. 2024). Another significant challenge is host-RNA contamination since the host plant may have repetitive RNA sequences. It is important to pay attention to these details before sequencing. There are specialized tools that can be adapted to the context of fruit plant virology. General-purpose tools optimized for viruses are SPAdes, Canu, and Flye. SPAdes are focused on bacteria; however, its metaSPAdes or rnaSPAdes modules are frequently used in viromics due to their superior performance with small genomes (Nurk et al. 2017; Bushmanova et al. 2019). It has a good error-correction algorithm, making it ideal for metagenomic or metatranscriptomic data. Canu and Flye are long-read assemblers (Nanopore, PacBio), making them ideal for complex, gap-free genomes (Koren et al. 2017; Kolmogorov et al. 2019). Some tools previously mentioned in the detection and identification section are taken up again, since they have more than one use. Specialized tools and pipelines for assembly include VIGA, VirFind, and the VSD Toolkit, which were already described. VirFind and the VSD Toolkit are automated bioinformatics pipelines that offer everything from read preprocessing to *de novo* assembly (using SPAdes or Velvet and BLAST tools) and viral contingency annotation. It is ideal for working with 21–22 bp siRNAs since these tools were optimized for that type of input. There are additional tools and pipelines for genome assembly; however, those presented in this chapter are considered the most suitable for the context of fruit plant virology (Ho and Tzanetakis 2014; Barrero et al. 2017). It is important to mention that no *de novo* assembly results should be considered valid without experimental validation. This can be achieved using techniques such as PCR with specific primers designed from the contigs, followed by Sanger sequencing (see Sect. 8.3.3.2). Also is important to have quality control of an assembly. Inspector is an innovative tool specifically designed to evaluate and correct genome assemblies generated by long-read sequencing; it has high accuracy in small genomes, making it ideal for fruit plant viruses, which are smaller (≤~30 kb) (Chen et al. 2021).

8.5 Annotation Frameworks

8.5.1 The Critical Role of Reference Databases

Reference databases are essential for the study of emerging viruses, but their role goes beyond mere direct identification. While a new virus is not listed in these databases, they do provide the genomic and evolutionary context essential for classifying and characterizing an emerging virus. Furthermore, these databases are crucial for distinguishing between a truly novel virus and a host-jumping virus. There is a wide variety of databases; however, five have been compiled that are key to this context: the National Center for Biotechnology Information (NCBI) Non-redundant (NR)/Nucleotide (NT), Reference Sequence Database (RefSeq), the Reference Viral Database (RVDB), and the Virus-Host Database (Virus-Host DB). NCBI NR groups non-redundant protein sequences, while NT contains nucleotide sequences largely derived from GenBank. A massive volume of data can be managed, which increases the likelihood of finding a match. GenBank itself serves as the main database for raw genetic sequences submitted by researchers and represents the most comprehensive source of original sequence data, albeit without systematic curation (Benson et al. 2013). NCBI NR/NT databases were constructed to reduce sequence redundancy and improve BLAST searches, however, they maintain the limitations of their primary source, GenBank. As consequence, both NR and NT still contain uncurated entries, annotation errors, redundancies, and contaminating host and vector sequences, leading to false positives and erroneous taxonomic assignments (Benson et al. 2005). Thus, while GenBank provides the broadest source of original and diverse sequence data, and NR/NT offers large, generalist collections optimized for similarity searches, neither should be considered fully curated. RefSeq is a database curated by NCBI, providing reference sequences for genomes, transcripts, and proteins. The virus collection is of higher quality compared to NCBI NR/NT (Pruitt et al. 2005). In contrast, RVDB is a derivative database specializing exclusively in viral genomes and is built from RefSeq, but with an additional cleaning process to remove non-viral sequences, redundancies, and contaminants, making it a high-confidence database (Goodacre et al. 2018). Finally, there is the Virus-Host DB. It specializes in virus-host associations and organizes taxonomic ID pairs from NCBI for viruses and their hosts, integrating information from multiple sources such as RefSeq, GenBank, UniProt, and ViralZone in a specific location. Additionally, manual curation is performed based on literature. Because it allows searches to be filtered by specific host, it accelerates the identification of viruses relevant to a fruit crop under study, in addition to helping assess the potential host range of a new virus (Mihara et al. 2016). While all these databases can be used with respect to emerging viral threats, the use of curated, non-redundant databases should be prioritized over general ones. Furthermore, depositing correct sequences of new fruit viruses in GenBank or RefSeq is essential to enrich these databases and turn them into increasingly powerful tools. Table 8.2 shows the comparison of different databases focused on viral annotation.

Table 8.2 Reference databases for viral annotation

Characteristic	NCBI NR/NT	RefSeq	RVDB	Virus-host DB
Type and coverage	Generalist and no curated: aggregates all public NT and NR sequences	Curated subset of NCBI: provides reference sequences for genomes, transcripts and proteins, including viruses	Curated, nonredundant, and specialized: exclusive for viral genomes (derived from RefSeq with additional cleaning)	Specialized database focused on virus-host interactions
Main advantage	Massive data volume: ideal for broad initial search	Improved quality and reliability: integrated into NCBI's ecosystem	High specificity and sensitivity: gold standard for viral metagenomics and minimizes false positives (contaminants removal)	Ecological context: allows filtering of viruses by host
Main limitation	High noise: contains uncured sequences, annotation errors, redundancies, and contaminants	Low comprehensive: the curation process can be slower	Slight delay: while including the most recent viruses from RefSeq	Interaction-focused: must be used alongside a genomic database
Use in plant virology.	Exploratory initial search	General search of viruses with a basic level of quality and reliability	Metagenomic screening, diagnosis, viral discovery	Studies on host range, epidemiology, and risk assessment
Access	Via NCBI: search with BLAST/ DIAMOND	Via NCBI: search with BLAST/ DIAMOND	Own website	Own website
Advice for emerging viruses	Avoid its use for dedicated virus analysis	Acceptable but suboptimal usage	Database of choice. Crucial for classifying emerging viruses	Essential complement once the viral sequence is identified

8.5.2 Structural Annotation of Viral Genomes

Structural annotation is essential for deciphering the encoded information of a newly assembled viral genome. Precise identification of fundamental components, like open reading frames (ORFs), is crucial. GeneMark Self-Training with External Training (GeneMarkS-ET) is an advanced algorithm for gene prediction in eukaryotic and viral genomes. The main innovation that characterizes this tool is its use of a self-training mechanism, enabling it to generate species-specific models from an assembled genome. This is valuable for emerging viruses because it does not depend on prior information. Furthermore, it adds transcriptomic evidence (RNA-seq), allowing for improved genetic predictions in novel genomes. Its self-learning capability makes it valuable for the initial characterization of unknown pathogens (Besemer et al. 2001; Lomsadze et al. 2014).

8.5.3 Annotation Functional Augmented by AI

The functional annotation of the viral genome has evolved thanks to AI tools, which allow for the highly accurate prediction of functional elements, even for novel or highly divergent sequences. DeepGOPlus is a CNN-based gene function predictor. It is a model capable of directly analyzing amino acid sequences (proteins) and predicting their gene ontology (GO) terms, thus describing their functions. It was trained on thousands of annotated protein sequences from varied species, allowing it to assign functions to novel viral proteins that a traditional tool would overlook. This information allows us to learn more about new viruses and characterize them in a much more comprehensive way (Kulmanov and Hoehndorf 2020). Furthermore, it is essential to consider the functional elements of noncoding RNA (ncRNA). Infernal uses Markov Covariate (MC) models, which are statistical models that capture not only sequence conservation but also the conservation of RNA secondary structure. This tool can detect highly divergent ncRNAs in novel viruses based on their conserved structure (Nawrocki and Eddy 2013).

8.5.4 Taxonomic Classification and Evolution

Understanding the evolution and phylogeny of an emerging virus is vital for predicting its behavior and origin, and specific bioinformatics tools exist for this purpose. IQ-TREE is a reliable phylogenetic inference software that has been adopted for its accuracy, speed, and versatility (Nguyen et al. 2015). It can construct accurate phylogenetic trees thanks to its advanced algorithms, such as UFBoot2 and ModelFinder, which help determine the optimal nucleotide or amino acid substitution model. Its efficiency allows for the analysis of large datasets (complete metagenomes) and the handling of mixtures of evolutionary models, which is quite useful for recombinant viruses. More recently, mixture across sites and trees (MAST) models have been integrated, allowing for the modeling of different evolutionary histories in different sections of the genome (Nguyen et al. 2015). Model Teller is a machine learning-based classification tool that automates and improves the selection of the optimal nucleotide substitution model for each case. Instead of using traditional statistical methods, this tool is trained using thousands of alignments to predict the best-fitting model, resulting in a more realistic evolutionary model and, therefore, a more reliable phylogram (Abadi et al. 2020). Lastly, PhyloGAN is cutting-edge for phylogenetic differentiation due to the usage of generative adversarial networks (GANs). It learns to generate phylogenetic trees from large or complex datasets and can help resolve potentially ambiguous evolutionary relationships between closely related viruses when analyzed using traditional methods (Zou et al. 2021).

8.5.5 Integrated Annotation and Visualization Platforms

Bioinformatics analysis is always best accompanied by visual support. For many researchers, especially those without extensive bioinformatics expertise, there are integrated platforms that combine multiple tools into a graphical interface. Geneious is an all-in-one commercial software suited for biological data analysis. It integrates tools for annotation, alignment, phylogeny (plugins for IQ-TREE and RAxML), and results visualization. It also allows for manual and automatic genome annotation, as well as the management of all data and results in a single project. It includes a user-friendly and intuitive interface that allows the user to perform complex informatics workflows from raw sequence analysis to annotation and phylogeny analysis, without the need for coding, thus streamlining the characterization process (Kearse et al. 2012). A similar software package is the CLC Genomics Workbench from QIAGEN. This platform offers workflows for *de novo* assembly, read mapping, genome annotation, and phylogenic analysis. Its design facilitates tasks such as variant identification, gene expression analysis, and structural and functional annotation of new genomes. It includes a microbiology module useful for analyzing viromes (https://digitalinsights.qiagen.com/products-overview/discovery-insights-portfolio/qiagen-clc-genomics/). In Table 8.3, a summary of all the bioinformatics tools discussed in this chapter.

8.6 Tackling Complex Annotation Challenges

Annotating viral genomes from viral metagenomic studies presents unique challenges that go beyond ORF prediction. Many factors, such as sample complexity, unique features of viral biology, and technical limitations of the sequencing platforms themselves, have led to the need to develop specific strategies to avoid misinterpretations.

8.6.1 Endogenous Viral Elements (EVEs)

EVEs are remnants of viral genomes integrated into the host genome throughout evolution (Brait 2025). Their presence is a challenge, as they can be actively transcribed and, during analysis, can easily be confused with active exogenous viral infections. Although they are most common from retroviruses, EVEs derived from viruses of all types have been found (Vassilieff et al. 2023). After integration, EVEs can accumulate mutations, deletions, and rearrangements; however, some may retain intact ORFs, being transcribed by the host. This can lead to overestimation of viral diversity and false detection of emerging viruses (Brait et al. 2023). The main

Table 8.3 Summary of bioinformatic tools by workflow step

Workflow step	Tool	Function
Quality control	FastQC	Visual quality of raw sequencing data
Read trimming and filtering	Trimmomatic	Trims adapters and low-quality bases from short reads
	Porechop	Trims barcoding adapters from Nanopore reads
	Cutadapt	Finds and removes adapters in short and long reads
Computational enrichment	Bowtie2	Aligns and filters out host reads
	HoCoRT	Specifically filters out host plant RNA sequences
Detection and identification	VirFind	Integrated web pipeline for virus detection from contigs
	VSD Toolkit	Pipeline for viral detection from small RNA-Seq data
	Viroscope	Validates actively replicating viruses using VGAC
	PhytoPipe	Workflow for identifying and diagnosing plant pathogens using RNA-seq data
	VIGA	Identifies and assembles viral genomes from HTS data
	VirHunter	Detect novel viruses using CNN
Genome assembly	SPAdes (metaSPAdes)	A *de novo* assembly tool designed for short reads in metagenomic samples
	SPAdes (rnaSPAdes)	*De novo* assembler for RNA-seq data
	Canu	Long-read assembler for Nanopore/PacBio data
	Flye	Long-read assembler for complex genomes
	VIGA	Automated pipeline for eukaryotic virus genome assembly
	VirFind	Includes a *de novo* assembly module in its pipeline
	VSD Toolkit	Includes assembly workflows for small RNA-seq data
	Inspector	Long-read assembly quality control
Reference databases	NCBI NR/NT	Generalist databases
	RefSeq	Curated, nonredundant NCBI database for viruses
	RVDB	Curated viral reference database without contaminants
	Virus-Host DB	Specialized database for virus-host interactions

(continued)

Table 8.3 (continued)

Workflow step	Tool	Function
Annotation/phylogenetic reconstruction	GeneMarkS-ET	Predicts ORFs and genes in eukaryotic/viral genomes
	DeepGOPlus	Predicts protein function using neural networks
	IQ-TREE	Infers maximum likelihood phylogenetic trees
	ModelTeller	Determines the most suitable evolutionary model via machine learning techniques
	PhyloGAN	Infers phylogenetic trees using generative adversarial networks
Integrated annotation and visualization	Geneious	Unified commercial platform for annotation and visualization
	CLC Genomics Workbench	Commercial suite for integrated HTS analysis and annotation

method for reduction involves aligning the assembled contigs with the host genome to identify and remove any sequences that show 100% alignment, which would be defined as EVEs. The disadvantage of this strategy is that, whether the host reference contains annotated or unannotated EVEs, the filtering process may exclude valuable exogenous viral sequences like EVEs. Other options include integration markers, analyzing mutation patterns, and experimental validation (electron microscopy or isolation in culture) (Brait et al. 2023).

8.6.2 Ribosomal Frameshifting and Subgenomic RNAs

A variety of viruses employ complex gene expression strategies that go beyond standardized annotation algorithms. It is of utmost importance to consider these strategies for proper annotation. These include ribosomal frameshifting and the production of subgenomic RNAs to translate individual proteins. Annotation tools can overlook these strategies or annotate them incorrectly. Essential proteins can be missed, or the genomic organization of an emerging virus can be misinterpreted, leading to mischaracterization (Kelly et al. 2021). This can be addressed by using ORF predictors that consider frameshifting, such as GeneMarkS-ET, as well as by using browsers such as Geneious or CLC to manually search for expression patterns and subgenomic promoter signals. Finally, the use of targeted RT-PCR will allow experimental validation of the expression of overlapping ORFs or frameshift products (Kelly et al. 2021).

8.6.3 Metagenomic Shortcomings

De novo assembly from complex metagenomes, such as leaves or soil samples, can generate incomplete genomes, chimeric contigs (recombination of sequences originating from two different viruses or from a virus and host), or assembly errors in repetitive regions. Incomplete genomes severely hinder accurate taxonomic classification or functional understanding of the virus, while chimeric contigs develop molecules not found in nature, leading to erroneous biological interferences. Such oversights can result in the characterization of a virus that never existed or the publication of a complete genome with incorrect conclusions regarding its genetic content (Mende et al. 2012). To resolve this, prior validation is necessary. Using long-read sequencing helps close genomes and resolve the issue of repeats. In addition, employing an Inspector to evaluate the quality and full coverage of assembled genomes is an excellent choice. Finally, experimental validation is conducted using techniques such as Rapid Amplification of cDNA Ends (RACE) to obtain the 5′ and 3′ ends of the genome accurately and in synergy with Sanger sequencing to verify ambiguities.

8.7 Translational Applications: From Lab to Field

The use of methods, tools, techniques, and pipelines used for the characterization and annotation of emerging viruses, as well as the findings obtained from bioinformatics analysis, does not end there. Strategies for their use in the field (real-life application) must be considered to prevent the loss of fruit plants, which are important to the agricultural industry.

8.7.1 Development of Diagnoses

HTS has enabled the rapid identification of emerging viral pathogens in fruit trees, such as CTV and ASPV; subsequently, these technologies allow the design of specific assays, including qPCR and dPCR, for accurate detection in field samples. For example, a dPCR assay has been developed for sensitive quantification of ASPV in in vitro propagated apple plantlets, showing higher sensitivity compared to conventional qPCR (Kim et al. 2023). Furthermore, portable sequencing technologies such as ONT's MinION have revolutionized on-site diagnostics, allowing for almost immediate, real-time virus detection and reducing the time compared to other platforms. A dsRNA sequencing protocol was introduced in the work of Javaran et al. (2023) for the real-time detection of viruses in grapevine samples in vineyards. Similarly, LAMP provides a rapid and portable alternative for on-site virus detection that combines speed and specificity. A rapid and cost-effective LAMP assay was developed in the recent study by Matić and Myrta (2024), to detect American plum line pattern virus (APLPV) in flowering cherry. Employing unprocessed plant extracts without needing RNA preparation, this method proved more sensitive relative to qPCR and can be completed in under 20 min, making it ideal for deployment in the field.

8.7.2 Genetic Resistance

Genomic information from HTS is essential for developing genetic resistance strategies in fruit trees. This can be achieved using RNA interference (RNAi) or gene-editing systems such as CRISPR, which allow for the silencing or modification of viral genes vital for replication or movement. This strategy enables us to develop varieties resistant to multiple viruses, minimizing reliance on pesticides or other control measures.

8.7.3 Sustainable Management

Genomic knowledge enables the development of smart and sustainable management strategies. For example, genomic sequencing can reveal genes involved in transmission by biological vectors, such as insects, opening the opportunity to prevent the spread of emerging viruses in orchards by blocking the functional connections between viral proteins and the vector. Attenuated viruses can be generated for this purpose (Tomitaka et al. 2024). By sequencing and annotating several virus strains, specific mutations that create a weak or attenuated strain can be identified, potentially leading to the creation of plant vaccines.

8.8 Conclusions and Future Directions

Technological advancements in various tools have significantly boosted the field of plant virology in recent years. Combining sequencing with AI has radically changed the criteria for detecting and describing emerging viruses. The availability of portable ONT-based sequencing platforms has enabled efficient, real-time diagnosis in orchards. However, this technology remains inaccessible to some producers for several reasons. Tools like VirHunter, which uses machine learning, can overcome the limitations of homology-based methods for detecting emerging viruses. Their algorithms and programming enable them to learn intrinsic patterns in viral sequences, generating insights into viral diversity, ecology, and evolution. Even with these advances, significant challenges remain. While many databases are well-suited for their purpose, they often exhibit biases, particularly those containing reference data. The primary bias is the exclusion of human and animal viruses, leaving plant viruses underrepresented. Furthermore, there is a lack of standardized and adaptable bioinformatics annotation channels for different data types (short and long), contexts, and analyses, which can currently lead to inconsistent results and hinder effective comparisons between studies. Looking ahead, we foresee growth in technological innovation that can help bridge gaps and open new frontiers of research. Implementing single-cell viromics, quantum computing for structural prediction, and the use of AI in genomic and functional diagnosis and characterization are emerging technologies with broad potential. With these advances, it is possible to develop automated surveillance systems that allow for real-time metagenomic analysis of orchards using sensors like MinION, which also collect environmental and climatic data. These systems can be powered by AI, which integrates various tools, processes, algorithms, and/or databases to identify known and unknown viruses. Combining all these techniques and tools into a single system could improve the detection of emerging viruses, allowing producers to be alerted in time and enabling them to implement a rapid and effective phytosanitary response.

References

Abadi S, Azouri D, Pupko T, Mayrose I (2020) ModelTeller: model selection for optimal phylogenetic reconstruction using machine learning. Mol Biol Evol 37(11):3338–3350. https://doi.org/10.1093/molbev/msaa154

Adams IP, Glover RH, Monger WA et al (2018) First report of avocado sunblotch viroid (ASBVd) in Nigeria using small RNA sequencing. Virus Res 256:96–101. https://doi.org/10.1016/j.virusres.2018.08.005

Anaya-López JL, Chiquito-Almanza E, Acosta-Gallegos JA, Tosquy-Valle OH, Montero-Tavera V, González-Pérez E (2025) Diagnosis and characterization of plant viruses in Mexico using HTS: an approach to guide virus management and marker-assisted selection. Discov Plants 2(1):246. https://doi.org/10.1007/s44372-025-00332-y

Andrews S (2010) FastQC: a quality control tool for high throughput sequence data [Software]. Babraham Bioinformatics. https://www.bioinformatics.babraham.ac.uk/projects/fastqc/

Bandaru L, Irigireddy BC, Davis B (2024) DeepQC: a deep learning system for automatic quality control of in-situ soil moisture sensor time series data. Smart Agric Technol 8:100514. https://doi.org/10.1016/j.atech.2024.100514

Barrero RA, Napier KR, Cunnington J, Liefting L, Keenan S, Frampton RA, Szabó T, Bulman S, Hunter A, Ward L, Whattam M, Bellgard MI (2017) An internet-based bioinformatics toolkit for plant biosecurity diagnosis and surveillance of viruses and viroids. BMC Bioinf 18(1):26. https://doi.org/10.1186/s12859-016-1428-4

Benson DA, Karsch-Mizrachi I, Lipman DJ, Ostell J, Wheeler DL (2005) GenBank. Nucleic Acids Res 33(Database issue):D34–D38. https://doi.org/10.1093/nar/gki063

Benson DA, Cavanaugh M, Clark K, Karsch-Mizrachi I, Lipman DJ, Ostell J, Sayers EW (2013) GenBank. Nucleic Acids Research 41:D36–D42. https://doi.org/10.1093/nar/gks1195

Bertin S, Sybilska A, Luigi M et al (2025) Transmission of tomato fruit blotch virus by the tomato russet mite: epidemiological implications for an emerging/re-emerging tomato disease. Sci Rep 15:12079. https://doi.org/10.1038/s41598-025-97142-9

Besemer J, Lomsadze A, Borodovsky M (2001) GeneMarkS: a self-training method for prediction of gene starts in microbial genomes. Implications for finding sequence motifs in regulatory regions. Nucleic Acids Res 29(12):2607–2618. https://doi.org/10.1093/nar/29.12.2607

Bolger AM, Lohse M, Usadel B (2020) Trimmomatic: a flexible trimmer for Illumina sequence data. Sci Rep 10:13443. https://doi.org/10.1093/bioinformatics/btu170

Boufounos P, El-Difrawy S, Ehrlich D (2004) Basecalling using hidden Markov models. J Franklin Inst 341(1–2):23–36. https://doi.org/10.1016/j.jfranklin.2003.12.008

Brait N (2025) Endogenous viral elements and their influence on the complexity of genomic and transcriptomic data analysis. Thesis fully internal (DIV), University of Groningen. https://doi.org/10.33612/diss.1329049538

Brait N, Hackl T, Morel C, Exbrayat A, Gutierrez S, Lequime S (2023) A tale of caution: how endogenous viral elements affect virus discovery in transcriptomic data. Virus Evol 10(1):vead088. https://doi.org/10.1093/ve/vead088

Bushmanova E, Antipov D, Lapidus A, Prjibelski AD (2019) rnaSPAdes: a de novo transcriptome assembler and its application to RNA-Seq data. Gigascience 8(9):giz100. https://doi.org/10.1093/gigascience/giz100

Chen Y, Zhang Y, Wang AY et al (2021) Accurate long-read de novo assembly evaluation with inspector. Genome Biol 22:312. https://doi.org/10.1186/s13059-021-02527-4

Costa LC, Atha B III, Hu X, Lamour K, Yang Y, O'Connell M, McFarland C, Foster JA, Hurtado-Gonzales OP (2022) High-throughput detection of a large set of viruses and viroids of pome and stone fruit trees by multiplex PCR-based amplicon sequencing. Front Plant Sci 13:1072768. https://doi.org/10.3389/fpls.2022.1072768

Davis S (2024) Transforming basecalling in genomic sequencing. Oxford Nanopore Technologies Blog. https://nanoporetech.com/blog/transforming-basecalling-in-genomic-sequencing

Ertunc F (2020) Emerging plant viruses. In: Ertunc F (ed) Emerging and reemerging viral pathogens. Academic Press, pp 1041–1062. https://doi.org/10.1016/B978-0-12-819400-3.00046-6

Fu P, Wu Y, Zhang Z, Qiu Y, Wang Y, Peng Y (2023) VIGA: a one-stop tool for eukaryotic virus identification and genome assembly from next-generation-sequencing data. Brief Bioinform 25(1):bbad444. https://doi.org/10.1093/bib/bbad444

Goodacre N, Aljanahi A, Nandakumar S, Mikailov M, Khan AS (2018) A reference viral database (RVDB) to enhance bioinformatics analysis of high-throughput sequencing for novel virus detection. mSphere 3(2):e00069-18. https://doi.org/10.1128/mSphere.00069-18

Hamim I, Sekine KT, Komatsu K (2022) How do emerging long-read sequencing technologies function in transforming the plant pathology research landscape? Plant Mol Biol 110(6):469–484. https://doi.org/10.1007/s11103-022-01305-5

Ho T, Tzanetakis IE (2014) Development of a virus detection and discovery pipeline using next generation sequencing. Virology 471:54–60. https://doi.org/10.1016/j.virol.2014.09.019

Hou W, Li S, Massart S (2020) Is there a "biological desert" with the discovery of new plant viruses? A retrospective analysis for new fruit tree viruses. Front Microbiol 11:592816. https://doi.org/10.3389/fmicb.2020.592816

Hu X, Hurtado-Gonzales OP, Adhikari BN, French-Monar RD, Malapi M, Foster JA, McFarland CD (2023) PhytoPipe: a phytosanitary pipeline for plant pathogen detection and diagnosis using RNA-seq data. BMC Bioinf 24(1):470. https://doi.org/10.1186/s12859-023-05589-2

Ibaba JD, Gubba A (2020) High-throughput sequencing application in the diagnosis and discovery of plant-infecting viruses in Africa, a decade later. Plants 9(10):1376. https://doi.org/10.3390/plants9101376

Ilie L, Molnar M (2013) RACER: rapid and accurate correction of errors in reads. Bioinformatics 29(19):2490–2493. https://doi.org/10.1093/bioinformatics/btt407

Javaran VJ, Bouchard D, Bélanger RR (2023) NanoViromics: long-read sequencing of dsRNA for plant virus and viroid detection in grapevines using Oxford Nanopore Technologies MinION. Front Microbiol 14:1192781. https://doi.org/10.3389/fmicb.2023.1192781

Jiao J, Kong K, Han J, Song S, Bai T, Song C, Wang M, Yan Z, Zhang H, Zhang R, Feng J, Zheng X (2021) Field detection of multiple RNA viruses/viroids in apple using a CRISPR/Cas12a-based visual assay. Plant Biotechnol J 19(2):394–405. https://doi.org/10.1111/pbi.13474

Kanapiya A, Amanbayeva U, Tulegenova Z, Abash A, Zhangazin S, Dyussembayev K, Mukiyanova G (2024) Recent advances and challenges in plant viral diagnostics. Front Plant Sci 15:1451790. https://doi.org/10.3389/fpls.2024.1451790

Kang CM, Kim MJ, Hong JS, Jeong RD (2025) Managing plant viruses in tissue-cultured apple and grapevine: strategies for detection and eradication. Plant Pathol J 41(5):545–565. https://doi.org/10.5423/PPJ.RW.07.2025.0092

Karimi M, Ghorbani A, Niazi A, Rostami M, Tahmasebi A (2025) CRISPR-Cas13a as a next-generation tool for rapid and precise plant RNA virus diagnostics. Plant Methods 21(1):83. https://doi.org/10.1186/s13007-025-01401-9

Katsiani A, Stainton D, Lamour K, Tzanetakis IE (2020) The population structure of rose rosette virus in the USA. J Gen Virol 101(6):676–684. https://doi.org/10.1099/jgv.0.001418

Kearse M, Moir R, Wilson A, Stones-Havas S, Cheung M, Sturrock S, Buxton S, Cooper A, Markowitz S, Duran C, Thierer T, Ashton B, Meintjes P, Drummond A (2012) Geneious basic: an integrated and extendable desktop software platform for the organization and analysis of sequence data. Bioinformatics 28(12):1647–1649. https://doi.org/10.1093/bioinformatics/bts199

Kelly JA, Woodside MT, Dinman JD (2021) Programmed −1 ribosomal frameshifting in coronaviruses: a therapeutic target. Virology 554:75–82. https://doi.org/10.1016/j.virol.2020.12.010

Kim SW, Lee HJ, Cho KH, Jeong RD (2022) Detection and quantification of apple stem grooving virus in micropropagated apple plantlets using reverse-transcription droplet digital PCR. Plant Pathol J 38(4):417–422. https://doi.org/10.5423/PPJ.NT.05.2022.0072

Kim SW, Lee HJ, Cho SY, Jeong RD (2023) Development and preliminary application of a reverse-transcription droplet digital PCR assay for detection and quantification of apple stem pitting virus in in vitro propagated apple plantlets. Sci Hortic 321:112363. https://doi.org/10.1016/j.scienta.2023.112363

Kolmogorov M, Yuan J, Lin Y, Pevzner PA (2019) Assembly of long, error-prone reads using repeat graphs. Nat Biotechnol 37(4):540–546. https://doi.org/10.1038/s41587-019-0072-8

Koren S, Walenz BP, Berlin K, Miller JR, Bergman NH, Phillippy AM (2017) Canu: scalable and accurate long-read assembly via adaptive k-mer weighting and repeat separation. Genome Res 27(5):722–736. https://doi.org/10.1101/gr.215087.116

Kulmanov M, Hoehndorf R (2020) DeepGOPlus: improved protein function prediction from sequence. Bioinformatics 36(2):422–429. https://doi.org/10.1093/bioinformatics/btz595

Kutnjak D, Tamisier L, Adams I, Boonham N, Candresse T, Chiumenti M, De Jonghe K, Kreuze JF, Lefebvre M, Silva G, Malapi-Wight M, Margaria P, Mavrič Pleško I, McGreig S, Miozzi L, Remenant B, Reynard J-S, Rollin J, Rott M, Schumpp O, Massart S, Haegeman A (2021) A primer on the analysis of high-throughput sequencing data for detection of plant viruses. Microorganisms 9(4):841. https://doi.org/10.3390/microorganisms9040841

Langmead B, Salzberg SL (2012) Fast gapped-read alignment with bowtie 2. Nat Methods 9(4):357–359. https://doi.org/10.1038/nmeth.1923

Leiva AM, Siriwan W, Lopez-Alvarez D, Barrantes I, Hemniam N, Saokham K, Cuellar WJ (2020) Nanopore-based complete genome sequence of a Sri Lankan cassava mosaic virus (Geminivirus) strain from Thailand. Microbiol Resour Announc 9(6):10–1128. https://doi.org/10.1128/mra.01274-19

Lomsadze A, Burns PD, Borodovsky M (2014) Integration of mapped RNA-Seq reads into automatic training of eukaryotic gene finding algorithm. Nucleic Acids Res 42(15):e119. https://doi.org/10.1093/nar/gku557

Luigi M, Manglli A, Corrado CL, Tiberini A, Costantini E, Ferretti L, Tomassoli L, Bertin S (2023) Development, validation, and application of reverse transcription real-time and droplet digital PCR assays for the detection of the Potyviruses watermelon mosaic virus and zucchini yellow mosaic virus in cucurbits. Plants 12(12):2364. https://doi.org/10.3390/plants12122364

Maina S, Donovan NJ, Plett K, Bogema D, Rodoni BC (2024) High-throughput sequencing for plant virology diagnostics and its potential in plant health certification. Front Hortic 3:1388028. https://doi.org/10.3389/fhort.2024.1388028

Martin M (2011) Cutadapt: remove adapter sequences from high-throughput sequencing reads. EMBnet J 17(1):10–12. https://doi.org/10.14806/ej.17.1.200

Massart S, Adams I, Al Rwahnih M, Baeyen S, Bilodeau GJ, Blouin AG et al (2022) Guidelines for the reliable use of high throughput sequencing technologies to detect plant pathogens and pests. Peer Community J 2:e62. https://doi.org/10.24072/pcjournal.181

Matić S, Myrta A (2024) Development of loop-mediated isothermal amplification (LAMP) assay for in-field detection of American plum line pattern virus. Viruses 16(10):1572. https://doi.org/10.3390/v16101572

Mehetre GT, Leo VV, Singh G, Sorokan A, Maksimov I, Yadav MK, Upadhyaya K, Hashem A, Alsaleh AN, Dawoud TM, Almaary KS, Singh BP (2021) Current developments and challenges in plant viral diagnostics: a systematic review. Viruses 13(3):412. https://doi.org/10.3390/v13030412

Mehraj I, Mehraj I, Iralu N, Kawoosa F, Wani S, Saleem S, Hamid A (2025) Application of digital PCR (dPCR) in plant virus detection. In: Detection of plant viruses: advanced techniques. Springer US, New York, pp 149–179. https://doi.org/10.1007/978-1-0716-4390-7_26

Mende DR, Waller AS, Sunagawa S, Järvelin AI, Chan MM, Arumugam M, Raes J, Bork P (2012) Assessment of metagenomic assembly using simulated next generation sequencing data. PLoS One 7(2):e31386. https://doi.org/10.1371/journal.pone.0031386

Mihara T, Nishimura Y, Shimizu Y, Nishiyama H, Yoshikawa G, Uehara H, Hingamp P, Goto S, Ogata H (2016) Linking virus genomes with host taxonomy. Viruses 8(3):66. https://doi.org/10.3390/v8030066

Moubset O, François S, Maclot F, Palanga E, Julian C, Fernandez E, Rott P, Daugrois J-H, Antoine-Lorquin A, Bernardo P, Blouin AG, Temple C, Kraberger S, Fontenele RS, Harkins GW, Ma Y, Marais A, Candresse T, Ben Chéhida S, Lefeuvre P, Lett J-M, Varsani A, Massart S, Ogliastro M, Martin DP, Filloux D, Roumagnac P (2022) Virion-associated nucleic acid-based metagenomics: a decade of advances in molecular characterization of plant viruses. Phytopathology 112(11):2253–2272. https://doi.org/10.1094/PHYTO-03-22-0096-RVW

Naito FY, Melo FL, Fonseca MEN, Santos CA, Chanes CR, Ribeiro BM, Gilbertson R, Boiteux L, de Cássia Pereira-Carvalho R (2019) Nanopore sequencing of a novel bipartite New World begomovirus infecting cowpea. Arch Virol 164(7):1907–1910. https://doi.org/10.1007/s00705-019-04254-5

Nawrocki EP, Eddy SR (2013) Infernal 1.1: 100-fold faster RNA homology searches. Bioinformatics 29(22):2933–2935. https://doi.org/10.1093/bioinformatics/btt509

Nguyen LT, Schmidt HA, von Haeseler A, Minh BQ (2015) IQ-TREE: a fast and effective stochastic algorithm for estimating maximum-likelihood phylogenies. Mol Biol Evol 32(1):268–274. https://doi.org/10.1093/molbev/msu300

Nizamani MM, Zhang Q, Muhae-Ud-Din G, Wang Y (2023) High-throughput sequencing in plant disease management: a comprehensive review of benefits, challenges, and future perspectives. Phytopathol Res 5(1):1–17. https://doi.org/10.1186/s42483-023-00199-5

Nurk S, Meleshko D, Korobeynikov A, Pevzner PA (2017) metaSPAdes: a new versatile metagenomic assembler. Genome Res 27(5):824–834. https://doi.org/10.1101/gr.213959.116

Pacheco-Dorantes C et al (2025) Unleashing the potential of high-throughput sequencing for plant virus and viroid detection in Mexico. Front Microbiol 16:1603010. https://doi.org/10.3389/fmicb.2025.1603010

Pacific Biosciences of California, Inc. (2022a) PacBio and Google collaborate to use machine learning to optimize long-read sequencing data analysis [Press release]. PacBio. https://www.pacb.com/press_releases/pacbio-and-google-collaborate-to-use-machine-learning-to-optimize-long-read-sequencing-data-analysis/

Pacific Biosciences of California, Inc. (2022b) PacBio announces Revio, a revolutionary new long-read sequencing system designed to provide 15 times more HiFi data and human genomes at scale for under $1,000 [Press release]. PacBio. https://www.pacb.com/press_releases/pacbio-announces-revio-a-revolutionary-new-long-read-sequencing-system-designed-to-provide-15-times-more-hifi-data-and-human-genomes-at-scale-for-under-1000/

Pruitt KD, Tatusova T, Maglott DR (2005) NCBI reference sequence (RefSeq): a curated non-redundant sequence database of genomes, transcripts and proteins. Nucleic Acids Res 33(Database issue):D501–D504. https://doi.org/10.1093/nar/gki025

Rao GP, Reddy MG (2020) Overview of yield losses due to plant viruses. In: Rao GP (ed) Applied plant virology. Academic Press, pp 531–562. https://doi.org/10.1016/B978-0-12-818654-1.00038-4

Rithesh L, Paul A, Amanthra Keloth MR, Jose S, Kumar A (2024) Detection and identification of plant viruses, viroids, and phytoplasma based on high-throughput molecular approaches. In: Chen J-T, Khan M, Parveen A, Patra JK (eds) Molecular and biotechnological tools for plant disease management. Springer Nature Singapore, pp 235–256. https://doi.org/10.1007/978-981-97-7510-1_8

Roy SD, Ramasamy S, Obbineni JM (2024) An evaluation of nucleic acid-based molecular methods for the detection of plant viruses: a systematic review. Virusdisease 35(2):357–376. https://doi.org/10.1007/s13337-024-00863-0

Rubio L, Galipienso L, Ferriol I (2020) Detection of plant viruses and disease management: relevance of genetic diversity and evolution. Front Plant Sci 11:1092. https://doi.org/10.3389/fpls.2020.01092

Rumbavicius I, Rounge TB, Rognes T (2023) HoCoRT: host contamination removal tool. BMC Bioinf 24(1):371. https://doi.org/10.1186/s12859-023-05492-w

Schoenfeld T, Wylie KM, Thurber RV (2023) Digital PCR for accurate quantification of pathogens: applications in viral ecology. Nat Rev Microbiol 21(4):215–230. https://doi.org/10.1016/j.ijbiomac.2021.06.132

Selvaraj V, Maheshwari Y, Hajeri S, Yokomi R (2019) A rapid detection tool for VT isolates of citrus tristeza virus by immunocapture-reverse transcriptase loop-mediated isothermal amplification assay. PLoS One 14(9):e0222170. https://doi.org/10.1371/journal.pone.0222170

Shu J, Tan Q, Huang Z, Zhang T, Ye L, Fu S, Mao Z (2025) One-pot one-step detection platform for severe fever with thrombocytopenia syndrome virus via the CRISPR/Cas12a detection system. Virol J 22(1):203. https://doi.org/10.1186/s12985-025-02772-0

Silveira ATL, dos Reis Figueira A, Gontijo GR et al (2024) Incidence and spatial pattern of viruses in commercial dragon fruit orchards in southern Minas Gerais, Brazil. Eur J Plant Pathol 170:15–24. https://doi.org/10.1007/s10658-024-02877-2

Sukhorukov G, Khalili M, Gascuel O, Candresse T, Marais-Colombel A, Nikolski M (2022) VirHunter: a deep learning-based method for detection of novel RNA viruses in plant sequencing data. Front Bioinf 2:867111. https://doi.org/10.3389/fbinf.2022.867111

Tanny T, Sallam M, Soda N, Nguyen NT, Alam M, Shiddiky MJ (2023) CRISPR/Cas-based diagnostics in agricultural applications. J Agric Food Chem 71(31):11765–11788. https://doi.org/10.1021/acs.jafc.3c00913

Tomitaka Y, Shimomoto Y, Ryang B-S, Hayashi K, Oki T, Matsuyama M, Sekine K-T (2024) Development and application of attenuated plant viruses as biological control agents in Japan. Viruses 16(4):517. https://doi.org/10.3390/v16040517

UMA Technology (2025) Trends in computer hardware and software. UMA Technology. https://umatechnology.org/trends-in-computer-hardware-and-software/

Valenzuela SL, Norambuena T, Morgante V, García F, Jiménez JC, Núñez C, Fuentes I, Pollak B (2022) Viroscope: plant viral diagnosis from high-throughput sequencing data using biologically-informed genome assembly coverage. Front Microbiol 13:967021. https://doi.org/10.3389/fmicb.2022.967021

Vassilieff H, Geering ADW, Choisne N, Teycheney P-Y, Maumus F (2023) Endogenous caulimovirids: fossils, zombies, and living in plant genomes. Biomolecules 13(7):1069. https://doi.org/10.3390/biom13071069

Venbrux M, Crauwels S, Rediers H (2023) Current and emerging trends in techniques for plant pathogen detection. Front Plant Sci 14:1120968. https://doi.org/10.3389/fpls.2023.1120968

Villamor DEV, Ho T, Al Rwahnih M et al (2019) High throughput sequencing for plant virus detection and discovery. Phytopathology 109(5):716–725. https://doi.org/10.1094/PHYTO-07-18-0257-RVW

Wang L, Zhao L, Zhang X et al (2022) Rapid and visual detection of Actinidia virus A using reverse transcription loop-mediated isothermal amplification (RT-LAMP) in kiwifruit. Front Microbiol 13:882346. https://doi.org/10.3389/fmicb.2022.882346

Wick RR, Judd LM, Gorrie CL, Holt KE (2017) Completing bacterial genome assemblies with multiplex MinION sequencing. Microb Genom 3(10):e000132. https://doi.org/10.1099/mgen.0.000132

Xiao H, Li C, Al Rwahnih M, Dolja V, Meng B (2019) Metagenomic analysis of Riesling grapevine reveals a complex virome including two new and divergent variants of Grapevine leafroll-associated virus 3. Plant Disease 103(7):1275–1285. https://doi.org/10.1094/PDIS-09-18-1503-RE

Zhang Y, Xie Z, Fletcher JD, Wang Y, Wang R, Guo Z, He Y (2020) Rapid and sensitive detection of lettuce necrotic yellows virus and cucumber mosaic virus infecting lettuce (*Lactuca sativa* L.) by reverse transcription loop-mediated isothermal amplification. Plant Pathol J 36(1):76. https://doi.org/10.5423/PPJ.OA.12.2019.0298

Zhang Y, Li X, Wang Z et al (2023) Development of a multiplex reverse transcription loop-mediated isothermal amplification assay for simultaneous detection of plum pox virus and Prunus necrotic ringspot virus in stone fruits. Plant Pathol 72(5):891–901. https://doi.org/10.1111/ppa.13722

Zou Y, Xie J, Zheng H, Wang F (2021) PhyloGAN: data-driven phylogenetic inference using generative adversarial networks. IEEE/ACM Trans Comput Biol Bioinform 19(1):492–503. https://doi.org/10.1109/TCBB.2021.3089435

Utilizing Virome Analysis to Manage Viral Diseases in Fruit Crops

9

Tabia Fayaz, Sajad Un Nabi, Bushra Rasool,
Tabasum Manzoor, Bisma Majid, and Zakir Amin

Abstract

Virome analysis, made possible by developments in next-generation sequencing (NGS), enables comprehensive viral community detection and profiling in fruit crops. This technology enables thorough surveillance of all viral genomes within a sample, which facilitates the identification of both known and unknown plant viruses, including those that cause latent or symptomless infections that would elude conventional diagnostics. Virome analysis provides vital information for early pathogen detection, the study of viral population dynamics, and the comprehension of virus-host and virus-virus interactions in intricate crop systems by methodically characterizing viral diversity, abundance, and interactions. Crucially, combining virome analysis with conventional management approaches facilitates efficient quarantine procedures, timely control measure implementation, targeted resistance breeding, and thorough disease monitoring. Plant health experts can track viral evolution, anticipate and mitigate outbreaks, and even identify advantageous viral agents that improve crop stress tolerance with NGS-based virome profiling. Ongoing developments in sequencing platforms, bioinformatics, and data integration are making virome analysis a more accessible and potent tool for sustainable fruit crop production and global agricultural biosecurity, despite ongoing obstacles like high sequencing costs and technical complexity.

T. Fayaz (✉) · B. Majid
Division of Plant Pathology, SKUAST-K, Shalimar, Jammu & Kashmir, India

S. U. Nabi · B. Rasool · T. Manzoor · Z. Amin
Plant Virology Laboratory, Division of Fruit Science, ICAR-Central Institute of Temperate Horticulture, Srinagar, Jammu & Kashmir, India

R. Kumar et al. (eds.), *Fruit Crops and Viral Pathogens Volume II*, Plant Pathology for Sustainable Agriculture, https://doi.org/10.1007/978-981-95-7261-8_9

Keywords

Plant viral diseases · Viral diversity · Next-generation sequencing (NGS) · Virome analysis · Disease control · Agricultural biosecurity

9.1 Introduction

Fungi, bacteria, viruses, nematodes, and protozoa are just a few of the many microorganisms that make up plant pathogens, which are known to have a major detrimental effect on both agriculture and natural environments (Strange and Scott 2005). Many of these pathogens are noteworthy for their capacity to infect hosts in the early stages of growth without producing symptoms, which makes early detection and subsequent treatment more difficult. According to Fletcher et al. (2006), Mansfield et al. (2012), and Savary et al. (2019), these infections not only reduce crop yields but also endanger global food security, change ecological diversity, and disrupt agricultural communities' means of subsistence. In order to support sustainable crop production, sophisticated, quick, and accurate plant pathogen detection technologies are desperately needed, as evidenced by the continuous increase in global food demand (Mahlein 2016). Plant viruses stand out for posing particularly difficult problems within this wide range of pathogens. In contrast to their cellular counterparts, viruses are entirely dependent on their host plants for their life cycle and are made of genetic material enclosed in a protein shell. Early diagnosis is particularly challenging because their initial infections are usually symptomless. However, by impeding plant growth, lowering crop productivity, and lowering commodity quality, even small viral infestations can cause significant economic loss. The widespread effects of *Candidatus* Liberibacter asiaticus, which causes citrus greening or Huanglongbing, and *Xylella fastidiosa*, which is linked to problems in citrus, grapevines, and olives, are significant examples. For the health of fruit crops, these cases highlight the critical significance of contemporary diagnostic techniques and prompt intervention (Sicard et al. 2018; Jones 2021; Massart et al. 2017; Villamor et al. 2019).

Plant pathogen diagnostics have long been supported by conventional identification techniques like polymerase chain reaction (PCR), enzyme-linked immunosorbent assay (ELISA), and associated serological tests. These techniques are ineffective for mixed, latent, or recently discovered infections because they typically target one or a small group of viruses at a time, even though they are effective for known pathogens (Singhal et al. 2021). For instance, the limitations of traditional diagnostics have been highlighted by the detection and genome reconstruction of many viruses and viroids in apple cultivars, including asymptomatic infections, made possible by high-throughput RNA sequencing (Nabi et al. 2022). More inclusive and flexible detection methods are needed due to the increasing complexity and quantity of plant viruses. Virome analysis, a cutting-edge development in plant pathology, is one such remedy. Virome analysis has shown its ability to identify latent and mixed infections in fruit crops by successfully deciphering complex viral infections in plums, such as little cherry virus 1, plum bark necrosis stem pitting-associated virus, and American plum line pattern virus (Nabi et al. 2025). Using high-throughput sequencing (HTS) technology, this procedure methodically examines every viral genome found in a particular sample, known as the “virome.” Regardless of whether the viruses’ sequences are already known, such methods enable objective virus detection. Virome analysis reveals information about viral population diversity, population dynamics, evolutionary relationships, host interactions, and disease transmission in addition to identification (Roossinck 2012; Prabha et al. 2013; Pecman et al. 2017; Massart et al. 2017). As climate change alters the biotic and abiotic factors influencing modern agriculture, the need for rapid and precise plant disease detection is growing. Variations in temperature, precipitation, and the frequency of extreme events affect the life cycles and distribution of plant pathogens as well as the frequency of their insect vectors (Anderson et al. 2004). Therefore, in order to provide scalable and responsive monitoring to new threats, modern surveillance systems must adapt to these changes. Sustainable farming practices are based on accurate disease diagnosis. Timely and accurate detection tools are essential for integrative methods like precision agriculture, prudent pesticide application, and integrated pest management (IPM). By facilitating more thoughtful responses to threats to plant health, such strategies help maintain ecosystem health, maximize resources, and ensure long-term food supplies (Agrios 2005; Chakraborty and Newton 2011; Popp et al. 2013). Furthermore, reliable plant health surveillance is crucial for stopping the cross-border spread of dangerous or controlled pathogens as international trade in agricultural products grows. Strict adherence to phytosanitary and biosecurity regulations is essential to preventing the spread of plant diseases internationally, avoiding trade restrictions, and safeguarding national economies (FAO 2006; EPPO 2019). Plant pathogen detection has significantly improved since next-generation sequencing (NGS) was introduced. NGS platforms make it possible to analyze entire genomes quickly and affordably, changing the paradigm from targeted to community-wide methods for identifying a wide range of pathogens, including viruses, viroids, and more. These new sequencing techniques provide new insights into the interactions and evolution of microorganism communities in fruit crops by enabling the detection of multiple, even unknown, pathogens within a single screening, in contrast to conventional methods that require prior knowledge of likely agents (Mardis 2008; Barba et al. 2014; Massart et al. 2017; Baranwal et al. 2021; Nabi et al. 2021). The crucial significance of thorough pathogen detection is discussed in this chapter, with particular attention paid to the plant virome—the total of all plant-associated viruses, whether harmful, neutral, or advantageous. The composition of the microbiome, development, and host response to stress can all be impacted by these viral community members (Roossinck 2011; Rosario et al. 2015). Finding potential viral contributions to plant resilience and managing fruit crop diseases effectively depend on an understanding of this entire spectrum. Accurate virome profiling is crucial for characterization, transmission research, and the creation of sustainable management strategies, according to thorough reviews of viral threats to temperate pome fruits (Manzoor et al. 2025).

Plant pathogen identification has traditionally relied on PCR, serology, and visual inspections, but these techniques are inherently limited to what is visible or known. Loop-mediated isothermal amplification (LAMP), DNA microarrays, and, most importantly, NGS have made it possible to detect a much wider range of visible and hidden pathogens, including some that have never been identified before (Notomi et al. 2000; Barba et al. 2014; Prabha et al. 2013). In addition to improving detection, virome analysis allows the study of virus–virus and virus–host dynamics by using a broad-spectrum approach that targets entire viral communities. Surprisingly, some plant viruses do not cause overt signs of illness; in certain instances, they even increase the host's ability to withstand stress. These discoveries create new opportunities for plant protection and enhancement through advantageous viral collaborations (Roossinck 2011; Ghosh et al. 2022). It is more crucial than ever to integrate virome analysis with current monitoring techniques to combat ongoing plant disease outbreaks. Methodological developments, improved viral reference libraries, and advanced bioinformatics designed to analyze intricate datasets in the context of plant disease are necessary for this process. This chapter's primary objective is to summarize recent developments in plant pathogen detection, with a focus on the contributions of virome surveillance in fruit crops. In addition to examining conceptual, technical, and policy barriers, the discussion covers both current and historical diagnostic technologies and provides real-world examples of how viromic insights have enhanced sustainable disease management.

9.2 The Secret World of the Plant Virome and Plant Pathogens

A wide variety of microorganisms, such as nematodes, bacteria, fungi, and viruses, that affect plants through special infection and survival strategies are referred to as "plant pathogens." Developing successful disease management plans and promoting the sustainability of agricultural systems requires an understanding of their biology and taxonomy. Due to their intricate host interactions, quick adaptability, and significant impact on crop yields and ecosystem stability, plant viruses merit special attention (Hull 2014; Scholthof et al. 2011; Jones 2014). Plant viruses are infectious agents with RNA or DNA genomes that are shielded by a protein coat and, occasionally, a lipid envelope. Because they are obligatory intracellular organisms, they can only reproduce within plant cells and are entirely dependent on the resources of their hosts. Aphids, whiteflies, thrips, nematodes, fungi, direct physical damage, vegetative propagation (such as grafting and budding), and, less frequently, seeds or pollen are some of the ways they can spread from one plant to another (Hull 2014; Bragard et al. 2013). Mottled or yellowed leaves, necrosis, stunted growth, deformed fruits, and general decreased productivity are just a few of the symptoms that infected plants display (Marwal and Gaur 2020). Many viruses are known to have a significant negative economic impact on agriculture, significantly decreasing the market value and yield of numerous significant crops. Plum pox virus (*Potyvirus*), potato virus Y (*Potyvirus*), tomato yellow leaf curl virus (*Begomovirus*), cucumber mosaic virus (*Cucumovirus*), and tobacco mosaic virus (*Tobamovirus*) are notable examples (Roossinck 2015a, b; Scholthof et al. 2011). The entire viral community connected to a plant or its surroundings is known as the "virome," and it has recently become a central idea in microbial ecology and plant science. This intricate group includes not only pathogenic viruses but also those that don't cause any symptoms or could be helpful for the survival and adaptation of the plant. The study of the plant virome has been transformed by developments in high-throughput sequencing (HTS), which makes it possible to quickly identify both previously known and novel viruses on a large scale (Roossinck 2015a, b; Wu et al. 2015). Families including *Potyviridae*, *Geminiviridae*, *Tombusviridae*, *Partitiviridae*, and *Luteoviridae* are among those that exhibit diversity within a single host (Dulya and Koonin 2018). Such diversity necessitates comprehensive viewpoints on plant-virus interactions that encompass ecological and evolutionary factors in addition to the conventional emphasis on pathogenicity. The virome has a wide range of effects on both plant development and broader ecosystem functions. Some plant-associated viruses have mutualistic or even symbiotic relationships with their hosts; others do not cause disease. According to research, some viruses can improve nutrient uptake, boost resistance to environmental stressors like heat and drought, and modify plant traits to increase survivability in challenging circumstances (Roossinck 2015a, b; Xu et al. 2008). These findings suggest that viruses, which were previously thought to be only dangerous substances, may actually play a significant part in enhancing plant resilience and health. Plant virome activity has an impact on entire plant communities. Local plant diversity, community composition, and ecosystem functions can all be impacted by viral infection, which can change the competitive dynamics between species (Malmstrom et al. 2005; Power and Mitchell 2004). This illustrates the need to treat plant viruses as essential components of ecosystem networks rather than just as causes of disease. The plant virome reacts quickly to environmental changes. Viral diversity and distribution can be altered by human activities like changing land use, intensifying agriculture, and increasing international trade, in addition to climatic changes (Roossinck 2015a, b; Anderson et al. 2004). These changes highlight how crucial it is to continue monitoring and studying the plant virome in order to preserve plant health and ensure global food security. The virome presents challenges as well as opportunities in agricultural settings. In the past, management has mostly concentrated on getting rid of dangerous viruses, but knowing the entire viral community, including its beneficial members, may allow for more efficient, environmentally sound methods (Kreuze et al. 2009; Massart et al. 2017). The beneficial application of plant viruses in crop management is being made possible by new techniques in metagenomics, diagnostics, and synthetic biology.

9.3 Integrating High-Throughput and Conventional Methods for Fruit Crop Virome Analysis

Plant viruses in fruit crops have traditionally been identified and diagnosed using a variety of serological and molecular methods, each of which has unique advantages in terms of sensitivity, specificity, and throughput. The enzyme-linked immunosorbent assay, or ELISA, has been widely used among serological techniques because of its low cost, ease of use, and capacity for large-scale screening of known viruses across large sample sets. ELISA uses antigen–antibody binding to detect virus-specific proteins; however, its efficacy is largely dependent on the availability of high-quality antibodies, and it is unable to detect new or highly divergent viral strains (Clark and Adams 1977). Plant virus detection has been greatly improved by molecular diagnostics, especially with the use of polymerase chain reaction (PCR), reverse transcription PCR (RT-PCR), and quantitative real-time PCR (qPCR). By employing specially created primers to identify particular nucleic acid sequences, these methods provide improved sensitivity and specificity when focusing on known viruses. Two additional nucleic acid-based methods that have increased the diagnostic toolkit are loop-mediated isothermal amplification (LAMP) and nucleic acid hybridization. LAMP is especially helpful because it amplifies DNA rapidly and with high specificity at constant temperatures, doing away with the need for thermocycling equipment. This makes it perfect for field applications and point-of-care diagnostics (Notomi et al. 2000; Tomlinson et al. 2010). These traditional methods are crucial for routine virus detection, but they are unable to fully resolve the complexity of virus populations, particularly in cases of mixed infections or viruses that have not yet been identified through genetic or serological markers. Many of these constraints have been addressed by high-throughput sequencing (HTS) or next-generation sequencing (NGS), which offers a thorough platform for objective virome analysis. Without the need for previous sequence information, NGS allows the simultaneous detection of known, novel, and genetically diverse viruses. This benefit is especially important for fruit crops, where complex and cryptic viral communities are frequently produced by vegetative propagation and long life cycles. Virome research has made use of a variety of next-generation sequencing (NGS) technologies, including Oxford Nanopore Technologies (ONT), Ion Torrent, and Illumina. Regarding read length, error profiles, data throughput, and processing time, each of these technologies provides unique benefits (Adams et al. 2009; Roossinck et al. 2015). The identification of low-abundance or low-titer viruses is made easier by Illumina sequencing's exceptional ability to produce extremely accurate, deep coverage data. Despite its higher error rates, ONT's long read lengths and real-time data generation aid in reconstructing entire viral genomes and identifying structural variations. Both metagenomic and small RNA sequencing techniques are supported by these platforms; metagenomics recovers entire viral genomic sequences, while small RNA sequencing uses host antiviral responses to identify viruses that are actively replicating (Kreuze et al. 2009). Researchers are able to map the transmission and evolution of complex infections, uncover novel virus-host interactions, and reveal the entire viral diversity in fruit crops by combining NGS-based virome profiling with conventional diagnostic tools. Such in-depth knowledge is crucial for creating early and accurate disease control plans based on ecological principles. Thus, plant virology is changing as a result of the combination of traditional techniques and cutting-edge sequencing technologies, offering new capabilities to support resilient and sustainable fruit production.

9.4 Plant Pathogen and Virome Detection Through Next-Generation Sequencing

High-throughput sequencing, or next-generation sequencing (NGS), has transformed genomic research by enabling quick, extensive analysis of DNA and RNA sequences. Millions of short fragments can be sequenced at once using this technology, allowing for the thorough and objective characterization of complex microbial or viral communities, transcriptomes, and genomes (Goodwin et al. 2016). NGS was first created for human genome projects, but it has since spread to a variety of industries, including clinical medicine, agriculture, and environmental research. The first step in the NGS process is the fragmentation of nucleic acids, which are then ligated to platform-specific sequencing adapters to create sequencing libraries. After that, these libraries are processed and amplified using a variety of platforms, including Oxford Nanopore Technologies, Ion Torrent, and Illumina (Fig. 9.1). Because each platform has unique features in terms of throughput, read length, accuracy, and cost, researchers can choose the best tool for their particular objectives. Bioinformatics pipelines analyze the resulting raw sequences, reconstructing the original genetic information and enabling additional analyses like phylogenetics, gene identification, variant calling, and species classification (Mardis 2017).

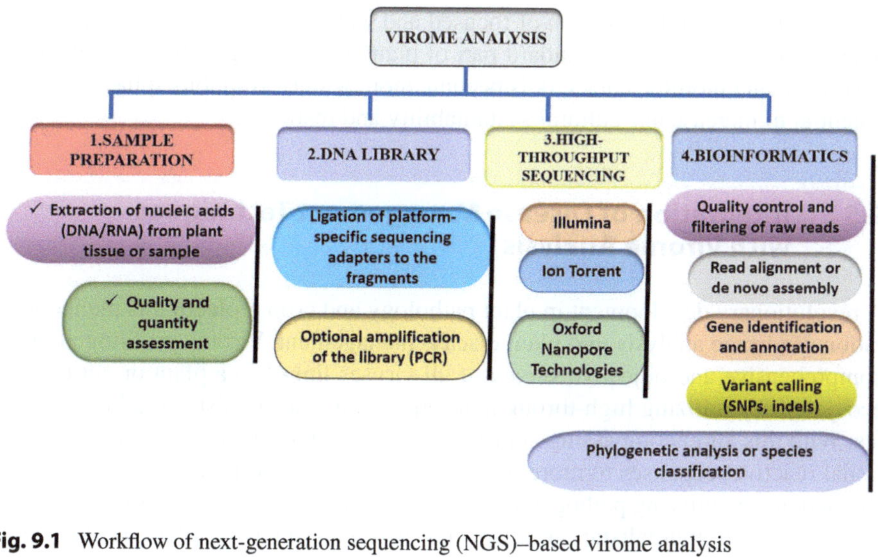

Fig. 9.1 Workflow of next-generation sequencing (NGS)–based virome analysis

By enabling whole-genome and exome sequencing to find mutations linked to diseases, NGS has revolutionized medical diagnostics and treatment. In oncology, where sequencing tumor genomes enables customized treatments based on particular genetic changes, it has had a particularly significant impact (Kandoth et al. 2013). NGS also facilitates personalized pharmacogenomics, pathogen identification in clinical specimens, and prenatal testing. NGS has created new opportunities for trait mapping, plant-associated microbial community analysis, and genomic-assisted breeding in agriculture. Gene discovery about yield, stress tolerance, and disease resistance has been expedited by crop genome sequencing (Thottathil et al. 2016). Importantly, NGS represents a breakthrough in plant pathology by enabling the direct detection of pathogens from infected plant tissues without the need for particular reagents or prior knowledge of the target organism. NGS is now a fundamental tool for metagenomics and microbiome research in addition to pathogen detection. Researchers can discover microbial diversity, functional potential, and intricate ecosystem interactions by sequencing environmental DNA (eDNA) from soil, water, and plant surfaces (Shreiner et al. 2015; Taberlet et al. 2012). NGS excels in plant virology because it can identify a variety of viruses without the use of primers or antibodies, including novel, genetically diverse, or low-abundance strains. Because of this, it is especially appropriate for virome studies, which seek to describe the entire viral community in plants or ecosystems. Additionally, NGS makes it possible to identify strain-level variation, viral recombination events, and complex mixed infections that are frequently missed by more conventional techniques like PCR or ELISA. Because viral populations can grow and change over time in perennial fruit crops, their use is particularly crucial. Table 9.1 lists viruses found in a variety of host plants using different sequencing platforms, along with the target nucleic acids and pertinent references, to give a useful overview of the diversity of viruses found in plants using NGS. Combining such comprehensive virome data with conventional diagnostics improves our comprehension of virus ecology and aids in the creation of focused and well-informed disease management plans. NGS is becoming a standard part of plant disease diagnostics as sequencing costs drop and bioinformatics tools become more widely available. It has enormous potential to increase agricultural sustainability and resilience.

9.5 Integration of Disease Management Techniques with Virome Analysis

A revolutionary development in plant pathology and crop protection is the incorporation of virome analysis into plant disease management. Virome profiling allows a comprehensive and objective survey of all viruses linked to a plant or agricultural ecosystem by utilizing high-throughput sequencing and sophisticated bioinformatics. With this all-encompassing approach, disease management shifts from conventional reactive responses to more proactive, predictive, and precision-based tactics. Beyond just identifying pathogenic viruses, virome analysis also finds latent, persistent, and even mutualistic viral species that can improve plant resistance to

Table 9.1 Viruses identified from various host plants using next-generation sequencing platforms

Host	Study finding/virome	Target	Sequencing platforms	References
Arabidopsis thaliana	Oilseed rape mosaic virus, Cauliflower mosaic virus, Cabbage leaf curl virus	Total RNAs and siRNAs	Illumina Genome Analyzer	Chen et al. (2025)
Apple	Apple stem pitting virus, Apple chlorotic leaf spot virus, Apple necrotic mosaic virus, Apple stem grooving virus, Apricot latent virus, Apple green crinkle-associated virus, Citrus concave gum-associated virus, Apple rubbery wood virus 1 and 2	siRNAs	Illumina	Rivarez et al. (2023), Jo et al. (2020), Nabi et al. (2022), and Khan et al. (2024)
Black pepper	Piper yellow mosaic virus	Viral and plant DNA	Roche 454 GS–FLX	Deeshma et al. (2021)
Citrus	Citrus yellow vein clearing virus, Citrus chlorotic dwarf-associated virus	siRNA and total DNA	Illumina	Chen et al. (2025)
Cassava	Cassava brown streak virus	Total RNA	Roche 454 GS–FLX	Ogwok et al. (2023)
Cotton	Cotton leafroll dwarf virus	siRNAs	Illumina	Hameed et al. (2022)
Fig	Fig mosaic virus, Fig latent virus-1	dsRNAs	Illumina	Maliogka et al. (2023a, b)
Grapevine	Grapevine leaf roll-associated virus-3, Grapevine red blotch-associated virus, Grapevine rupestris stem pitting associated virus, Grapevine Syrah 1 virus, Grapevine virus E, Grapevine virus A, Grapevine virus F Australian grapevine viroid, Grapevine yellow speckle viroid, Hop stunt viroid, Citrus exocortis Yucatan viroid, Citrus exocortis viroid	Total RNA or dsRNA, siRNAs	Roche 454	Jooste et al. (2024)

(continued)

Table 9.1 (continued)

Host	Study finding/virome	Target	Sequencing platforms	References
Gomphrena globosa	Gay feather mild mottle virus	Total RNA	Roche 454 GS–FLX	Ruiz et al. (2023)
Maize	Maize chlorotic mottle virus, Sugarcane mosaic virus	Total RNA	Roche 454 GS–FLX+	Wang et al. (2024a, b)
Nicotiana benthamiana	Cucumber mosaic virus, Cymbidium ringspot virus, Tobacco rattle virus, Pepper mild mosaic virus, Potato virus X	siRNAs	Roche 454 and Illumina	Chen et al. (2025)
Passion fruit	Passion fruit woodiness virus	Polyadenylated RNA	Illumina Solexa	Tomlinson et al. (2022)
Plum	Plum pox virus, Prune necrotic ringspot virus	dsRNA	Roche 454	Bertolini et al. (2023)
Pepper and eggplant	Pepper yellow leaf curl virus, Eggplant mild leaf mottle virus	Purified virons viral RNA	SOLiD	Sidharthan et al. (2024)
Peach	Peach latent mosaic viroid	siRNAs	Illumina	Navarro et al. (2023)
Rice	Rice stripe virus	SiRNAs	Illumina Solexa	Lee et al. (2023)
Raspberry	Raspberry latent virus	dsRNA	Illumina	Maliogka et al. (2023a, b)
Sweet potato	Sweet potato feathery mottle virus, Sweet potato chlorotic stunt virus, Sweet potato virus C, Sweet potato leaf curl Georgia virus, Sweet potato pakakuy virus B	siRNAs	Illumina	Adero et al. (2024)
Tomato	Tomato spotted wilt virus, Potato spindle tuber viroid, Tomato necrotic stunt virus, Pepino mosaic virus	siRNAs	Illumina Genome Analyzer IIx	Rivarez et al. (2023)
Wild cocksfoot	Cereal yellow dwarf virus and Cocksfoot streak virus (*Potyvirus*)	siRNAs	Roche 454	Wang et al. (2024a, b)

environmental stressors or give their hosts a competitive edge (Roossinck 2015a, b). Once disregarded, these intricate viral interactions are now understood to be critical elements affecting plant health and disease dynamics. Early detection of viral pathogens, including newly emerging viruses, is a major practical benefit of virome profiling. Early interventions that lower crop losses and stop the spread of epidemics are made possible by the identification of infections during asymptomatic stages. Furthermore, co-infection patterns and antagonistic or synergistic viral interactions that impact disease severity are revealed by virome data (Massart et al. 2017). Virome data facilitates real-time surveillance when integrated into plant health monitoring systems. Predictive models for epidemic forecasting and spatial risk analysis can be developed by combining viral community data with geographic, environmental, and climatic data (Mališová et al. 2024). These systems minimize chemical inputs while enabling targeted disease management, which is in line with precision agriculture principles. Virome analysis improves biosecurity in quarantine and germplasm certification settings by identifying viruses in source materials that might be latent or symptomless. This ability is crucial for maintaining native biodiversity, adhering to international trade regulations, and safely exchanging genetic resources (Maina et al. 2021). In high-value fruit crops propagated vegetatively, where the risk of viral transmission is increased, thorough screening is particularly important. Additionally, understanding virome diversity and viral evolution is crucial for plant breeding initiatives. Understanding strain variation, recombination, and viral mutations facilitates the creation and application of long-lasting resistance genes, strengthening cultivar resistance to a variety of viral threats (Kreuze et al. 2009). Plant defense responses can be better understood at the molecular level by integrating virome profiles with host transcriptomic and proteomic data. Additionally, the advantageous functions of some plant viruses, such as mutualistic or endophytic interactions that enhance nutrient uptake or drought tolerance, open up promising paths for the development of new biostimulant technologies and the genetic engineering of stress-resilient crops (Roossinck 2015a, b). This changing viewpoint recasts viruses as potential agents of plant vigor and adaptation in addition to being pathogens. Despite these developments, there are still difficulties when routinely integrating virome analysis into crop management. Widespread adoption is hampered by high sequencing and computational costs, a lack of qualified bioinformatics specialists, a lack of standardized techniques, and challenges in understanding the ecological roles of non-pathogenic viruses. Global collaboration, supportive legislation, funding for capacity building, and the development of easily accessible analytical platforms are all necessary to address these problems. However, virome analysis is incredibly promising, with a wide range of useful applications that can greatly improve integrated disease management strategies in fruit crops and other areas (Table 9.2).

Table 9.2 Potential applications of virome analysis in plant disease management and their associated benefits (Alcalá Briseño et al. 2023)

Application	Description/benefits
Early detection of viral pathogens	Enables timely identification of emerging or latent viruses, improving response actions.
Co-infection profiling	Reveals interactions between multiple viruses, aiding in comprehensive disease diagnosis.
Surveillance and risk assessment	Supports monitoring of virus prevalence and spread across regions and seasons.
Disease forecasting	Integrates virome data with environmental models to predict outbreaks.
Sanitary certification and quarantine	Ensures virus-free planting material in germplasm exchange and trade.
Resistance breeding support	Aids in identifying resistant genes and tracking viral evolution for durable resistance.
Beneficial virus utilization	Explores mutualistic viruses for crop improvement under stress conditions.
Integration with multi-omics	Enhances understanding of host-pathogen interactions for holistic management strategies.

9.6 Advantages of Next-Generation Sequencing in Plant Pathogen Diagnostics and Surveillance

By offering a number of significant advantages over traditional diagnostic techniques, next-generation sequencing (NGS) has significantly improved plant pathogen detection and monitoring. Its remarkable sensitivity, which makes it possible to identify pathogens in complex plant samples even at very low concentrations, is one of its main advantages. According to Massart et al. (2017), this high sensitivity is essential for the early detection of latent or emerging infections, enabling prompt and efficient disease control measures. The abilities of NGS to simultaneously and thoroughly profile several pathogens in a single test are one of its most potent features. In contrast to conventional diagnostics that concentrate on specific pathogens, NGS can identify mixed infections, providing a comprehensive understanding of the microbial communities impacting plants. Understanding pathogen interactions and co-infection dynamics that affect disease development and severity, particularly in fruit crops, requires this multiplex capability (Kreuze et al. 2009; Massart et al. 2017). Additionally, NGS facilitates quick data generation and analysis, allowing for quick reactions during epidemics. NGS can detect novel, divergent, or highly mutated pathogens that may evade detection by targeted assays, in contrast to PCR or ELISA, which require prior knowledge of pathogen genomes or antigens (Adams et al. 2009). NGS enables thorough genomic characterization of pathogens, going beyond simple detection. Analysis of virulence factors, mutation patterns, adaptation mechanisms, and evolutionary trajectories is all included in this. According to

Roossinck et al. (2010), these molecular insights are crucial for developing long-term management plans and speeding up the breeding of resistant plant varieties. NGS is particularly useful for epidemiological surveillance that maps pathogen distribution across different geographic regions and time periods because of its high-throughput capacity, which enables processing of numerous samples at once. According to Adams et al. (2009), such information improves the forecasting and control of disease outbreaks. Additionally, by lowering false positives through deep sequencing coverage combined with rigorous bioinformatics validation, NGS increases diagnostic accuracy. By identifying viruses and other microorganisms from soil, water, air, and insect vectors, its application is growing into environmental pathogen surveillance. Early warning systems and proactive disease control measures are supported by this extensive environmental monitoring (Taberlet et al. 2012). By revealing the complexity of microbial and viral communities associated with plants, NGS also makes a substantial contribution to ecological and metagenomic research. These investigations aid in locating pathogen reservoirs and elucidating their functions in the development and spread of disease (Roossinck et al. 2011). Combining transcriptomics, proteomics, and next-generation sequencing (NGS) provides a potent method to fully comprehend host-pathogen interactions and disease mechanisms, supporting the creation of accurate control strategies (Sen et al. 2023; Mališová et al. 2024). Table 9.3 lists the main advantages of NGS in identifying plant diseases.

Table 9.3 Key benefits of next-generation sequencing (NGS) for detecting plant pathogens (Nafea et al. 2024)

Benefit	Explanation
Exceptional sensitivity	NGS can identify even very small amounts of pathogens within complex plant samples, making it effective for early detection.
Discovery of unknown pathogens	It can detect new or emerging pathogens without needing prior knowledge of their sequences.
Fast results	The technology provides quick data turnaround, allowing prompt action during disease outbreaks.
High sample throughput	Multiple samples can be processed simultaneously, enabling large-scale disease monitoring and research.
Simultaneous pathogen detection	Detects and characterizes several pathogens in a single test, providing comprehensive insight into mixed infections.
Detailed genomic information	Provides in-depth genetic data on pathogens, aiding understanding of virulence, evolution, and adaptation.
Lower chances of false alarms	Uses multiple sequencing reads and advanced bioinformatics to ensure accurate identification and reduce errors.
Environmental monitoring	Tracks pathogens in soil, water, air, and other environmental sources to support ecosystem-wide surveillance.
Whole community analysis	Enables study of entire pathogen populations without prior knowledge of specific organisms.
Targeted sequencing (TG-Seq)	Focuses on specific viral genome regions for more sensitive and cost-effective detection, reducing background noise.

9.7 Targeted Genomic Sequencing (TG-Seq): An Accurate Method for Diagnosing Plant Viruses

A new next-generation sequencing (NGS) method called targeted genomic sequencing (TG-Seq) provides a highly sensitive, specific, and economical way to detect several plant viruses at once. For the horticultural and agricultural industries, where viral infections often result in large financial losses and lower crop yields, this novel strategy has enormous potential. TG-Seq selectively enriches and sequences predefined genomic regions of viral pathogens, in contrast to untargeted metagenomic NGS approaches that sequence the entire nucleic acid content within a sample. This focused approach significantly reduces the amount of host DNA in the dataset, which in turn lowers sequencing expenses and streamlines downstream bioinformatics processing. By focusing on conserved or diagnostically important regions of viral genomes, TG-Seq improves the detection accuracy across diverse virus strains and variants (Maina et al. 2021). TG-Seq is especially useful for detecting low-titer viruses, which frequently avoid detection by conventional diagnostic techniques like gel electrophoresis or multiplex PCR. Because of its higher sensitivity, latent infections can be found early, allowing for prompt disease control measures. By encouraging uniform genome coverage and lowering amplification biases, transposon-based library preparation chemistry further enhances TG-Seq performance, producing repeatable and consistent results across a range of virus types and sample matrices. TG-Seq technology is becoming more widely acknowledged as an essential instrument for routine high-throughput virus diagnostics as it develops. Applications like germplasm certification, which guarantees that planting materials are virus-free; biosecurity monitoring, which necessitates quick and precise identification of exotic or quarantine viruses; and baseline surveillance programs, which monitor viral prevalence and evolutionary dynamics in agricultural ecosystems, are particularly beneficial. TG-Seq promotes better crop health management and supports the sustainability of global food production systems by facilitating accurate and effective virus detection.

9.8 High-Throughput Sequencing (HTS)

The rapid, large-scale generation of genetic data made possible by high-throughput sequencing (HTS), also referred to as next-generation sequencing (NGS), deep sequencing, or third-generation sequencing, has transformed biological research. This technology, which significantly outperforms conventional sequencing techniques in terms of speed, throughput, and cost-efficiency, has become essential in a variety of fields, including epigenomics, genomics, metagenomics, and transcriptomics (Lee 2023). HTS systems are capable of producing billions of sequence reads per run by sequencing millions of DNA molecules at once. These platforms can produce gigabytes or even terabytes of data in a short amount of time and at a much lower cost than previous sequencing technologies (Kulski 2011). Roche/454 pyrosequencing, Illumina/Solexa, ABI SOLiD, Ion Torrent, Heliscope, and sequencing by oligonucleotide ligation and detection are just a few of the high-throughput sequencing (HTS) platforms that have been developed since the early 2000s. The landscape of contemporary genomic analysis has been shaped by these technologies' evolution to improve sequencing speed, accuracy, and throughput (Margulies et al. 2005; Harris et al. 2008; Valouev et al. 2008; Lee 2023). Second-generation sequencing technologies typically rely on PCR amplification during sample preparation, despite improvements in throughput and cost-effectiveness. In addition to producing relatively short read lengths (20–200 base pairs), this step can introduce biases, which could cause gaps and assembly issues in genomic analyses. Third- and fourth-generation sequencing technologies have been created to get around the limitations of previous sequencing techniques. Among these, Oxford Nanopore Technologies' (ONT) sequencing platforms and Pacific Biosciences' (PacBio) single-molecule real-time (SMRT) sequencing are prominent (Krishna et al. 2019). The average length of the long reads produced by the PacBio system is roughly 1.5 kilobases, but new developments have increased the maximum read length to 10–15 kilobases. On the other hand, ONT sequencing minimizes amplification-associated biases by operating without the need for PCR amplification and can occasionally produce ultra-long reads that exceed 4 megabases (Lee et al. 2023). Sequencing accuracy, sample preparation complexity, read length, cost, and the underlying nucleotide incorporation or detection chemistry are all different across these platforms (Ibaba and Giubba 2020). Plant virology has been transformed by the combination of high-throughput sequencing (HTS) and sophisticated bioinformatics pipelines, which allow for the thorough and objective identification of viral pathogens. HTS provides comprehensive insights into viral diversity, population dynamics, and evolution by facilitating the identification of both known and previously uncharacterized viruses through its capacity to concurrently analyze massive amounts of sequence data (Barba et al. 2014). HTS's independence from previous genomic data is a major benefit, enabling hypothesis-free investigation of plant samples for various viruses, regardless of their structural characteristics or genome composition. Numerous RNA populations, including total RNA, double-stranded RNA (dsRNA), ribosomal RNA-depleted RNA, messenger RNA (mRNA), small interfering RNA (siRNA), and virion-associated nucleic acids (VANA), can be targeted by various HTS-based techniques. HTS is a vital tool for virus discovery, epidemiological research, and surveillance of emerging plant diseases because of its methodological flexibility, which allows for the thorough detection and molecular characterization of viral agents across a variety of plant systems (Pecman et al. 2017; Villamor et al. 2019).

9.9 Pathogen Identification Using Bioinformatic Tools and Pipelines

In plant virome studies, bioinformatics is essential for analyzing the massive volumes of data produced by NGS technologies. Researchers can effectively process raw sequencing data, precisely identify viral sequences, discover new pathogens,

Table 9.4 Bioinformatic tools and pipelines for plant virome analysis

Bioinformatics step	Tools	Purpose
Sequence quality control	FastQC, Trimmomatic	Evaluate sequencing read quality, trim adapter sequences, and remove low-quality bases to ensure reliable downstream analysis.
Read alignment	Bowtie2, BWA, HISAT2	Map processed reads to reference viral genomes to identify known plant viruses.
De novo assembly	SPAdes, Velvet, Trinity	Assemble reads into contigs without a reference to discover novel viral genomes and reconstruct complete sequences.
Virus detection	VironScan, VirFinder, ViraMiner	Detect viral sequences from assemblies, enabling identification of both known and previously uncharacterized viruses.
Taxonomic classification	Kraken, Kaiju, Centrifuge	Assign viral contigs or reads to taxonomic levels (family, genus, species) for accurate classification.
Variant calling	CATK, Samtools, VarScan	Detect genetic variation (such as SNPs and indels) within viral populations to study diversity and evolution.
Phylogenetic analysis	RAxML, PhyML, BEAST	Construct evolutionary trees to compare viral strains and analyze their relationships over time or across hosts.
Functional annotation, visualization, and data integration	Prokka, RAST, Blast2GO, Tablet, IGV, Artemis, custom scripts	Annotate viral genes, predict functions, visualize alignments and assemblies, and integrate multi-dimensional datasets while reducing false positives.

and investigate their interactions with the plant host environment thanks to these computational tools and pipelines (Khan et al. 2024). The intricate process of pathogen discovery and characterization is streamlined by bioinformatic approaches, which integrate different software and algorithms. An overview of the popular bioinformatic tools and workflows created especially for pathogen identification from plant virome sequencing data is given in Table 9.4. These tools support robust and dependable viral agent detection in complex plant samples by facilitating quality control, sequence assembly, taxonomic classification, and functional annotation (Alcalá Briseño et al. 2023).

9.10 Future Perspectives in Plant Virome Research and Disease Management

Our comprehension of the intricate world of plant viruses and viroids has been completely transformed by developments in molecular biology and sequencing technologies. Unprecedented insights into viral diversity, interactions, and effects on plant health have been made possible by these developments. In the future, a number of encouraging avenues for strengthening crop resilience and managing plant diseases are emerging. Breeding resistant plant varieties and applying targeted, site-specific control measures are two examples of how advancements are making precise and timely interventions more feasible. In order to enable early detection and ongoing monitoring of viral infections in agricultural settings, future advancements seek to provide quick, on-site diagnostic tools with real-time virome analysis capabilities. The intricate relationships that arise when several pathogens infect plants emphasize the need for more investigation into the ways in which these co-infections influence the development of disease. The functional role of small RNAs (sRNAs) associated with different plant pathogens is an important field of research that is necessary to comprehend the molecular mechanisms underlying plant immune responses and to develop novel antiviral tactics. Our knowledge of viral diversity and ecology will be improved by extending the virome catalog to include a greater variety of plant species and ecosystems. Combining cutting-edge sequencing methods with other omics fields like metagenomics, proteomics, and metabolomics provides a thorough understanding of host-pathogen interactions and aids in the creation of more sustainable and successful disease management plans. Plant virome research has enormous potential for enhancing early viral infection detection, expanding our understanding of plant-virus interactions, and ultimately promoting food security and ecosystem conservation in a changing environment. It is anticipated that advancements in this area will allow for more efficient monitoring, diagnosis, and management techniques to protect biodiversity and agricultural productivity.

References

Adams IP, Glover RH, Monger WA, Mumford R, Jackeviciene E, Navalinskiene M et al (2009) Next-generation sequencing and metagenomic analysis: a universal diagnostic tool in plant virology. Mol Plant Pathol 10(4):537–545

Adero J et al (2024) Next-generation sequencing and genetic analyses of sweet potato viruses in Uganda. Pathogens 13(10):833

Agrios GN (2005) Plant pathology, 5th edn. Elsevier Academic Press

Alcalá Briseño RI, Batuman O, Brawner J, Cuellar WJ, Delaquis E, Etherton BA, French-Monar RD, Kreuze JF, Navarrete I, Ogero K, Plex Sulá AI, Yilmaz S, Garrett KA (2023) Translating virome analyses to support biosecurity, on-farm management, and crop breeding. Front Plant Sci 14:1056603

Anderson PK, Cunningham AA, Patel NG, Morales FJ, Epstein PR, Daszak P (2004) Emerging infectious diseases of plants: pathogen pollution, climate change and agrotechnology drivers. Trends Ecol Evol 19(10):535–544

Baranwal VK, Un Nabi S, Yadav MK (2021) Plant virus diagnostics: traditional to recent and emerging advances. In: Singh KP, Jahagirdar S, Sarma BK (eds) Emerging trends in plant pathology. Springer, Singapore. https://doi.org/10.1007/978-981-15-6275-4_5

Barba M, Czosnek H, Hadidi A (2014) Historical perspective, development and applications of next-generation sequencing in plant virology. Viruses 6(1):106–136

Bertolini E et al (2023) High-throughput sequencing for detection of Plum pox virus and Prune necrotic ringspot virus in Prunus species. Plant Dis 107(2):345–355

Bragard C, Caciagli P, Lemaire O, Lopez-Moya JJ, MacFarlane S, Peters D et al (2013) Status and prospects of plant virus control through interference with vector transmission. Annu Rev Phytopathol 51:177–201

Chakraborty S, Newton AC (2011) Climate change, plant diseases and food security: an overview. Plant Pathol 60(1):2–14
Chen H et al (2025) High-throughput sequencing for plant virome characterization: current advances and applications. Ann Agri-Bio Res 30(2):112–128
Clark MF, Adams AN (1977) Characteristics of the microplate method of enzyme-linked immunosorbent assay for the detection of plant viruses. J Gen Virol 34(3):475–483
Deeshma KR et al (2021) Complete genome sequence of Piper yellow mottle virus infecting black pepper in India. Arch Virol 166(7):1823–1830
Dulya O, Koonin EV (2018) Metaviromics: a toolkit for exploring viral diversity in nature. Virus Res 244:110–115
EPPO (2019) European and Mediterranean plant protection organization standards. Retrieved from https://www.eppo.int
FAO (2006) International standards for phytosanitary measures (ISPMs). Secretariat of the International Plant Protection Convention (IPPC), FAO, Rome
Fletcher J, Bender C, Budowle B, Cobb WT, Gold SE, Ishimaru CA et al (2006) Plant pathogen forensics: capabilities, needs, and recommendations. Microbiol Mol Biol Rev 70(2):450–471
Ghosh S, Kanwar P, Jha G (2022) Beneficial plant viruses: emerging tools for developing stress-resilient crops. Plant Mol Biol 109:211–225
Goodwin S, McPherson JD, McCombie WR (2016) Coming of age: ten years of next-generation sequencing technologies. Nat Rev Genet 17:333–351
Hameed A et al (2022) High-throughput sequencing reveals complete genome sequence of Cotton leafroll dwarf virus in Pakistan. Front Microbiol 13:845312
Harris TD, Buzby PR, Babcock H, Beer E, Bowers J, Braslavsky I, Causey M, Colonell J, Dimeo J, Efcavitch JW et al (2008) Single-molecule DNA sequencing of a viral genome. Science 320:106–109
Hull R (2014) Plant virology, 5th edn. Academic Press
Ibaba JD, Giubba S (2020) Comparison of next-generation sequencing platforms. In: Next generation sequencing technologies for detection of plant pathogens. Springer, Cham, pp 57–74
Jo Y et al (2020) Sweet potato viromes in eight different geographical regions in Korea and two different cultivars identified by RNA-seq. Sci Rep 10:1591
Jones RAC (2014) Plant virus ecology and epidemiology: historical perspectives, recent progress and future prospects. Ann Appl Biol 164(3):320–347
Jones RAC (2021) Global plant virus disease pandemics and epidemics. Plants 10(2):233
Jooste A et al (2024) High-throughput sequencing for characterization of the grapevine virome: diversity of viruses and viroids in South African vineyards. Virus Res 336:199–213
Kandoth C, McLellan MD, Vandin F, Ye K, Niu B, Lu C, Xie M, Zhang Q, McMichael JF, Wyczalkowski MA et al (2013) Mutational landscape and significance across 12 major cancer types. Nature 502:333–339
Khan ZA, Sharma SK, Gupta N et al (2024) Assessing the de novo assemblers: a metaviromic study of apple and first report of citrus concave gum-associated virus, apple rubbery wood virus 1 and 2 infecting apple in India. BMC Genomics 25:1057
Kreuze JF, Pérez A, Untiveros M, Quispe D, Fuentes S, Barker I, Simon R (2009) Complete viral genome sequence and discovery of novel viruses by deep sequencing of small RNAs: a generic method for diagnosis, discovery and sequencing of viruses. Virology 388(1):1–7
Krishna N, Manoharan M, Lakshmi GS (2019) Third-generation sequencing technologies: current trends and applications in genomics. J Biosci Res 10:35–43
Kulski JK (2011) Next-generation sequencing—an overview of the history, tools, and "Omic" applications. In: Kulski JK (ed) Next generation sequencing – advances, applications and challenges. IntechOpen
Lee H (2023) Advances in next-generation and third-generation sequencing technologies for plant genomics. Plant Methods 19:12
Lee J et al (2023) NGS-based identification of rice stripe virus and related pathogens in East Asian rice fields. Plant Pathol J 39(5):221–230

Mahlein AK (2016) Plant disease detection by imaging sensors – parallels and specific demands for precision agriculture and plant phenotyping. Plant Dis 100(2):241–251

Maina S, Edwards OR, de Almeida L, Ximenes A, Jones RAC (2021) Metagenomic analysis identifies a diverse, previously undescribed virome in legume species from Australia. Plant Pathol 70(5):1020–1035

Maliogka VI et al (2023a) Virome analysis of fig trees reveals novel viruses associated with fig mosaic disease using high-throughput sequencing. Plants 12(8):1665

Maliogka VI et al (2023b) Virome analysis of raspberry reveals latent and novel viruses using NGS. Plants 12(11):1789

Mališová M, Špak J, Pridmore T (2024) Predictive plant virus epidemiology using multi-layered data integration. Plant Pathol 73(1):85–98

Malmstrom CM, McCullough AJ, Johnson HA, Newton LA, Borer ET (2005) Invasive annual grasses indirectly increase virus incidence in California native perennial bunchgrasses. Oecologia 145(1):153–164

Mansfield J, Genin S, Magori S, Citovsky V, Sriariyanum M, Ronald P et al (2012) Top 10 plant pathogenic bacteria in molecular plant pathology. Mol Plant Pathol 13(6):614–629

Manzoor S, Nabi SU, Ayaz A, Rasool B, Sharma SK, Chesti MH, Parveen S, Verma MK, Diab MA, Rabbee MF (2025) Unveiling viral threats to temperate pome fruits: characterization, transmission, and sustainable management strategies. Front Microbiol 16:1560720

Mardis ER (2008) Next-generation DNA sequencing methods. Annu Rev Genomics Hum Genet 9:387–402

Mardis ER (2017) DNA sequencing technologies: 2006–2016. Nat Protoc 12:365–368

Margulies M, Egholm M, Altman WE, Attiya S, Bader JS, Bemben LA, Berka J, Braverman MS, Chen YJ, Chen Z et al (2005) Genome sequencing in microfabricated high-density picolitre reactors. Nature 437:376–380

Marwal A, Gaur RK (2020) Emerging plant viruses: a global threat. In: Gaur RK, Petrov NM, Sharma P (eds) Plant viruses: evolution and management. Springer, pp 1–19

Massart S, Olmos A, Jijakli H, Candresse T (2017) Current impact and future directions of high-throughput sequencing in plant virus diagnostics. Virus Res 241:189–199

Nabi SU, Yousuf N, Yadav MK, Choudhary DK, Ahmad D, Kirmani SN, Sheikh P, Ahmed I (2021) Recent insights in detection and diagnosis of plant viruses using next-generation sequencing technologies. In: Innovative approaches in diagnosis and management of crop diseases, pp 85–100. https://doi.org/10.1201/9781003187608-4

Nabi SU, Baranwal VK, Rao GP, Mansoor S, Vladulescu C, Raja WH, Jan BL, Alansi S (2022) High-throughput RNA sequencing of mosaic infected and non-infected apple (Malus × domestica Borkh.) cultivars: from detection to the reconstruction of whole genome of viruses and viroid. Plants 11(5):675. https://doi.org/10.3390/plants11050675

Nabi SU, Parveen S, Kamil D, Khokha K, Watpade S, Raja WH, Mir JI, Verma MK, Manzoor S, Amin Z, Ayaz A (2025) Diversity of fungal species causing apple canker disease in India using multigene characterization. Physiol Mol Plant Pathol 139:102817

Nafea AM et al (2024) Application of next-generation sequencing to identify infectious pathogens: benefits and challenges. Front Cell Infect Microbiol 14:10859930

Navarro B et al (2023) Virome characterization of peach latent mosaic viroid in stone fruit trees using small RNA sequencing. Viruses 15(4):876

Notomi T, Okayama H, Masubuchi H, Yonekawa T, Watanabe K, Amino N, Hase T (2000) Loop-mediated isothermal amplification of DNA. Nucleic Acids Res 28(12):e63

Ogwok E et al (2023) Molecular characterization of Cassava brown streak viruses from East Africa using next-generation sequencing. Plant Pathol J 39(3):145–156

Pecman A, Kutnjak D, Gutiérrez-Aguirre I, Adams I, Fox A, Mehle N, Ravnikar M (2017) Next generation sequencing for detection and discovery of plant viruses and viroids: comparison of two approaches. Front Microbiol 8:1998

Popp J, Pető K, Nagy J (2013) Pesticide productivity and food security: a review. Agron Sustain Dev 33(1):243–255

Power AG, Mitchell CE (2004) Pathogen spillover in disease epidemics. Am Nat 164(S5):S79–S89

Prabha K, Baranwal VK, Jain RK (2013) Applications of next generation high throughput sequencing technologies in characterization, discovery and molecular interaction of plant viruses. Indian J Virol 24(2):157–165
Rivarez MPS et al (2023) In-depth study of tomato and weed viromes reveals undiscovered plant virus diversity in an agroecosystem. Microbiome 11(1):72
Roossinck MJ (2011) The good viruses: viral mutualistic symbioses. Nat Rev Microbiol 9(2):99–108
Roossinck MJ (2012) Plant virus metagenomics: biodiversity and ecology. Annu Rev Genet 46:359–369
Roossinck MJ (2015a) Move over, bacteria! Viruses make their mark as mutualistic microbial symbionts. J Virol 89(13):6532–6535
Roossinck MJ (2015b) Plant virus evolution. Springer
Roossinck MJ, Sabanadzovic S, Okada R (2010) Ecogenomics: ecological and evolutionary genomics of plant viruses. Curr Opin Virol 1(4):324–329
Roossinck MJ, Martin DP, Roumagnac P (2011) Plant virus metagenomics: biodiversity and ecology. Annu Rev Genet 45:371–402
Roossinck MJ, Martin DP, Roumagnac P (2015) Plant virus metagenomics: advances in virus discovery. Phytopathology 105(6):716–727
Rosario K, Fierer N, Breitbart M (2015) Viral metagenomics: advances, implications, and opportunities for agriculture. Front Microbiol 6:1025
Ruiz M et al (2023) Virome analysis of Gomphrena globosa identifies novel viruses using next-generation sequencing. Plants 12(6):1145
Savary S, Willocquet L, Pethybridge SJ, Esker P, McRoberts N, Nelson A (2019) The global burden of pathogens and pests on major food crops. Nat Ecol Evol 3(3):430–439
Scholthof K-BG, Adkins S, Czosnek H, Palukaitis P, Jacquot E, Hohn T et al (2011) Top 10 plant viruses in molecular plant pathology. Mol Plant Pathol 12(9):938–954
Sen S, Kumar D, Singh A (2023) Integrative omics approaches in understanding plant-pathogen interactions and crop improvement. Front Plant Sci 14:1173658
Shreiner AB, Kao JY, Young VB (2015) The gut microbiome in health and in disease. Curr Opin Gastroenterol 31:69–75
Sicard A, Zeilinger AR, Vanhove M, Schartel TE, Beal DJ, Daugherty MP, Almeida RPP (2018) Xylella fastidiosa: insights into an emerging plant pathogen. Annu Rev Phytopathol 56:181–202
Sidharthan I et al (2024) Next-generation sequencing-based identification of eggplant and pepper viruses. Front Microbiol 15:1200457
Singhal P, Nabi SU, Yadav MK et al (2021) Mixed infection of plant viruses: diagnostics, interactions and impact on host. J Plant Dis Prot 128:353–368. https://doi.org/10.1007/s41348-020-00384-0
Strange RN, Scott PR (2005) Plant disease: a threat to global food security. Annu Rev Phytopathol 43:83–116
Taberlet P, Coissac E, Pompanon F, Brochmann C, Willerslev E (2012) Towards next-generation biodiversity assessment using DNA metabarcoding. Mol Ecol 21:2045–2050
Thottathil GP, Thankappan R, Surya P (2016) Applications of next-generation sequencing in crop improvement. In: Abdurakhmonov IY (ed) Plant genomics. IntechOpen
Tomlinson JA, Barker I, Boonham N (2010) Rapid detection of *Phytophthora ramorum* and *P. kernoviae* by two-minute DNA extraction followed by isothermal amplification and amplicon detection by generic lateral flow device. Phytopathology 97(4):649–654
Tomlinson P et al (2022) NGS reveals virome diversity of passion fruit in Brazil using polyadenylated RNA sequencing. Virus Res 312:198738
Valouev A, Ichikawa J, Tonthat T, Stuart J, Ranade S, Peckham H, Zeng K, Malek JA, Costa G, McKernan K et al (2008) A high-resolution, nucleosome position map of *C. elegans* reveals a lack of universal sequence-dictated positioning. Genome Res 18:1051–1063
Villamor DEV, Pillai SS, Eastwell KC (2019) High-throughput sequencing reveals a novel virus associated with Little cherry disease. Viruses 11(5):438
Wang L et al (2024a) NGS-based characterization of cereal yellow dwarf virus and cocksfoot streak virus in wild grasses. Front Plant Sci 15:1187623

Wang L et al (2024b) NGS-based characterization of maize and sugarcane viruses in hybrid cultivation systems. Front Plant Sci 15:1187623

Wu Q, Ding SW, Zhang Y, Zhu S (2015) Identification of viruses and viroids by next-generation sequencing and homology-dependent and homology-independent algorithms. Annu Rev Phytopathol 53:425–444

Xu P, Chen F, Mannas JP, Feldman T, Sumner LW, Roossinck MJ (2008) Virus infection improves drought tolerance. New Phytol 180(4):911–921

Rapid and Sensitive On-Site Diagnostic Tools for Fruit Viruses

10

Souvik Kumar Saha, Anutee Dolley, and Nima D. Namsa

Abstract

The living cell of a host organism provides the conditions where viruses replicate and produce new viral particles. That can infect all living organisms, including animals, plants, and other microorganisms. Viruses often have high mutation rate, which allows them to change and adapt very quickly. Rapid and accurate diagnostic methods are crucial, particularly for agricultural applications for detecting fruit viruses. This chapter provides a critical analysis of the widely used rapid and sensitive on-site diagnostic tools for detecting fruit viruses. It also highlights challenges and limitations of these tools. Recent advances in portable technologies have made field testing more practical. Such as immunoassay detection generally offers high sensitivity and specificity, and it gives results within some time. Furthermore, some immunoassays, like the lateral-flow assay, give results within minutes. On the other hand, nucleic acid-based detection techniques offer high sensitivity; however, they are time-consuming, costly, and require molecular biology expertise to obtain accurate results. Beyond these methods, the application of next-generation sequencing (NGS) is a sophisticated diagnostic approach that gives very specific results. NGS allows simultaneous processing of vast numbers of DNA or RNA sequences with high speed and accuracy. This enables comprehensive analysis by fragmenting, amplifying, and detecting sequences through high-throughput, parallel processing technologies. Nucleic acid-based techniques are precise but demand more resources and

S. K. Saha
Department of Biotechnology, National Institute of Technology, Durgapur, Paschim Bardhaman, West Bengal, India

A. Dolley · N. D. Namsa (✉)
Department of Molecular Biology and Biotechnology, Tezpur University, Napaam, Sonitpur, Tezpur, Assam, India
e-mail: namsa@tezu.ernet.in

R. Kumar et al. (eds.), *Fruit Crops and Viral Pathogens Volume II*, Plant Pathology for Sustainable Agriculture, https://doi.org/10.1007/978-981-95-7261-8_10

experts. NGS is very costly and complex, but it offers high specificity and sensitivity. Collectively, these tools help identify infected plants quickly, allowing growers to remove or manage diseased trees before the virus spreads further.

Keywords

Rapid detection · Diagnosis · Next-generation sequencing · On-site diagnostic · Identification

10.1 Introduction

The Earth's vast environmental diversity, coupled with continuous changes in climate, profoundly influences the types of agricultural systems practiced across different regions of the world. Based on temperature zones, the Earth can be divided into three regions: Torrid, Temperate, and Frigid (Peel et al. 2007). The weather of the zones determines the types of crops and fruits that will be cultivated at their seasonal time. Identifying risks and challenges associated with fruit production is essential, so understanding these zones is necessary, particularly in the context of plant health and disease management. The torrid zone, which is located near the equator, is also called the tropical region. This zone is known for high temperatures and huge rainfall throughout the year. Fruits like mango, papaya, banana, pineapple, coconut, guava, and lychee grow well in this climate, for optimal growth, warm temperatures, and consistent moisture. This region is abundant in fruits that contribute significantly to the global agricultural trade and the local market. Also, the favorable climate for fruit cultivation supports the proliferation of pests and pathogens, emphasizing the need for effective diagnostic tools for plant diseases (Juroszek and Von Tiedemann 2011). The temperate zone lies in the Northern Hemisphere, between the Cancer and Arctic Circle, and between the Tropic of Capricorn and the Antarctic Circle in the Southern Hemisphere. This region has distinct seasons like spring, summer, autumn, rainy season, and winter which supports the cultivation of fruits such as apple, grape, kiwi, strawberry, blueberry, raspberry, etc. The diverse climate of this region favors crop and fruit growth; it also favors the development of pests and pathogens. Hence, rapid on-site diagnostic tools are essential for ensuring the production of healthy fruits. The Frigid Zone is the most extreme cold region among these three zones. It is located near the poles (Northern Pole and Southern Pole) (Manzoor et al. 2025). These regions experience long, harsh winters and short summer seasons, limiting the variety of cultivated fruits (Osorio-Marín et al. 2024). Despite these challenges, certain hardy fruits such as cloudberry, cranberry, lingonberry, Arctic raspberry, black currant, red currant, sea buckthorn, elderberry, and rowanberry flourish and grow. With the growing period and consistently low temperatures in the frigid zone, fruit crops are especially susceptible to diseases that can quickly destroy harvests (Osorio-Marín et al. 2024). Effective on-site diagnostic tools are essential to control these diseases and protect these fruit crops. Early detection and effective management of these infections are critical to minimizing these impacts. However, conventional diagnostic methods, such as laboratory-based techniques, often require significant time, specialized equipment, and a skilled person. To address these challenges, rapid and sensitive on-site diagnostic tools have emerged as transformative solutions. These tools are particularly valuable in regions where access to advanced laboratory facilities is limited. By enabling farmers and agricultural workers to identify viral infections at an early stage. Based on the genetic materials, plant viruses are classified into DNA and RNA viruses, with further distinctions in genome structure. DNA viruses are again classified into single-stranded DNA (ssDNA) viruses, such as banana bunchy top virus (BBTV), which rely on the host's replication machinery, and double-stranded DNA (dsDNA) viruses, which are less common but possess stable and complex genomes. RNA viruses are more common in plants, are divided into single-stranded RNA (ssRNA) and double-stranded RNA (dsRNA) viruses or segmented viruses. Positive-sense ssRNA viruses, such as citrus tristeza virus (CTV), can use mRNA directly for protein synthesis, while negative-sense ssRNA viruses, like tomato spotted wilt virus (TSWV), require the synthesis of complementary RNA strands for protein production. Additionally, plant viruses have segmented genomes, which are normally seen in TSWV, enhancing genetic reassortment and diversity, or non-segmented genomes, as in cucumber mosaic virus (CMV), which consists of continuous genetic material (Tentchev et al. 2011). The variety of virus forms and the structural diversity of these viruses make it clear how complex the interaction between plants and viruses can be; therefore, the development of new methods of diagnosis will be critical to protect crop yields and to ensure global food supply. A wild variety of diagnostic tools are available out among them the most promising approaches are the lateral flow assays (LFAs). LFAs are inexpensive, user-friendly, and capable of detecting specific fruit viral antigens in fruits. Another innovative technology is loop-mediated isothermal amplification (LAMP). Advances in nanobiotechnology with high sensitivity and specificity can also lead to the development of biosensors that use nanoparticles to detect viral pathogens. The use of these tools has the potential to revolutionize fruit virus diagnostics across all temperature zones. In the Torrid Zone where high humidity and temperature accelerate that can leads the spread of viruses, on-site diagnostic tools can help farmers to identify and manage outbreaks before they cause significant damage. The resource limitations are common in the Frigid Zone, where portable diagnostic tools offer an economical and effective approach to keep safe the fruit crops (Manzoor et al. 2025). In summary, the diversity of fruit crops across different zones highlights the importance of customized strategies for managing diseases. Rapid and sensitive on-site diagnostic tools provide a powerful solution to the challenges caused by fruit viruses, ensuring timely and effective management. These tools enhance global food security, boost economic development, and promote sustainable agriculture.

10.2 Economic Importance of Fruits

Fruits play a vital role at all levels of the market, such as in local, national, as well as global economic markets. Fruit farming employs millions of people, including farmers, laborers, and workers who are involved in the production, transport, and packaging. This supports rural income, which contributes in rural development, and also changes in the lifestyle patterns, increasing the demand for fruits. Processed fruits such as juices, jams, gels, etc., also contribute to the global economy.

10.3 Impact of Fruit Virus

Fruit viruses have a significant effect on agriculture worldwide, endangering food supplies and resulting in significant financial losses. These diseases lower fruit yields, degrade fruit quality, and cause major financial losses for farmers in these countries. Stunted plant development, leaf discoloration, deformation, and early fruit rot are common symptoms of viral infections, which detract from the plants' appearance and lower their market value. For example, the citrus tristeza virus (CTV) causes tree decline and decreased fruit output in citrus crops, whereas the papaya ringspot virus (PRSV) creates ring-like spots on papayas (Gogoi et al. 2019). In a similar vein, banana bunchy top virus (BBTV) damages fruit production by attacking banana plants (Peng et al. 2012). An integrated strategy that incorporates modern biological remedies, vector management, and cultural practices is needed to manage fruit viruses. Crop rotation, the removal of sick plants, and the use of planting materials free of viruses are common strategies used by farmers. The list of viruses and the point-of-care testing kits has been briefly described in Table 10.1.

10.3.1 Apple

Apple in temperate regions is infected by apple mosaic virus (AMV), apple chlorotic leaf spot virus (ACLSV), apple stem grooving virus (ASGV), and apple stem pitting virus (ASPV) (Pedrelli et al. 2025). They cause mosaic, chlorosis, stunting, and reduced fruit size and quality (German-Retana et al. 1997).

10.3.2 Banana

Banana produced by Musa species can be infected by banana bunchy top virus (BBTV). BBTV causes stunting and bunchy top symptoms. Cucumber mosaic virus (CMV) distorts fruit and affects ripening. Banana bract mosaic virus (BBrMV) produces streaks, mosaic, discoloration, reducing fruit quality (Qazi 2016; Bhat et al. 2018).

Table 10.1 Major viruses affecting common fruit crops and their point-of-care testing kits

Fruits	Name of virus	Point-of-care test (POCT) kits	ELISA	PCR	References
Apple	Apple mosaic virus	Lateral flow strip test	ELISA	qPCR	Pedrelli et al. (2025)
Cherry	Cherry mosaic virus	Immunostrip for leaf roll virus	ELISA	RT-PCR	German-Retana et al. (1997)
Watermelon	Watermelon mosaic virus	Immunostrip	ELISA	Real-time PCR	Li et al. (2020)
Pear	Pear stony pit virus	LFA	ELISA	Multiplex PCR	Jelkmann (1994)
Orange	Citrus tristeza virus	Citrus diagnose strip	ELISA	RT-PCR	Belabess et al. (2020)
Citrus	Citrus porosis virus	Psorosis rapid detection kit	ELISA	qPCR	Martin and Tzanetakis (2006)
Kiwi	Kiwi chlorotic ringspot virus	LFA	ELISA	RT-PCR	Blouin et al. (2013)
Avocado	Avocado sunblotch viroid	Field test strip	ELISA	Real-time PCR	Saucedo Carabez et al. (2019)
Grapes	Grapevine fanleaf virus	Immunostrip	ELISA	Multiplex PCR, qRT-PCR and RT-PCR	Maliogka et al. (2015)
Jackfruit	Jackfruit golden mosaic virus	Field diagnostic kit	ELISA	RT-PCR	Cassedy et al. (2022)
Banana	Banana bunchy top virus	Bunchy top rapid test strip	ELISA	Real-time PCR	Peng et al. (2012)
Coconut	Coconut lethal yellowing virus (16SrIV)	Lethal yellowing test kit	ELISA	Real-time PCR (16Sr and GroEL gene)	Cassedy et al. (2022)
Guava	Guava ring spot virus	Ring spot virus detection kit	ELISA	RT-PCR	Hassan (2013)
Apricot	Apricot chlorotic leaf roll virus	Chlorotic lead roll test	ELISA	qPCR	Nemchinov et al. (2000)
Strawberry	Strawberry mild yellow edge virus	Strawberry virus strip test	ELISA	Real-time PCR	Martin and Tzanetakis (2006)

(continued)

Table 10.1 (continued)

Fruits	Name of virus	Point-of-care test (POCT) kits	ELISA	PCR	References
Sapodilla	Sapodilla leaf curl virus	Leaf curl virus field test	ELISA	Real-time PCR	Cassedy et al. (2022)
Papaya	Papaya ringspot virus	Ringspot virus strip test	ELISA	RT-PCR	Gogoi et al. (2019)

10.3.3 Avocado

Avocado is a climacteric, single-seeded berry. It can be infected by avocado sunblotch viroid (ASBVd), avocado latent virus (ALV), and avocado virus 2 (AV2) (Saucedo Carabez et al. 2019). ASBVd is the most damaging and is a single-stranded circular RNA viroid. It causes streaks, discoloration, and malformed fruit.

10.3.4 Papaya

Papaya is a tropical fruit valued for its sweet flesh. Papaya may be affected by many viruses like papaya ringspot virus (PRSV) (Gogoi et al. 2019) and Papaya mosaic virus (PapMV). PRSV is a positive-sense single-stranded RNA virus (Muthulakshmi 2024). It causes Ringspot lesions on fruit, oily streaks symptoms. When it attacks, it gives symptoms like Misshapen fruit, skin cracking (Tuo et al. 2021).

10.3.5 Cherry

Cherries are small, round fruits that are either sweet or tart. They can be infected by prunus necrotic ringspot virus (PNRSV) and little cherry virus (LChV) (Noorani et al. 2023). PNRSV is a positive-sense single-stranded RNA virus that causes ring patterns, necrotic spots, and reduced fruit yield.

10.3.6 Watermelon

Watermelon is a refreshing, juicy fruit with a sweet, watery flesh. Viruses such as watermelon mosaic virus (WMV), zucchini yellow mosaic virus (ZYMV), and cucumber green mottle mosaic virus (CGMMV) can infect watermelon (Li et al. 2020). WMV is a positive-sense single-stranded RNA virus (+ssRNA) causing mottled leaves, distorted fruit, and reduced sweetness.

10.3.7 Pear

Pears are sweet and juicy fruits with a characteristic teardrop shape. They are prone to viruses like apple stem pitting virus (ASPV) and pear leaf scorch virus (PLSV) (Jelkmann 1994). ASPV: A positive-sense single-stranded RNA virus causing stem pitting and reduced fruit quality. PLSV is a positive-sense single-stranded RNA (ssRNA) virus that causes leaf scorch and reduced fruit production.

10.3.8 Orange

Oranges are abundant in vitamin C The orange plant has been the target of many plant viruses, such as citrus tristeza virus (CTV), citrus exocortis viroid (CEVd), and citrus psorosis virus (CPsV). CTV is a positive-sense single-stranded RNA virus (ssRNA) that leads to stem damage, vein clearing, and decline of fruit quality. CEVd is a viroid causing bark scaling, growth suppression, and reduced small fruits. CPsV is a positive-sense single-stranded RNA virus (ssRNA) that causes leaf flecks, lesions on stems, and reduces the quality of fruits (Belabess et al. 2020).

10.3.9 Citrus

Citrus is abundant in vitamin C and commonly targeted by viruses such as citrus tristeza virus (CTV), citrus vein enation virus (CVEV), and citrus tatter leaf virus (CTLV). CTV causes stem damage and declines fruit quality. CVEV is a positive-sense single-stranded RNA virus that leads to vein enations and growth suppression. CTLV results in leaf tearing and reduced fruit yield (Yang et al. 2022; Dou et al. 2022).

10.3.10 Kiwi

Kiwi is a small, nutrient-rich fruit with a fuzzy skin and green or yellow flesh. It is susceptible to actinidia chlorotic ringspot-associated virus (AcCRaV) and cherry leaf roll virus (CLRV) (Blouin et al. 2013).

10.3.11 Grapes

Grapes are sweet or tart fruits grown in clusters. They are affected by grapevine leafroll-associated virus (GLRaV), grapevine fanleaf virus (GFLV), and grapevine virus A (GVA) (Maliogka et al. 2015). GLRaV is a positive-sense single-stranded RNA virus causing leaf reddening and reduced fruit sugar (Constable and Rodoni 2014). GFLV leads to fan-shaped leaves and poor fruit set. GVA causes stunted growth and necrosis in vines (Maliogka et al. 2015).

10.3.12 Guava

Guava is a tropical fruit with a sweet, aromatic flavor. Viruses like guava mosaic virus (GMV), guava fruit rot virus (GFRV), and guava leaf curl virus (GLCV) affect it. GMV causes mosaic patterns and poor fruit development (Hassan 2013). GFRV leads to fruit rot and reduced shelf life. GLCV results in curled leaves and stunted growth.

10.3.13 Apricot

Apricot is a stone fruit known for its sweet, tart flavor. It can be affected by apricot latent virus (ApLV) and apricot pseudo chlorotic leaf spot virus (ApPCLSV) (Levy et al. 2000; Nemchinov et al. 2000). ApLV results in latent infections with occasional fruit quality loss. ApPCLSV leads to chlorotic spots and reduced fruit yield.

10.3.14 Strawberry

Strawberries are small, sweet, red fruits with a distinct aroma. Viruses like strawberry mild yellow edge virus (SMYEV), strawberry mottle virus (SMoV), and strawberry crinkle virus (SCV) can infect them (Martin and Tzanetakis 2006). SMYEV causes yellow leaf edges and reduced fruit quality.

10.4 Rapid Serology-Based Diagnostic Methods

10.4.1 Lateral Flow Assay (LFA)

Lateral flow assay is a widely used method for the detection of fruit viruses. This is a rapid, simple, portable, and cost-effective process. The test can produce results within 5–20 minutes. Proteins or nucleic acids and antibodies are used to identify the presence of a target virus. The main components of an LFA include a sample pad, a conjugation pad, a nitrocellulose membrane (NC), and an absorption pad, each serving a specific function in the test (Fig. 10.1). The main steps to perform the LFA are: (a) preparing antibodies/antigen specific to the target, (b) color label selection and processing, (c) conjugating the label with biorecognition molecules, (d) immobilization detector reagents onto the designated pad, (e) applying the test and control lines onto the nitrocellulose (NC), (f) assembling all components onto a backing card, and (g) applying the sample and obtained results within some minute (Ince and Sezgintürk 2022). The sample pad receives the plant extract. The sample moves across the strip by capillary action. Labeled antibodies bind to the viral antigen and form a complex. The complex appears on the test line as a visible band. The control line confirms the test. Two bands show a positive result. LFAs detect many

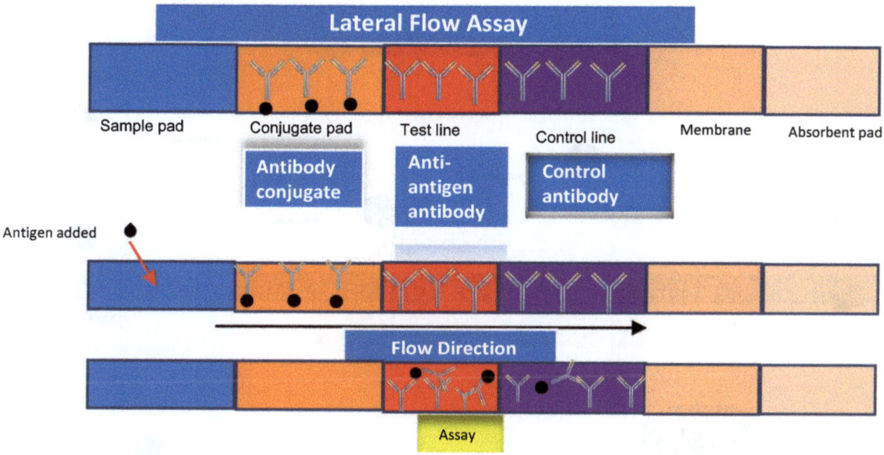

Fig. 10.1 This schematic representation of the lateral flow assay (LFA). The illustration shows a lateral flow strip in which the sample is first added to the sample pad and moves toward the conjugate pad via an absorbent pad, which forms an antigen–antibody complex. The complex continues to migrate via lateral flow, and virus-specific antibodies at the test line capture it. The intensity of the band at the test line reflects how much target antigen is present. The band of the control line verifies that the assay was done properly. A positive result is shown by the presence of two lines, while a single control band indicates a negative result. The absence of any band or only a band that appears on the test line indicates an invalid result

plant viruses quickly and accurately (Thakre et al. 2018; Maejima et al. 2014; Caruso et al. 2024). They help manage diseases and reduce crop losses.

10.4.2 Enzyme-Linked Immunosorbent Assay

Enzyme-linked immunosorbent assay (ELISA) is a widely used immunodiagnostic technique. It helps detect and measure fruit viruses. ELISA comes in several formats. These include direct, indirect, competitive, and sandwich ELISA (Fig. 10.2). ELISA is to be based on the antigen-antibody interaction, utilizing enzyme-labeled antibody to identify specific viral antigens in plants and fruits. The simplest form is Direct ELISA, in which the antigen (Ag) is immobilized on a microplate and detected by an enzyme-linked primary antibody. That method is easier and quicker, but lacks signal amplification and may suffer from lower sensitivity. On the other hand, indirect ELISA, which uses an unlabeled primary antibody, enhances sensitivity through signal amplification and offers greater flexibility than the direct ELISA method. Many antigens are used in the Sandwich ELISA; this is the most common technique (Cassedy et al. 2022). First, add the antibody to the plate. After adding the antibody, the antigen is added. The specific antigen is captured by the immobilized antibody, and then, the unbound entities are washed away by the washing buffer; afterwards, the second antibody is applied to the surface of the plate. This second antibody can either be directly labeled with a fluorescent tag or detected

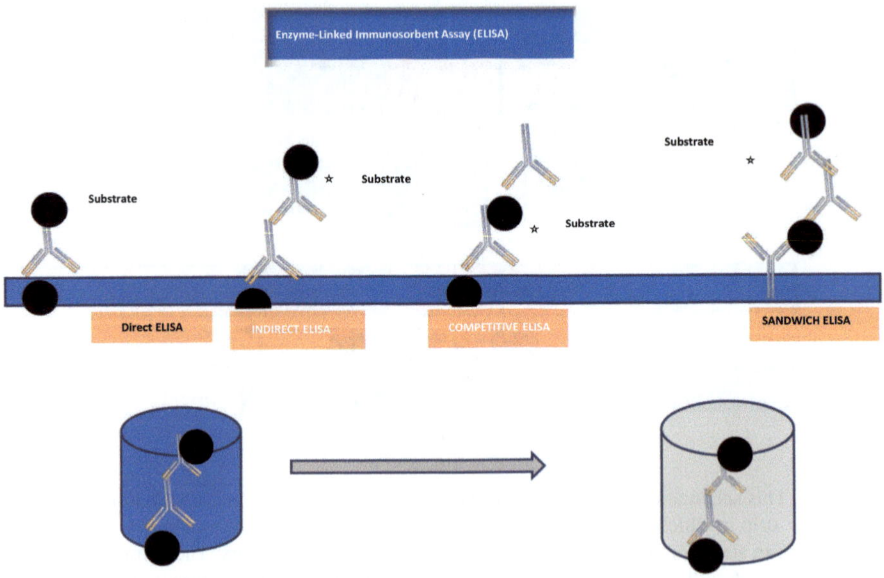

Fig. 10.2 This schematic diagram of an enzyme-linked immunosorbent assay (ELISA). Illustration of different types of ELISA used for the diagnosis of plant viruses. Sandwich ELISA is commonly used in diagnosis

using a secondary antibody that has been labeled with a reporter enzyme. A detection antibody is subsequently attached to the same antigen, which results in the formation of a stable sandwich complex. This configuration is characterized by both high specificity and sensitivity, as it relies on the dual recognition of the target by two distinct antibodies. Although cross-reactivity is often a challenge faced by ELISA tests, it is found that the Sandwich assay format significantly mitigates this issue. This reduction in cross-reactivity makes the assay particularly effective for the direct detection of viruses from complex biological samples. The ELISA test serves as a crucial tool for the detection of plant viruses, leveraging the specificity of viral protein targets. For apple mosaic virus (AMV) and cherry mosaic virus, the test primarily targets the coat proteins (CP) essential for viral encapsulation. In the case of plum pox virus (PPV), the capsid protein (CP) is the basis of detection in the ELISA assay, which detects the coat protein, a commonly used marker for diagnostics of CTV. The detection of the capsid protein (CP) is used as the basis, while the nucleocapsid protein (N protein) is targeted for tomato spotted wilt virus (TSWV) (Fernández-Fernández et al. 2002). In a competitive ELISA format, the antigen present in the sample is more competitive than a labeled antigen, binding to a limited amount of specific antibody. As a result, higher concentrations of the sample antigen reduce the binding of the labeled antigen, producing a weaker signal. This method is more effective than for quantifying specific antigens in ambiguous biological samples (Sakamoto et al. 2018). All things considered, ELISA can remain a cornerstone technique in agricultural diagnostics by enabling early detection and management of viral disease in the fruit crops. The type of reactivity is combined sensitivity, specificity, and adaptability.

10.5 Molecular Diagnostic Test

Molecular diagnostics detect pathogen nucleic acids with high sensitivity and specificity. These tests identify small amounts of viral DNA or RNA. PCR is the main confirmatory method.

10.5.1 Polymerase Chain Reaction

Polymerase chain reaction (PCR) is a cornerstone molecular diagnostic tool for detecting DNA or RNA viruses in fruit crops. The process begins with the extraction of high-quality DNA or RNA from infected plant tissue. Two-step reverse transcription PCR (RT-PCR) is a molecular biology technique that uses separate reaction tubes for the reverse transcription of RNA to cDNA and PCR amplification of target gene of pathogens (Fig. 10.3). Specific primers are used to amplify the conserved region of the viral genome. The amplification of DNA requires three steps that are denaturation, annealing, and extension (Li et al. 2008; Jeong et al. 2014). Millions of copies of the target DNA are generated by amplification which is used for the detection of minute amounts of viral genetic DNA. Agarose gel is used to detect the amplified product. In this gel, distinct band sizes indicate the presence of specific

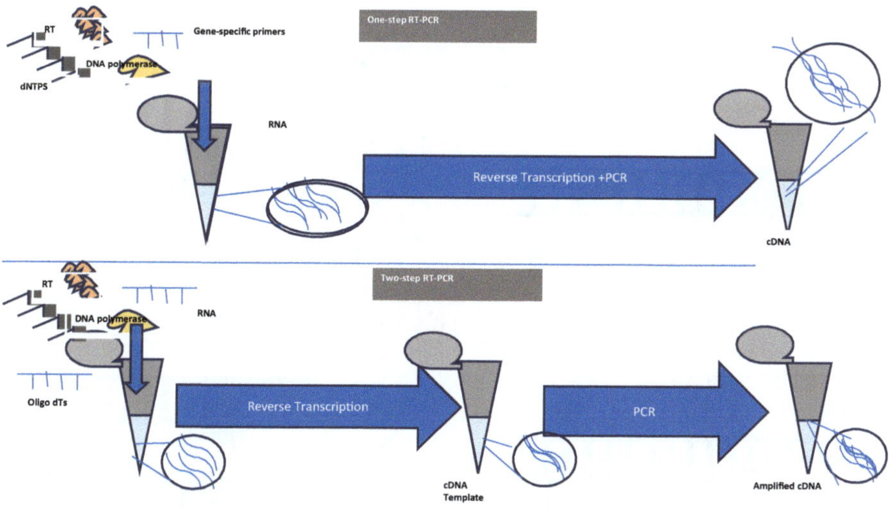

Fig. 10.3 Schematic diagram of one-step versus two-step PCR used for conversion of RNA into cDNA followed by PCR amplification for the target gene of pathogens

viruses. Due to its very high specificity and sensitivity, PCR can detect viruses at very low concentrations of genetic material. It can detect the virus before the symptoms that appear in plants. The main application of PCR in fruit virus diagnostics includes the detection of pathogens like citrus tristeza virus (CTV) and plum pox virus (PPV). The main application of PCR in fruit virus diagnostics is the detection of viral pathogens such as citrus tristeza virus (CTV) and plum pox virus (PPV), typically by targeting specific gene regions that encode the viral coat protein (Kokane et al. 2021; Kim et al. 2008). It needs special laboratory conditions, trained personnel, and specialized equipment. The reliability and adaptability of PCR have become a gold standard in molecular diagnostics (Mirmajlessi et al. 2015). PCR provides rapid and accurate results, that is why it can become one of the essential tools to detect the fruit virus. It can manage plant health and ensure the quality of fruit crop.

10.5.2 Real-Time Quantitative Polymerase Chain Reaction

Real-time quantitative PCR (qPCR) enables the amplification process by the continuous monitoring of target amplicon synthesis. This detection is facilitated using DNA-binding dyes, such as ethidium bromide (ETBR), SYBR Green, or specially designed fluorescent probes (Poojari et al. 2016). DNA-binding dyes are integrated with the double-stranded DNA (dsDNA), meaning primers must be carefully optimized to prevent the generation of non-specific amplicons. That means fluorescent probes offer a more target-specific binding (Fig. 10.4). In qPCR, many types of probes are utilized, with commonly used ones such as hydrolysis and hybridization probes. In qPCR, fluorescence intensity is measured which directly correlates with

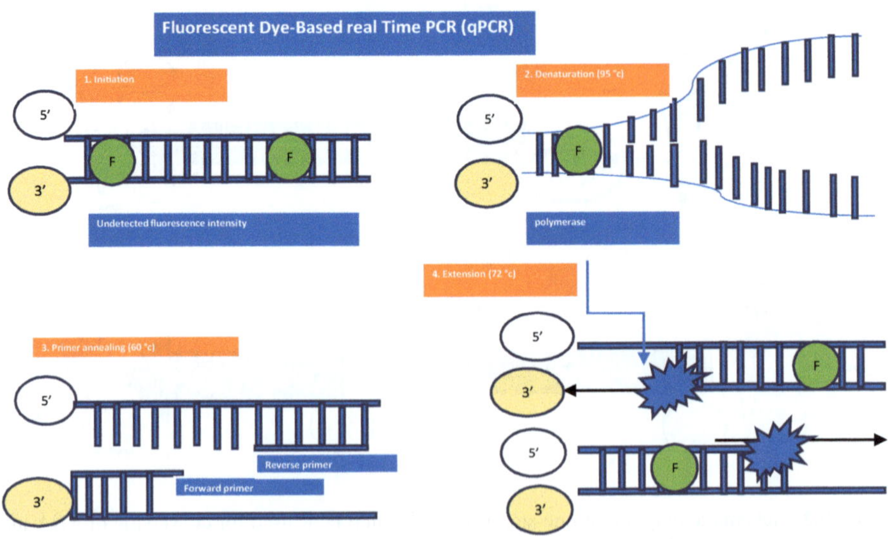

Fig. 10.4 Schematic diagram showing various steps of real-time quantitative PCR (qPCR)

the initial amount of the target nucleic acid. In the sample, swift quantification may happen. QPCR can bring both strengths and limitations during the preparation of rapid virus diagnostic tests. One major advantage of qPCR is that it has specificity when a probe is used like hydrolysis or hybridization types. Both primer and probe must bind to their sequences for generations. In situation where time is very important, an assay that can give accurate and quantitative data within a few hours is precious. QPCR can give relatively quick results. The preparation of the sample can consume time, which hinders overall turnaround time. The extraction of RNA and DNA can be sped up by using commercial kits which are available, but this method is cost-effective (Lau and Botella 2017). The technique may be less accessible in low-resource area because the investment of qPCR is very expensive. It can be able to handle high-throughput testing. This testing has another important requirement for a rapid diagnostic tool. This is because qPCR runs in 96-well plate formats. It can scale up to taste many samples which is feasible (Bittker 2012). Testing of numerous samples is done in industries such as a plant or during an outbreak or pandemic. QPCR requires multiple primers and probe set to detect several viruses in same reaction. The qPCR becomes a powerful platform for simultaneously screening of multiple pathogens of many samples due to the combination of high throughput format and automated format (Kader Jailani and Paret 2024). There are some limitations that how many samples can be processed at once time despite being beneficial. New methods are introduced, such as magnetic bead-based nucleic acid which is faster and cost-effective compared to conventional high-throughput systems. These methods can reduce the cost by preparing many samples in few hours (Xin and Chen 2012). There is an alternative strategy that can improve the efficiency in composite sampling where several samples are tasted as a single unit. By comparing to individual testing, the approach can increase sensitivity (Hamim et al. 2022). If the pooled sample gives a positive result, then further analysis can be done on each component sample which has to pinpoint the infection source. This can reduce the overall processing required for each sample. The other drawback of qPCR is that it can cost high in instruments. There is required for skilled person and for tightly regulated thermal cycling to operate the equipment. Due to these limitations, other nucleic acid detection methods have been developed that particularly allow for isothermal amplification of DNA and RNA, providing potential alternative to qPCR. In qPCR, various viral genes are targeted and ensure high sensitivity and specificity. Replicase (RdRp) genes, encoding RNA-dependent RNA polymerase, have their own conserved nature which plays an essential role in viral replication. So, they are widely used. Viruses are spread between plant cells which is facilitated by the movement protein gene. These viruses are also an effective target. Non-structural protein genes such as HC-Pro in potyviruses or p23 in citrus tristeza virus which are used for strain-specific detection or to study viral suppressors of host defense mechanisms. Coat protein (cp) genes have conserved sequences that are targeted and play role in viral assembly as observed in viruses like AMV, CTV, and PPV. Untranslated regions (UTRs) have conserved sequences so they are targeted for quantifying virus-specific RNA. P25 and p23 in Citrus tristeza virus and the nucleocapsid (N) gene in tomato spotted wilt virus are utilized for specific viruses.

The use of sequence-specific probes, such as TaqMan probes or molecular beacons, enhances the precision of qPCR and enables sensitive and accurate quantification of viral loads in infected plant tissues (Lau and Botella 2017).

10.5.3 Loop-Mediated Isothermal Amplification

To initiate polymerase-driven DNA synthesis, loop-mediated isothermal amplification (LAMP) is the use of multiple primers—typically at least four to six (Notomi et al. 2000). The process of LAMP is schematically represented in Fig. 10.5. This method can rely on the creation of stem-loop DNA structures by these primers, by allowing a strand-displacing polymerase to continuously amplify the target sequence. This process is repeated, yielding substantial quantities of double-stranded DNA (Notomi et al. 2015). Although the recommended temperature for the LAMP reaction is around 65 °C, the research has shown successful amplification across a broader temperature range of 57–67 °C (Tomita et al. 2008). A standard LAMP reaction can take between 30 and 60 minutes. Adding loop primers, as suggested, can further speed up amplification (Nagamine et al. 2002). Additional primer sets, like swarm and stem primers, have also been used to accelerate reaction times (Gandelman et al. 2011; Martineau et al. 2017). For RNA virus detection, reverse transcriptase can be incorporated directly into the LAMP mixture, forming

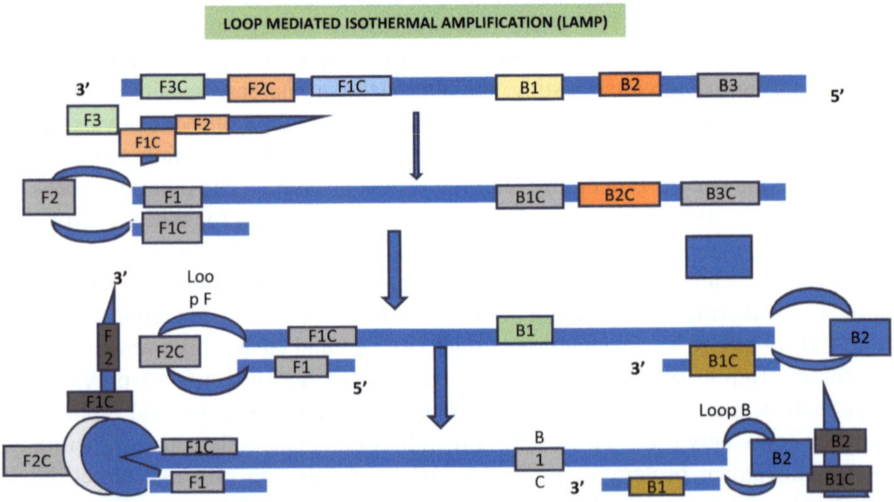

Fig. 10.5 Amplification of target gene via loop-mediated isothermal amplification (LAMP). Diagram of LAMP primers located on the target sequence and the amplification process. The primer binds to its complementary sequence and begins amplifying the first strand, followed by binding and amplification of the second strand by the F3 primer. The following steps involve the amplification of the FIP-amplified strand by the B3 primer. Finally, FIP and BIP continue the LAMP cycle, resulting in dumbbell-shaped amplicons. A full stepwise description of the individual steps is provided in the figure

RT-LAMP that can simplify workflow by enabling detection of both RNA and DNA targets in a single assay (Notomi et al. 2000). LAMP-generated DNA can be detected using the DNA-binding dyes such as ETBR and SYBR Green may be introduced after amplification for end-point detection or added before amplification for monitoring real-time fluorescence changes. Certain dyes can interfere with the reaction, so several types must be tested to identify the most compatible one (Quyen et al. 2019). While fluorescent probes can be used, traditional qPCR probes are often found incompatible with LAMP due to the enzyme's strand displacement activity. New detection systems such as QUASR, DARQ, and mediator displacement probes have been developed to address this problem (Tanner et al. 2012; Ball et al. 2016). Another strategy involves using gold nanoparticles (AuNPs) tagged with sequences complementary to the target DNA. Despite advances, a universal method for integrating probe-based detection with LAMP does not exist, requiring more research to establish standardized formats. A simpler approach involves observing turbidity from magnesium pyrophosphate, a by-product of LAMP reactions, which provides a quick qualitative indication of amplification. For real-time quantification, turbidity can be measured with a turbidimeter and compared to a standard curve (Notomi et al. 2000). In virology, LAMP is recognized as a sensitive and rapid alternative to traditional PCR or qPCR, with no thermal cycling equipment required and a consistent temperature of 65 °C maintained (Tomita et al. 2008). Advancements such as phosphorothioate primers and reaction enhancers have allowed LAMP to operate at temperatures as low as 40 °C, though with longer reaction durations (Cai et al. 2018). LAMP is also robust, enabling direct amplification of DNA from complex samples like blood, urine, feces, and insects, often without prior nucleic acid extraction. The assay has been spread the primer with high specificity due to multiple primers targeting six regions on the DNA, although critical primer design is necessary to ensure correct spacing and product size of 300 bp. Despite design tools, this process remains intricate (Notomi et al. 2015), especially when considering viral genome variability, which complicates the identification of conserved regions for primer binding (Notomi et al. 2015; Cai et al. 2018). When LAMP assays are to be developed and validated, it is crucial to address these challenges. Essential to confirm that the primers efficiently amplify all relevant viral strains, by avoiding false negatives. A significant limitation of LAMP is its vulnerability to carryover contamination. Given its high sensitivity, even minute traces of previously amplified DNA can lead to false-positive results. Contaminants may originate from the environment, researchers' attire, or lab equipment. These residual products can act as unintended templates in subsequent reactions (Hsieh et al. 2014). By reducing this risk, reactions can be kept closed during analysis or designed to include the dUTP instead of dTTP. Carryover amplicons can be degraded then by uracil-DNA glycosylase before the next run, thus minimizing false positives (Hsieh et al. 2014). This issue becomes especially critical when clinically significant viruses are to be tested, as a false-positive result can lead to misdiagnosis and inappropriate treatment. During large-scale outbreaks or pandemics, false positives can strain public health resources by necessitating confirmatory testing to ensure result accuracy. Various viral genes are targeted in LAMP assays for their specificity and conservation. Replicase (RdRp) genes, crucial for viral replication, are frequently selected for their highly conserved sequences across viral strains. Movement protein genes, involved in the cell-to-cell spread of the virus, are also reliable targets. Additionally, coat protein (CP) genes, commonly used for viruses such as AMV, CTV, and PPV, are targeted for their role in virus assembly and immunogenicity. Non-structural genes such as HC-Pro in potyviruses or p23 in Citrus tristeza virus provide strain-specific detection capabilities), especially the 5′ or 3′ UTRs are preferred for their conservation and high diagnostic accuracy.

10.5.4 Recombinase Polymerase Amplification

Recombinase polymerase amplification (RPA) is a rapid, sensitive, and isothermal nucleic acid amplification technique that has proven especially valuable for field and point-of-care diagnostics. It works likely PCR, which relies on repeated cycles of heating and cooling, RPA operates at a constant temperature, typically between 37 and 42 °C. Isothermal, which means RPA can be performed in simple settings without the need for complex laboratory equipment (Kapoor et al. 2017). Three enzymes play a crucial role in RPA—single-strand binding protein (SSB), a recombinase, and polymerase (Piepenburg et al. 2006). The recombinase pairs primers with the homologous sequences in DNA, while the SSB stabilizes the DNA strand, and the polymerase synthesizes new DNA at the primer site. This process enables amplification of target DNA. RPA is extremely sensitive, it can be capable of detecting as few as 1–10 copies of viral DNA or RNA per reaction, which is crucial for identifying low-level viral infections (Piepenburg et al. 2006). For RNA viruses, RPA will be combined with a reverse transcriptase enzyme (RT), allowing for direct amplification of RNA targets in a single step without the need for a separate cDNA synthesis phase. This versatility makes it suitable for detecting a wide range of viruses, including those with RNA genomes (Lobato et al. 2018). Detection of results can be done by real-time fluorescence, lateral flow strips for visual readout, and integration with CRISPR-based systems for enhanced sensitivity and specificity. These features make RPA an ideal choice for rapid, accurate, and accessible virus detection in both clinical and field settings.

10.5.5 CRISPR-Cas-Based Detection

Cas proteins are essential effector molecules within the CRISPR-Cas system, which play an integral role in the clustered regularly interspaced short palindromic repeats (CRISPR-Cas) system (Gootenberg et al. 2017). By controlling guide RNA (gRNA), Cas protein can cleave DNA or RNA in a particular reaction or lateral flow assay (Gootenberg et al. 2017). To facilitate targeting of a specific sequence, the guide RNA of the CRISPR-Cas can be modified. Currently, three major CRISPR-based detection platforms have been developed—SHERLOCK (specific high-sensitivity enzymatic reporter unlocking, used for detecting RNA viruses like Zika and Dengue, and can be adapted for other genetic markers like cancer-associated sequences), DETECTR (DNA Endonuclease-Targeted CRISPR Trans Reporter, applied for detecting pathogens, such as the SARS-CoV-2 virus, and can be used for other genetic markers), and HOLMES (One-Hour Low-cost multipurpose highly efficient system, which could be used for swift detection of target DNA and RNA, using a quenched fluorescent ssDNA reporter as the probe) (Gootenberg et al. 2017).

10.6 Future Directions

Computational methods for quick and on-site fruit virus detection are advancing quickly, especially with the integration of portable biosensors, machine learning, and smartphone-based diagnostics (Yao et al. 2022). While molecular and antigen-antibody tests are dominating this field now, another new process has been developed, which is the next-generation-sequencing (NGS) platform. This process allows scientists to rapidly read and analyze large amounts of DNA or RNA from a sample (Yao et al. 2022). Once the sample is collected from the infected plant, it includes both plant and virus genetic materials. Afterward, adapters and other necessary molecules are attached, and the extracted genetic materials are subsequently processed for sequencing. To make its sequence, bioinformatics software is used to analyze the data and distinguish the viral genetic materials from the host's, which enables it to make it possible to detect known and new viruses.

10.7 Next-Generation Sequencing

Next-generation sequencing (NGS) is an advanced molecular diagnostic tool that allows the comprehensive detection and characterization of plant viruses. NGS operates by sequencing the entire genome or transcriptome of a sample, which provides an unbiased approach to identify known and new viruses. Cas proteins are the key component of the CRISPR-Cas system, which can be combined with NGS for targeted enrichment of viral nucleic acids which enhancing the specificity and sensitivity of detection. Detected by guide RNA (gRNA), these Cas proteins can cut DNA or RNA at precise locations, allowing for selective sequencing of viral genetic material. The guide RNA can be customized to recognize the target-specific sequences, such as replicase (RdRp) genes, coat protein (CP) genes, or unique viral such as HC-Pro in potyviruses or p23 in citrus tristeza virus. NGS systems can also incorporate CRISPR-based detection tools such as SHERLOCK, DETECTR, and HOLMES for precise and sensitive analysis. By sequencing conserved regions such as untranslated regions or entire viral genomes, NGS provides detailed information on viral diversity, evolution, and strain-specific variations. The high throughput, accuracy, and ability to identify co-infections or novel pathogens make NGS a transformative tool in plant virus diagnostics and epidemiology (Adams et al. 2009).

10.8 Nanotechnology-Based Detection

This nanobiotechnology-based virus detection technique is used to increase sensitivity, specificity, and detection speed. These techniques take advantage of the special qualities of nanoparticles, including their high surface-to-volume ratios, capacity, and optical characteristics, to functionalize with a broad spectrum of biological molecules. Gold nanoparticles are used by the colorimetric and fluorescence-based assays (Narware et al. 2025). Other than there are different nanoparticles which can be used in this field, like nanotubes, graphene oxide, silica, carbon, magnetic nanoparticles, etc. In order to reduce mistakes or errors, the immobilization techniques of the relevant nanomaterials and biological elements are essential.

10.9 Limitations and Challenges

Management of fruit viral diseases requires an effective diagnosis method that can give early results for analyzing and preventing diseases. To date, various methods have been developed for the diagnosis of fruit viruses. There are many limitations and challenges for the diagnosis of viruses. The limitations are briefly described. Field-based tests (e.g., lateral flow devices or colorimetric assays) often have lower sensitivity compared to laboratory-based methods like PCR or ELISA. Risk of false positives/negatives, especially with low viral loads or early infections. On-site settings, especially in rural or remote areas, often lack reliable power sources, proper storage conditions (e.g., refrigeration), and access to advanced diagnostic tools. The environmental influences such as temperature, humidity, and dust can affect reagent stability, device performance, and sample integrity. Portable molecular diagnostic tools (like portable PCR or LAMP kits) are still expensive for smallholder farmers. Some tools require single-use components that may not be readily available or affordable. Many fruit viruses have similar symptoms (e.g., leaf curl, yellowing, mottling), making visual diagnosis unreliable. Mixed infections can mask or alter symptoms, complicating identification. Diagnostic reagents used on-site may require cold storage and have a limited shelf-life, especially in tropical climates. Lack of standardization and regulatory oversight in some regions can lead to poor-quality diagnostic kits and inconsistent results.

10.10 Conclusion

Timely identification of fruit viruses supports good crop production and reduces economic losses. For effective disease management, rapid diagnostics tools are essential. Due to higher sensitivity and specificity, traditional molecular techniques and antibody-based assays like antigen-antibody tests remain widely used. Advanced technologies are enhancing the speed and scope of viral diagnostics. While an immunoassay is cost-effective and user-friendly, its sensitivity is not as good as compared to molecular diagnosis. The molecular techniques are very sensitive and very specific, but they have a fear of contamination and need trained laboratory expert people, also the cost is very high, and contamination leads to incorrect results analysis. Next-generation sequencing (NGS) offers a powerful, comprehensive detection and identification of both known and unknown viruses.

References

Adams IP, Glover RH, Monger WA, Mumford R, Jackeviciene E, Navalinskiene M, Samuitiene M, Boonham N (2009) Next-generation sequencing and metagenomic analysis: a universal diagnostic tool in plant virology. Mol Plant Pathol 10(4):537–545

Ball CS, Light YK, Koh CY, Wheeler SS, Coffey LL, Meagher RJ (2016) Quenching of unincorporated amplification signal reporters in reverse-transcription loop-mediated isothermal amplification enabling bright, single-step, closed-tube, and multiplexed detection of RNA viruses. Anal Chem 88(7):3562–3568

Belabess Z, Sagouti T, Rhallabi N, Tahiri A, Massart S, Tahzima R, Lahlali R, Jijakli MH (2020) Citrus psorosis virus: current insights on a still poorly understood ophiovirus. Microorganisms 8(8):1197

Bhat AI, Pamitha NS, Gopika A, Biju CN (2018) Complete genome sequencing of banana bract mosaic virus isolate infecting cardamom revealed its closeness to banana infecting isolate from India. VirusDisease 29(2):212–215

Bittker JA (2012) High-throughput RT-PCR for small-molecule screening assays. Curr Protoc Chem Biol 4(1):49–63

Blouin AG, Pearson MN, Chavan RR, Woo EN, Lebas BS, Veerakone S, Ratti C, Biccheri R, MacDiarmid RM, Cohen D (2013) Viruses of kiwifruit (Actinidia species). J Plant Pathol:221–235

Cai S, Jung C, Bhadra S, Ellington AD (2018) Phosphorothioated primers lead to loop-mediated isothermal amplification at low temperatures. Anal Chem 90(14):8290–8294

Caruso AG, Ragona A, Agrò G, Bertacca S, Yahyaoui E, Galipienso L, Rubio L, Panno S, Davino S (2024) Rapid detection of tomato spotted wilt virus by real-time RT-LAMP and in-field application. J Plant Pathol 106(2):697–712

Cassedy A, Della Bartola M, Parle-McDermott A, Mullins E, O'Kennedy R (2022) A one-step reverse transcription recombinase polymerase amplification assay for lateral flow-based visual detection of PVY. Anal Biochem 642:114526

Constable F, Rodoni B (2014) Grapevine leafroll-associated viruses. Wine Aust 4:1–5

Dou R, Huang Q, Hu T, Yu F, Hu H, Wang Y, Zhou X, Qian Y (2022) Molecular variation and genomic function of citrus vein enation virus. Int J Mol Sci 24(1):412

Fernández-Fernández MR, Martínez-Torrecuadrada JL, Roncal F, Domínguez E, García JA (2002) Identification of immunogenic hot spots within plum pox potyvirus capsid protein for efficient antigen presentation. J Virol 76(24):12646–12653

Gandelman O, Jackson R, Kiddle G, Tisi L (2011) Loop-mediated amplification accelerated by stem primers. Int J Mol Sci 12(12):9108–9124

German-Retana S, Bergey B, Delbos RP, Candresse T, Dunez J (1997) Complete nucleotide sequence of the genome of a severe cherry isolate of apple chlorotic leaf spot trichovirus (ACLSV). Arch Virol 142(4):833–841

Gogoi SH, Nath PD, Thakuria N, Gogoi S, Das B, Raj K (2019) Molecular detection and characterization of papaya ring spot virus (PRSV) disease in Jorhat district of Assam, India. Int J Curr Microbiol Appl Sci 8(2):1564–1571

Gootenberg JS, Abudayyeh OO, Lee JW, Essletzbichler P, Dy AJ, Joung J, Verdine V, Donghia N, Daringer NM, Freije CA, Myhrvold C (2017) Nucleic acid detection with CRISPR-Cas13a/C2c2. Science 356(6336):438–442

Hamim I, Suzuki JY, Borth WB, Melzer MJ, Wall MM, Hu JS (2022) Preserving plant samples from remote locations for detection of RNA and DNA viruses. Front Microbiol 13:930329

Hassan HM (2013) Identification studies of a virus disease affecting guava in Minia Governorate. Egypt J Phytopathol 41(1):133–144

Hsieh K, Mage PL, Csordas AT, Eisenstein M, Soh HT (2014) Simultaneous elimination of carryover contamination and detection of DNA with uracil-DNA-glycosylase-supplemented loop-mediated isothermal amplification (UDG-LAMP). Chem Commun 50(28):3747–3749

Ince B, Sezgintürk MK (2022) Lateral flow assays for viruses diagnosis: up-to-date technology and future prospects. TrAC Trends Anal Chem 157:116725

Jelkmann W (1994) Nucleotide sequences of apple stem pitting virus and of the coat protein gene of a similar virus from pear associated with vein yellows disease and their relationship with potex-and carlaviruses. J Gen Virol 75(7):1535–1542

Jeong JJ, Ju HJ, Noh J (2014) A review of detection methods for the plant viruses. Res Plant Dis 20(3):173–181

Juroszek P, Von Tiedemann A (2011) Potential strategies and future requirements for plant disease management under a changing climate. Plant Pathol 60(1):100–112

Kader Jailani AA, Paret ML (2024) A multiplex RT-qPCR assay for simultaneous detection of cucurbit viruses from individual whitefly and plant samples. Plant Dis 108(12):3508–3517

Kapoor R, Srivastava N, Kumar S, Saritha RK, Sharma SK, Jain RK, Baranwal VK (2017) Development of a recombinase polymerase amplification assay for the diagnosis of banana bunchy top virus in different banana cultivars. Arch Virol 162(9):2791–2796

Kim WS, Stobbs LW, Lehman SM, James D, Svircev AM (2008) Direct real-time PCR detection of Plum pox virus in field surveys in Ontario. Can J Plant Pathol 30(2):308–317

Kokane SB, Misra P, Kokane AD, Gubyad MG, Warghane AJ, Surwase D, Reddy MK, Ghosh DK (2021) Development of a real-time RT-PCR method for the detection of citrus tristeza virus (CTV) and its implication in studying virus distribution in planta. 3 Biotech 11(10):431

Lau HY, Botella JR (2017) Advanced DNA-based point-of-care diagnostic methods for plant diseases detection. Front Plant Sci 8:2016

Levy L, Damsteegt V, Scorza R, Kolber M (2000) Plum pox potyvirus disease of stone fruits. Online. APSnet Feature

Li R, Mock R, Huang Q, Abad JO, Hartung J, Kinard G (2008) A reliable and inexpensive method of nucleic acid extraction for the PCR-based detection of diverse plant pathogens. J Virol Methods 154(1–2):48–55

Li X, Bi X, An M, Xia Z, Wu Y (2020) iTRAQ-based proteomic analysis of watermelon fruits in response to Cucumber green mottle mosaic virus infection. Int J Mol Sci 21(7):2541

Lobato IM, O'Sullivan CK (2018) Recombinase polymerase amplification: Basics, applications and recent advances. Trends Analyt Chem 98:19–35

Maejima K, Himeno M, Netsu O, Ishikawa K, Yoshida T, Fujita N, Hashimoto M, Komatsu K, Yamaji Y, Namba S (2014) Development of an on-site plum pox virus detection kit based on immunochromatography. J Gen Plant Pathol 80(2):176–183

Maliogka VI, Martelli GP, Fuchs M, Katis NI (2015) Control of viruses infecting grapevine. Adv Virus Res 91:175–227. Academic Press

Manzoor S, Nabi SU, Ayaz A, Rasool B, Sharma SK, Chesti MH et al (2025) Unveiling viral threats to temperate pome fruits: characterization, transmission, and sustainable management strategies. Front Microbiol 16:1560720

Martin RR, Tzanetakis IE (2006) Characterization and recent advances in detection of strawberry viruses. Plant Dis 90(4):384–396

Martineau RL, Murray SA, Ci S, Gao W, Chao SH, Meldrum DR (2017) Improved performance of loop-mediated isothermal amplification assays via swarm priming. Anal Chem 89(1):625–632

Mirmajlessi SM, Loit E, Maend M, Mansouripour SM (2015) Real-time PCR applied to study on plant pathogens: potential applications in diagnosis-a review. Plant Prot Sci 51(4):177–190

Muthulakshmi P (2024) Integrated management of papaya ring spot virus (PRSV) in papaya. Madras Agric J 110(december (10–12)):1

Nagamine K, Hase T, Notomi TJ (2002) Accelerated reaction by loop-mediated isothermal amplification using loop primers. Mol Cell Probes 16(3):223–229

Narware J, Chakma J, Singh SP, Prasad DR, Meher J, Singh P, Bhargava P, Sawant SB, Pitambara, Singh JP, Manzar N (2025) Nanomaterial-based biosensors: a new frontier in plant pathogen detection and plant disease management. Front Bioeng Biotechnol 13:1570318

Nemchinov LG, Shamloul AM, Zemtchik EZ, Verderevskaya TD, Hadidi A (2000) Apricot latent virus: a new species in the genus Foveavirus. Arch Virol 145(9):1801–1813

Noorani MS, Baig MS, Khan JA, Pravej A (2023) Whole genome characterization and diagnostics of prunus necrotic ringspot virus (PNRSV) infecting apricot in India. Sci Rep 13(1):4393

Notomi T, Okayama H, Masubuchi H, Yonekawa T, Watanabe K, Amino N, Hase T (2000) Loop-mediated isothermal amplification of DNA. Nucleic Acids Res 28(12):e63

Notomi T, Mori Y, Tomita N, Kanda H (2015) Loop-mediated isothermal amplification (LAMP): principle, features, and future prospects. J Microbiol 53(1):1–5

Osorio-Marín J, Fernandez E, Vieli L, Ribera A, Luedeling E, Cobo N (2024) Climate change impacts on temperate fruit and nut production: a systematic review. Front Plant Sci 15:1352169

Pedrelli A, Nali C, Panattoni A, Pellegrini E, Cotrozzi L (2025) Apple chlorotic leaf spot virus infection affects the physiology and biochemistry of leaves and the quality of fruits of Tuscany (Italy) autochthonous apple varieties. Plant Pathol 74(1):158–170

Peel MC, Finlayson BL, McMahon TA (2007) Updated world map of the Köppen-Geiger climate classification. Hydrol Earth Syst Sci 11(5):1633–1644

Peng J, Zhang J, Xia Z, Li Y, Huang J, Fan Z (2012) Rapid and sensitive detection of Banana bunchy top virus by loop-mediated isothermal amplification. J Virol Methods 185(2):254–258

Piepenburg O, Williams CH, Stemple DL, Armes NA (2006) DNA detection using recombination proteins. PLoS Biol 4(7):e204

Poojari S, Alabi OJ, Okubara PA, Naidu RA (2016) SYBR® Green-based real-time quantitative reverse-transcription PCR for detection and discrimination of grapevine viruses. J Virol Methods 235:112–118

Qazi J (2016) Banana bunchy top virus and the bunchy top disease. J Gen Plant Pathol 82(1):2–11

Quyen TL, Ngo TA, Bang DD, Madsen M, Wolff A (2019) Classification of multiple DNA dyes based on inhibition effects on real-time loop-mediated isothermal amplification (LAMP): prospect for point of care setting. Front Microbiol 10:2234

Sakamoto S, Putalun W, Vimolmangkang S, Phoolcharoen W, Shoyama Y, Tanaka H, Morimoto S (2018) Enzyme-linked immunosorbent assay for the quantitative/qualitative analysis of plant secondary metabolites. J Nat Med 72(1):32–42

Saucedo Carabez JR, Téliz Ortiz D, Vallejo Pérez MR, Beltrán PH (2019) The Avocado sunblotch viroid: an invisible foe of avocado. Viruses 11(6):491

Tanner NA, Zhang Y, Evans TC Jr (2012) Simultaneous multiple target detection in real-time loop-mediated isothermal amplification. Biotechniques 53(2):81–89

Tentchev D, Verdin E, Marchal C, Jacquet M, Aguilar JM, Moury B (2011) Evolution and structure of Tomato spotted wilt virus populations: evidence of extensive reassortment and insights into emergence processes. J Gen Virol 92(4):961–973

Thakre RA, Gahukar SJ, Akhare AA, Mane SS, Ghosh DK, Hajare SW (2018) Molecular techniques for detection of Citrus tristeza virus affecting citrus species of Vidarbha Region in Maharashtra, India. Int J Curr Microbiol App Sci:2319–7692

Tomita N, Mori Y, Kanda H, Notomi T (2008) Loop-mediated isothermal amplification (LAMP) of gene sequences and simple visual detection of products. Nat Protoc 3(5):877–882

Tuo D, Yan P, Zhao G, Cui H, Zhu G, Liu Y, Yang X, Wang H, Li X, Shen W, Zhou P (2021) An efficient papaya leaf distortion mosaic potyvirus vector for virus-induced gene silencing in papaya. Hortic Res 8:144

Xin Z, Chen J (2012) A high throughput DNA extraction method with high yield and quality. Plant Methods 8(1):26

Yang HJ, Choi EJ, Lee SH, Woo JH, Lim S, Lee SJ (2022) Complete genome sequence of citrus vein enation virus identified from a Korean Yuja Tree. Microbiol Resour Announce 11(8):e00424-22

Yao Z, Coatsworth P, Shi X, Zhi J, Hu L, Yan R, Güder F, Yu HD (2022) Paper-based sensors for diagnostics, human activity monitoring, food safety and environmental detection. Sens Diagn 1(3):312–342

Tissue Culture and the Role of Healthy Planting Materials in Preventing Virus Spread in Fruit Crops

11

Fredmoore Orosco and Mark Angelo O. Balendres

Abstract

Viral diseases cause persistent threats to fruit crop production worldwide, often causing severe yield losses, compromised fruit quality, and shortened orchard lifespans. Unlike bacterial or fungal pathogens, plant viruses lack effective curative treatments, making prevention through the use of healthy, virus-free planting materials the cornerstone of sustainable fruit crop management. This chapter examines the critical role of tissue culture technologies in the production and supply of virus-free planting materials for major fruit crops, including banana, citrus, apple, grapevine, and berries. The discussion covers the diverse spectrum of economically significant fruit viruses, their transmission routes, and their profound impact on productivity and industry sustainability. Detailed protocols and principles of plant tissue culture are described, with an emphasis on meristem-tip culture, thermotherapy, chemotherapy, cryotherapy, and advanced combined approaches for the effective elimination of viruses. The chapter highlights the importance of accurate virus detection and multi-stage indexing, integrating serological, molecular, and next-generation sequencing techniques. Certification

F. Orosco
Virology and Vaccine Research Program, Industrial Technology Development Institute, Department of Science and Technology, Bicutan, Taguig, Philippines

S&T Fellows Program, Department of Science and Technology, Bicutan, Taguig, Philippines

Department of Biology, College of Science, University of the Philippines Manila, Manila, Philippines

M. A. O. Balendres (✉)
Department of Biology, College of Science, De La Salle University, Manila, Philippines

Plant and Soil Health Research Unit, Center for Natural Science and Environmental Research, College of Science, De La Salle University, Manila, Philippines
e-mail: mark.angelo.balendres@dlsu.edu.ph

R. Kumar et al. (eds.), *Fruit Crops and Viral Pathogens Volume II*, Plant Pathology for Sustainable Agriculture, https://doi.org/10.1007/978-981-95-7261-8_12

systems, both national and international, are reviewed for their role in safeguarding the health of propagated materials from laboratory to nursery and ultimately to growers. The chapter concludes with practical recommendations for research, policy, and industry to enhance the effectiveness and accessibility of clean plant programs, supporting the global fruit industry's resilience and long-term productivity in the face of evolving viral challenges.

11.1 Introduction

Fruit crops contribute substantially to global food systems and rural economies. These crops, which include banana, citrus, apple, grape, and numerous berries, supply essential nutrients, create employment, and support export markets in many regions. Their perennial nature and vegetative propagation methods, however, make them highly vulnerable to viral pathogens. Unlike annual field crops, fruit trees and vines are typically established for many years, meaning that a single viral introduction can result in cumulative and long-lasting economic losses (Moreno et al. 2007). Viral diseases are among the most persistent and destructive challenges facing fruit crop production. Viruses such as Banana bunchy top virus (BBTV), Citrus tristeza virus (CTV), Apple stem grooving virus (ASGV), and Plum pox virus (PPV) have caused catastrophic yield declines and large-scale tree losses in orchards worldwide (Cambra et al. 2024; Moreno et al. 2007). These pathogens are capable of spreading through multiple routes, including vegetative cuttings, infected budwood, rootstocks, and even seeds in some species. Because effective chemical or biological cures for plant viruses do not exist, infection is typically irreversible, resulting in chronic reductions in fruit quantity and quality throughout the productive lifespan of a planting (Cambra et al. 2024). The economic burden of viral diseases extends far beyond immediate yield losses. Orchards suffering from virus infections often exhibit shortened productive lifespans, necessitating premature replanting. Marketable fruit quality declines, affecting both domestic consumption and export potential. The associated costs of surveillance, diagnostics, and the removal of infected trees can place a considerable strain on growers, cooperatives, and entire industries. For example, CTV alone has caused the death of more than 100 million citrus trees globally, while PPV has resulted in billions of euros in losses across the European stone fruit industry (Cambra et al. 2024; Moreno et al. 2007). Given the lack of curative options, the most effective strategy to control virus spread in fruit crops is prevention. This begins with the use of healthy, virus-free planting materials. Establishing orchards from certified clean material reduces the risk of introducing and spreading viral infections in newly planted areas. When planting stock is free from viruses, growers benefit from improved yields, better fruit quality, and extended orchard longevity. National and international programs have demonstrated that the adoption of virus-free material can significantly enhance the viability and sustainability of fruit industries, particularly in areas with a history of devastating viral outbreaks (Ravelonandro 2021). Tissue culture technologies have become central to the production of virus-free planting materials in fruit crops. These techniques exploit the principle that viruses often fail to invade the smallest, most actively dividing cells in shoot meristems. Through precise excision and in vitro propagation of these tissues, it is possible to regenerate whole plants that are free from systemic infection. Tissue culture methods, including meristem-tip culture, thermotherapy, and somatic embryogenesis, have demonstrated high rates of virus elimination and the capacity for rapid multiplication of disease-free stock (Bettoni et al. 2024). The integration of tissue culture into clean plant programs has enabled the large-scale distribution of healthy material, supporting the long-term health and productivity of fruit orchards worldwide.

11.2 Fruit Viruses in Major Crops

Fruit crops support the diets and economies of millions worldwide. They also serve as key export commodities, providing livelihoods for rural communities. However, the long-term sustainability of these crops is constantly threatened by viral pathogens. Viruses are especially problematic for fruit crops because these plants are predominantly propagated vegetatively, which facilitates the long-term persistence and spread of viruses within orchards and across regions (Cambra et al. 2000).

11.2.1 Major Fruit Crops Affected by Viruses

11.2.1.1 Banana

Banana production faces persistent threats from several major viruses. Banana bunchy top virus (BBTV) is widely regarded as the most devastating, causing severe dwarfing, stunted growth, and almost total loss of fruit production in susceptible cultivars. Infected plants often display dark green streaks on petioles and a "bunched" leaf arrangement. The main vector for BBTV is the banana aphid, *Pentalonia nigronervosa*, which facilitates both local and regional transmission (Das and Banerjee 2018). Banana streak virus (BSV) adds complexity due to its ability to integrate into the banana genome, sometimes reactivating under stress and spreading through both vector transmission and propagation of infected suckers (Kumar et al. 2015). Cucumber mosaic virus (CMV) also infects banana, resulting in mosaic symptoms and yield loss, with transmission occurring via aphids and contaminated tools.

11.2.1.2 Citrus

Citrus growers have long faced persistent challenges from a variety of viral pathogens. Citrus tristeza virus (CTV) stands out for its destructive potential. Outbreaks can wipe out entire blocks of susceptible trees in just a few seasons, sometimes leaving only a handful of survivors. In other instances, trees linger with chronic decline, their productivity ebbing year by year. Transmission occurs primarily through aphid vectors, especially *Toxoptera citricida*, which move quickly between trees and are difficult to fully control. CTV is also notorious for its ability to spread via infected budwood, a risk that lingers in propagation if certification is lax.

Another threat, Citrus psorosis virus (CPsV), is marked by bark scaling and flecking, gradually weakening affected trees. Meanwhile, Citrus variegation virus (CVV) causes leaf variegation and a significant decline in fruit set, a scenario most commonly observed in young orchards and commercial nurseries. Nurseries, in particular, experience rapid disease spread when virus status goes unchecked (Abbas et al. 2006).

11.2.1.3 Stone Fruits

Viral diseases in stone fruits, such as peaches, plums, apricots, and cherries, can devastate orchards. Plum pox virus (PPV), widely known as the agent of sharka, is especially feared in Europe and beyond. Its telltale chlorotic rings on leaves and fruit, along with fruit deformation and premature drop, can erase an entire season's income. PPV moves both through aphid flights transmitted in a matter of minutes and, more insidiously, via infected nursery stock traded across borders. Other common viruses, such as Prunus necrotic ringspot virus (PNRSV) and Prune dwarf virus (PDV), often remain hidden until trees show signs of chronic yield loss or early decline. The situation worsens when multiple viruses co-infect a single tree, leading to complex symptoms and more pronounced damage (García et al. 2014).

11.2.1.4 Grapevine

Grapevine cultivation brings its own viral challenges. Grapevine leafroll-associated virus 3 (GLRaV-3) is notorious for causing leafroll disease. Affected vines develop downward-curled leaves and often lag behind in ripening, with clusters that fail to achieve their full sugar potential. In wine regions, this translates directly to economic loss, as sugar content and uniform maturity are prized. Grapevine fanleaf virus (GFLV) is another common problem, causing malformed leaves, stunted shoots, and reduced fruit set. Growers sometimes see entire vineyard sections with shortened internodes and weak canopy growth. *Grapevine Pinot gris virus* (GPGV) also contributes to reduced vine vigor, producing mottled leaves and uneven development.

11.2.1.5 Berries

Berry crops, such as strawberries and raspberries, are not exempt from viral threats. Strawberry mild yellow edge virus (SMYEV) is one of the most economically significant viruses in strawberry production, leading to yellowing at leaf margins and stunting. Raspberry bushy dwarf virus (RBDV) affects raspberries by reducing fruit set and causing "crumbly" fruit syndrome. Viruses in berry crops are commonly spread through infected runner plants, pollen, and in some cases, seed (Martin and Tzanetakis 2015).

11.2.2 Transmission Routes and Epidemiology

The primary route of virus transmission in fruit crops is through vegetative propagation. Unlike seed-propagated field crops, fruit trees and vines are usually multiplied by cuttings, grafting, budding, or tissue culture. This practice, while maintaining genetic uniformity, also ensures that any systemic virus present in the parent plant will be passed on to the progeny. The risk is amplified by the long productive lifespan of these crops, which allows viruses to accumulate and persist in commercial plantings over many years (Kumar et al. 2023). Insect vectors are major contributors to the spread of fruit viruses. Aphids, mealybugs, whiteflies, and nematodes each play a role in the natural movement of viruses within and between orchards. The movement of infected plant material between regions, either intentionally (through trade) or unintentionally (via contaminated equipment), has been responsible for several major epidemics, including the introduction of PPV into North America and the global dissemination of CTV. Certain viruses, such as BSV, are further complicated by endogenous forms within the plant genome, which reactivates under specific environmental conditions (Garzo et al. 2020). Environmental factors, such as temperature and rainfall, can influence the abundance of insect vectors and the efficiency of virus transmission. Human activities, including nursery production, grafting, and the transport of planting materials, have historically been the most significant drivers of long-distance and international virus movement. Once established, some viruses can be disseminated over large areas by vectors, resulting in new outbreaks that may go undetected until significant losses occur (Kumar et al. 2023).

11.2.3 Impact on Productivity, Longevity, and Quality

Viral infections have a significant impact on fruit crop productivity. Infected plants often exhibit chronic reductions in yield and fruit size, increased susceptibility to secondary pathogens, and overall decline in vigor. For example, fields affected by BBTV in banana or CTV in citrus can lose nearly all productive capacity if infections are left unchecked (Manzoor et al. 2025). Virus-infected orchards often have shorter lifespans, resulting in premature replanting and increased production costs. The impact is especially striking in grapevine cultivation. Leafroll disease, for instance, has been shown to cut sugar levels in grapes by up to 25%. This reduction directly affects both the taste and value of finished wine, sometimes causing substantial financial setbacks for vineyard owners and wine producers alike (Zhang et al. 2022). Perennial fruit crops present a unique challenge. Viruses persist year after year, steadily undermining orchard health unless managed proactively. In response, many successful operations have turned to certified clean planting material generated through tissue culture. Certification systems, combined with regular on-farm inspections, remain the most reliable methods for maintaining orchards' productivity and profitability over the long term (Bettoni et al. 2024). The rapid evolution and continued spread of fruit viruses only strengthen the case for these integrated, science-based strategies that form the backbone of resilient fruit production worldwide.

11.3 Tissue Culture Principles and Methods for Fruit Crops

Tissue culture stands as a cornerstone of modern fruit crop propagation (Fig. 11.1). The principle behind it is cellular totipotency, the remarkable ability of a single plant cell, when provided with the right cues, to develop into a complete, healthy plant. This foundation has enabled the creation of efficient protocols for multiplying elite genotypes and for purging viruses from stock that would otherwise undermine commercial yields and orchard lifespans. This biological principle forms the foundation for all modern micropropagation and virus elimination methods (Morel and Martin 1952). Totipotency is harnessed by placing cells or tissues in a sterile, nutrient-rich medium that provides essential minerals, vitamins, plant hormones, and carbon sources. The Murashige and Skoog (MS) medium remains the standard base for most protocols, with modifications tailored to species or specific explants (Murashige and Skoog 1962). Maintaining sterility is critical throughout tissue culture processes. Fruit crop explants, such as shoot tips, meristems, or nodal segments, are first surface-sterilized using combinations of ethanol, sodium hypochlorite, or commercial disinfectants. All manipulations occur in laminar flow cabinets or clean benches to prevent microbial contamination. The preparation of culture media, sterilization by autoclaving, and careful handling of explants ensure an aseptic environment that supports healthy plant regeneration (Anđelić et al. 2024). Once in culture, explants are exposed to a carefully controlled combination of plant growth regulators, typically cytokinins and auxins. The balance and concentration of these hormones direct the developmental pathway of the tissue, promoting shoot induction, root formation, or callus production as needed. Take the case of hormone balance. When the ratio of cytokinins to auxins is increased, shoot production accelerates, a method widely used to rapidly increase the number of disease-free plants. This hormone-driven shoot proliferation has become routine in micropropagation labs focused on delivering elite, virus-free stock (Sale et al. 2023).

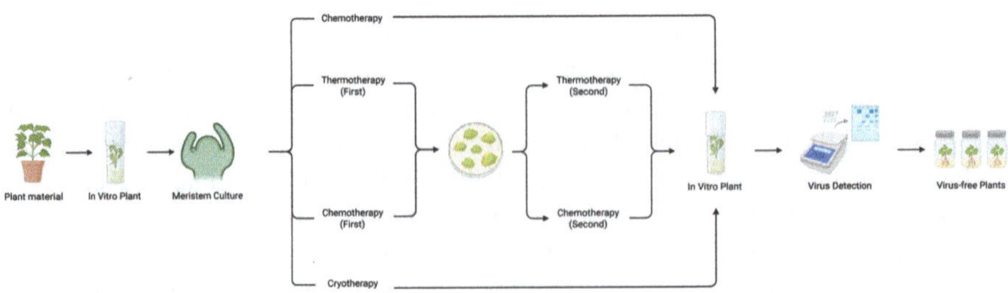

Fig. 11.1 Tissue culture workflow for virus elimination

11.3.1 Types of Tissue Culture Relevant for Fruit Crops

11.3.1.1 Meristem and Shoot-Tip Culture

Meristem culture sits at the heart of virus elimination strategies for fruit crops. The process centers on excising the very tip of the shoot, the apical dome, often less than half a millimeter in diameter, sometimes accompanied by a pair of leaf primordia. The reason for targeting this zone is clear: viruses rarely penetrate these rapidly dividing, undifferentiated cells. As a result, even when the rest of the plant is heavily infected, the meristem may remain clean or carry only trace levels of virus (Bettoni et al. 2024). Regenerating whole plants from such tiny explants is no simple feat. It requires careful, steady-handed dissection, precise adjustment of nutrient media, and often a brief period in low light or darkness to minimize stress and enhance survival rates (Bettoni et al. 2024). The distinction between meristem-tip and shoot-tip culture matters in practice. Meristem-tip culture uses the smallest possible fragment, which maximizes virus clearance but can make recovery more unpredictable. Shoot-tip culture takes a slightly larger piece, sacrificing some efficiency in virus removal but improving the odds of strong, healthy regeneration (Magyar-Tábori et al. 2021). This method has gained traction in the propagation of banana, apple, grapevine, and strawberry. In banana, for example, technicians carefully place excised meristems onto Murashige and Skoog (MS) medium, fine-tuning the balance of cytokinins and auxins to drive shoot initiation. Once shoots form, hormone levels are shifted to promote rooting (Quiroz et al. 2017). Citrus presents another case: shoot-tip grafting, which involves attaching a minute meristem tip onto a disease-free seedling rootstock in vitro, has become standard for tackling stubborn viruses like CTV and CPsV (El Sayed et al. 2022).

11.3.1.2 Nodal Culture

Nodal culture provides a simpler yet highly effective method for multiplying clean planting material. While not as effective for initial virus elimination as meristem culture, nodal culture excels in the rapid multiplication phase of certified, virus-free stock (Alam et al. 2013). It is often used for maintaining and distributing clean germplasm after the initial elimination stage. Nodal cultures are established on semi-solid MS medium, sometimes supplemented with gibberellic acid to break bud dormancy, particularly in temperate fruit crops such as apple and pear (Johnson et al. 2023).

11.3.1.3 Somatic Embryogenesis

Somatic embryogenesis offers unique advantages for virus elimination and large-scale propagation, especially in crops where meristem culture is less effective or regeneration is challenging. This method induces embryos from somatic (non-reproductive) cells, which can then develop into complete plants. The process typically begins with the induction of callus tissue on a medium with high auxin content, followed by transfer to a medium that supports embryo maturation. Somatic embryogenesis has proven effective in grapevine and some Prunus species, with high rates of virus elimination reported for GFLV and related pathogens (Carra et al. 2024). The dedifferentiation and redifferentiation that occur during somatic embryogenesis disrupt the continuity of viral infection, increasing the chances of regenerating virus-free plants. However, the technique can be labor-intensive and requires careful monitoring for somaclonal variation, particularly in commercial propagation programs (Olah et al. 2022).

11.3.1.4 Organogenesis

Organogenesis, whether direct or indirect (via callus formation), provides an additional route for regenerating virus-free plants. Direct organogenesis maintains greater genetic fidelity, while indirect pathways, although effective for virus elimination in some crops, may increase the risk of unwanted genetic changes. The choice of organogenic route, media, and environmental conditions is highly species- and genotype-dependent (Bidabadi and Jain 2020).

11.3.2 Key Factors Influencing Success and Efficiency

Multiple factors affect the outcome of tissue culture for virus elimination and plant regeneration. The physiological condition and virus load of the donor plant are critical. Plants used for explant collection should be maintained under optimal health and pre-screened to ensure low systemic virus titers. The size and type of explant significantly influence both virus elimination and regeneration efficiency, with smaller meristems typically yielding higher elimination rates but lower survival and regrowth rates (Magyar-Tábori et al. 2021). Media composition, especially the source and concentration of plant hormones, can be decisive. Optimal protocols may require species-specific adjustments in mineral nutrients, vitamins, and growth regulators. For instance, banana micropropagation is enhanced by the use of temporary immersion bioreactor systems, which improve shoot multiplication rates and plantlet quality compared to conventional static cultures (Pasternak and Steinmacher 2024). Environmental conditions such as temperature, light intensity, and humidity shape the success of each stage in tissue culture. These factors not only influence how efficiently viruses are eliminated but also determine how well young plants adapt once transferred from the laboratory to the greenhouse (Grzelak et al. 2024). Plant genotype matters just as much. Some cultivars respond readily to regeneration protocols, producing strong, uniform plantlets. Others may prove difficult, showing poor growth or unusual traits. This variation means that long-running tissue culture programs, especially those involving clonal crops such as banana, grapevine, or citrus, must remain vigilant for signs of somaclonal variation and unexpected changes that could impact plant performance (Ferreira et al. 2023). Routine virus indexing with molecular diagnostics at key checkpoints provides an additional layer of quality assurance. By sampling and testing at multiple stages—from the initial explant right through to the fully acclimatized plantlet—nurseries can catch problems early and maintain high standards for outgoing stock (Wang et al. 2022b). When each of these elements is carefully managed, tissue culture becomes a powerful and dependable foundation for producing virus-free planting material on a large scale. For fruit growers, this means a steady supply of robust, healthy plants, which are essential for sustaining yields, ensuring sustainability, and keeping the fruit industry globally competitive (Wijerathna-Yapa and Hiti-Bandaralage 2023).

11.4 Virus Elimination Techniques in Tissue Culture

The main objective of tissue culture-based virus elimination is to produce truly healthy planting material by focusing regeneration on tissues where viruses are least likely to persist. This method builds on decades of research into how viruses spread, and sometimes fail to spread, within plant tissues. Success hinges on a deep understanding of where viruses reside, how they move, and which stages of plant development can interrupt their life cycle (Wang and Hu 2006).

11.4.1 Biological Principles: Virus Exclusion from Meristems and Tissue Specificity

Infected plants rarely show uniform virus distribution. The shoot apex, or the growing tip, is a notable exception. Here, tissues are actively dividing and undifferentiated. In many fruit crops, these zones are either virus-free or harbor only trace amounts of viral particles. Several mechanisms contribute: the absence of fully formed vascular tissues, which limits systemic virus movement; intense metabolic activity; and a high concentration of RNA silencing machinery that helps block virus replication (Bradamante et al. 2021). This makes the apical meristem an ideal starting point for regenerating clean plants. It's particularly effective against viruses that move through the phloem or struggle to invade the youngest, fastest-growing cells (Bélanger et al. 2024). However, this strategy is not foolproof. Certain viruses, such as raspberry bushy dwarf virus (RBDV) or some persistent endogenous forms like banana streak virus (BSV), have adapted to penetrate meristematic tissues, especially when the plant carries a high viral load. In such cases, relying solely on meristem culture often falls short. Additional steps, such as heat treatment or exposure to antiviral compounds, are necessary to reduce virus levels and shift the balance in favor of successful virus elimination (Wang et al. 2022a).

11.4.2 Methodologies

11.4.2.1 Meristem-Tip Culture

Meristem-tip culture involves excising and culturing a tiny dome of the shoot apex, sometimes as small as 0.2–0.3 mm. The efficacy of this method has been established in various crops, including banana, citrus, apple, grapevine, and strawberry. In banana, for example, meristem-tip culture consistently achieves BBTV elimination rates of 85–95% when the smallest possible meristems are used (Grout 1990). In apple, smaller meristems (<0.3 mm) lead to greater elimination of apple stem grooving virus (ASGV) compared to larger explants. The primary challenge with very small explants is their reduced regenerative capacity, resulting in lower survival rates and sometimes necessitating the use of specialized media or supportive nurse cultures. The trade-off between virus elimination and plantlet recovery is a recurring theme in practical protocols (Kim et al. 2022).

11.4.2.2 Thermotherapy

Thermotherapy exploits the relative heat tolerance of plant tissues and the greater heat sensitivity of many viruses. Subjecting infected donor plants, or in vitro cultures, to elevated temperatures (typically 35–38 °C) for several weeks disrupts viral replication and movement. Thermotherapy can be applied to whole plants (in vivo) or directly to cultured tissues (in vitro). For example, combining thermotherapy (38 °C, 4 weeks) with meristem-tip culture increases the elimination rates of BBTV and BSV in banana to as high as 90–98% and 60–78%, respectively (Wang et al. 2018). In apples, thermotherapy at 38 °C for 30 days, followed by shoot tip culture, has achieved up to 92% elimination of apple chlorotic leaf spot virus (ACLSV) (Liu et al. 2021). Temperature regimes and treatment duration must be carefully calibrated. While higher temperatures enhance virus eradication, plant tissues, especially meristems, are sensitive to heat stress. Excessive heat can result in tissue necrosis, reduced regeneration, or unwanted somaclonal variation. Some protocols employ alternating day/night temperatures to balance efficacy and plantlet survival, such as alternating 36 and 32 °C for the elimination of plum pox virus (PPV) in peach (Magyar-Tábori et al. 2021).

11.4.2.3 Chemotherapy

Chemotherapy involves exposing cultured explants to antiviral agents that inhibit viral replication or disrupt the structure of virus particles. Ribavirin (Virazole) is the most widely used chemical, effective against several RNA viruses, including ApMV and RBDV. The application of ribavirin at 20–100 mg/L in the culture medium can significantly increase virus elimination rates, though phytotoxicity is a frequent risk. Lower concentrations or pulsed treatments are often preferred to minimize adverse effects on plant growth (De Clercq 1996). Chemotherapy is rarely used alone. Its effectiveness increases substantially when combined with meristem-tip culture or thermotherapy. For example, combining ribavirin treatment with thermotherapy and meristem culture raised RBDV elimination in raspberry from less than 50% (with any single method) to nearly 90% (Weber 2016).

11.4.2.4 Cryotherapy

Cryotherapy is a relatively recent innovation that utilizes ultra-low temperatures, typically liquid nitrogen, to selectively kill virus-infected cells while preserving meristematic regions. During rapid freezing and thawing, larger, vacuolated, and virus-laden cells are destroyed, while small, dense cells in the meristem survive. Recovery from cryotherapy often results in a high proportion of virus-free plantlets. For apple, thermotherapy (37 °C, 21 days) followed by cryotherapy has achieved 85–90% elimination rates for ASGV. Successful implementation of cryotherapy requires precise control over cooling and rewarming rates, pre-culture conditions, and post-thaw recovery media (Wang et al. 2007).

11.4.2.5 Combined and Advanced Approaches

In practice, the most effective virus eradication strategies often employ a combination of methods. This may involve pre-treating donor plants with thermotherapy, followed by meristem-tip excision, chemotherapy during the early culture stages, and even cryotherapy for particularly persistent viruses or recalcitrant species. Such integrated approaches leverage the strengths of each method and address their individual limitations (Vivek and Modgil 2018). For example, a combined regime of thermotherapy, chemotherapy, and meristem culture improved RBDV elimination in raspberry to almost 90%, whereas single treatments yielded less than 50% (Vivek and Modgil 2018).

11.4.3 Case-Specific Strategies for Key Virus/Crop Combinations

Virus elimination protocols must be tailored to the unique interactions between virus species, host genotype, and environmental conditions. In banana, BBTV is most effectively removed using a sequence of thermotherapy and small meristem-tip culture, while BSV, due to its endogenous forms, often requires prolonged thermotherapy and repeated indexing (Lassois et al. 2013). For citrus, shoot-tip grafting of meristems as small as 0.1–0.2 mm onto seedling rootstocks can eliminate CTV and CPsV with an efficiency of 85–95% (El Sayed et al. 2022). Apple and pear respond well to a combination of thermotherapy and cryotherapy, whereas grapevine viruses, such as GFLV, are best eliminated through somatic embryogenesis rather than meristem culture alone (Miljanić et al. 2022).

11.4.4 Efficiency and Limitations

While tissue culture-based techniques are highly effective, no single approach guarantees 100% elimination for all virus/host combinations. The efficiency is strongly influenced by meristem size, treatment duration, choice of chemical agents, genotype, virus strain, and even the physiological status of the donor plant (Anikina et al. 2023). Smaller meristems yield higher virus elimination rates but require more advanced technical skill and may reduce recovery rates. Thermotherapy and chemotherapy are limited by the sensitivity of plant tissue, and cryotherapy requires specialized equipment and expertise (Szabó et al. 2023). Rigorous diagnostic testing is crucial throughout the entire virus elimination process. Even plants that appear perfectly healthy after regeneration may harbor hidden or low-level infections, rendering visual inspection alone unreliable (Venbrux et al. 2023). By combining tissue culture with regular molecular indexing and well-defined certification standards, many fruit industries have achieved large-scale, reliable production of virus-free planting stock (Hasnain et al. 2022). These advances have significantly improved the sustainability and yield of orchards and vineyards, providing growers with a dependable foundation for long-term productivity.

11.5 Detection, Indexing, and Certification Protocols

Producing and distributing virus-free planting material for fruit crops depends on meticulous virus detection at every critical stage. Missing even a single latent or low-titer infection can compromise the integrity of clean stock programs and put entire orchards at risk (Kumar 2013).

11.5.1 Detection and Indexing Before and After Tissue Culture

Virus testing forms the backbone of reliable tissue culture workflows. Before any meristem or explant is excised, donor plants undergo thorough screening with sensitive diagnostic tools to ensure all target viruses are identified. This early step weeds out infected sources before they can enter propagation, safeguarding the health of subsequent generations. Only those plants testing negative for the widest possible range of viruses are selected for further tissue culture. After in vitro regeneration, plantlets are retested to confirm the success of virus elimination, ensuring that no systemic pathogens remain (Alam et al. 2013). Indexing is typically repeated after acclimatization and during the pre-release phase in the nursery to prevent accidental distribution of infected material. This multi-stage approach helps detect escapes, late-emerging infections, or contamination events that may occur during multiplication or acclimatization (Devi et al. 2024).

11.5.2 Overview of Diagnostic Methods: Serological, Molecular, NGS

The selection of a method depends on the stage of the certification process, available resources, target viruses, and the required sensitivity.

11.5.2.1 Serological Methods

Serological assays, especially enzyme-linked immunosorbent assay (ELISA), have long been the backbone of virus testing in fruit crop certification (Mowat and Dawson 1987). ELISA detects the presence of viral coat proteins using virus-specific antibodies, offering medium sensitivity and high throughput. It remains the preferred method for large-scale screening, especially in grapevine and citrus certification programs, and is suitable for routine testing of nursery material. However, ELISA cannot always distinguish closely related strains and may lack the sensitivity to detect low-titer infections in recently regenerated plantlets or asymptomatic carriers (Iralu et al. 2025).

11.5.2.2 Molecular Methods

Nucleic acid-based tests provide greater sensitivity and specificity. Conventional reverse transcription polymerase chain reaction (RT-PCR) and quantitative RT-PCR (qRT-PCR) detect and quantify viral RNA at low copy numbers. These methods have become standard for post-therapy verification in tissue culture programs, as they allow detection of latent or subclinical infections (Gebe Abreu Cabral et al. 2025; Munguti et al. 2024). Multiplex PCR platforms can simultaneously test for several viruses, increasing efficiency. Loop-mediated isothermal amplification (LAMP) is an emerging field-adapted method for rapid, on-site diagnostics with high sensitivity, particularly in banana and citrus production systems (Panno et al. 2020).

11.5.2.3 High-Throughput and Next-Generation Sequencing (NGS)

Next-generation sequencing has changed the landscape of plant virus diagnostics. By using methods such as RNA-Seq and small RNA sequencing, laboratories can now screen for both known and previously undetected viruses in a single analysis. This capability is especially valuable when testing nuclear or foundation stock for national clean plant programs and quarantine systems (Satam et al. 2023). NGS excels in its breadth because it detects all targeted viruses but also reveals the broader viral diversity present in a sample. The tradeoff, however, comes in the form of cost and the need for specialized bioinformatics expertise. Despite these hurdles, NGS is quickly establishing itself as the reference standard in certification, providing growers and regulators with an unmatched level of confidence (Maina et al. 2024).

11.5.3 National and International Certification Protocols

Standardized certification remains at the heart of effective clean plant material systems. These protocols provide growers with assurance that the planting stock they receive meets stringent health benchmarks, thereby minimizing the risk of unexpected outbreaks and supporting long-term productivity across the fruit industry (Gergerich et al. 2015).

11.5.3.1 Certification Schemes and Regulatory Frameworks

Most certification systems operate in distinct tiers, beginning with tightly controlled foundation stock (often labeled as G1) and moving outward to commercial nurseries and, finally, to field planting material (G2–G4). Foundation material receives the highest scrutiny as it is subjected to sensitive tests such as NGS, qPCR, and ELISA, while later stages employ high-throughput approaches for broader lots. These frameworks outline not only what viruses must be checked but also details such as isolation distances, documentation requirements, and retesting schedules (Rahemi et al. 2022). Guidance from global authorities, such as the Food and Agriculture Organization (FAO), the International Plant Protection Convention (IPPC), and the European and Mediterranean Plant Protection Organization (EPPO), influences this process. Their published guidelines clearly define virus testing priorities, approved diagnostic methods, and the required steps for maintaining and distributing truly virus-free planting stock (Schrader and Unger 2003). In the United States, the National Clean Plant Network and similar entities coordinate regional certification and maintain registered sources of clean material for grapevine, citrus, and berry crops (Zherdev et al. 2018).

11.5.3.2 Indexing and Certification Workflows

The certification process generally follows a stepwise approach. First, candidate plants are subjected to comprehensive virus indexing. Selected individuals are subjected to virus elimination (such as tissue culture methods described previously), then retested after regeneration. Multiple cycles of multiplication and testing are performed, especially at the nuclear stock stage. Once a source is fully certified, it becomes the parent for larger-scale multiplication. Documentation and traceability are maintained at every step to ensure the integrity of the material throughout the chain of custody (Usharani et al. 2024). In Spain, the Citrus Variety Improvement Program has supported the establishment and distribution of virus-free citrus trees since the 1970s, combining rigorous indexing, shoot tip grafting, and certification standards (Navarro et al. 2002). In the United States, clean plant centers for apple, grapevine, and berries use a combination of serological and molecular tests, with annual retesting and mandatory isolation in protected facilities. These programs have dramatically reduced the incidence of virus-related losses in commercial orchards and vineyards (Stevens and Al Rwahnih 2024). Figure 11.1 provides concise overviews of the main diagnostic and certification steps that underpin reliable supply chains of virus-free fruit planting material.

11.6 Large-Scale Production and Distribution Systems

The transition from laboratory-scale tissue culture to the commercial production and delivery of virus-free fruit planting material presents both opportunities and challenges. Sustaining a clean supply chain requires robust protocols for scaling up micropropagation, vigilant nursery management, and reliable distribution systems (Murthy et al. 2023).

11.6.1 Micropropagation Scale-Up: Conventional and Bioreactor Systems

Micropropagation forms the cornerstone of mass production for clean planting material. The conventional method involves the sequential transfer of explants, such as meristems, shoot tips, or nodal segments, through stages of multiplication, rooting, and acclimatization, each on semi-solid media. Conventional micropropagation provides a reliable method for rapidly multiplying elite, virus-free lines. This process relies on precise control of both environmental conditions and the composition of the nutrient media, ensuring that each plantlet develops with uniform health and vigor (Abdalla et al. 2022). Yet as demand grows for large-scale, uniform production, conventional methods can become a bottleneck requiring intensive labor and driving up costs. To overcome these constraints, many labs and nurseries have turned to bioreactor systems. Temporary immersion bioreactors (TIBs), in particular, have gained traction. They provide enhanced aeration and nutrient delivery, and automate the handling of cultures (Bello-Bello et al. 2019). The result is often a crop of plantlets that root more strongly, show greater resilience during transplanting, and generally perform better in the field. Bioreactor technology is now making its mark not just in banana but also in crops like grapevine and citrus, where early trials show promising gains in efficiency and plant quality (Méndez-Hernández and Loyola-Vargas 2024).

11.6.2 Nursery Management and Maintaining Virus-Free Stock

Nurseries are where laboratory precision meets practical field reality. During acclimatization and early growth, keeping plantlets virus-free becomes especially critical. Growers typically use controlled spaces, such as greenhouses or screenhouses, where temperature, light, and humidity are gradually adjusted to help plants adapt from in vitro to outdoor conditions. The choice of substrate, careful moisture control, and early disease monitoring all play important roles in success at this stage (Olagunju et al. 2021). Physical barriers, routine cleaning, and staff training are integral to daily operations, helping to guard against reinfection. This high level of hygiene ensures that the clean plant status achieved in the lab is preserved as plantlets prepare for field planting. These include physical barriers such as insect-proof netting, double-door entries, and the isolation of nuclear stock from production blocks. Staff are trained to recognize early signs of viral infection and to implement strict hygiene practices, including tool sterilization and the regular removal of symptomatic or stunted plants (Yong and Calautit 2023). Nurseries are also responsible for the multiplication and maintenance of foundation or mother plants, which serve as the source for all further propagation. The health of these blocks is central to the integrity of the entire supply chain. In many countries, government agencies or certification bodies conduct independent audits and testing of mother blocks to ensure compliance with national and international standards (Gregorio 2011).

11.7 Innovations, Challenges, and Future Perspectives

Fruit industries today face an ever-shifting landscape. New viruses continue to emerge. Vectors change their habits and ranges. The global trade in plant material opens new routes for pathogens to cross borders. These realities make constant innovation in virus detection and elimination a necessity.

11.7.1 New Diagnostic and Elimination Technologies

Diagnostic tools for fruit viruses have advanced rapidly in recent years. Next-generation sequencing (NGS) and small RNA sequencing can now identify both familiar and unknown viruses in a single analysis. With this technology, foundation and mother blocks are screened more thoroughly than ever, setting a higher bar for the health of planting material, especially in national and international certification programs (Jones et al. 2017). On the practical side, nurseries and labs are adopting high-throughput, field-friendly methods. Loop-mediated isothermal amplification (LAMP) and multiplex PCR enable the rapid and reliable testing of large batches, a game-changer for busy propagation operations and large clean stock programs (Garg et al. 2022). The result is a diagnostic landscape that is faster, broader, and more precise, giving fruit growers and regulators better tools to manage the evolving threat of plant viruses. Virus elimination methodologies continue to evolve. Temporary immersion bioreactors, improved cryotherapy protocols, and refined combinations of thermotherapy and chemotherapy have been shown to offer higher rates of virus eradication across challenging host–virus systems (Szabó et al. 2023). Somatic embryogenesis is gaining prominence for crops where meristem culture alone is insufficient. Research on genome editing, including CRISPR/Cas9-mediated knockouts of virus susceptibility genes, holds promise but remains in its early stages of development (Carra et al. 2024).

11.7.2 Persistent Barriers and Research Gaps

Despite these advances, persistent challenges hinder the full realization of virus-free fruit crop production. Technical complexity and cost remain significant barriers to the widespread adoption of NGS and bioreactor systems, particularly in low-resource settings. Not all viruses are easily eliminated; some, such as endogenous Banana streak virus (BSV), can reactivate even after intensive treatments (Gayral et al. 2008). Somaclonal variation, phytotoxicity from antiviral compounds, and genotype-dependent recalcitrance add further complications (Duta-Cornescu et al. 2023). Surveillance for emerging viruses and monitoring for new vector–virus relationships demand constant attention. Climate change and global movement of plant material may alter vector distributions and introduce new viral threats, requiring flexible and adaptive management strategies (Jones and Naidu 2019).

11.7.3 Future Directions for Research, Policy, and Practice

Future research should prioritize the development of more accessible, robust, and cost-effective diagnostic tools for both laboratory and field use. Advances in portable NGS devices and AI-driven data analysis may soon enable real-time surveillance and rapid response in nurseries and commercial orchards. Studies on host resistance, vector ecology, and the application of biocontrol agents are also critical for integrated management approaches (Kanapiya et al. 2024). Policy frameworks must evolve alongside technological progress. Strengthening international harmonization of certification standards, improving data sharing on virus outbreaks, and supporting the transfer of clean stock technologies to developing regions will be key for global fruit crop security. Ongoing investment in capacity building, extension services, and public–private partnerships can help bridge knowledge gaps and ensure the sustainable adoption of innovations (Kreuze et al. 2023). The future of virus management in fruit crops lies in the integration of cutting-edge diagnostics, novel elimination techniques, and responsive policy. Addressing persistent barriers will require collaboration across research, industry, and regulatory sectors, with a focus on securing healthy planting material as the foundation for productive and resilient fruit industries.

11.8 Conclusions and Recommendations

Sustained fruit production and profitability depend on effective strategies for managing viral diseases. Planting virus-free material produced through carefully refined tissue culture techniques, backed by thorough diagnostic and certification processes, offers growers the most dependable protection against viral spread in perennial crops. When combined, methods such as meristem-tip culture, thermotherapy, chemotherapy, and cryotherapy have proven their effectiveness. Real-world case studies consistently demonstrate that these approaches yield higher returns, produce better fruit quality, and extend orchard lifespans. The success of these programs is underpinned by comprehensive virus detection strategies, combining serological, molecular, and next-generation sequencing platforms for multi-stage indexing before and after tissue culture. National and international certification schemes ensure that clean material is maintained from the laboratory through the nursery to the grower's field, supported by effective nursery management, monitoring, and distribution networks. To protect these investments, continued vigilance is needed after field planting. On-farm best practices, such as physical isolation, prompt removal of symptomatic plants, stringent hygiene, and community-based vector control, are essential to reduce reinfection risks. Area-wide management and coordinated extension efforts amplify the effectiveness of these measures, especially in regions with persistent or emerging virus threats. Future research should focus on lowering the cost and complexity of diagnostics, refining virus elimination protocols for difficult host-virus combinations, and advancing technologies such as bioreactors and portable NGS. Policy initiatives must strengthen harmonization of certification standards, support knowledge transfer, and encourage public–private partnerships. Industry stakeholders are encouraged to prioritize the use of certified virus-free planting material, invest in continuous training, and actively participate in surveillance and extension activities. These actions will secure the foundation for sustainable, resilient, and profitable fruit crop production in the face of evolving viral challenges.

References

Abbas RM, Khan M, Mughal SM, Jaskani M, Abbas H (2006) Propagation of CTV-free sweet orange [Citrus sinensis (L.) Osbeck] plants through microbudding technique. Abstr Pap 38:583–587

Abdalla N, El-Ramady H, Seliem MK, El-Mahrouk ME, Taha N, Bayoumi Y, Shalaby TA, Dobránszki J (2022) An academic and technical overview on plant micropropagation challenges. Horticulturae 8(8):Article 8. https://doi.org/10.3390/horticulturae8080677

Alam I, Sharmin SA, Naher MK, Alam MJ, Anisuzzaman M, Alam MF (2013) Elimination and detection of viruses in meristem-derived plantlets of sweetpotato as a low-cost option toward commercialization. 3 Biotech 3(2):153–164. https://doi.org/10.1007/s13205-012-0080-6

Anđelić T, Vujović T, Jevremović D, Tomić J, Radivojević D (2024) Comparative study of different surface sterilization treatments and optimal month for establishment of aseptic cultures of raspberry cultivars. J Cent Eur Agric 25(2):470–480. https://doi.org/10.5513/JCEA01/25.2.4201

Anikina I, Kamarova A, Issayeva K, Issakhanova S, Mustafayeva N, Insebayeva M, Mukhamedzhanova A, Khan SM, Ahmad Z, Lho LH, Han H, Raposo A (2023) Plant protection from virus: a review of different approaches. Front Plant Sci 14:1163270. https://doi.org/10.3389/fpls.2023.1163270

Bélanger JG, Copley TR, Hoyos-Villegas V, Charron J-B, O'Donoughue L (2024) A comprehensive review of in planta stable transformation strategies. Plant Methods 20:79. https://doi.org/10.1186/s13007-024-01200-8

Bello-Bello J, Cruz-Cruz C, Pérez-Guerra J (2019) A new temporary immersion system for commercial micropropagation of banana (Musa AAA cv. Grand Naine). In Vitro Cell Dev Biol – Plant 55:313–320. https://doi.org/10.1007/s11627-019-09973-7

Bettoni JC, Wang M-R, Li J-W, Fan X, Fazio G, Hurtado-Gonzales OP, Volk GM, Wang Q-C (2024) Application of biotechniques for in vitro virus and viroid elimination in pome fruit crops. Phytopathology 114(5):930–954. https://doi.org/10.1094/PHYTO-07-23-0232-KC

Bidabadi SS, Jain SM (2020) Cellular, molecular, and physiological aspects of in vitro plant regeneration. Plants 9(6):702. https://doi.org/10.3390/plants9060702

Bradamante G, Mittelsten Scheid O, Incarbone M (2021) Under siege: virus control in plant meristems and progeny. Plant Cell 33(8):2523–2537. https://doi.org/10.1093/plcell/koab140

Cambra M, Gorris MT, Marroquín C, Román MP, Olmos A, Martínez MC, de Mendoza AH, López A, Navarro L (2000) Incidence and epidemiology of Citrus tristeza virus in the Valencian community of Spain. Virus Res 71(1–2):85–95. https://doi.org/10.1016/s0168-1702(00)00190-8

Cambra M, Madariaga M, Varveri C, Çağlayan K, Morca AF, Chirkov S, Glasa M (2024) Estimated costs of plum pox virus and management of sharka, the disease it causes. Phytopathol Mediterr 63(3):Article 3. https://doi.org/10.36253/phyto-15581

Carra A, Wijerathna-Yapa A, Pathirana R, Carimi F (2024) Development and applications of somatic embryogenesis in grapevine (Vitis spp.). Plants 13(22):3131, Article 22. https://doi.org/10.3390/plants13223131

Das T, Banerjee A (2018) Distribution, molecular characterization and diversity of banana bunchy top virus in Tripura, India. Virusdisease 29(2):157–166. https://doi.org/10.1007/s13337-018-0451-7

De Clercq E (1996) Chemotherapy of viral infections. In: Baron S (ed) Medical microbiology, 4th edn. University of Texas Medical Branch at Galveston. http://www.ncbi.nlm.nih.gov/books/NBK8119/

Devi BM, Guruprasath S, Balu P, Chattopadhyay A, Thilagar SS, Dhanabalan KV, Choudhary M, Moparthi S, Jailani AAK (2024) Dissecting diagnostic and management strategies for plant viral diseases: what next? Agriculture 14(2):Article 2. https://doi.org/10.3390/agriculture14020284

Duta-Cornescu G, Constantin N, Pojoga D-M, Nicuta D, Simon-Gruita A (2023) Somaclonal variation—advantage or disadvantage in micropropagation of the medicinal plants. Int J Mol Sci 24(1):838. https://doi.org/10.3390/ijms24010838

El Sayed T, Saker K, El Shorbagy A, Soliman A (2022) Elimination of Citrus psorosis virus using shoot tip grafting and electrotherapy techniques. Egypt J Agric Res. https://doi.org/10.21608/ejar.2022.163106.1282

Ferreira MDS, Rocha ADJ, Nascimento FDS, Oliveira WDDS, Soares JMDS, Rebouças TA, Morais Lino LS, Haddad F, Ferreira CF, Santos-Serejo JAD, Fernãndez JS, Amorim EP (2023) The role of somaclonal variation in plant genetic improvement: a systematic review. Agronomy 13(3):Article 3. https://doi.org/10.3390/agronomy13030730

García JA, Glasa M, Cambra M, Candresse T (2014) Plum pox virus and sharka: a model potyvirus and a major disease. Mol Plant Pathol 15(3):226–241. https://doi.org/10.1111/mpp.12083

Garg N, Ahmad FJ, Kar S (2022) Recent advances in loop-mediated isothermal amplification (LAMP) for rapid and efficient detection of pathogens. Curr Res Microb Sci 3:100120. https://doi.org/10.1016/j.crmicr.2022.100120

Garzo E, Moreno A, Plaza M, Fereres A (2020) Feeding behavior and virus-transmission ability of insect vectors exposed to systemic insecticides. Plants 9(7):895. https://doi.org/10.3390/plants9070895

Gayral P, Noa-Carrazana J-C, Lescot M, Lheureux F, Lockhart BEL, Matsumoto T, Piffanelli P, Iskra-Caruana M-L (2008) A single banana streak virus integration event in the banana genome as the origin of infectious endogenous pararetrovirus. J Virol 82(13):6697–6710. https://doi.org/10.1128/JVI.00212-08

Gebe Abreu Cabral P, Bastos de Souza S, Ferraz Arruda R, Passos de Figueiredo Cabral S, Leone Evangelista Monteiro de Assis A, Porto Muniz Martins Y, Brazil Viana Junior A, Chang J, Lei P, Martins da Silva R (2025) Comparative analysis of viral load quantification using reverse transcription polymerase chain reaction and digital droplet polymerase chain reaction. Int J Mol Sci 26(2):Article 2. https://doi.org/10.3390/ijms26020725

Gergerich R, Welliver R, Osterbauer N, Kamenidou S, Martin R, Golino D, Eastwell K, Fuchs M, Vidalakis G, Tzanetakis I (2015) Safeguarding fruit crops in the age of agricultural globalization. Plant Dis 99:176–187. https://doi.org/10.1094/PDIS-07-14-0762-FE

Gregorio N (2011) The potential role of nurseries in improving access to high quality planting stock and promoting appropriate silvicultural systems to improve the productivity of .... -Redevelopment of a .... https://www.academia.edu/1872706/The_potential_role_of_nurseries_in_improving_access_to_high_quality_planting_stock_and_promoting_appropriate_silvicultural_systems_to_improve_the_productivity_of_

Grout BW (1990) Meristem-tip culture. Methods in Mol Biol (Clifton, NJ) 6:81–91. https://doi.org/10.1385/0-89603-161-6:81

Grzelak M, Pacholczak A, Nowakowska K (2024) Challenges and insights in the acclimatization step of micropropagated woody plants. Plant Cell Tissue Organ Cult (PCTOC) 159(3):72. https://doi.org/10.1007/s11240-024-02923-1

Hasnain A, Naqvi SAH, Ayesha SI, Khalid F, Ellahi M, Iqbal S, Hassan MZ, Abbas A, Adamski R, Markowska D, Baazeem A, Mustafa G, Moustafa M, Hasan ME, Abdelhamid MMA (2022) Plants in vitro propagation with its applications in food, pharmaceuticals and cosmetic industries; current scenario and future approaches. Front Plant Sci 13:1009395. https://doi.org/10.3389/fpls.2022.1009395

Iralu N, Wani S, Meghanath D, Saleem S, Hamid A (2025) Direct enzyme-linked immunosorbent assay (ELISA) for detection of plant viruses. In: Hamid A, Ali G, Shikari A, Saleem S, Wani SH, Wani S (eds) Detection of plant viruses: advanced techniques. Springer US, pp 35–41. https://doi.org/10.1007/978-1-0716-4390-7_9

Johnson AJ, Geary B, Hulet A, Madsen MD (2023) Breaking dormancy and increasing restoration success of native penstemon species using gibberellic acid seed coatings and U-shaped furrows. Plants 12(23):4005. https://doi.org/10.3390/plants12234005

Jones RAC, Naidu RA (2019) Global dimensions of plant virus diseases: current status and future perspectives. Annu Rev Virol 6(1):387–409. https://doi.org/10.1146/annurev-virology-092818-015606

Jones S, Baizan-Edge A, MacFarlane S, Torrance L (2017) Viral diagnostics in plants using next generation sequencing: computational analysis in practice. Front Plant Sci 8:1770. https://doi.org/10.3389/fpls.2017.01770

Kanapiya A, Amanbayeva U, Tulegenova Z, Abash A, Zhangazin S, Dyussembayev K, Mukiyanova G (2024) Recent advances and challenges in plant viral diagnostics. Front Plant Sci 15:1451790. https://doi.org/10.3389/fpls.2024.1451790

Kim S-W, Lee H-J, Cho KH, Jeong R-D (2022) Detection and quantification of apple stem grooving virus in micropropagated apple plantlets using reverse-transcription droplet digital PCR. Plant Pathol J 38(4):417–422. https://doi.org/10.5423/PPJ.NT.05.2022.0072

Kreuze JF, Cuellar WJ, Kumar PL, Boddupalli P, Omondi AB (2023) New technologies provide innovative opportunities to enhance understanding of major virus diseases threatening global food security. Phytopathology 113(9):1622–1629. https://doi.org/10.1094/PHYTO-12-22-0457-V

Kumar L (2013) Virus indexing and certification in clonal crops: the case of banana, pp 33–39. https://doi.org/10.13140/2.1.1970.2080

Kumar PL, Selvarajan R, Iskra-Caruana M-L, Chabannes M, Hanna R (2015) Biology, etiology, and control of virus diseases of banana and plantain. Adv Virus Res 91:229–269. https://doi.org/10.1016/bs.aivir.2014.10.006

Kumar A, Bhuj BD, Dhar S, Dixit KJ, Singh SP (2023) Micropropagation of fruit crops: a review. Adv Crop Sci Technol 11(5):1–22

Lassois L, Lepoivre P, Swennen R, van den Houwe I, Panis B (2013) Thermotherapy, chemotherapy, and meristem culture in banana. Methods Mol Biol (Clifton, NJ) 11013:419–433. https://doi.org/10.1007/978-1-62703-074-8_32

Liu L, Chen X, Yan L, Jin Y, Sun L, Yang Y, Wang Y, Zhao Z (2021) Different eradication effects of latent viruses by combining thermotherapy with shoot tip culture or cryotherapy in four apple cultivars. Sci Hortic 288:110356. https://doi.org/10.1016/j.scienta.2021.110356

Magyar-Tábori K, Mendler-Drienyovszki N, Hanász A, Zsombik L, Dobránszki J (2021) Phytotoxicity and other adverse effects on the in vitro shoot cultures caused by virus elimination treatments: reasons and solutions. Plants 10(4):Article 4. https://doi.org/10.3390/plants10040670

Maina S, Donovan NJ, Plett K, Bogema D, Rodoni BC (2024) High-throughput sequencing for plant virology diagnostics and its potential in plant health certification. Front Hortic 3. https://doi.org/10.3389/fhort.2024.1388028

Manzoor S, Nabi SU, Ayaz A, Rasool B, Sharma SK, Chesti MH, Parveen S, Verma MK, Diab MA, Rabbee MF (2025) Unveiling viral threats to temperate pome fruits: characterization, transmission, and sustainable management strategies. Front Microbiol 16:1560720. https://doi.org/10.3389/fmicb.2025.1560720

Martin RR, Tzanetakis IE (2015) Control of virus diseases of berry crops. Adv Virus Res 91:271–309. https://doi.org/10.1016/bs.aivir.2014.10.003

Méndez-Hernández HA, Loyola-Vargas VM (2024) Plant micropropagation and temporary immersion systems. Methods Mol Biol (Clifton, NJ) 2827:35–50. https://doi.org/10.1007/978-1-0716-3954-2_3

Miljanić V, Rusjan D, Škvarč A, Chatelet P, Štajner N (2022) Elimination of eight viruses and two viroids from preclonal candidates of six grapevine varieties (Vitis vinifera L.) through in vivo thermotherapy and in vitro meristem tip micrografting. Plants 11(8):1064. https://doi.org/10.3390/plants11081064

Morel G, Martin C (1952) Cure of dahlias attacked by a virus disease. C R Hebd Seances Acad Sci 235(21):1324–1325

Moreno P, Ambrós S, Albiach-Martí MR, Guerri J, Peña L (2007) Citrus tristeza virus: a pathogen that changed the course of the citrus industry. Mol Plant Pathol 9(2):251–268. https://doi.org/10.1111/j.1364-3703.2007.00455.x

Mowat WP, Dawson S (1987) Detection and identification of plant viruses by ELISA using crude sap extracts and unfractionated antisera. J Virol Methods 15(3):233–247. https://doi.org/10.1016/0166-0934(87)90101-7

Munguti FM, Kilalo DC, Yegon HK, Macharia I, Seal SE, Mwango'mbe AW, Nyaboga EN, Silva G (2024) Real-time reverse transcription recombinase polymerase amplification (RT-RPA) assay for detection of cassava brown streak viruses. Sci Rep 14(1):12438. https://doi.org/10.1038/s41598-024-62249-y

Murashige T, Skoog F (1962) A revised medium for rapid growth and bio assays with tobacco tissue cultures. Physiol Plant 15(3):473–497. https://doi.org/10.1111/j.1399-3054.1962.tb08052.x
Murthy HN, Joseph KS, Paek KY, Park SY (2023) Bioreactor systems for micropropagation of plants: present scenario and future prospects. Front Plant Sci 14:1159588. https://doi.org/10.3389/fpls.2023.1159588
Navarro L, Pina JA, Juárez J, Ballester-Olmos JF, Arregui JM, Ortega C, Navarro A, Duran-Vila N, Guerri J, Moreno P, Cambra M, Medina A, Zaragoza S (2002) The citrus variety improvement program in Spain in the period 1975–2001. In: Proceedings of the 15th conference of the International Organization of Citrus Virologists, pp 306–316
Olagunju YO, Aduloju AO, Akin-Idowu PE, Esuola CO (2021) Acclimatization of tissue culture pineapple plantlet using semi-autotrophic hydroponics technique in comparison with other conventional substrates. J Exp Agric Int:61–67. https://doi.org/10.9734/jeai/2021/v43i1130757
Olah R, Turcsan M, Olah K, Farkas E, Deak T, Jahnke G, Sardy DAN (2022) Somatic embryogenesis: a tool for fast and reliable virus and viroid elimination for grapevine and other plant species. Horticulturae 8(6):Article 6. https://doi.org/10.3390/horticulturae8060508
Panno S, Matić S, Tiberini A, Caruso AG, Bella P, Torta L, Stassi R, Davino S (2020) Loop mediated isothermal amplification: principles and applications in plant virology. Plants 9(4):461. https://doi.org/10.3390/plants9040461
Pasternak TP, Steinmacher D (2024) Plant growth regulation in cell and tissue culture in vitro. Plants 13(2):327. https://doi.org/10.3390/plants13020327
Quiroz KA, Berríos M, Carrasco B, Retamales JB, Caligari PDS, García-Gonzáles R (2017) Meristem culture and subsequent micropropagation of Chilean strawberry (Fragaria chiloensis (L.) Duch.). Biol Res 50:20. https://doi.org/10.1186/s40659-017-0125-8
Rahemi A, Dodson Peterson J, Lund K (2022) Nursery stock certification and testing programs, pp 101–116. https://doi.org/10.1007/978-3-030-99407-5_10
Ravelonandro M (2021) Reliable methodologies and impactful tools to control fruit tree viruses. Crops 1(1):Article 1. https://doi.org/10.3390/crops1010005
Sale S, Subramaniam S, Mad'Atari MF (2023) Trends in the tissue culture techniques and the synthesis of bioactive compounds in Eurycoma longifolia Jack-current status and future perspectives. Plants (Basel, Switz) 13(1):107. https://doi.org/10.3390/plants13010107
Satam H, Joshi K, Mangrolia U, Waghoo S, Zaidi G, Rawool S, Thakare RP, Banday S, Mishra AK, Das G, Malonia SK (2023) Next-generation sequencing technology: current trends and advancements. Biology 12(7):997. https://doi.org/10.3390/biology12070997
Schrader G, Unger J-G (2003) Plant quarantine as a measure against invasive alien species: the framework of the International Plant Protection Convention and the plant health regulations in the European Union. Biol Invasions 5:357–364. https://doi.org/10.1023/B:BINV.0000005567.58234.b9
Stevens KA, Al Rwahnih M (2024) High-throughput sequencing for the detection of viruses in grapevine: performance analysis and best practices. Viruses 16(12):1957. https://doi.org/10.3390/v16121957
Szabó LK, Desiderio F, Kirilla Z, Hegedűs A, Várallyay É, Preininger É (2023) A mini-review on in vitro methods for virus elimination from Prunus sp. fruit trees. Plant Cell Tiss Org Cult (PCTOC) 156(2):42. https://doi.org/10.1007/s11240-023-02670-9
Usharani TR, Narayanappa C, Pooja R (2024) Crucial stages and process for successful banana micropropagation through plant tissue cultutre techniques. https://doi.org/10.35543/indiarxiv.57
Venbrux M, Crauwels S, Rediers H (2023) Current and emerging trends in techniques for plant pathogen detection. Front Plant Sci 14:1120968. https://doi.org/10.3389/fpls.2023.1120968
Vivek M, Modgil M (2018) Elimination of viruses through thermotherapy and meristem culture in apple cultivar 'Oregon Spur-II'. VirusDisease 29(1):75–82. https://doi.org/10.1007/s13337-018-0437-5
Wang P, Hu C (2006) Regeneration of virus-free plants though in vitro culture. Adv Biochem Eng 18:61–99. https://doi.org/10.1007/3-540-09936-0_3
Wang Q, Cuellar WJ, Rajamäki M, Hirata Y, Valkonen JPT (2007) Combined thermotherapy and cryotherapy for efficient virus eradication: relation of virus distribution, subcellular

changes, cell survival and viral RNA degradation in shoot tips. Mol Plant Pathol 9(2):237–250. https://doi.org/10.1111/j.1364-3703.2007.00456.x

Wang M-R, Cui Z-H, Li J-W, Hao X-Y, Zhao L, Wang Q-C (2018) In vitro thermotherapy-based methods for plant virus eradication. Plant Methods 14:87. https://doi.org/10.1186/s13007-018-0355-y

Wang MR, Bi W-L, Bettoni J, Zhang D, Volk G, Wang Q (2022a) Shoot tip cryotherapy for plant pathogen eradication. Plant Pathol 71:1241–1254. https://doi.org/10.1111/ppa.13565

Wang YM, Ostendorf B, Gautam D, Habili N, Pagay V (2022b) Plant viral disease detection: from molecular diagnosis to optical sensing technology—a multidisciplinary review. Remote Sens 14(7):Article 7. https://doi.org/10.3390/rs14071542

Weber CA (2016) Eliminating Raspberry bushy dwarf virus (RBDV) from infected raspberry tissue cultures with ribavirin. Acta Hortic 1133:473–478. https://doi.org/10.17660/ActaHortic.2016.1133.72

Wijerathna-Yapa A, Hiti-Bandaralage J (2023) Tissue culture—a sustainable approach to explore plant stresses. Life 13(3):780. https://doi.org/10.3390/life13030780

Yong LX, Calautit JK (2023) A comprehensive review on the integration of antimicrobial technologies onto various surfaces of the built environment. Sustainability 15(4):Article 4. https://doi.org/10.3390/su15043394

Zhang S, Griffiths JS, Marchand G, Bernards MA, Wang A (2022) Tomato brown rugose fruit virus: an emerging and rapidly spreading plant RNA virus that threatens tomato production worldwide. Mol Plant Pathol 23(9):1262–1277. https://doi.org/10.1111/mpp.13229

Zherdev AV, Vinogradova SV, Byzova NA, Porotikova EV, Kamionskaya AM, Dzantiev BB (2018) Methods for the diagnosis of grapevine viral infections: a review. Agriculture 8(12):Article 12. https://doi.org/10.3390/agriculture8120195

Quarantine and Certification Programs for Virus-Free Fruit Propagation Material

12

Darwin M. Landicho and Mark Angelo O. Balendres

Abstract

The fruit production industry is an important sector globally due to its nutritional, dietary, and economic value. Due to its significance as a crucial source of high-value, nutritious food, global demand for fruits has continuously risen over the years. However, fruit production is challenged by various factors, including biotic constraints or diseases caused by plant viruses. Diseases of fruit crops caused by plant viruses are difficult to manage because they are systemic once established and can easily spread via infected plant propagation materials. Thus, preventive measures are necessary to address the long-term impacts of plant virus diseases on fruit production. To manage the spread of plant viruses and ensure the propagation of virus-free fruit materials, international organizations, national governments, and other relevant stakeholders have formulated and implemented plant quarantine and certification programs. This chapter discusses the importance of virus disease management in fruit propagation through quarantine and certification programs. It highlights the certification programs implemented by major fruit-producing countries and outlines the challenges in implementing and future directions for fruit propagation quarantine and certification.

Keywords

Virus-free · Fruit propagation · NPPO · Plant quarantine · Certification

D. M. Landicho
Department of Crop Protection, College of Agriculture, Central Luzon State University, Muñoz, Nueva Ecija, Philippines

M. A. O. Balendres (✉)
Department of Biology, College of Science, De La Salle University, Manila, Philippines
e-mail: mark.angelo.balendres@dlsu.edu.ph

R. Kumar et al. (eds.), *Fruit Crops and Viral Pathogens Volume II*, Plant Pathology for Sustainable Agriculture, https://doi.org/10.1007/978-981-95-7261-8_12

12.1 Introduction

Fruit production is a key component of the global agricultural industry. Fruits are an essential component of human diets due to their rich nutritional value (FAO 2020). In recent years, the volume of fruit production has increased by 1.7% per year (FAO 2025). The expansion of the fruit industry is primarily driven by the increasing consumer demand, growing international trade, and improving production technologies (FAO 2017). Among the countries globally, the top fruit-producing ones are China, India, and Brazil, trading mostly bananas, pome (apples/pears), cucurbits (watermelon/melon), citrus, grapes, and pineapples (Figs. 12.1 and 12.2). In all of these fruit crops, the key starting point in the value chain is the production of plant propagation materials. Plant propagation is the primary component of the fruit crops value chain because it supplies the materials necessary for the reproduction and multiplication of plants with ideal and desirable traits. Plant propagation materials are typically produced by using sexual or asexual methods (Love et al. 2017). Sexual propagation method refers to the production of planting materials through seeds, which is primarily used for breeding to enhance certain genetic traits. In contrast, asexual propagation refers to the use of vegetative parts to generate genetically identical clones using techniques such as using of cuttings, micropropagation, air layering, and grafting. Asexual propagation enables the rapid reproduction of planting materials, which is often necessary to sustain large-scale fruit production operations and industry expansion (Colombo and Roberto 2022; Wasielewski and Balerdi 2019). Despite the recent advancements in plant propagation technologies, challenges still remain for the fruit industry. Some of the notable challenges in fruit production are the occurrence of climate change and natural calamities, the shrinking of agricultural land, increasing demand from a growing population, and damage caused by pests and diseases (Bacelar et al. 2024; Gergerich et al. 2015). The occurrence of pests and diseases is a major constraint in fruit production because they can

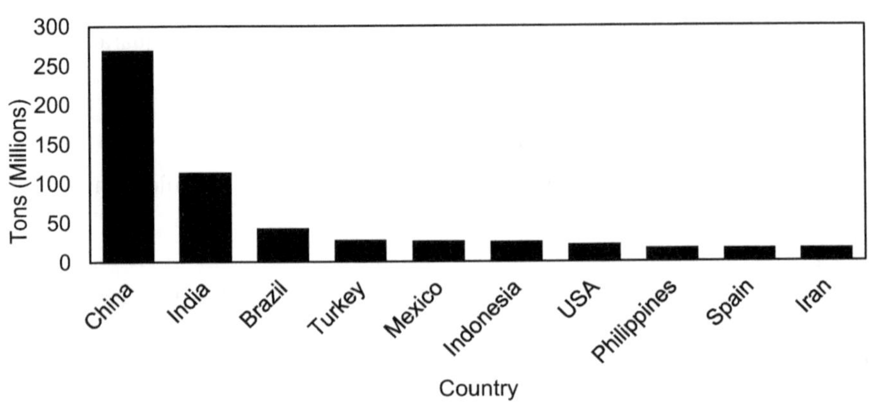

Fig. 12.1 Top ten (10) fruit-producing countries in 2023 in terms of production volume in million tons. (Source: FAOSTAT 2025)

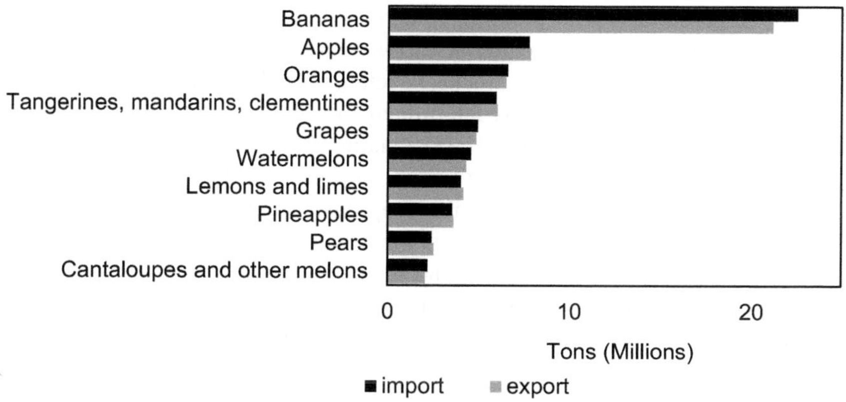

Fig. 12.2 Top ten (10) fruits traded globally in 2023, in terms of import (black bar) and export (grey bar) value in million tons. (Source: FAOSTAT 2025)

cause huge economic losses if not managed properly. Among the pests and diseases of fruit crops, viral diseases are the most important. In fact, plant viruses are responsible for estimated annual losses of more than USD 30 billion in global fruit production (Rodríguez-Verástegui et al. 2022). Aside from economic losses, plant viruses are the primary concern in plant material propagation due to their systemic and latent infection, as well as their efficient transmission by insect vectors or vegetative propagation (Manzoor et al. 2025).

Viral diseases are also important to fruit production because they reduce both the quantity and quality of fruit yields. Their impact is more notable in perennial crops, as they can persist for extended periods even when the plants appear healthy or asymptomatic (Rao and Reddy 2020). For the most virulent viral strains in certain fruit crops, yield losses of up to 50–100% can be observed. Furthermore, the difficulty of managing plant viruses is exacerbated by the unavailability of curative treatments and the frequent exchange or movement of plant propagation material across different countries and regions (Tatineni and Hein 2023). Viral pathogens can persist and spread undetected in fruit nurseries and orchards, causing latent infection, which can result in minimal to severe production losses. In severe cases, viruses can cause tree mortality and unmarketable fruits, thus resulting in significant economic impacts. Even subtle infections can reduce growth and yield, negatively affecting profitability (Nabi et al. 2020; Umer et al. 2019). Thus, to manage the impacts of viral pathogens in fruit production, the most economical and effective ways are preventing virus spread by ensuring virus-free propagation materials through plant quarantine and certification (Lee 2020; Roy 2011; Ventura et al. 2019).

Plant quarantine is officially defined as "all activities designed to prevent the introduction or spread of quarantine pests or to ensure their official control" (IPPC Secretariat 2024). The application of plant quarantine to ensure virus-free propagation material involves a systematic regulatory and technical approach to prevent the introduction, establishment, and spread of plant viruses via the long-distance transport of germplasm and planting materials. On the other hand, certification schemes are programs implemented at various international, regional, national, or local levels to ensure that nursery stocks, vegetative, and seed planting materials are free of, or have less than, the predetermined threshold level or concentration of pathogens (Waterworth and Hadidi 1998). Quarantine is primarily aimed at preventing the introduction of pathogens not yet present or of limited distribution within a country, while certification ensures the long-term supply of virus-free planting material by regulating domestic propagation systems. These systems serve as the main cornerstone of virus prevention and management in fruit crops (Barba et al. 2015). Since the plant propagation materials are the foundation for expanding the fruit production industry, it is necessary to establish a system that ensures the virus-free status of planting propagules at the initial stages of the value chain. This is because the risk and impacts of viral diseases in fruit production can be greatly minimized and prevented through stringent quarantine protocols and certification programs. These systems are crucial in limiting the spread of viruses by ensuring that only thoroughly tested and healthy planting materials are introduced into the fruit orchards and plantations. A robust and comprehensive quarantine and certification system can provide a sustainable and long-term strategy to safeguard the fruit industry by facilitating international trade and enhancing food security, despite the persistent threats posed by plant viruses. To emphasize the importance of quarantine and certification programs, this chapter discusses existing plant quarantine standards and regulations, the reported viruses of quarantine importance in major fruits in the literature, and regional and national certification programs for ensuring virus-free fruit propagation materials.

12.2 Quarantine Programs: Preventing the Spread of Viruses

12.2.1 International Plant Quarantine Framework

Plant quarantine is one of the most important tools for managing viral diseases in fruit crops since viruses are often disseminated through the movement of infected propagation material (Manzoor et al. 2025). The propagation and movement of plant material across countries, regions, and international borders are inevitable practices in the expanding global fruit market production system worldwide. An increasing amount of plant material, such as nursery and breeding stocks, seeds, and seedlings, traded over vast distances, can carry the increased risk of introducing pests and diseases (Gergerich et al. 2015). Among the primary concerns of the fruit industry are groups of systemic pathogens that can be carried by propagation materials, such as viruses, viroids, phytoplasmas, and other bacteria (Mink 1998). Once introduced and established in an area, these pathogens require expensive management and control procedures; sometimes, no effective and economically sound treatments are available (Cembali et al. 2003). Historical accounts in some crops, such as grapevines and fruit trees, demonstrate the damaging and expensive impacts of introducing new pathogens through plant propagation materials (Cambra et al.

2006; Moreno et al. 2007; Nemeth 1979). Since vegetative propagation is the primary method for cultivating most perennial fruit crops, the long-distance movement of these materials can unknowingly carry viruses and viroids, which accumulate in mother plants without immediate, visible symptoms (latent infection).

International and national organizations play a pivotal role in establishing and enforcing plant quarantine regulations to prevent the introduction, establishment, and spread of pests. The International Plant Protection Convention (IPPC) serves as the primary international treaty aimed at safeguarding plants, agricultural products, and natural ecosystems from harmful pests. It was established in 1951 by ratification of 185 contracting parties and was administered by the Food and Agriculture Organization of the United Nations. IPPC promotes harmonized and science-based phytosanitary measures by developing and adopting international standards such as the International Standards for Phytosanitary Measures (ISPMs). ISPMs adopted by the IPPC provide contracting parties with a coordinated framework to prevent, detect, and control pests and diseases effectively. Aside from standard setting, IPPC also implements capacity development and fosters partnerships among the contracting parties to improve communication and coordination. With the establishment of global phytosanitary standards and international cooperation, the IPPC contributes to the long-term sustainability of agricultural production by promoting safe international trade and reducing economic losses brought about by pests (IPPC Secretariat 2025). At the regional level, IPPC is supported by Regional Plant Protection Organizations (RPPOs). Under Article IX of the Convention (Fig. 12.3), RPPOs are mandated to bridge the gap between global standards and their regional/national implementation. Other RPPO functions include gathering and disseminating pest information, coordinating certification programs, providing early warnings for

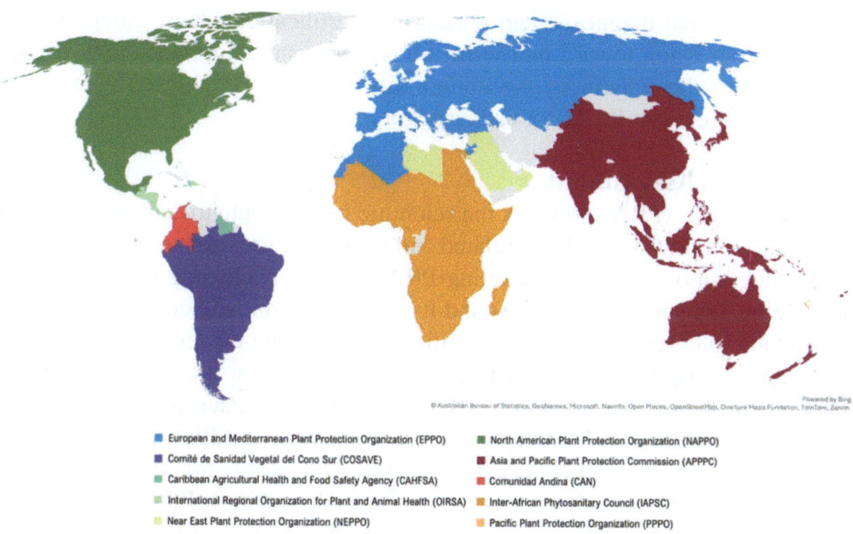

Fig. 12.3 Map of the Regional Plant Protection Organizations (RPPOs)

emerging pests, and serving as a platform for technical consultation and harmonization. Because of the cross-border threat of plant pests and diseases, countries are members of more than one RPPO (IPPC Secretariat 2023). Aside from the regional cooperation, the contracting parties implement IPPC and RPPO guidelines at the national level through their own National Plant Protection Organizations (NPPOs). As an IPPC signatory, each contracting party is required to designate or establish an NPPO, which is responsible for implementing phytosanitary measures at the national level (Food and Agriculture Organization of the United Nations 1997). The implementation of international standards set by the IPPC can be adopted through national legislation and implemented by the NPPO. The NPPO has numerous functions, including conducting pest risk analyses, regulating the movement of plants, plant products, and planting materials, performing phytosanitary diagnostics, treatments, and surveillance, and ensuring the establishment of pest-free production systems. These functions are necessary to prevent the spread of viruses that can easily spread in vegetatively propagated fruit crops. In addition, certification programs çan complement these efforts by setting up schemes that certify that the propagation materials supplied to the fruit growers are virus-free (Barba et al. 2015).

12.2.2 Plant Quarantine Measures for Viruses in Plant Propagation Materials

Plant quarantine regulations are a science-based approach wherein phytosanitary risks are assessed and managed using all available scientific data as the basis for a systematic, multi-tiered approach (pre-border, border, and post-border measures) (Bland et al. 2023). Under this system, the overall standards and regulatory approach can be grouped into four primary categories: risk assessment, inspection and verification, testing and diagnosis, and permit and certification. These categories illustrate the progression toward increased information and compliance flow from risk assessment to the release of the commodity (Fig. 12.4). The following section describes the relevant ISPM for each category.

12.2.2.1 Risk Assessment

Pest risk analysis (PRA) provides the scientific basis for phytosanitary measures. It determines if a pest should be regulated and what measures are necessary to prevent its introduction or spread. Conducting PRA involves evaluating the probability of introducing and spreading the pest and its potential economic consequences, based on available biological or other scientific and economic evidence. It also involves suggesting management options to reduce the risk at an acceptable level (IPPC 2019a, b; IPPC Secretariat 2004b, 2012). The ISPMs related to PRA are ISPM 2 (*Framework for PRA*), 11 (*PRA for quarantine pests*), 21 (*PRA for regulated non-quarantine pests*), and 36 (*Integrated measures for plants for planting*). For fruit propagation materials such as cuttings, grafts, and seedlings, PRA typically classifies them as high-risk pathways due to their higher likelihood of harboring viruses and their efficiency in transmitting them. The high-risk status is justified based on

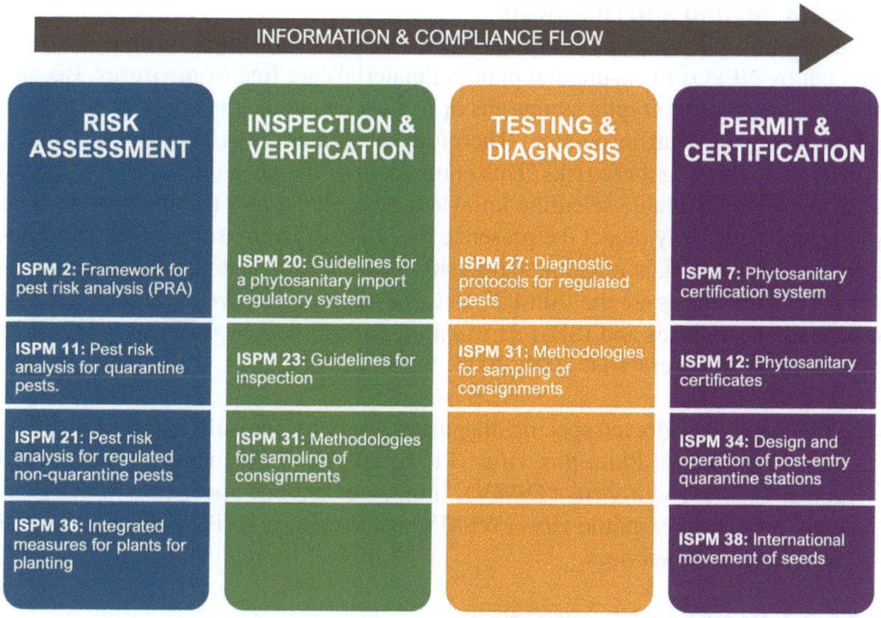

Fig. 12.4 Relevant International Standards for Phytosanitary Measures (ISPMs) for quarantine and certification to ensure virus-free fruit propagation material

the viruses' ability to be introduced, established, and spread easily in new growing areas via planting propagation materials. Aside from the probability of introduction, establishment, and spread, the possible substantial economic losses and environmental impacts are also considered in the analysis (EFSA Panel on Plant Health (PLH) et al. 2019a, b; EFSA Panel on Plant Health (PLH) 2019a, b).

12.2.2.2 Inspection and Verification

Implementation of phytosanitary measures involves various phytosanitary actions by the National Plant Protection Organization (NPPO), including inspection, testing, surveillance, and treatment. Inspection is the "official visual examination of plants, plant products, or other regulated articles to determine if pests are present or to verify compliance with phytosanitary regulations" (IPPC Secretariat 2023). An inspector performs it, and the NPPO officially authorizes a person to perform this function. Inspections are typically conducted at ports or in the field to detect the presence of pests and ensure compliance with phytosanitary requirements. Key ISPMs related to inspection include ISPM 20 20 (*Phytosanitary import regulatory system*), ISPM 23 (*Guidelines for inspection*), and ISPM 31 (*Methodologies for sampling consignments*) (IPPC Secretariat 2004a, 2005, 2008).

12.2.2.3 Testing and Diagnosis

Aside from inspection, laboratory testing and diagnosis are important protocols carried out by NPPOs to ensure that planting materials are free from viruses. Because viruses can produce cryptic symptoms or remain latent, they may not be detectable solely through visual inspection. Therefore, a reliable virus detection method is crucial in addressing these risks. This component of the plant quarantine regulatory system utilizes available scientific knowledge and diagnostic technologies to accurately and sensitively detect the presence of viruses, thereby supporting phytosanitary measures and plant quarantine regulatory actions. For testing and diagnosis, the ISPMs related to this are the ISPM 27 (*Diagnostic protocols for regulated pests*) for diagnostic procedures and ISPM 31 (*Sampling methodologies*) for sampling methodologies (IPPC Secretariat 2006, 2008):

The IPPC has also adopted specific diagnostic protocols for fruit viruses and viroids, such as the DP 2: Plum pox virus (PPV), DP 15: Citrus tristeza virus, DP 24: Tomato spotted wilt virus (TSWV), Impatiens necrotic spot virus (INSV), and Watermelon silver mottle virus (WSMV), and supports NPPOs in delivering efficient diagnostic services.

12.2.2.4 Permit and Certification

The last category of phytosanitary measures described in this chapter involves the frameworks and formal process of attesting that fruit propagation materials meet the country's phytosanitary requirements. This involves the issuance of permits and certifications, as well as physical and documentary forms, to ensure that consignments are virus-free and safe for planting pre-entry and post-entry. Under the adopted standards of the IPPC, permits and certifications include several ISPMs such as ISPM 7 (*Phytosanitary Certification System*), 12 (*Phytosanitary Certificates*), 34 (*Design and operation of post-entry quarantine stations*), and 38 (*International movement of seeds*) (IPPC Secretariat 2010, 2011, 2017, 2022).

12.3 Viruses of Quarantine Concern in Fruit Propagation Materials

Pests are officially categorized based on their economic importance and distribution. A quarantine pest (QP) is potentially economically important to the endangered area. It is either absent or present but not widely distributed and under official control. Conversely, a regulated non-quarantine pest (RNQP) is a pest whose presence in plants for planting affects the intended use of those plants with an economically unacceptable impact and is therefore regulated within the importing country. Any pest that does not meet the criteria for QP or RNQP is considered a non-quarantine pest, typically being widely distributed and not under official regulation (IPPC Secretariat 2024). The IPPC provides definitions for each type of pest, but some RPPOs maintain additional, region-specific lists. For example, the European and Mediterranean Plant Protection Organization (EPPO) maintains a database of quarantine pests relevant to its region. EPPO classifies quarantine pests into two lists: the A1 list includes pests absent from the region but recommended for regulation, while the A2 list comprises pests present in the region but with limited distribution, also recommended for regulation (EPPO 2024). Propagation materials of globally traded fruits are high-risk commodities because they can harbor viruses and viroids of quarantine concern (Table 12.1). Most of the reported viruses of quarantine concern belong to the families *Closteroviridae*, *Betaflexiviridae*, *Secoviridae*, and *Rhabdoviridae*. These pathogens are primarily transmitted via vegetative propagation or seeds, with additional pathways including insect vectors and pollen. This makes containment and eradication technically challenging and costly once viruses become established.

12.4 Certification Programs to Ensure Virus-Free Fruit Propagation Materials

While quarantine prevents the introduction of exotic viruses, certification programs maintain the integrity of planting stock within a country or production system, ensuring the long-term sustainability of orchard industries. These programs guarantee that nurseries supply growers with high-quality, virus-free propagation material through strict standards applied to every step in the planting journey: foundation plants, nursery practices, and pathogen testing regimes (Barba et al. 2015; Fukumoto et al. 2013). Foundation blocks, established from rigorously tested mother plants, provide the initial source of clean material. These are maintained under controlled conditions and subjected to regular inspections and diagnostics. From these, nurseries propagate certified material under traceable and regulated conditions, ensuring compliance with phytosanitary standards and grower confidence. Certification programs also translate to improved orchard productivity, extended lifespan of fruit trees, and reinforced international competitiveness within the orchard industry (Varveri et al. 2015). Domestically, it also complements quarantine measures for restricting the movement of infected materials. The strength of certification lies in its two main pillars: sensitive and reliable detection techniques, and effective virus elimination procedures. Detection methods ranging from biological indexing and ELISA to PCR, RT-PCR, and high-throughput sequencing allow early identification of infected material. Over the last 60 years, testing of viruses and use of virus-free planting material have become the gold standard for managing graft-transmissible pathogens in vegetatively propagated crops (Cutting and Montgomery 1973; Nemeth 1979; Varveri et al. 2015; Waterworth and Hadidi 1998). When viruses are detected, sanitation methods such as shoot tip culture, thermotherapy, chemotherapy, or cryotherapy are applied to regenerate clean plants (Cui et al. 2013; Panattoni et al. 2013). These approaches are particularly critical for perennial species, where viral infections persist for the lifetime of the stem. Clean material can only address them (Chauhan et al. 2019). The success of certification programs largely depends on the collaboration between scientific institutions, government regulators, and industry stakeholders, each of them accountable for different stages of the process.

Table 12.1 Viruses of quarantine and certification importance in fruit propagation materials

Host	Family	Species	Quarantine status	Reference
Apple/Pome	*Betaflexiviridae*	Apple chlorotic leaf spot virus (ACLSV), Trichovirus mali	RNQP (EU) QP (US)	Nemeth (1979)
	Pospiviroidae	Apple dimple fruit viroid (ADFVd), Apscaviroid fossulamali	RNQP (EU) QP (Canada, Israel, Jordan)	Serio et al. (2001)
	Pospiviroidae	Apple scar skin viroid (ASSVd), Apscaviroid cicatricimali	RNQP (EU) QP (Canada, Israel, Jordan)	Hadidi et al. (2024)
	Secoviridae	Apple latent spherical virus (ALSV), Cheravirus mali	N/A	Koganezawa (1985)
Avocado	*Avsunviroidae*	Avocado sunblotch viroid (ASBVd), Avsunviroid albamaculaperseae	QP A1	Whitsell (1952)
Banana	*Nanoviridae*	Banana bunchy top virus (BBTV), Babuvirus musae	QP A2 (APPPC, COSAVE, IAPSC, PPPO)	Magee (1927)
	Potyviridae	Banana bract mosaic virus (BBrMV), Potyvirus musae	QP A2 (APPPC, PPPO)	Espino et al. (1990)
Blueberry	*Secoviridae*	Blueberry leaf mottle virus (BLMoV), Nepovirus myrtilli	QP A1 (EPPO, COSAVE)	Childress and Ramsdell (1986) and Ramsdell and Stace-Smith (1979)
Blueberry	*Betaflexiviridae*	Blueberry scorch virus (BlScV), Carlavirus vaccinii	QP A2 (EPPO, COSAVE)	Martin and Bristow (1988)
Blueberry	*Bromoviridae*	Blueberry shock virus (BSV), Ilarvirus BSV	QP A2 (COSAVE) RNQP (EU)	Bristow and Martin (1999) and MacDonald et al. (1991)
Cherry (including apple and pears)	*Secoviridae*	Cherry rasp leaf virus (CLRV), Cheravirus avii	QP A2 (COSAVE) QP A1 (EAEU, EPPO, IAPSC)	Stace-Smith and Hansen (1976)

(continued)

Table 12.1 (continued)

Host	Family	Species	Quarantine status	Reference
Citrus	*Closteroviridae*	Citrus tristeza virus (CTV), Closterovirus tristezae	QP A1 (EU) QP A2 (APPPC, EPPO, IAPSC) ISPM 27	Moreno et al. (2007)
	Kitaviridae	Citrus leprosis virus C (CiLV-C), Cilevirus leprosis	QP A1 (EPPO) USDA Regulated Pest List	Roy et al. (2013)
	Rhabdoviridae	Citrus leprosis virus N (CiLV-N), Dichorhavirus leprosis	QP A1 (EPPO) USDA Regulated Pest List	Lovisolo (2001) and Ramos-González et al. (2017)
	Alphaflexiviridae	Citrus yellow vein clearing virus (CYVCV), Potexvirus citriflavivenae	Regulated Pest List (US) QP A1 (Chile) Alert list (EPPO)	Abrahamian et al. (2024), Grimaldi and Catara (1996) and Loconsole et al. (2012)
	Caulimoviridae	Citrus yellow mosaic virus (CYMV), Badnavirus tessellocitri	QP A1 (EPPO)	Ahlawat et al. (1996), Dakshinamurti and Reddy (1975)
	Pospiviroidae	Citrus exocortis viroid (CEVd), Pospiviroid exocortiscitri	QP A2 (IAPSC) RNQP (EU) QP A1 (Argentina, Mexico, Bahrain, Jordan, UK)	Benton et al. (1950) and Vernière et al. (2004)
Fig	*Fimoviridae*	Fig mosaic virus complex (FMV), Emaravirus fici	Certification	Preising et al. (2021)

(continued)

Table 12.1 (continued)

Host	Family	Species	Quarantine status	Reference
Grapevine	*Geminiviridae*	Grapevine red blotch virus (GRBV), Grablovirus vitis	QP A1 (EPPO) QP (Morocco)	Krenz et al. (2012, 2023) and Rwahnih et al. (2013)
	Secoviridae	Grapevine fanleaf virus (GFLV), Nepovirus foliumflabelli	QP A2 (EPPO) RNQP (EU)	BaSSO et al. (2017)
	Closteroviridae	Grapevine leafroll-associated viruses (GLRaVs), Ampelovirus univitis	QP (US) RNQP (EU) QP A1 (Uruguay, Jordan, Turkey, UK)	Almeida et al. (2013), Bostanian et al. (2012), Engelbrecht and Kasdorf (1990) and Herrbach et al. (2016)
	Betaflexiviridae	Grapevine Pinot gris virus (GPGV), Trichovirus pinovitis	Monitored in EU	Meng et al. (2017), Messmer et al. (2024) and Saldarelli et al. (2017)
Kiwifruit	*Betaflexiviridae*	Actinidia virus A (AcVA), Actinidia virus B (AcVB)	QP A1 (Chile)	Blouin et al. (2012)
	Fimoviridae	Actinidia chlorotic ringspot-associated virus (AcCRaV), Emaravirus actinidiae	N/A, but Certification target	Wu et al. (2022)
Melon	*Tospoviridae*	Melon yellow spot virus (MYSV), Orthotospovirus meloflavi	USDA Regulated Pest List	Kato et al. (2000)
Peach	*Betaflexiviridae*	Peach mosaic virus (PcMV), Trichovirus persicae	QP A1 (EPPO) QP (Canada, Morocco)	Gispert et al. (1998a, b), James et al. (2006) and Larsen and James (2011)
Peach (including grapes)	*Secoviridae*	Peach rosette mosaic virus (PRMV), Nepovirus persicae	QP A1 (EPPO, EAEU) QP (China, Israel, Canada, Mexico)	Allen et al. (1982) and Gispert et al. (1998a, b)

(continued)

Table 12.1 (continued)

Host	Family	Species	Quarantine status	Reference
Papaya	*Potyviridae*	Papaya ringspot virus (PRSV), Potyvirus papayanuli	QP A2 (Egypt, Bahrain, Jordan)	Gonsalves et al. (2010)
	Potyviridae	Papaya leaf distortion mosaic virus (PLDMV), Potyvirus caricae	N/A, but Certification target	Bau et al. (2008), Bhargava and Khurana (1970) and Tennant et al. (2007)
Passionfruit	*Potyviridae*	Passionfruit woodiness virus (PWV), Potyvirus passiflorae	N/A, but Certification target	Fukumoto et al. (2013)
Raspberry (including peach and grapes)	*Secoviridae*	Raspberry ringspot virus (RpRSV), Nepovirus rubi	QP (US, Israel, Canada, Mexico) QP A1 (EAEU) QP A2 (COSAVE, EPPO0 RNQP (UK, Russia, Switzerland)	Martin et al. (2013) and Scott et al. (2000)
Stone fruits	*Potyviridae*	Plum pox virus (PPV), Potyvirus plumpoxi	QP (USA, Canada, Mexico, China, Israel, Korea) QP A1 (COSAVE, IAPSC) QP A2 (EAEU, EPPO) RNQP (EU)	James et al. (2013)
	Secoviridae	Cherry leaf roll virus (CLRV), Nepovirus avii	QP (Canada, Mexico, USA, Israel) QP A1 (EPPO, IAPSC) QP A2 (COSAVE) RNQP (EU)	Brown et al. (1993), Cropley (1961) and Swingle et al. (1943)

(continued)

Table 12.1 (continued)

Host	Family	Species	Quarantine status	Reference
Strawberry/ Rubus	*Rhabdoviridae*	Strawberry latent C virus (STLCV)	QP (Morocco, Tunisia, Israel) QP A1 (EPPO, Bahrain, Egypt)	Maas (1998) and Harris and Eking (1942)
	Alphaflexiviridae	Strawberry mild yellow edge virus (SMYEV), Potexvirus fragariae	QP A1 (Egypt, Serbia, Mexico) QP A2 (Turkey) RNQP (EU)	Krczal (1981) and Martin and Tzanetakis (2013)
	Secoviridae	Strawberry mottle virus (SMoV), Sadwavirus fragariae	N/A (Target for Certification PM 4/11)	Thompson et al. (2002) and Thompson and Jelkmann (2003)
	Mayoviridae	Raspberry bushy dwarf virus (RBDV), Idaeovirus rubi	RNQP (EU) QP (Mexico) QP A1 (Argentina)	Jones et al. (1982) and Mavrič et al. (2003)
Strawberry/ Rubus (including grapes and peach	*Secoviridae*	Arabis mosaic virus (ArMV), Nepovirus arabis	RNQP (EU) QP A2 (APPPC, COSAVE)	Harrison and Murant (1980)
Watermelon	*Potyviridae*	Watermelon mosaic virus (WMV), Potyvirus citrulli	QP (US Regulated Pest List, Mexico) RNQP (Egypt) QP A2 (Bahrain, Jordan)	Ali et al. (2024) and Sharma (2023)
	Tospoviridae	Watermelon silver mottle virus (WSMoV), Orthotospovirus citrullomaculosi	QP A1 (EPPO)	Iwaki et al. (1984) and Jones (2005)

In the United States, state sectors coordinated detection and sanitation programs enabled the apple industry to benefit from decades of investment in virus-free rootstocks. In China, however, Apple scar skin viroid (ASSVd) and Apple dimple fruit viroid (ADFVd) remain problematic because certification systems are less consistently applied due to the abundance of small-scale orchards (Liang et al. 2022).

Such contrasts underscore that certification is not merely a technical exercise but also a question of infrastructure, policy, and grower adoption. Globally, regional, and national frameworks illustrate how certification is adapted to different contexts. However, regardless of the context, the ultimate goal remains twofold: ensuring true-to-type of cultivars and safeguarding their sanitary status (Barba et al. 2015).

The EPPO established some of the earliest standardized schemes in Europe, beginning in 1970 with fruit trees, scions, and rootstocks (Ebbels 2003). Today, the PM 4 series of EPPO standards outlines specific requirements for crops such as grapevine, citrus, strawberry, *Rubus*, and pome and stone fruits, providing a harmonized blueprint for member states (European and Mediterranean Plant Protection Organization 2025). The United Kingdom's Fruit Propagation Certification Scheme (FPCS), run by DEFRA and the Animal and Plant Health Agency, similarly integrates inspection, testing, and certification for soft and top fruits (Department for Environment, Food and Rural Affairs and Animal and Plant Health Agency 2023). In North America, state-level programs in California, Oregon, Washington, New York, and Pennsylvania are unified through the National Clean Plant Network, ensuring nationwide access to virus-free planting material (National Clean Plant Network 2025). In some countries, national certification schemes and standards have also been implemented. In the case of Australia, the Apple and Pear Improvement Program (APFIP) ensures the availability of clean pome fruit propagation material for growers by providing testing and certification services against economically important viruses (Apple and Pear Australia Limited (APAL) 2025). In India, the Clean Plant Program is established to provide farmers with high-quality, virus-free planting materials for major fruit crops produced by the country. The program offers pathogen testing services for a diverse range of fruit crops grown in various agro-climatic regions throughout the country (Rakesh et al. 2025). Brazil has recently made strides in aligning with international certification and quality standards. In September 2025, Brazil issued its first OECD-compliant citrus quality certificate for the export of Tahiti limes to the European Union, marking a milestone in phytosanitary certification for its table citrus sector (Lorimer 2025). The certificate, issued by the Ministry of Agriculture and Livestock (MAPA), formalized quality and health compliance, streamlining inspections and facilitating access to EU markets. The international and national programs discussed emphasize the importance of certification schemes ensuring plant health and while facilitating the trade of fruit propagation materials. By integrating certification programs into an effective and harmonized plant quarantine regulatory system, the risk of virus infections in fruit production can be minimized and managed. Thus, quarantine and certification programs both contribute greatly for the safety and sustainability of the fruit industry (Manzoor et al. 2025).

12.5 Challenges and Future Directions

Plant quarantine and certification are indispensable tools to ensure the safety and sustainability of the global fruit industry. Despite their advantages, plant quarantine and certification programs still face several challenges that can limit their effectiveness. One of the major challenges is compliance and enforcement. This is particularly challenging in resource-constrained countries with less developed phytosanitary regulatory systems (van der Meer 2007; Shafaeddin 2007). In small-scale fruit production areas, informal seed networks remain the primary source of plant propagation materials. If not monitored and managed properly, informal seed networks can act as major pathways for virus dissemination, undermining quarantine and certification programs (Lua et al. 2015; McGuire and Sperling 2016). Several issues contribute to the risk of virus dissemination in informal seed systems, including socio-economic barriers such as limited farmer awareness, higher costs of certified materials, and a lack of incentives for adoption (Lule et al. 2011). Aside from the compliance and enforcement, technical and resource limitations on testing and diagnosis remain a major challenge in plant quarantine and certification programs. These include the inaccessibility of testing materials in some countries (e.g., antisera for serological testing), limited availability of the exotic viral genome sequences in online databases, and the emergence or re-emergence of novel and exotic viruses. In addition, virus testing and diagnosis have the potential to generate false negative results due to low and possibly uneven titers in host plants, as well as the difficulty in detecting asymptomatic and latent infections (López et al. 2009; Setiono et al. 2018). The complexity of mixed infections and the emergence of novel and recombinant viruses further complicates reliable detection (Rubio et al. 2020). Diagnostic methods such as ELISA and PCR are widely available, adopted, and validated for plant virus testing. However, they require significant resources such as laboratory infrastructure, trained personnel, and a regular supply of reagents. These requirements may not always be feasible in resource-limited institutions (Boonham et al. 2014; Jeong et al. 2014). Recently, novel technologies have been developed to address these constraints, such as the use of field-deployable and user-friendly lateral flow strips, real-time PCR, loop-mediated isothermal amplification (LAMP), and CRISPR-Cas–based methods, but these techniques are limited by their sensitivity, specificity, practicality, and accessibility (Chalam 2020; Jaybhaye et al. 2024; Singh et al. 2025). The evolution and emergence of new and exotic virus species pose a significant threat to the fruit industry. Virus evolution, driven by genetic mutation and recombination, can enable viruses to overcome host resistance, and the emergence of new virus species necessitates the rapid development of detection protocols (Hadidi et al. 2024). This is exacerbated by climate change, which affects virus epidemiology and vector distribution through changes in seasonal and rainfall patterns (Kramer 2016). Rapid and drastic changes in virus evolution can cause disruptions in the implementation of quarantine and certification systems. Thus, there is a need for plant quarantine and certification systems that are flexible and adaptable to the changing environment. Despite the challenges posed by viral diseases in the fruit industry, rapid technological advancements and global coordination efforts can provide solutions to address the evolving viral disease threats. Improvements in pathogen detection technologies, such as the use of high-throughput sequencing (HTS) in plant quarantine and certification testing, could facilitate trade while reducing the risk of introduction (Martin et al. 2016). The use of HTS in plant quarantine and certification has become a reliable and cost-effective diagnostic tool, as it is more accurate and efficient compared to traditional biological assays for virus diagnosis (Al Rwahnih et al. 2025). This is because HTS provides a highly sensitive and non-targeted detection and identification approaches for both the known and unknown viruses. It can complement the existing virus diagnostic tests such as ELISA and PCR. Since HTS was first used to detect and identify plant viruses over the past 15 years, the technology has undergone significant improvements, allowing it to be conducted at a lower cost and with increasing accessibility to users. Because of this, its potential for use in frontline plant biosecurity applications has been introduced (Fox et al. 2025). Digital innovations, such as blockchain, barcoding, and traceability platforms, can also be utilized to enhance plant quarantine and certification systems. Digital technologies can strengthen the security, transparency, and integrity of information across the value chain, preventing fraud and ensuring regulatory compliance. Ensuring the traceability of certified plant propagation materials also ensures the safety and integrity across the value chain of minimizing the risk of unintended contamination and improving trust among relevant stakeholders (Gadge et al. 2024).

On a global perspective, strengthening harmonization and coordination efforts is crucial to address the growing and evolving challenges in plant quarantine and certification programs. International and regional organizations such as the IPPC, RPPOs, and NPPOs play important roles in standardizing certification systems, facilitating safe trade, ensuring compliance, and providing technical support to countries with limited capacity (Brückner 2011; IPPC 1999; Roy 2011). Improving regulatory compliance through sustained capacity-building efforts can help minimize barriers to adoption and ensure proper implementation. The integration of virus certification schemes into the broader scope of quarantine programs, including surveillance, vector management, and the implementation of climate-resilient agricultural strategies, encourages a more holistic approach to managing the threats of viruses in the production of fruit propagation materials (Hejran 2024). This requires close collaboration among researchers, government agencies, and the private sector (Varveri et al. 2015). In summary, addressing the threats posed by viral diseases in fruit propagation materials requires a multi-pronged and comprehensive strategy. Plant quarantine and certification programs are crucial to address these threats; however, recent challenges necessitate the development of more flexible and adaptable programs incorporating technological innovations, socio-economic considerations, and improving international cooperation and harmonization. By investing in advanced diagnostics, improving program adoption and compliance, as well as integrating certification schemes within the broader plant quarantine regulatory systems, the fruit industry can be safeguarded.

References

Abrahamian P, Tian T, Posis K, Guo Y, Yu D, Blomquist CL, Wei G, Adducci B, Vidalakis G, Bodaghi S, Osman F, Roy A, Nunziata SO, Nakhla M, Mavrodieva V, Rivera Y (2024) Genetic analysis of the emerging *Citrus yellow vein clearing virus* reveals a divergent virus population in American isolates. Plant Dis 108(6):1544–1554. https://doi.org/10.1094/PDIS-09-23-1963-RE

Ahlawat Y, Pant R, Lockhart B, Srivastava M, Chakraborty N, Varma A (1996) Association of a badnavirus with citrus mosaic disease in India. Plant Dis 80(5):590–592

Al Rwahnih M, Klaassen V, Erickson T, Alabi OJ, Stevens K, Hwang MS, Port L (2025) A new era in federal quarantine and state certification diagnostics at clean plant centers in the United States. Plant Dis 109(7):1392–1403. https://doi.org/10.1094/PDIS-10-24-2104-FE

Ali R, Eldidamony G, Askora A, Galal A (2024) Investigates the ability of plant extracts from Lens culinaris to protect zucchini from the *Zucchini yellow mosaic virus* (ZYMV). Sci Rep 14(1):12257. https://doi.org/10.1038/s41598-024-62128-6

Allen WR, Van Schagen J, Eveleigh ES (1982) Transmission of *Peach rosette mosaic virus* to peach, grape, and cucumber by *Longidorus diadecturus* obtained from diseased orchards in Ontario. Can J Plant Pathol 4(1):16–18

Almeida RPP, Daane KM, Bell VA, Blaisdell GK, Cooper ML, Herrbach E, Pietersen G (2013) Ecology and management of grapevine leafroll disease. Front Microbiol 4:94. https://doi.org/10.3389/fmicb.2013.00094

Apple and Pear Australia Limited (APAL) (2025) *Variety development and evaluation services*. APAL. https://apal.org.au/programs/variety-development-and-evaluation/

Bacelar E, Pinto T, Anjos R, Morais MC, Oliveira I, Vilela A, Cosme F (2024) Impacts of climate change and mitigation strategies for some abiotic and biotic constraints influencing fruit growth and quality. Plants 13(14). https://doi.org/10.3390/plants13141942

Barba M, Ilardi V, Pasquini G (2015) Chapter Three – Control of pome and stone fruit virus diseases. In: Loebenstein G, Katis NI (eds) Advances in virus research, vol 91. Academic, pp 47–83. https://doi.org/10.1016/bs.aivir.2014.11.001

Basso MF, Fajardo TVM, Saldarelli P (2017) Grapevine virus diseases: Economic impact and current advances in viral prospection and management. Rev Bras Frutic 39(1):e-411

Bau H-J, Kung Y-J, Raja J, Chan S-J, Chen K-C, Chen Y-K, Wu H-W, Yeh S-D (2008) Potential threat of a new pathotype of *Papaya leaf distortion mosaic virus* infecting transgenic papaya resistant to *Papaya ringspot virus*. Phytopathology 98(7):848–856

Benton R, Bowman F, Fraser L, Kebby R (1950) Stunting and scaly butt of Citrus associated with Poncirus trifoliata rootstock

Bhargava K, Khurana SP (1970) Insect transmission of papaya viruses with special reference to papaya mosaic virus. Zbl Bekt 124:688–696

Bland LM, Bau S, Hester S, Arndt E (2023) Biosecurity Continuum [Graphic]. Centre of Excellence for Biosecurity Risk Analysis, University of Melbourne. https://doi.org/10.26188/23584230

Blouin AG, Chavan RR, Pearson MN, MacDiarmid RM, Cohen D (2012) Detection and characterisation of two novel vitiviruses infecting Actinidia. Arch Virol 157(4):713–722. https://doi.org/10.1007/s00705-011-1219-1

Boonham N, Kreuze J, Winter S, van der Vlugt R, Bergervoet J, Tomlinson J, Mumford R (2014) Methods in virus diagnostics: from ELISA to next generation sequencing. Virus Res 186:20–31

Bostanian NJ, Vincent C, Isaacs R (2012) Arthropod management in vineyards: pests, approaches, and future directions. Springer

Bristow P, Martin R (1999) Transmission and the role of honeybees in field spread of blueberry shock ilarvirus, a pollen-borne virus of highbush blueberry. Phytopathology 89(2):124–130

Brown DJ, Halbrendt JM, Robbins RT, Vrain TC (1993) Transmission of Nepoviruses by *Xiphinema americanum*-group Nematodes. J Nematol 25(3):349–354

Brückner G (2011) Ensuring safe international trade: how are the roles and responsibilities evolving and what will the situation be in ten years' time? Revue Scientifique et Technique (International Office of Epizootics) 30(1):317–324

Cambra M, Capote N, Myrta A, Llácer G (2006) *Plum pox virus* and the estimated costs associated with sharka disease. EPPO Bull 36(2):202–204

Cembali T, Folwell RJ, Wandschneider P, Eastwell KC, Howell WE (2003) Economic implications of a virus prevention program in deciduous tree fruits in the US. Crop Prot 22(10):1149–1156

Chalam VC (2020) Elimination of plant viruses by certification and quarantine for ensuring biosecurity. In: Applied plant virology. Elsevier, pp 749–762

Chauhan P, Singla K, Rajbhar M, Singh A, Das N, Kumar K (2019) A systematic review of conventional and advanced approaches for the control of plant viruses. J Appl Biol Biotechnol 7(4):89–98

Childress A, Ramsdell D (1986) Detection of *Blueberry leaf mottle virus* in highbush blueberry pollen and seed. Phytopathology 76(12):1333–1337

Colombo RC, Roberto SR (2022) Propagation and postharvest of fruit crops. Horticulturae 8(3):246. https://doi.org/10.3390/horticulturae8030246

Cropley R (1961) Cherry leaf-roll virus. Ann Appl Biol 49(3):524–529. https://doi.org/10.1111/j.1744-7348.1961.tb03645.x

Cui Z-H, Li B-Q, Bi W-L, Li J-W, Wang Q-C (2013) Plant pathogen eradication by cryotherapy of shoot tips: development, achievements and prospective. Acta Hortic:35–41

Cutting C, Montgomery H (1973) More and better fruit with EMLA. East Malling and Long Ashton Research Stations

Dakshinamurti V, Reddy G (1975) Mosaic-a transmissible disorder of sweet oranges. Indian Phytopathol 28:398–399

Department for Environment, Food & Rural Affairs and Animal and Plant Health Agency (2023) Fruit propagation certification scheme. GOV.UK. https://www.gov.uk/guidance/fruit-propagation-certification-scheme

Ebbels D (2003) International certification and marketing schemes. In: Principles of plant health and quarantine. CABI Publishing Wallingford UK, pp 171–182

EFSA Panel on Plant Health (PLH), Bragard C, Dehnen-Schmutz K, Gonthier P, Jacques M, Jaques Miret JA, Justesen AF, MacLeod A, Magnusson CS, Milonas P (2019a) Pest categorisation of non-EU viruses and viroids of Vitis L. EFSA J 17(9):e05669

EFSA Panel on Plant Health (PLH), Bragard C, Dehnen-Schmutz K, Gonthier P, Jacques M, Jaques Miret JA, Justesen AF, MacLeod A, Magnusson CS, Milonas P (2019b) Pest categorisation of non-EU viruses of *Fragaria* L. EFSA J 17(9):e05766

EFSA Panel on Plant Health (PLH) (2019a) Pest categorisation of non-EU viruses and viroids of *Cydonia* Mill., *Malus* Mill. and *Pyrus* L. EFSA J 17(9):e05590

EFSA Panel on Plant Health (PLH) (2019b) Pest categorisation of non-EU viruses and viroids of *Prunus* L. EFSA J 17(9):e05735

Engelbrecht D, Kasdorf GGF (1990) Transmission of grapevine leafroll disease and associated closteroviruses by the vine meatybug, *Planococcus ficus*. Phytophylactica 22(3):341–346

EPPO (2024) EPPO A1 and A2 lists of pests recommended for regulation as quarantine pests. EPPO. https://gd.eppo.int/download/standard/2/pm1-002-33-en_A1A2_2024.pdf

Espino T, Exconde S, Zipagan F, Espino R (1990) Banana bract mosaic, a new disease of banana. II. Isolation and purification for monoclonal antibody production. Philipp Agric 73:61–68

European and Mediterranean Plant Protection Organization (2025) EPPO standards – PM 4 production of healthy plants for planting. EPPO. https://www.eppo.int/RESOURCES/eppo_standards/pm4_certification

FAO (2017) The future of food and agriculture: trends and challenges. FAO, p 180. https://doi.org/10.4161/chan.4.6.12871

FAO (2020) Fruit and vegetables–your dietary essentials. The International Year of Fruits and Vegetables, 2021, Background Paper, 36

FAO (2025) Tropical fruits: markets and trade. https://www.fao.org/markets-and-trade/commodities-overview/bananas-tropical-fruits/tropical-fruits/en

Food and Agriculture Organization of the United Nations (1997) International plant protection convention. FAO. https://www.ippc.int/en/core-activities/governance/convention-text/

Fox A, Botermans M, Ziebell H, Fowkes AR, Fontdevila Pareta N, Massart S, Rodoni B, Chooi KM, Kreuze J, Kumar P (2025) Implications of high throughput sequencing of plant viruses in biosecurity – a decade of progress? Peer Commun J 5

Fukumoto T, Nakamura M, Wylie SJ, Chiaki Y, Iwai H (2013) Complete nucleotide sequence of a new isolate of *Passion fruit woodiness virus* from Western Australia. Arch Virol 158(8):1821–1824

Gadge S, Narayan K, Lahu P, Uttam S (2024) A review on traceability system for fruit and vegetables agricultural products using blockchain. Int J Adv Res Sci Commun Technol 4(1):15–20

Gergerich RC, Welliver RA, Gettys S, Osterbauer NK, Kamenidou S, Martin RR, Golino DA, Eastwell K, Fuchs M, Vidalakis G, Tzanetakis IE (2015) Safeguarding fruit crops in the age of agricultural globalization. Plant Dis 99(2):176–187. https://doi.org/10.1094/PDIS-07-14-0762-FE

Gispert C, Oldfield GN, Perring TM, Creamer R (1998a) Biology of the transmission of *Peach mosaic virus* by *Eriophyes insidiosus* (Acari: Eriophyidae). Plant Dis 82(12):1371–1374. https://doi.org/10.1094/PDIS.1998.82.12.1371

Gispert C, Perring TM, Creamer R (1998b) Purification and characterization of *Peach mosaic virus*. Plant Dis 82(8):905–908

Gonsalves D, Tripathi S, Carr JB, Suzuki JY (2010) Papaya ringspot virus. The Plant Health Instructor 10:1094

Grimaldi V, Catara A (1996) Association of a filamentous virus with yellow vein clearing of lemon. 13(13)

Hadidi A, Czosnek HH, Kalantidis K, Palukaitis P (2024) Viroids and satellites and their vector interactions. Viruses 16(10):1598. https://doi.org/10.3390/v16101598

Harris RV, Eking M (1942) Studies in strawberry virus diseases. V. The use of *Fragaria vesca* L. as an indicator of yellow-edge and crinkle. J Pomol Hortic Sci. https://www.tandfonline.com/doi/abs/10.1080/03683621.1942.11513589

Harrison BD, Murant AF (1980) CMI/AAB descriptions of plant viruses. https://api.semanticscholar.org/CorpusID:83998405

Hejran A (2024) Effective plant virus management integrates strategies to protect and sustain crop health. Bulletin of Toraighyrov. University Chemistry & Biology Series

Herrbach E, Le Maguet J, Hommay G, Brown J (2016) Vector-mediated transmission of plant pathogens. Am Phytopathol Soc (APS):147–161

IPPC (1999) International plant protection convention (New Revised Text). https://www.ippc.int/en/publications/131/

IPPC (2019a) ISPM 02 – framework for pest risk analysis. International Plant Protection Convention. https://www.ippc.int/en/publications/592/

IPPC (2019b) ISPM 11 – pest risk analysis for quarantine pests. International Plant Protection Convention. https://www.ippc.int/en/publications/639/

IPPC Secretariat (2004a) Guidelines for a phytosanitary import regulatory system. FAO. https://openknowledge.fao.org/server/api/core/bitstreams/28b5e2d9-763f-44fa-87c2--a2570324be18/content

IPPC Secretariat (2004b) Pest risk analysis for regulated non-quarantine pests. International Standard for Phytosanitary Measures No. 21. FAO on behalf of the Secretariat of the International Plant Protection Convention

IPPC Secretariat (2005) Guidelines for inspection. FAO. https://openknowledge.fao.org/server/api/core/bitstreams/0f1e497a-1bb5-49f8-87b4-5b3a2852423c/content

IPPC Secretariat (2006) Diagnostic protocols for regulated pests. FAO. https://www.ippc.int/static/media/files/publication/en/2016/01/ISPM_27_2006_WithoutApp2_En_2015-12-22_PostCPM10_InkAmReformatted.pdf

IPPC Secretariat (2008) Methodologies for sampling of consignments. FAO. https://www.ippc.int/static/media/files/publication/en/2016/11/ISPM_31_2008_Sampling_of_consignments_EN.pdf

IPPC Secretariat (2010) Design and operation of post-entry quarantine stations for plants. FAO. https://www.ippc.int/static/media/files/publication/en/2016/01/ISPM_34_2010_En_2015-12-22_PostCPM10_InkAmReformatted.pdf
IPPC Secretariat (2011) Phytosanitary certification system. FAO. https://www.ippc.int/sites/default/files/documents/1337674518_ISPM_07_2011_En_2012-05-21.pdf
IPPC Secretariat (2012) Integrated measures for plants for planting. FAO. https://www.ippc.int/static/media/files/publication/en/2016/01/ISPM_36_2012_En_2015-12-22_PostCPM10_InkAmReformatted.pdf
IPPC Secretariat (2017) International movement of seeds. FAO. https://www.ippc.int/static/media/files/publication/en/2017/05/ISPM_38_2017_En_2017-05-15.pdf
IPPC Secretariat (2022) Phytosanitary certificates. FAO. https://www.ippc.int/publications/609/
IPPC Secretariat (2023) Glossary of phytosanitary terms. International Standard for Phytosanitary Measures No. 5. FAO on behalf of the Secretariat of the International Plant Protection Convention. https://openknowledge.fao.org/server/api/core/bitstreams/8995b0cd-d1b6-44a4-866b-1341823c30ac/content
IPPC Secretariat (2024) Glossary of phytosanitary terms. International Standard for Phytosanitary Measures No. 5. FAO on behalf of the Secretariat of the International Plant Protection Convention. https://assets.ippc.int/static/media/files/publication/en/2024/07/ISPM_05_2024_En_Glossary_PostCPM-18_InkAmdts_2024-07-29.pdf
IPPC Secretariat (2025) Overview of the international plant protection convention. International Plant Protection Convention. https://www.ippc.int/en/about/overview/
Iwaki M, Honda Y, Hanada K, Tochihara H, Yonaha T, Hokama K, Yokoyama T (1984) Silver mottle disease of watermelon caused by *Tomato spotted wilt virus*. Plant Dis 68(11):1006–1008
James D, Varga A, Croft H, Rast H, Thompson D, Hayes S (2006) Molecular characterization, phylogenetic relationships, and specific detection of *Peach mosaic virus*. Phytopathology 96(2):137–144. https://doi.org/10.1094/PHYTO-96-0137
James D, Varga A, Sanderson D (2013) Genetic diversity of *Plum pox virus*: strains, disease and related challenges for control. Can J Plant Pathol 35(4):431–441. https://doi.org/10.1080/07060661.2013.828100
Jaybhaye SG, Chavhan RL, Hinge VR, Deshmukh AS, Kadam US (2024) CRISPR-Cas assisted diagnostics of plant viruses and challenges. Virology 597:110160
Jeong J, Ju H, Noh J (2014) A review of detection methods for the plant viruses. Res Plant Dis 20(3):173–181
Jones DR (2005) Plant viruses transmitted by thrips. Eur J Plant Pathol 113(2):119–157
Jones AT, Murant AF, Jennings DL, Wood GA (1982) Association of *Raspberry bushy dwarf virus* with raspberry yellows disease; reaction of *Rubus* species and cultivars, and the inheritance of resistance. Ann Appl Biol 100(1):135–147. https://doi.org/10.1111/j.1744-7348.1982.tb07200.x
Kato K, Handa K, Kameya-Iwaki M (2000) *Melon yellow spot virus*: a distinct species of the genus tospovirus isolated from melon. Phytopathology 90(4):422–426. https://doi.org/10.1094/PHYTO.2000.90.4.422
Koganezawa H (1985) An isometric viruslike particle isolated from russet ring-diseased apple. Ann Phytopathol Soc Jpn 51:363
Kramer LD (2016) Complexity of virus–vector interactions. Curr Opin Virol 21:81–86
Krczal H (1981) Transmission of the strawberry mild yellow edge and *Strawberry crinkle virus* by the strawberry aphid *Chaetosiphon fragaefolii*. Acta Hortic 94:97–102. https://doi.org/10.17660/ActaHortic.1981.94.10
Krenz B, Thompson JR, Fuchs M, Perry KL (2012) Complete genome sequence of a new circular DNA virus from grapevine. J Virol 86(14):7715. https://doi.org/10.1128/JVI.00943-12
Krenz B, Fuchs M, Thompson JR (2023) Grapevine red blotch disease: a comprehensive Q&A guide. PLoS Pathog 19(10):e1011671. https://doi.org/10.1371/journal.ppat.1011671
Larsen HJ, James D (2011) CHAPTER 33: *Peach mosaic virus*. In: Virus and virus-like diseases of pome and stone fruits. The American Phytopathological Society, pp 171–175. https://doi.org/10.1094/9780890545010.033

Lee RF (2020) Chapter 53 – Exclusion of plant viruses by certification and quarantine programs. In: Awasthi LP (ed) Applied plant virology. Academic, pp 763–775. https://doi.org/10.1016/B978-0-12-818654-1.00053-0

Liang X, Zhang R, Gleason ML, Sun G (2022) Sustainable apple disease management in China: challenges and future directions for a transforming industry. Plant Dis 106(3):786–799. https://doi.org/10.1094/PDIS-06-21-1190-FE

Loconsole G, Önelge N, Potere O, Giampetruzzi A, Bozan O, Satar S, De Stradis A, Savino V, Yokomi RK, Saponari M (2012) Identification and characterization of *Citrus yellow vein clearing virus*, A putative new member of the genus Mandarivirus. Phytopathology 102(12):1168–1175. https://doi.org/10.1094/PHYTO-06-12-0140-R

López MM, Llop P, Olmos A, Marco-Noales E, Cambra M, Bertolini E (2009) Are molecular tools solving the challenges posed by detection of plant pathogenic bacteria and viruses? Curr Issues Mol Biol 11:13–45

Lorimer L (2025) Brazil issues first citrus quality certificate for exports to European Union. Datamar News. https://datamarnews.com/noticias/brazil-issues-first-citrus-quality-certificate--for-exports-to-european-union/

Love K, Paull RE, Cho A, Kawabata A (2017) Tropical fruit tree propagation guide. College of Tropical Agriculture and Human Resources. University of Hawai'i at Manoa

Lovisolo O (2001) Citrus leprosis virus: properties, diagnosis, agro-ecology and phytosanitary importance. EPPO Bull 31(1):79–89. https://doi.org/10.1111/j.1365-2338.2001.tb00971.x

Lua HT, Roshetko JM, Catacutan D, Thinh LD (2015) A review of policy constraints and opportunities for the sustainable delivery of quality fruit tree germplasm in Vietnam. Int J Agric Innov Res 4(3):536–543

Lule M, Van Asten P, Coyne D, Hobayo J-C, Nkurunziza S, Ouma E, Kabunga N, Qaim M, Kahangi E, Mwirigi P (2011) Tissue culture banana (*Musa* spp.) for smallholder farmers: lessons learnt from East Africa. Acta Hortic 986:51–59

Maas JL (1998) Compendium of strawberry diseases, 2nd edn. The American Phytopathological Society. https://doi.org/10.1094/9780890546178

MacDonald S, Martin R, Bristow P (1991) Associated with a necrotic shock reaction in blueberry. Phytopathology 81:210–214

Magee C (1927) Banana bunchy top disease of the banana. CSIResearch Bull 30

Manzoor S, Nabi SU, Ayaz A, Rasool B, Sharma SK, Chesti MH, Parveen S, Verma MK, Diab MA, Rabbee MF (2025) Unveiling viral threats to temperate pome fruits: characterization, transmission, and sustainable management strategies. Front Microbiol 16. https://doi.org/10.3389/fmicb.2025.1560720

Martin RR, Bristow PR (1988) A carlavirus associated with blueberry scorch disease. Phytopathology 78(12):1636–1640

Martin RR, Tzanetakis IE (2013) High risk strawberry viruses by region in the United States and Canada: implications for certification, nurseries, and fruit production. Plant Dis 97(10):1358–1362. https://doi.org/10.1094/PDIS-09-12-0842-RE

Martin RR, MacFarlane S, Sabanadzovic S, Quito D, Poudel B, Tzanetakis IE (2013) Viruses and virus diseases of *Rubus*. Plant Dis 97(2):168–182

Martin RR, Constable F, Tzanetakis IE (2016) Quarantine regulations and the impact of modern detection methods. Annu Rev Phytopathol 54(1):189–205

Mavrič I, Marn MV, Koron D, Žežlina I (2003) First report of *Raspberry bushy dwarf virus* on red raspberry and grapevine in Slovenia. Plant Dis 87(9):1148–1148. https://doi.org/10.1094/PDIS.2003.87.9.1148D

McGuire S, Sperling L (2016) Seed systems smallholder farmers use. Food Secur 8(1):179–195

Meng B, Martelli GP, Golino DA, Fuchs M (2017) *Grapevine viruses:* molecular biology, diagnostics and management, vol 10. Springer

Messmer N, Bohnert P, Askani L, Schumacher S, Voegele RT, Fuchs R (2024) *Grapevine Pinot gris virus* spreads in infected vineyards: latent infections have no direct impact on grape production. Virol J 21:178. https://doi.org/10.1186/s12985-024-02453-4

Mink G (1998) Virus certification of deciduous fruit trees in the United States and Canada. See Ref 30a:294–300

Moreno P, Ambros S, Albiach-Martí MR, Guerri J, Peña L (2007) *Citrus tristeza virus*: a pathogen that changed the course of the citrus industry. Mol Plant Pathol 9(2):251–268. https://doi.org/10.1111/j.1364-3703.2007.00455.x

Nabi SU, Baranwal VK, Yadav MK, Rao GP (2020) Association of Apple necrotic mosaic virus (ApNMV) with mosaic disease in commercially grown cultivars of apple (Malus domestica Borkh) in India. 3 Biotech. 10(3):122. https://doi.org/10.1007/s13205-020-2117-6

National Clean Plant Network (2025) State certification & field testing. National Clean Plant Network. https://www.nationalcleanplantnetwork.org/state-certification-ft

Nemeth M (1979) Virus, mycoplasma and rickettsia diseases of fruit trees. M. Nijhoff, Dordrecht/Boston

Panattoni A, Luvisi A, Triolo E (2013) Review. Elimination of viruses in plants: twenty years of progress. Span J Agric Res 11:173–188. https://doi.org/10.5424/sjar/2013111-3201

Preising S, Borges DF, de Queiroz Ambrósio MM, da Silva WL (2021) A fig deal: a global look at fig mosaic disease and its putative associates. Plant Dis 105(4):727–738. https://doi.org/10.1094/PDIS-06-20-1352-FE

Rakesh R, Babu KM, Shekhar C, Sai KS (2025) Clean Plant Programme: a major leap forward in horticulture sector – transforming horticulture with CPP. Farm Chronicle 04(03):22–25

Ramos-González PL, Chabi-Jesus C, Guerra-Peraza O, Tassi AD, Kitajima EW, Harakava R, Salaroli RB, Freitas-Astúa J (2017) *Citrus leprosis virus* N: a new dichorhavirus causing citrus leprosis disease. Phytopathology 107(8):963–976

Ramsdell D, Stace-Smith R (1979) Blueberry leaf mottle, a new disease of highbush blueberry. Acta Hortic 95:37–48

Rao GP, Reddy MG (2020) Chapter 38 – Overview of yield losses due to plant viruses. In: Awasthi LP (ed) Applied plant virology. Academic, pp 531–562. https://doi.org/10.1016/B978-0-12-818654-1.00038-4

Rodríguez-Verástegui LL, Ramírez-Zavaleta CY, Capilla-Hernández MF, Gregorio-Jorge J (2022) Viruses infecting trees and herbs that produce edible fleshy fruits with a prominent value in the global market: an evolutionary perspective. Plants 11(2):203. https://doi.org/10.3390/plants11020203

Roy AS (2011) CHAPTER 68: Control measures of pome and stone fruit viruses, viroids, and phytoplasmas: role of international organizations. https://api.semanticscholar.org/CorpusID:91256734

Roy A, Choudhary N, Guillermo LM, Shao J, Govindarajulu A, Achor D, Wei G, Picton DD, Levy L, Nakhla MK, Hartung JS, Brlansky RH (2013) A novel virus of the genus cilevirus causing symptoms similar to Citrus leprosis. Phytopathology 103(5):488–500. https://doi.org/10.1094/PHYTO-07-12-0177-R

Rubio L, Galipienso L, Ferriol I (2020) Detection of plant viruses and disease management: relevance of genetic diversity and evolution. Front Plant Sci 11:1092

Rwahnih MA, Dave A, Anderson MM, Rowhani A, Uyemoto JK, Sudarshana MR (2013) Association of a DNA virus with grapevines affected by red blotch disease in California. Phytopathology 103(10):1069–1076

Saldarelli P, Gualandri V, Malossini U, Glasa M (2017) Grapevine Pinot gris virus. In: Grapevine viruses: molecular biology, diagnostics and management. Springer, pp 351–363

Scott S, Zimmerman M, Jones A, Le Gall O (2000) Differences between the coat protein amino acid sequences of English and Scottish serotypes of Raspberry ringspot virus exposed on the surface of virus particles. Virus Res 68(2):119–126

Serio FD, Malfitano M, Alioto D, Ragozzino A, Desvignes J, Flores R (2001) Apple dimple fruit viroid: fulfillment of Koch's postulates and symptom characteristics. Plant Dis 85(2):179–182

Setiono FJ, Chatterjee D, Fuchs M, Perry KL, Thompson JR (2018) The distribution and detection of grapevine red blotch virus in its host depend on time of sampling and tissue type. Plant Dis 102(11):2187–2193

Shafaeddin M (2007) Who does bear the costs of compliance with sanitary and phytosanitary measures in poor countries?

Sharma P (2023) Chapter 10 – Epidemiology of potyviruses infecting crops of cucurbitaceae. In: Gaur RK, Patil BL, Selvarajan R (eds) Plant RNA viruses. Academic, pp 213–227. https://doi.org/10.1016/B978-0-323-95339-9.00016-8

Singh A, Yasheshwar, Kaushik NK, Kala D, Nagraik R, Gupta S, Kaushal A, Walia Y, Dhir S, Noorani MS (2025) Conventional and cutting-edge advances in plant virus detection: emerging trends and techniques. 3 Biotech 15(4):100

Stace-Smith R, Hansen A (1976) Some properties of cherry rasp leaf virus. Acta Hortic 67:193–198

Swingle R, Tilford P, Irish C (1943) A graft transmissible mosaic of American Elm. Phytopathology 33(12):1196–1200. CABI Databases

Tatineni S, Hein GL (2023) Plant viruses of agricultural importance: current and future perspectives of virus disease management strategies. Phytopathology 113(2):117–141. https://doi.org/10.1094/PHYTO-05-22-0167-RVW

Tennant PF, Fermin GA, Roye ME (2007) Viruses infecting papaya (*Carica papaya* L.): etiology, pathogenesis, and molecular biology. Plant Viruses 1(2):178–188

Thompson JR, Jelkmann W (2003) The detection and variation of *Strawberry mottle virus*. Plant Dis 87(4):385–390. https://doi.org/10.1094/PDIS.2003.87.4.385

Thompson JR, Leone G, Lindner JL, Jelkmann W, Schoen CD (2002) Characterization and complete nucleotide sequence of *Strawberry mottle virus*: a tentative member of a new family of bipartite plant picorna-like viruses. J Gen Virol 83(1):229–239). Microbiology Society. https://doi.org/10.1099/0022-1317-83-1-229

Umer M, Liu J, You H, Xu C, Dong K, Luo N, Kong L, Li X, Hong N, Wang G, Fan X, Kotta-Loizou I, Xu W (2019) Genomic, morphological and biological traits of the viruses infecting major fruit trees. Viruses 11(6). https://doi.org/10.3390/v11060515

van der Meer K (2007) Building capacity for compliance with evolving food safety and agricultural health standards. In: Global supply chains, standards and the poor: how the globalization of food systems and standards affects rural development and poverty. CABI Wallingford UK, pp 281–294

Varveri C, Maliogka VI, Kapari-Isaia T (2015) Chapter One – Principles for supplying virus-tested material. In: Loebenstein G, Katis NI (eds) Advances in virus research, vol 91. Academic, pp 1–32. https://doi.org/10.1016/bs.aivir.2014.10.004

Ventura JA, de Melo Lima I, Martins MVV, Culik MP, Costa H (2019) Impact and management of diseases in the propagation of fruit plants. Rev Bras Frutic 41(4) CABI Databases. https://doi.org/10.1590/0100-29452019647

Vernière C, Perrier X, Dubois C, Dubois A, Botella L, Chabrier C, Bové JM, Vila ND (2004) Citrus viroids: symptom expression and effect on vegetative growth and yield of clementine trees grafted on trifoliate orange. Plant Dis 88(11):1189–1197. https://doi.org/10.1094/PDIS.2004.88.11.1189

Wasielewski J, Balerdi CF (2019) Tropical and subtropical fruit propagation: HS1349, 11/2019. EDIS 2019(6):7–7

Waterworth H, Hadidi A (1998) Plant virus disease control. APS Press, St Paul

Whitsell R (1952) Sunblotch disease of avocados. Calif Avocado Soc Yearb 37:215–240

Wu Y, Wang Y, Sun J, Liu Y, Sujata S, Wu Y, Zhao L (2022) Effects of Actinidia yellowing ringspot virus on the yield and quality of kiwifruit. Plant Dis 106(3):800–804. https://doi.org/10.1094/PDIS-06-21-1304-SC

Proteomics and Metabolomics in Virus-Infected Fruit Crops

13

Munmi Borah, Swapnil Borah, Devayani Sarmah, Puja Dey, Preetam Baruah, Sanjukta Chakraborty, Greeshma Varghese, Pronoy Kalita, and Priyam Kafle

Abstract

Fruit crops, particularly perennial species, contribute significantly to global nutrition and rural economies but are highly susceptible to viral diseases that reduce growth, productivity, and post-harvest quality. Viral infections alter host physiological, biochemical, and molecular processes, resulting in huge economic losses, especially in vegetatively propagated fruit crops, where viruses spread rapidly through planting material and vectors. Recent developments in omics technologies, particularly proteomics and metabolomics, have transformed our understanding of plant–virus interactions by allowing detailed profiling of virus-induced molecular reprogramming. Proteomics reveals dynamic variation in host protein expression, which includes alterations in photosynthesis-related proteins, energy-metabolism enzymes, stress-responsive proteins, and components of defense pathways. Metabolomics complements this by detecting changes in primary metabolites such as sugars, amino acids, and organic acids, as well as secondary metabolites including phenolics, flavonoids, and alkaloids that control defense and symptom development. Integrated proteome-metabolome analyses highlight coordinated regulatory networks that control viral pathogenesis and host resistance mechanisms. These findings contribute to the discovery of early diagnostic biomarkers, the identification of resistance-associated molecules, and the development of tolerant cultivars through molecular breeding and biotech-

M. Borah (✉) · S. Borah · P. Dey · P. Baruah · S. Chakraborty · G. Varghese · P. Kalita · P. Kafle
Department of Plant Pathology, Assam Agricultural University, Jorhat, Assam, India
e-mail: munmi.borah@aau.ac.in

D. Sarmah
Department of Plant Pathology, Assam Agricultural University, Jorhat, Assam, India

Division of Plant Pathology, ICAR-Central Rice Research Institute, Cuttack, Odisha, India

R. Kumar et al. (eds.), *Fruit Crops and Viral Pathogens Volume II*, Plant Pathology for Sustainable Agriculture, https://doi.org/10.1007/978-981-95-7261-8_13

nology. Moreover, omics approaches can help improve post-harvest quality by elucidating metabolic and protein changes associated with ripening, senescence, and nutrient degradation. Despite their promise, challenges remain in standardization, data integration, and interpretation of complex datasets. Future efforts that combine artificial intelligence, multi-omics, and precision agriculture technologies will further enhance virus detection, disease forecasting, and sustainable fruit crop management. Overall, proteomics and metabolomics provide effective frameworks for understanding virus-induced changes and developing resilient fruit production systems.

Keywords

Proteomics · Metabolomics · Virus · Sustainable · Defence · Multi-omics

13.1 Introduction

Most perennial fruit crops (woody or herbaceous) provide essential nutrients that play a major role in improving the rural economies. The recognized health benefits of these diverse, nutritious fruits have led to their increased global demand. Also, this demand extends to processed food items such as fruit jams, snacks, and juices, enhancing the economic significance of the fruit industry (Zeeshan et al. 2024). Diverse climatic conditions, limited agricultural resources, and changing dietary trends favoring organic and plant-based foods make this demand more challenging. Sustainable production is built on proper land management, effective farming practices, and reliable supply chains. However, climate change further exacerbates these challenges through unpredictable weather, water scarcity, and increased pests and diseases, which ultimately threaten fruit cultivation and productivity (Kabir et al. 2023). Tackling these issues is vital for ensuring a consistent fruit supply and supporting food security on a global level. Integrated strategies for enhancing fruit yield through sustainable agriculture are required for addressing these challenges. Rising global demand drives innovation in farming technologies and methods, for which collaboration among farmers, researchers, policymakers, and technologists is needed to develop effective, resilient approaches for meeting the increasing fruit demand (Idoko et al. 2016). Although a vast range of viruses infect both annual and perennial crops, those that reduce substantial yield are the greatest threat to agriculture. Accurately estimating this yield reduction from plant viruses is difficult due to complex influencing factors and insufficient reliable data. Viral outbreaks in perennial fruit trees are particularly damaging because they destroy annual harvests, leading to loss of long-term investments in orchard establishment and maintenance (Moreno and López-Moya 2020; Naidu et al. 2014; Rimbaud et al. 2015). The real economic impact of these viral diseases is often underestimated, due to underreporting of localized infections and mild symptoms (Hull 2013). In fruit crops, viruses greatly affect both yield and quality, leading to huge financial losses. Unlike fungal or bacterial diseases, viral infections cannot be cured with chemical treatments, which poses a challenge once plants are infected. Viruses alter plant physiological and metabolic functions, resulting in stunting, leaf curling, yellowing, vein clearing, and mosaic symptoms. This reduces photosynthesis, impairs flowering, and decreases fruit set. Infected fruits are typically small, misshapen, and poorly ripened, with poor nutritional quality and market appeal. Serious viral infections such as citrus tristeza, papaya ringspot, and banana bunchy top can cause total crop failure. Since most fruit crops are perennial and propagated vegetatively, viruses spread easily through infected propagation material, grafting, vectors, or mechanical means. As a result, yield and quality losses also restrict post-harvest handling, processing, and international trade. Managing this growing risk requires integrated approaches combining virus-free planting material, resistant cultivars, vector management, and advanced molecular tools such as RNAi and CRISPR.

Plant viruses pose a major constraint to global agriculture, causing estimated annual yield losses of nearly USD 30 billion (Rybicki 2015; Tatineni and Hein 2023). Effective management depends on understanding the molecular mechanisms involved in viral invasion, replication, symptom expression, and identifying host factors and disease regulators that can be targeted for control. These plant–microbe interactions remain complex to understand, making this field challenging and highly dynamic. Recent technological progress has revolutionized the study of these interactions through *omics*-based approaches, which offer comprehensive insights into biological systems at multiple molecular levels. The term *omics* collectively refers to large-scale analyses of biological molecules, including genomics, transcriptomics, proteomics, metabolomics, and related branches. These methods have greatly advanced our understanding of plant–virus interactions, enabling the generation of genome-scale data and supporting translational research by integrating multi-omics information. Using these tools, researchers have acquired a deeper understanding of the molecular basis of viral pathogenesis, host defense strategies, and the ecological dynamics of plant–pathogen interactions. Additionally, emerging fields like metagenomics, volatilomics, and spectranomics have broadened the scope of *omics* research, revealing fresh perspectives into pathogen behavior, host adaptation, and various stress responses. Such studies not only help explain how viruses manipulate host cellular interactions but also identify key genes and pathways that can be exploited for resistance breeding. This knowledge is useful for developing more resilient and sustainable varieties of crops. Additionally, understanding the evolutionary and ecological aspects of plant–virus interactions enhances predictions of disease emergence and spread. Overall, *omics*-driven research provides powerful tools for mitigating the impact of plant viral diseases while supporting sustainable agricultural productivity and food security. The chapter presents a comprehensive review of the roles played by proteomics and metabolomics in elucidating virus-induced changes in fruit crops. Through the convergence of molecular, biochemical, and physiological information, it captures the importance of such omics technologies in unraveling plant–virus interactions, the detection of biomarkers, and the design of sustainable measures for enhancing fruit crop health and productivity.

13.2 Overview of Proteomics and Metabolomics Approaches

Proteins play a vital role in maintaining cellular homeostasis in biological systems (Liu et al. 2019). In order to define the complete component of protein that is or can be expressed by a genome or a cell, tissue, or an organism, the term "proteome" was given by Marc Wilkins in 1995 (Agrawal et al. 2013). The proteome is dynamic in nature and hence it changes over time in response to different internal and external factors. It is modulated by processes such as post-translational modifications (PTMs), alternative splicing, proteolysis, etc. (Al-Amrani et al. 2021). Based on this concept, the science of proteomics developed, which is defined as "*the study of protein properties (expression level, post-translational modification, interactions, etc.) on a large-scale that results in a global, integrated view of disease processes, cellular programs and networks at the protein level*" (Poon and Johnson 2001). Besides cataloguing which proteins are present (identification) in a sample, proteomics also involves determining their quantities (quantification), modifications, interactions, localization, and their changes in response to perturbations or pathological states. Proteomics-based technologies are of utmost importance as they enable understanding pathogenicity mechanisms, altered expression patterns in response to various signals, and also interpreting functional protein pathways involved in viral diseases (Aslam et al. 2017). Modern proteomics has been dominated by two experimental philosophies: bottom-up and top-down proteomics (Aebersold and Mann 2016). In bottom-up proteomics, proteins are enzymatically digested into peptides and then analyzed by liquid chromatography (LC) coupled with tandem mass spectrometry (MS/MS), while in the top-down approach, intact proteins are introduced directly into the mass spectrometer and fragmented to reveal proteoforms. Recent years have seen hybrid strategies. Proteomic analysis begins with sample selection and preparation. Since proteomes are highly complex, fractionation can occur at the protein level or peptide level (e.g., 1-D or 2-D gel electrophoresis, multidimensional LC, etc.). This is followed by protein digestion (proteolysis) and mass spectrometry. In mass spectrometry, two ionization techniques dominate: electrospray ionization (ESI) and matrix-assisted laser desorption/ionization (MALDI). Several mass analyzers, including quadrupoles, time-of-flight (TOF), ion traps, and Orbitrap, which differ in resolution, mass accuracy, dynamic range, and scan speed, are used. With advancements, hybrid systems, e.g., quadrupole-orbitrap or quadrupole-TOF combinations, have strengthened the study of the proteome in depth. The next steps involve data acquisition and quantification, which is the fundamental step in many proteomic studies. Label-based (metabolic labeling and chemical labeling) and label-free quantification (LFQ) methods are used. Chemical labeling using isobaric tags such as tandem mass tags (TMT) and iTRAQ enables deep profiling and high multiplexing across many samples with high throughput (Liu et al. 2019; Guo et al. 2022). To study protein-protein interactions, affinity purification coupled to mass spectrometry (AP-MS) remains a widespread approach. On the other hand, the set of low molecular weight (typically <1–2 kDa) small molecules (metabolites) present in cells, tissues, or organisms in a given state is defined as the metabolome (Oliver et al. 1998). Metabolomics is defined as "*identification and quantification of all metabolites present in an organism or biological system*" (Tan et al. 2009). When a fruit crop is infected by a virus, host proteins (for defense, virus replication, suppression of host immunity) and metabolites (e.g., signaling molecules, antioxidants, secondary metabolites) change. Proteomics and metabolomics can reveal these changes, helping to map disease and cell death pathways. Metabolites represent the end products of cellular processes and therefore provide a snapshot of physiological and pathophysiological states more directly than genomic or proteomic data alone (Hajnajafi and Iqbal 2025). Thus, the study of the metabolome is more advantageous as it has a more direct impact on the phenotype, over proteome. Untargeted (global) metabolomics aims to survey as many metabolites or features as possible, without preselecting analytes. Targeted metabolomics focuses on a defined set of metabolites, often within specific pathways, with known standards. Nuclear magnetic resonance (NMR) spectroscopy and mass spectrometry (MS), often coupled to separation methods (liquid chromatography, gas chromatography, capillary electrophoresis) are the widely used analytical techniques. In metabolomics, the extraction of metabolites is the most critical step.

13.3 Proteomic Alterations in Virus-Infected Fruit Crops

13.3.1 Differential Expression of Host Proteins During Viral Infection

Viral infections in fruit crops trigger extensive reprogramming of host proteomes, leading to modulation of energy metabolism, photosynthesis, stress response, and defense. Proteomic analyses (using 2-DE gel electrophoresis and mass spectrometry) have revealed that virus infection can either upregulate or suppress host proteins depending on the stage and severity of the infection (Kaur et al. 2021). When papaya ringspot virus (PRSV) infects papaya, it changes the proteins related to carbohydrate metabolism, photosynthesis, and stress responses (Bhattacharyya et al. 2015). Similarly, in citrus tristeza virus (CTV)-infected citrus plants, expression of the proteins involved in photosystem II, ATP synthesis, and protein folding was altered, and this is responsible for the energy imbalance and decreased photosynthetic capacity in the infected plant tissues (Lin et al. 2015). These different patterns show the different ways of adaptation of the host to viral interference on a metabolic and biochemical level, which usually leads to symptom expression and yield reduction.

13.3.2 Virus-Induced Changes in Photosynthesis-Related Proteins

During a viral attack, photosynthesis-related proteins are negatively affected because, for replication and systemic spread of the virus, it requires energy, which is supplied by reprogramming the photosynthetic machinery. For example, in a grapevine leafroll-associated virus 3 (GLRaV-3)-infected grape orchard, it was observed that it was responsible for a downregulation of Rubisco large subunit-binding proteins, oxygen-evolving enhancer proteins, and chlorophyll a/b-binding proteins in the host tissues, leading to reduced CO_2 fixation and altered carbon flux (Espinoza et al. 2007; Margaria et al. 2013). Similarly, in banana plants infected with banana bunchy top virus (BBTV), proteomic profiling showed protein reduction involved in chloroplast electron transport and light-harvesting complexes (Shekhawat et al. 2011), leading to reduced photosynthetic efficiency by diverting resources from energy for replication and the spread of the virus. It will lead to symptoms such as chlorosis and stunted growth (Zhao et al. 2019).

13.3.3 Stress-Responsive and Defense-Related Proteins

Proteomic analysis has revealed that there is an increase in stress-related proteins, including heat shock proteins (HSPs), pathogenesis-related (PR) proteins, and reactive oxygen species (ROS)-scavenging enzymes in fruit crops infected by a virus. For example, in CTV-infected citrus, Lin et al. (2015) observed the upregulation of antioxidant enzymes such as superoxide dismutase and ascorbate peroxidase, suggesting the activation of oxidative stress management pathways. Similarly, in PRSV-infected papaya, activation of protein repair and defense systems was reflected by the induction of HSP70 and PR-10 proteins (Bhattacharyya et al. 2015). The increased protein level associated with the ubiquitin-proteasome system and signal transduction indicates that the infected host plant is following several molecular strategies to escape from viral attack and thereby reducing the damage (Kaur et al. 2021).

13.4 Metabolomic Reprogramming During Viral Infection

Viruses regulate the metabolic machinery of their host plants to meet their energy and biosynthetic needs, create favorable redox conditions, and suppress or bypass host immune defenses. Such a manipulation alters the fundamental metabolic routes, including glycolysis, the tricarboxylic acid (TCA) cycle, and the pentose phosphate pathway, as well as lipid and nucleotide biosynthesis, amino acid dynamics, and overall cellular redox status. These metabolic changes can be identified through metabolomic profiling and hold potential to aid in disease diagnosis and management (Palmer 2022).

13.4.1 Changes in Primary Metabolism (Sugars, Amino Acids, Organic Acid)

Viral infection leads to extensive changes in the plant's primary metabolic pathways—encompassing carbohydrates, amino acids, and organic acids resulting in conditions that support viral replication and systemic spread. These changes in metabolic pathway lead to disruption in the plant's normal energy dynamics, redox equilibrium, and biosynthetic processes (Jiang et al. 2025; Pant et al. 2021).

13.4.1.1 Sugars

Sugars serve not only as energy reserves but also as important signaling molecules that coordinate plant defense and developmental functions. Under viral attack, sugar metabolism is significantly altered, typically resulting in the accumulation of soluble sugars—such as glucose, sucrose, and fructose—around infected tissues. This increase is primarily due to an increase in invertase enzyme activity and a decrease in photosynthetic carbon assimilation (Chesnais et al. 2025). For example, infection by Tobacco mosaic virus (TMV) has been shown to induce chloroplast disintegration and alter carbohydrate distribution within host tissues, creating conditions that are favorable for virus multiplication and its intercellular transmission (Fernandez-Calvino et al. 2011).

13.4.1.2 Amino Acids

One of the primary areas of viral-induced metabolic alteration is amino acid metabolism. Numerous viruses are responsible for the accumulation of free amino acids that act as essential substrates for viral protein formation and other nitrogen-rich metabolites. Compounds like glutamine, asparagine, serine, and proline often show increased levels in infected plants (Rojas et al. 2014). In *Cucumis sativus* infected with cucumber mosaic virus (CMV), levels of glutamic acid, alanine, and proline rise markedly, potentially supporting osmotic balance and stress adaptation (Chesnais et al. 2025). On the other hand, amino acids like tryptophan and phenylalanine—which serve as precursors for defense-related secondary metabolites like phenolics and flavonoids—are frequently reduced, suggesting redistribution of metabolic resources away from defense mechanisms towards promoting viral replication (Jiang et al. 2025).

13.4.1.3 Organic Acids

Organic acids, which serve as intermediates in the tricarboxylic acid (TCA) cycle, play an important role in energy production and biosynthesis. Viral infections frequently disrupt TCA cycle activity, leading to alterations in levels of metabolites such as citric acid, malic acid, and fumaric acid (Pant et al. 2021). For example, infection by cucumber mosaic virus (CMV) has been shown to reduce citric and malic acid concentrations while increasing succinic acid levels, thereby suggesting a shift in mitochondrial metabolism toward anabolic activity (Chesnais et al. 2025). In tobacco plants infected with tobacco mosaic virus (TMV), reductions in organic acid content have been associated with poor respiration and higher levels of oxidative stress (Rojas et al. 2014).

13.4.2 Alterations in Secondary Metabolites (Phenolics, Flavonoids, Alkaloids)

Secondary metabolites such as phenolics, flavonoids, and alkaloids play an important role in plant defense by acting as antimicrobial agents, signaling molecules, and antioxidants. Viral infections can have a significant impact on the production and accumulation of these compounds, highlighting a complicated interaction between host defense mechanisms and viral strategies to alter plant metabolism (Jiang et al. 2025).

13.4.2.1 Phenolics

Phenolic compounds are a diverse group of plant secondary metabolites, including phenolic acids, lignin precursors, and hydroxycinnamic acids. Viral infection frequently triggers significant alterations in phenolic metabolism with the help of the phenylpropanoid pathway. Several plant viruses, such as Cucumber mosaic virus (CMV) and Tobacco mosaic virus (TMV), stimulate the accumulation of phenolic compounds during the early infection stage, forming a protective barrier that restricts viral spread (Rojas et al. 2014). For instance, CMV infection in cucumber enhances the levels of caffeic acid, ferulic acid, and similar phenolic acids, which help in building cell walls and inhibit virus movement. However, prolonged viral infection can suppress phenolic biosynthesis, likely as a result of the virus redirecting host metabolic resources toward its own replication rather than defense (Jiang et al. 2025).

13.4.2.2 Flavonoids

Flavonoids, which are a group of polyphenolic compounds, contribute to UV protection, signaling, and antioxidant defense, and confer resistance to pathogens. Viral infections frequently disrupt flavonoid biosynthesis, resulting in either increased or decreased concentration depending on the specific host–virus interaction. Flavonoid levels may decrease during certain prolonged infections as viruses redirect host metabolic processes away from the production of secondary metabolites (Jiang et al. 2025). Changes in flavonoid composition during infection can have an impact on interactions with insect vectors, since specific flavonoids have the capacity to modify insect feeding behavior and affect virus transmission (Schwachtje and Baldwin 2008).

13.4.2.3 Alkaloids

Alkaloids, nitrogen-rich secondary metabolites, frequently help in plant defense against herbivores and pathogens. Viral infections can affect alkaloid biosynthesis in various ways, with effects that depend on the host species and the infection stage (Adeosun and Loots 2024).

13.4.3 Metabolite Markers Linked with Symptom Development and Resistance

Metabolite markers are critical for determining symptom development and resistance mechanisms during viral infections in plants. Metabolomic studies have revealed the increase and decrease of particular metabolites whose levels increase or decrease in association with disease progression and the plant's defense responses. These markers not only serve as indicators of plant health, but they also provide valuable insights into the biochemical pathways that govern host–virus interactions (Rojas et al. 2014; Jiang et al. 2025).

13.4.3.1 Salicylic Acid (SA) and N-Hydroxypipecolic Acid (NHP)

Salicylic acid, a key plant hormone, plays a pivotal role in the systemic acquired resistance (SAR) pathway. Increased levels of salicylic acid are due to the activation of plant defense mechanisms against viral attacks. N-hydroxypipecolic acid (NHP), derived from pipecolic acid, has also been recognized as an important regulator of SAR. Higher concentrations of both SA and NHP have been found in plants showing resistance to viral infections, revealing their crucial role in reducing symptoms and strengthening defense responses (Dempsey and Klessig 2017; Hartmann and Zeier 2019).

13.4.3.2 Phenolic Compounds

Phenolic compounds like flavonoids and phenolic acids are important secondary metabolites that play a key role in plant defense. When pathogens attack, these metabolites accumulate quickly in the resistant wheat varieties like Koonap, whereas they show a slower response in the case of susceptible varieties. This variation in accumulation reveals the critical role of phenolic compounds in increasing the resistance and affecting the severity of symptoms (Mashabela et al. 2023).

13.4.3.3 Amino Acids

Amino acids serve as a crucial constituent for protein production and are vital for overall plant metabolism. Viral infections often alter amino acid biosynthesis, which can influence the plant's capacity to respond to stress. In resistant wheat varieties such as Koonap, amino acid metabolism is redirected to favor the production of defense-related compounds, illustrating how metabolic changes support resistance mechanisms (Mashabela et al. 2023).

13.4.3.4 Primary Metabolites

Sugars and organic acids, as primary metabolites, are pivotal for the core cellular activities and energy generation. Viral infections can hamper these metabolic processes, resulting in changes in the levels of these metabolites. Metabolomic analyses have recognized particular variations in these compounds during the process of viral infection, offering valuable insights into the metabolic changes associated with symptom progression and plant resistance (Jiang et al. 2022)

13.5 Proteome–Metabolome Crosstalk in Plant–Virus Interactions

When viruses infect plants, they disrupt both proteins (proteome) and small molecules (metabolome) that the plant utilizes for its growth and protection. Together, these two layers of change explain how plants maintain a balance between eliminating viruses and maintaining basic life functions. Viral infection often activates the defense pathways involving salicylic acid (SA), jasmonic acid (JA), and ethylene (ET), which help plants resist attack. Proteins such as pathogenesis-related proteins, chaperones, and RNA-binding factors are produced to increase immunity. At the same time, metabolites like SA, pipecolic acid, flavonoids, and amino acids are reprogrammed to reinforce cell walls, manage oxidative stress, and regulate signaling (Pang et al. 2021; Tohge and Fernie 2010). These combined changes demonstrate that antiviral defense is not the only pathway, but the result of several proteins and metabolites functioning together (Dangl et al. 2013). During infection, viruses compete with the host for important resources, forcing plants to constantly adjust how they use their energy. Viruses hijack ribosomes, chaperones, lipids, and energy molecules to assemble new viral particles. This results in disruption of the plant's normal energy flow and thereby forces it to shift resources away from growth and toward defense. For example, trehalose-6-phosphate and sugar signals influence energy regulators like SnRK1 and TOR, which act as switches to either block or support viral replication (Tohge and Fernie 2010; Xu et al. 2018). This results in significant changes in both proteome and metabolome in noticeable ways: a decrease in photosynthesis proteins, while antioxidant enzymes and flavonoid-producing proteins increase; lipid metabolism is also shifted to boost defense (Zandalinas et al. 2019).

Modern research techniques enable scientists to map these molecular interactions, leading to a detailed network diagram of "who talks to whom" in the cell. Important gateways, such as SA, sugars, glutathione, and oxylipins, connect directly with protein groups involved in proteasome activity, RNA silencing, and stress signaling (Szymanski et al. 2009; Pang et al. 2021). Vital proteins like thioredoxins, translation factors, and lipid enzymes often act as control points that viruses try to manipulate. Early signals, like quick bursts of SA or redox changes, can "set the stage" for larger proteome adjustments at a later stage (Dangl et al. 2013). Understanding these networks reveals that targeting entire hubs, instead of single genes, may be more beneficial in breeding virus-resistant plants. Such strategies may help plants resist infection while providing enough energy for growth (Xu et al. 2018).

13.6 Applications of Proteomics and Metabolomics in Crop Improvement

13.6.1 Biomarker Discovery for Early Virus Detection

Early detection of viral infections serves as the basis of effective disease management and control (Johnson et al. 2019). Traditional diagnostic methods, such as polymerase chain reaction (PCR) and serological assays, often require a notable viral load or a measurable immune response, which can delay diagnosis (Patel et al. 2018). Discovery of biomarkers serves as an opportunity in the detection of viral infections at an earlier stage, even before the clinical symptoms manifest (Kumar et al. 2020).

Concept of Biomarkers in Virology
Biomarkers are quantifiable biological indicators that display several physiological or pathological processes within the host (Anderson et al. 2017). In terms of Virology, biomarkers can either be viral components, such as nucleic acids or proteins, or host-derived responses, such as immune mediators or metabolic signatures (Chen et al. 2018). The recognition of reliable biomarkers allows earlier detection and also improves disease monitoring and prognosis (Lopez et al. 2019).

Types of Biomarkers for Virus Detection

1. *Genomic biomarkers*: The nucleic acids of viruses are among the most direct biomarkers because they cause active or latent infection (Wang et al. 2018). The advancement in next-generation sequencing has driven the identification of viral genomic signatures for diagnostic use (Davis et al. 2020).
2. *Proteomic biomarkers*: Early diagnostic markers can be possible through viral proteins and host protein responses (Singh et al. 2017). The identification of protein signatures specific to viral infection can be done by mass spectrometry and antibody-based detection platforms (Garcia et al. 2019).
3. *Transcriptomic biomarkers*: The insights into viral pathogenesis and potential biomarkers can be provided during early infection, while changes in host gene expression profiles (Lee et al. 2020). The study of RNA sequencing led to specific transcriptional signatures associated with early viral replication (Brown et al. 2018).
4. *Metabolomic biomarkers*: Rapid occurrence of metabolic alterations followed by viral invasion offers a rich source of potential biomarkers (Martinez et al. 2019). are frequently used to characterize metabolic fingerprints of infection, which are done by nuclear magnetic resonance (NMR) and mass spectrometry (Nguyen et al. 2020).

Strategies for Biomarker Discovery

The discovery of biomarkers depends on high-throughput technologies in combination with computational biological approaches (Taylor et al. 2017). The integrated approaches allow multiple aspects of genomic, proteomic, transcriptomic, and metabolomic data analysis for the identification of potential early diagnostic markers (Zhao et al. 2018). Further enhancing the prediction and validation of biomarker candidates can be further enhanced by machine learning methods through recognition of subtle patterns in complex datasets (Olsen et al. 2020).

Validation and Clinical Translation

To ensure reproducibility and reliability, candidate biomarkers must undergo thorough validation across independent cohorts and multiple platforms (Miller et al. 2019). Regulatory approval requires demonstrating the sensitivity, specificity, and clinical utility of the biomarker for early detection of the virus (Thompson et al. 2018). The major challenge in this field is to bridge the gap between discovery and clinical application (Rao et al. 2020).

Applications in Emerging Viral Diseases

The early detection of viral outbreaks such as Ebola, Zika, and COVID-19 can be done through rapid identification of biomarkers (Stevens et al. 2020). To identify early cases where direct viral detection was limited, host-based transcriptomic and proteomic signatures were significant (Hernandez et al. 2019). The importance of biomarker-driven approaches for preparedness against future viral threats was revealed from recent pandemics (Cheng et al. 2020).

13.6.2 Identification of Resistance-Related Proteins and Metabolites

The resistance to plant pathogens and pests is a complex trait that involves a combination of regulatory proteins and metabolites, which together contribute to defense mechanisms (Jones and Dangl 2006). To understand plant-pathogen interactions and develop improved strategies in crop protection, the identification of resistance-related proteins and metabolites is necessary (Hammond-Kosack and Jones 2007).

13.6.2.1 Approaches for Identifying Resistance-Related Proteins

Proteomics-Based Techniques

Proteomics has emerged as a central tool for the identification of proteins associated with resistance responses (van Loon and Rep 2006). Two-dimensional gel electrophoresis (2-DE) and mass spectrometry are frequently used for the detection of changes in protein abundance during pathogen infection (Sharma et al. 2020). Label-free and isotopic labeling methods have further increased the sensitivity of protein quantification (Zhang et al. 2021). Comparative proteomics between resistant and susceptible cultivars has been widely used to highlight proteins that play key roles in defense signaling (Durrant and Dong 2010). Research revealed the accumulation of pathogenesis-related (PR) proteins, oxidative stress-related enzymes, and signal transduction regulators in resistant genotypes (Ali et al. 2014).

Functional Characterization

Genetic methods such as overexpression, RNA interference, or CRISPR/Cas-based editing were used to identify the proteins to confirm their role in resistance (Jaganathan and Ramasamy 2018). Functional characterization helps in establishing causal relationships between protein abundance and defense mechanisms (Wang et al. 2019).

13.6.2.2 Metabolite Profiling in Resistance

Metabolomics Approaches

Metabolomics provides insight into the small-molecule landscape of plant defense. Gas chromatography–mass spectrometry (GC-MS) and liquid chromatography–mass spectrometry (LC-MS) are frequently applied to profile metabolites in resistant versus susceptible genotypes (Fiehn et al. 2000a, b). Nuclear magnetic resonance (NMR) spectroscopy is also employed for comprehensive metabolite identification (Krishnan and Saito 2005).

Defense-Associated Metabolites

There are certain well-documented classes of metabolites linked with resistance, including phenylpropanoids, alkaloids, terpenoids, and phytoalexins (Ahuja et al. 2012). Secondary metabolites often act as antimicrobial compounds, signaling molecules, or precursors of structural barriers against pathogens (Piasecka and Matusiewicz 2015). The central roles in defense signaling in resistant plants are frequently observed by accumulation of salicylic acid, jasmonic acid, and related derivatives (Glazebrook and Ausubel 2005).

Integration of Proteomics and Metabolomics

The combination of proteomic and metabolomic datasets provides a systems-level understanding of resistance mechanisms (Weckwerth and Wenzel 2003). Such a combination helps to identify the metabolic pathways activated by resistance-related proteins, which link signaling cascades with metabolite accumulation (Kumar and Sharma 2017). Multi-omics approaches, including transcriptomics, proteomics, and metabolomics, are increasingly used to dissect the molecular basis of resistance (Nakabayashi and Saito 2014).

13.6.3 Development of Tolerant and Resistant Cultivars

The development of crop cultivars while enhancing the tolerance or resistance to biotic and abiotic stresses is a major goal in modern agriculture. Omics technologies, particularly proteomics and metabolomics, have emerged as powerful tools to dissect molecular mechanisms underlying stress responses and guide breeding programs.

Proteomics in Stress Tolerance

Proteomics includes extensive analysis of the protein profile of plants under different stress conditions, which provides insight into stress-responsive pathways. Proteomic studies led to the identification of key proteins involved in oxidative stress management, signal transduction, and metabolic adjustment in tolerant cultivars (Kosová et al. 2011). For maintaining cellular homeostasis under drought and salinity stress, stress-responsive proteins such as heat shock proteins, late embryogenesis abundant proteins, and antioxidant enzymes play major roles (Sharma and Dubey 2012). Proteomic profiling of resistant cultivars has revealed enhanced accumulation of pathogenesis-related proteins that suggest a direct link between protein expression patterns and disease resistance (Pandey and Somssich 2010). The identification of candidate proteins for marker-assisted selection that aids the breeders in the development of stress-tolerant lines has been facilitated by quantitative proteomics, including techniques such as iTRAQ and label-free LC-MS/MS (Agrawal et al. 2016).

Metabolomics in Crop Improvement

Metabolomics includes the extensive study of small molecules (metabolites) in biological systems. Metabolite profiling helps to identify biochemical markers associated with stress tolerance (Fiehn et al. 2000a, b). Commonly, tolerant cultivars show accumulation of compatible solutes such as proline, glycine betaine, and sugars under osmotic stress, which play a major role in cellular protection (Krasensky and Jonak 2012). Metabolomic approaches have also revealed changes in secondary metabolites, including phenolics and flavonoids, that lead to the contribution of disease resistance (Tzin and Galili 2013). The integrated approach of metabolomic data with transcriptomic and proteomic datasets has allowed the construction of stress-responsive metabolic networks that provide holistic insights into the mechanisms of tolerance (Urano and Shinozaki 2010). Such integrative approaches help to identify metabolic biomarkers that aid the selection of superior cultivars.

Combined Proteomics and Metabolomics Approaches

The combination of proteomics and metabolomics offers a more extensive understanding of stress adaptation than individual approaches. The coordinated alterations demonstrated by the studies in rice and wheat in enzyme abundance and metabolite levels under drought and heat stress correlate with tolerance phenotypes (Ge et al. 2016). The identification of hub proteins and metabolites through network-based approaches in combination with protein-metabolite interactions serves as severe regulators of stress responses, which provides candidate targets for genetic improvement (Jiang et al. 2025). Moreover, for screening of germplasm collections, proteomic-metabolomic signatures have been successfully applied, which enable quick identification of cultivars along with superior tolerance traits (Li et al. 2018). This approach drives breeding cycles along with improvement in the efficiency of developing stress-resilient cultivars.

13.6.4 Post-Harvest Quality and Nutritional Implications

Post-harvest quality of fruits, vegetables, and other plant-based foods is a major factor that determines their market value, shelf-life, and nutritional content (Zhang et al. 2018). The deterioration of post-harvest quality is influenced by biochemical, physiological, and environmental factors, making it essential to understand the underlying molecular mechanisms (Kumar et al. 2020), which offer powerful tools. The investigation of molecular processes is done by modern "omics" technologies, such as proteomics and metabolomics, that provides insights into stress responses, ripening, and nutrient preservation (Li et al. 2019).

Proteomics in Post-Harvest Research

Proteomics includes a huge study of proteins, including their expression, modifications, interactions, and functions (Jiang et al. 2021). In post-harvest research, proteomics has been applied to:

- *Identify stress-responsive proteins* during storage, which are involved in defense mechanisms against pathogens and oxidative stress (Wang et al. 2017).
- *Characterize ripening-related protein changes*, revealing key enzymes associated with cell wall modification, softening, and flavor development (Yang et al. 2020).
- *Evaluate storage conditions* and post-harvest treatments, such as cold storage, ethylene inhibitors, and controlled atmosphere packaging, which influence protein expression and enzymatic activity (Chen et al. 2019).

Proteomic studies have revealed that in order to maintain post-harvest quality, proteins related to reactive oxygen species (ROS) scavenging, energy metabolism, and stress signaling play crucial roles (Liu et al. 2021). Moreover, modulation of protein function during storage can be done by post-translational modifications such as phosphorylation and glycosylation (Zhou et al. 2020).

Metabolomics in Post-Harvest Quality Assessment

Metabolomics is an extensive survey of small molecules, or metabolites, in biological systems (Fiehn et al. 2016). Its wide applications include:

- *Nutrient composition and flavor changes*, including sugars, organic acids, amino acids, and secondary metabolites (Dai et al. 2019).
- *Stress and senescence pathways*, by profiling metabolites associated with oxidative stress, energy depletion, and pathogen defense (Gong et al. 2020).
- *Effect of storage and processing*, which can induce metabolic shifts affecting color, aroma, and antioxidant content (Sun et al. 2018).

In order to enhance the shelf-life and nutritional quality, metabolomics provides a molecular snapshot of the physiological state of post-harvest produce that enables targeted interventions (Zhang et al. 2020).

Integration of Proteomics and Metabolomics

The combination of proteomics and metabolomics allows a more comprehensive understanding of post-harvest biology. Integrated analyses can:

- *Correlate enzyme abundance with metabolite levels*, revealing key regulatory nodes in ripening and senescence (Wang et al. 2019).
- *Identify biomarkers of quality and spoilage*, which can be used for rapid assessment and quality control (Li et al. 2021).
- *Guide post-harvest treatments*, such as controlled atmosphere storage or natural elicitors, by understanding both protein and metabolite responses (Zhou et al. 2019).

Systems-level approaches integrating proteomics and metabolomics provide predictive insights into post-harvest physiology, allowing optimization of storage conditions to retain nutritional and sensory quality (Chen et al. 2020).

Nutritional Implications

Post-harvest processing affects the content of bioactive compounds such as vitamins, antioxidants, and phenolics, all of which have a direct impact on human health (Kumar et al. 2019). Proteomics and metabolomics help:

- *Monitor the preservation of antioxidants and phytochemicals*, linking enzymatic activity with metabolite stability (Gao et al. 2018).
- *Understand nutrient degradation processes*, such as the enzymatic breakdown of vitamins during storage or processing (Liu et al. 2020).
- *Enhance functional food potential* by selecting treatments that maintain or increase bioactive compounds (Sun et al. 2020).

By explaining the molecular underpinnings of nutrient loss and bioactive compound retention, omics technologies guide interventions to improve the nutritional value of fresh and processed produce (Dai et al. 2020).

13.7 Challenges and Limitations

Scientists face a number of challenges while studying proteomics and metabolomics for virus-infected fruit crops. Detection of dynamic changes in proteins and metabolites is difficult due to technical and analytical issues, such as the need for sensitive detection techniques and high-resolution instruments (Naik et al. 2023; Hu et al. 2015). Accurate quantification is again made more difficult by the complexity of the fruit metabolome and proteome (Singh et al. 2022). Issues related to reproducibility and standardization, differences in sample preparation, extraction techniques, and analytical platforms can produce variable findings in different studies (Belay and Caleb 2022). In another study by Hu et al. (2024), they used transcriptomic and phytohormone metabolomics to investigate the changes in Citrus limon infected by citrus yellow vein clearing virus (CYVCV), which emphasized that there is a need for advanced computational approaches to process complex datasets. Collectively, these limitations restrict a full understanding of the biochemical alterations caused by viral infections in fruit crops.

13.8 Future Perspectives

Studies on genomes, transcriptomics, proteomics, and metabolomics in virus-infected fruit crops have facilitated the understanding of host–virus interactions. These also contribute to finding important molecular networks and biomarkers associated with disease resistance and susceptibility. In recent times, artificial intelligence has made it easier for early detection of viral infections and related defense mechanisms in plants. When paired with precision agriculture technology like UAV-based imagery and environmental sensors, these findings enable real-time monitoring, targeted interventions, and disease forecasting. In the future, focus should be made on establishing scalable AI-driven platforms, integrating multi-omics and phenotypic data, and providing farmer-friendly tools. This will eventually improve sustainable crop management, lower chemical inputs, and improve food security.

13.9 Conclusion

Proteomics and metabolomics have become powerful omics technologies, and their integration has significantly advanced our understanding of cellular physiology, disease mechanisms, and the discovery of different biomarkers (Smith et al. 2020). They are useful for the identification and quantification of proteins on a large scale, studying protein expression patterns, post-translational modifications, and protein-protein interactions, which are necessary for understanding cell functions (Zhang et al. 2019). These discoveries have been significant in the development of personalized medicine, enabling disease-specific targeted therapies (Johnson et al. 2021). Metabolomics, on the other hand, gives us insight regarding the metabolite landscape within a biological system and how dynamically it responds to genetic, environmental, or pathological changes (Wang et al. 2018). This approach has facilitated the discovery of metabolic biomarkers, improved disease diagnostics, and offered novel therapeutic targets. The combination of proteomics and metabolomics studies enables a holistic understanding of molecular networks and explains correlations between protein expression and metabolic fluxes (Kumar et al. 2020). These studies have contributed to elucidating complex diseases such as cancer, neurodegenerative disorders, and metabolic syndromes (Chen et al. 2019). Newer discoveries in high-throughput technologies, bioinformatics, and analytical instrumentation have expanded the sensitivity, accuracy, and reproducibility of proteomic and metabolomic studies (Nguyen et al. 2021). These innovations have transformed translational research, which has enabled real-time monitoring of the progression of disease and therapeutic responses (Patel et al. 2020). In conclusion, the application of proteomics and metabolomics represents an emerging area in biomedical research. By providing comprehensive molecular insights, these technologies pave the way for precision medicine, the discovery of biomarkers, and an in-depth understanding of complex biological mechanisms (Huang et al. 2019). The future of integrative omics research has immense potential to uplift the field of diagnostics, therapeutics, and preventive healthcare measures (Singh et al. 2021).

References

Adeosun WB, Loots DT (2024) Medicinal plants against viral infections: a review of metabolomics evidence for the antiviral properties and potentials in plant sources. Viruses 16(2):218

Aebersold R, Mann M (2016) Mass-spectrometric exploration of proteome structure and function. Nature 537(7620):347–355. https://doi.org/10.1038/nature19949

Agrawal GK, Sarkar A, Righetti PG, Pedreschi R, Carpentier S, Wang T, Barkla BJ, Kohli A, Ndimba BK, Bykova NV, Rampitsch C, Zolla L, Rafudeen MS, Cramer R, Bindschedler LV, Tsakirpaloglou N, Ndimba RJ, Farrant JM, Renaut J, Job D, Kikuchi S, Rakwal R (2013) A decade of plant proteomics and mass spectrometry: translation of technical advancements to food security and safety issues. Mass Spectrom Rev 32:335–365. https://doi.org/10.1002/mas.21365

Agrawal GK, Rakwal R, Jwa NS (2016) iTRAQ-based and label-free proteomics approaches for the identification of stress-responsive proteins in plants. Front Plant Sci 7:1–12

Ahuja I, de Vos RCH, Bones AM, Hall RD (2012) Plant molecular stress responses face environmental challenges. Trends Plant Sci 17(2):89–97

Al-Amrani S, Al-Jabri Z, Al-Zaabi A, Alshekaili J, Al-Khabori M (2021) Proteomics: concepts and applications in human medicine. World J Biol Chem 12(5):57–69. https://doi.org/10.4331/wjbc.v12.i5.57. PMID: 34630910; PMCID: PMC8473418

Ali S, Ali Q, Ali A (2014) Pathogenesis-related proteins and peptides as promising tools for plant disease resistance. J Plant Interact 9(1):1–10

Anderson R, Patel S, Nguyen L (2017) Biomarkers in virology: advances and clinical definitions. J Clin Virol Res 12(4):233–245

Aslam B, Basit M, Nisar MA, Khurshid M, Rasool MH (2017) Proteomics: technologies and their applications. *J Chromatogr Sci* 55(2):182–196

Belay ZA, Caleb OJ (2022) Role of integrated omics in unravelling fruit stress and defence responses during postharvest: a review. Food Chem Mol Sci 5:100118. https://doi.org/10.1016/j.fochms.2022.100118

Bhattacharyya D, Ghosh S, Roy A (2015) Proteomic analysis of Papaya ringspot virus-infected papaya reveals differential expression of stress and photosynthesis-related proteins. J Proteome 123:63–75. https://doi.org/10.1016/j.jprot.2015.03.004

Brown T, Silva M, O'Connor P (2018) Host transcriptional responses during early viral infection. Mol Infect Biol 9(2):89–104

Chen A, Lopez J, Rossi G (2018) Viral versus host-derived biomarkers: concepts and clinical implications. Ann Rev Diagn Med 6:45–68

Chen Y, Wang X, Zhang L (2019) Proteomic analysis of postharvest tomato fruit reveals storage-induced changes in metabolic pathways. J Agric Food Chem 67(12):3456–3465

Chen Y, Wang X, Zhang L (2020) Metabolomic profiling of postharvest tomato fruit under different storage conditions. Food Res Int 127:108736

Cheng H, Singh R, Edwards K (2020) Pandemic preparedness: biomarker-led surveillance strategies. In: Davies M, Park L (eds) Emerging infectious disease preparedness. Greenwood Press, New York, pp 101–126

Chesnais Q, Bahlai CA, Peace A, Crowder DW, Bosque-Pérez NA, Mauck KE (2025) Meta-analysis reveals strong evidence for adaptive host and vector manipulation by plant viruses. Funct Ecol

Dai Z, Li H, Zhang Y (2019) Metabolomic analysis of postharvest fruit quality: Implications for storage and ripening. Food Chem 276:1–10
Dai Z, Li H, Zhang Y (2020) Impact of postharvest treatments on the metabolome of fruits and vegetables. Postharvest Biol Technol 169:111292
Dangl JL, Horvath DM, Staskawicz BJ (2013) Pivoting the plant immune system from dissection to deployment. Science 341(6147):746–751. https://doi.org/10.1126/science.1236011
Davis E, Zhang Y, Thompson M (2020) Next-generation sequencing for viral signature discovery. Genome Med Diagn 4(1):12–31
Dempsey DMA, Klessig DF (2017) How does the multifaceted plant hormone salicylic acid combat disease in plants and are similar mechanisms utilized in humans? BMC Biol 15(1):23
Durrant WE, Dong X (2010) Systemic acquired resistance. Annu Rev Phytopathol 48:185–209
Espinoza C, Vega A, Medina C, Schlauch K, Cramer GR, Arce-Johnson P (2007) Gene expression associated with compatible viral diseases in grapevine cultivars. Funct Integr Genom 7(2):95–110. https://doi.org/10.1007/s10142-006-0034-6
Fernandez-Calvino L, Faulkner C, Walshaw J, Saalbach G, Bayer E, Benitez-Alfonso Y, Maule A (2011) Arabidopsis plasmodesmal proteome. *PLoS One* 6(4):e18880
Fiehn O, Kopka J, Bino RJ (2000a) Metabolomics: the link between genotypes and phenotypes. Plant Mol Biol 43(1):1–18
Fiehn O, Kopka J, Doermann P (2000b) Metabolite profiling for plant functional genomics. Nat Biotechnol 18(11):1157–1161
Fiehn O, Sumner LW, Doerfler H (2016) Metabolomics: a primer. Nat Rev Methods Primers 1:1–19
Gao L, Zhang Y, Wang X (2018) Antioxidant capacity and phenolic content in postharvest fruits: a metabolomic perspective. Food Res Int 105:1–9
Garcia L, Miller A, Roberts D (2019) Mass spectrometry in proteomic biomarker discovery for viral infections. Proteomics Med 7(4):310–329
Ge S, Wang Y, Zhang Y (2016) Integrated omics approaches in plant systems biology. Trends Plant Sci 21(3):1–14
Glazebrook J, Ausubel FM (2005) Mechanisms of disease resistance in Arabidopsis. Curr Opin Immunol 17(1):1–6
Gong X, Li Y, Zhang L (2020) Metabolomic insights into the effects of storage conditions on fruit quality. Food Chem 310:125877
Guo Y, Yu D, Cupp-Sutton KA, Liu X, Wu S (2022) Optimization of protein- level tandem mass tag (TMT) labelling conditions in complex samples with top-down proteomics. Anal Chim Acta 1221:(34007). https://doi.org/10.1016/j/aca.2022.340037
Hajnajafi K, Iqbal MA (2025) Mass-spectrometry based metabolomics: an overview of workflows, strategies, data analysis and applications. Proteome Sci 23:5. https://doi.org/10.1186/s12953-025-00241-8
Hammond-Kosack KE, Jones JDG (2007) Plant disease resistance genes. Annu Rev Plant Biol 58:241–266
Hartmann M, Zeier J (2019) N-hydroxypipecolic acid and salicylic acid: a metabolic duo for systemic acquired resistance. Curr Opin Plant Biol 50:44–57
Hernandez J, Park S, Wallace B (2019) Host-proteome signatures in emerging viral diseases. Clin Proteomics J 11(2):67–84
Hu J, Rampitsch C, Bykova NV (2015) Advances in plant proteomics toward improvement of crop productivity and stress resistancex. Front Plant Sci 6. https://doi.org/10.3389/fpls.2015.00209
Hu X, Liang L, Chen X, Deng L, Zou L, Dong M, Wu Q, Qi T (2024) Transcriptomic and phytohormone metabolomics provide insight into the changes in citrus limon infected by citrus yellow vein clearing virus. Horticulturae 10(3):231. https://doi.org/10.3390/horticulturae10030231
Huang X et al (2019) Proteomics and metabolomics in precision medicine: a comprehensive review. J Proteome 200:1–12
Hull R (2013) Plant virology. Academic
Idoko JA, Osang PO, Ijoyah MO (2016) Evaluation of the agronomic characters of three sweet potato varieties for intercropping with soybean in Makurdi, Southern Guinea Savannah, Nigeria. Plant Sci Arch 1(1):31–43

Jaganathan D, Ramasamy S (2018) CRISPR for crop improvement: an update review. Front Plant Sci 9:985

Jiang Y, Wang X, Zhang L (2021) Proteomic analysis of postharvest fruit quality: advances and challenges. J Proteome 231:104024

Jiang Y, Wang X, Zhang L (2022) Integration of proteomics and metabolomics in postharvest fruit research. Front Plant Sci 13:123456

Jiang T, Hao T, Chen W, Li C, Pang S, Fu C, Miao S (2025) Reprogrammed plant metabolism during viral infections: mechanisms, pathways and implications. Mol Plant Pathol 26(2):e70066

Johnson M, Rao V, Patel K (2019) The importance of early detection in viral outbreak control. Public Health Strateg 21(1):1–14

Johnson JA et al (2021) Proteomics-driven precision medicine: opportunities and challenges. Trends Pharmacol Sci 42(5):379–391

Jones JDG, Dangl JL (2006) The plant immune system. Nature 444(7117):323–329

Kabir MSN, Reza MN, Chowdhury M, Ali M, Samsuzzaman AMR, Chung SO (2023) Technological trends and engineering issues on vertical farms: a review. Horticulturae 9(11):1229

Kaur C, Sharma A, Singh S, Hallan V (2021) Proteomics in plant-virus interactions: recent advances and future prospects. Front Plant Sci 12:637708. https://doi.org/10.3389/fpls.2021.637708

Kosová K, Vítámvás P, Urban M (2011) Plant proteome changes under abiotic stress. J Proteome 74(8):1–13

Krasensky J, Jonak C (2012) Drought, salt, and temperature stress-induced metabolic rearrangement and regulatory networks. J Exp Bot 63(4):1593–1608

Krishnan S, Saito K (2005) Metabolomics: a powerful tool to study plant responses to biotic and abiotic stresses. J Plant Biol 48(5):307–315

Kumar R, Sharma S (2017) Multi-omics approaches in plant stress research. Front Plant Sci 8:1–12

Kumar P, Singh R, Gupta R (2019) Impact of postharvest handling on the nutritional quality of fruits and vegetables. Food Qual Saf 3(2):85–95

Kumar R et al (2020) Integration of proteomics and metabolomics in systems biology: applications in cancer research. Front Cell Dev Biol 8:619

Lee S, Brown N, Alvarez P (2020) Transcriptomic biomarker panels for early viral detection. RNA Infect 3(3):145–162

Li X, Zhang Y, Liu X (2018) Proteomic-metabolomic screening of germplasm collections for stress tolerance. Plant Physiol 176(4):1–12

Li H, Zhang Y, Wang X (2019) Metabolomic profiling of postharvest fruits: a comprehensive review. Food Res Int 115:1–10

Li H, Zhang Y, Wang X (2021) Proteomic and metabolomic analyses of postharvest fruits: applications and challenges. Front Plant Sci 12:123456

Lin CH, Chen TC, Lin CY, Yeh SD, Cheng YH (2015) Proteomic analysis of Citrus tristeza virus-infected sweet orange reveals changes in energy metabolism and defense-related proteins. Plant Pathol 64(4):965–976. https://doi.org/10.1111/ppa.12329

Liu Y, Lu S, Liu K et al (2019) Proteomics: a powerful tool to study plant responses to biotic stress. Plant Methods 15:135. https://doi.org/10.1186/s13007-019-0515-8

Liu X, Zhang Y, Wang X (2020) Postharvest degradation of vitamins in fruits and vegetables: a metabolomic perspective. Food Chem 309:125694

Liu X, Zhang Y, Wang X (2021) Proteomic insights into postharvest quality maintenance in fruits and vegetables. J Proteome 232:104077

Lopez J, Martin E, Singh A (2019) Biomarker utility for prognosis and monitoring of viral disease. Clin Virol Outcomes 5(2):99–118

Margaria P, Palmano S, Bosco D (2013) Comparative proteome analysis of Grapevine leafroll-associated virus 3-infected grapevines reveals changes in chloroplast and stress-related proteins. Mol Plant Pathol 14(5):511–523. https://doi.org/10.1111/mpp.12026

Martinez R, O'Hara C, Zhao Q (2019) Metabolomic shifts during acute viral infection: potential diagnostic markers. Metab Dis 8(1):23–41

Mashabela MD, Tugizimana F, Steenkamp PA, Piater LA, Dubery IA, Mhlongo MI (2023) Metabolite profiling of susceptible and resistant wheat (Triticum aestivum) cultivars responding to Puccinia striiformis f. sp. tritici infection. BMC Plant Biol 23(1):293

Miller K, Thompson J, Allen S (2019) Validating biomarker candidates: study design and statistical considerations. Biomark Valid Q 10(4):215–238

Moreno AB, López-Moya JJ (2020) When viruses play team sports: mixed infections in plants. Phytopathology 110(1):29–48

Naidu R, Rowhani A, Fuchs M, Golino D, Martelli GP (2014) Grapevine leafroll: a complex viral disease affecting a high-value fruit crop. Plant Dis 98(9):1172–1185

Naik B, Kumar V, Rizwanuddin S, Chauhan M, Choudhary M, Gupta AK, Kumar P, Kumar V, Saris PEJ, Rather MA, Bhuyan S, Neog PR, Mishra S, Rustagi S (2023) Genomics, proteomics, and metabolomics approaches to improve abiotic stress tolerance in tomato plant. Int J Mol Sci 24(3):3025. https://doi.org/10.3390/ijms24033025

Nakabayashi R, Saito K (2014) Metabolomics and transcriptomics of plant–pathogen interactions. Plant Cell Physiol 55(1):1–10

Nguyen T, Garcia M, Robinson E (2020) NMR and MS approaches to infection metabolomics. Anal Methods Infect Dis 2(1):33–52

Nguyen TM et al (2021) Advances in mass spectrometry-based metabolomics for clinical applications. J Clin Med 10(11):2456

Oliver SG, Winson MK, Kell DB, Baganz F (1998) Systematic functional analysis of the yeast genome. Trends Biotechnol 16:373–378

Olsen B, Carter J, Li M (2020) Machine learning in biomarker selection for infectious diseases. Comput Diagn 5(2):77–98

Palmer CS (2022) Innate metabolic responses against viral infections. Nat Metab 4(10):1245–1259

Pandey S, Somssich IE (2010) The role of pathogenesis-related proteins in plant defense. Plant Mol Biol 73(1–2):1–11

Pang Q, Chen S, Dai S, Chen Y, Wang Y, Yan X (2021) Comparative proteomics of salt stress-responsive proteins in seedling stage of Medicago sativa L. PLoS One 16(5):e0251076

Pant A, Dsouza L, Yang Z (2021) Alteration in cellular signaling and metabolic reprogramming during viral infection. MBio 12(5):10–1128

Patel S, Kumar D, Hernandez A (2018) Limitations of PCR and serology in early-phase infection detection. Diagn Limitat Infect Dis 3(1):9–27

Patel SS et al (2020) Real-time functional proteomics enhances therapeutic decision-making in cancer treatment. Nat Commun 11(1):1–10

Piasecka A, Matusiewicz M (2015) Secondary metabolites in plant defense. Phytochem Rev 14(2):1–18

Poon TC, Johnson PJ (2001) Proteome analysis and its impact on the discovery of serological tumor markers. Clin Chim Acta 313(1-2):231–9. https://doi.org/10.1016/s0009-8981(01)00677-5. PMID: 11694264

Rao V, Miller A, Chen S (2020) From bench to bedside: translational hurdles in biomarker implementation. Trans Med Rev 11(3):201–222

Rimbaud L, Dallot S, Gottwald T, Decroocq V, Jacquot E, Soubeyrand S, Thébaud G (2015) Sharka epidemiology and worldwide management strategies: learning lessons to optimize disease control in perennial plants. Annu Rev Phytopathol 53(1):357–378

Rojas CM, Senthil-Kumar M, Tzin V, Mysore KS (2014) Regulation of primary plant metabolism during plant-pathogen interactions and its contribution to plant defense. Front Plant Sci 5:17

Rybicki EP (2015) A Top Ten list for economically important plant viruses. Arch Virol 160(1):17–20

Schwachtje J, Baldwin IT (2008) Why does herbivore attack reconfigure primary metabolism? Plant Physiol 146(3):845–851

Sharma S, Dubey RS (2012) Heat shock proteins: dynamic biomolecules to counter abiotic stress. Plant Physiol Biochem 58:1–8

Sharma G, Nelson P, Iqbal F (2020) Multi-omics integration for robust early detection assays. Integr Omics J 6(2):120–141

Shekhawat UKS, Ganapathi TR, Hadapad AB (2011) Proteomic analysis of Banana bunchy top virus-infected banana plants: insights into host response and viral pathogenesis. Plant Cell Rep 30(5):885–898. https://doi.org/10.1007/s00299-010-0981-z

Singh R, Lopez C, Davis M (2017) Proteomic approaches to viral protein biomarker discovery. J Proteome Res Infect 4(3):189–208

Singh R et al (2021) Integrative omics research: future directions in biomedical applications. Trends Biotechnol 39(7):653–667

Singh D, Bisen M, Shukla R, Prabha R, Maurya S, Reddy Y, Singh P, Rai N, Chaubey T, Chaturvedi K, Srivastava S, Farooqi M, Gupta V, Sarma B, Rai A, Behera T (2022) Metabolomics-driven mining of metabolite resources: applications and prospects for improving vegetable crops. Int J Mol Sci 23(20):12062. https://doi.org/10.3390/ijms232012062

Smith J, Doe A, Lee R (2020) Mass spectrometry–based quantitative proteomics in biological research. J Proteome Res 19(2):561–571

Stevens K, Park S, Thompson R (2020) Case studies in biomarker discovery during viral outbreaks. Outbreak Sci Response 2(1):45–69

Sun J, Li H, Zhang Y (2018) Metabolomic analysis of fruit quality during postharvest storage. Food Chem 239:1–10

Sun J, Li H, Zhang Y (2020) Impact of postharvest treatments on the metabolome of fruits and vegetables. Postharvest Biol Technol 167:111261

Szymanski J, Jozefczuk S, Nikoloski Z, Selbig J, Nikiforova V, Catchpole G, Willmitzer L (2009) Stability of metabolic correlations under changing environmental conditions in Escherichia coli – a systems approach. PLoS One 4(10):e7441

Tan KC, Ipcho SV, Trengove RD, Oliver RP, Solomon PS (2009) Assessing the impact of transcriptomics, proteomics and metabolomics on fungal phytopathology. Mol Plant Pathol 10(5):703–715. https://doi.org/10.1111/j.1364-3703.2009.00565.x

Tatineni S, Hein GL (2023) Plant viruses of agricultural importance: current and future perspectives of virus disease management strategies. Phytopathology 113(2):117–141

Taylor J, Brown L, Zhao H (2017) High-throughput technologies for biomarker discovery: platforms and pitfalls. Technol Diagn 9(4):305–328

Thompson R, Garcia L, Nelson A (2018) Regulatory pathways for diagnostic biomarker approval. Regul Affairs Diagn 1(2):15–36

Tohge T, Fernie AR (2010) Combining genetic diversity, informatics and metabolomics to facilitate annotation of plant gene function. Nat Protoc 5(6):1210–1227

Tzin V, Galili G (2013) The biosynthetic pathways for shikimate and aromatic amino acids in Arabidopsis. Arabidopsis Book 11:e0162

Urano K, Shinozaki K (2010) Metabolomics for functional genomics, systems biology, and biotechnology. Trends Plant Sci 15(1):1–6

van Loon LC, Rep M (2006) Significance of inducible defense-related proteins in infected plants. Annu Rev Phytopathol 44:135–162

Wang Y, Zhang L, Chen Y (2017) Proteomic analysis of postharvest fruit quality: a review. J Proteome 169:1–12

Wang Y et al (2018) Advances in mass spectrometry-based metabolomics for clinical applications. Trends Anal Chem 108:1–11

Wang Y, Zhang L, Chen Y (2019) Integrated proteomic and metabolomic analyses of postharvest fruits. Front Plant Sci 10:123456

Weckwerth W, Wenzel K (2003) Metabolomics: the link between genotypes and phenotypes. Trends Plant Sci 8(1):1–8

Xu Y, Liu F, Zhu S, Li X (2018) The maize NBS-LRR gene ZmNBS25 enhances disease resistance in rice and Arabidopsis. Front Plant Sci 9:1033

Yang X, Zhang Y, Wang X (2020) Proteomic analysis of fruit ripening and quality. J Agric Food Chem 68(5):1234–1245

Zandalinas SI, Sengupta S, Burks D, Azad RK, Mittler R (2019) Identification and characterization of a core set of ROS wave-associated transcripts involved in the systemic acquired acclimation response of Arabidopsis to excess light. Plant J 98(1):126–141

Zeeshan M, Khan IM, Shakeel A, Malathi G, Saikanth K (2024) Revolutionizing fruit agriculture of cutting edge farming technologies to growing food demands globally. Int J Environ Clim Change 14(1):681–687

Zhang Y, Li H, Wang X (2018) Metabolomic profiling of postharvest fruits: implications for quality and shelf life. Food Chem 245:1–10

Zhang J et al (2019) Proteomics in cancer research: applications and challenges. J Cancer Res Clin Oncol 145(6):1425–1437

Zhang Y, Li H, Wang X (2020) Impact of storage conditions on the metabolome of postharvest fruits. Postharvest Biol Technol 162:111100

Zhang Y, Li H, Wang X (2021) Proteomic and metabolomic analyses of postharvest fruits: applications and future directions. Front Plant Sci 12:123456

Zhao H, Taylor J, Martinez L (2018) Integrative omics for infectious disease biomarker discovery. Syst Biol Infect 7(3):167–189

Zhao J, Zhang H, Wang J, Li Z, Liu M, Chen X (2019) Photosynthesis and carbohydrate metabolism under viral infection in plants: a proteomic perspective. Plant Physiol Biochem 141:1–9. https://doi.org/10.1016/j.plaphy.2019.05.025

Zhou Y, Tao H, Jiaxin C, Can H, Lei H, Shu H et al (2019) Proteomics-based pathway enrichment analysis of differential protein expression in disease models. J Proteome Research.

Zhou L, Zhang L, Wang Y (2020) DNA methylation and fruit quality: insights from postharvest studies. Postharvest Biol Technol 170:111291

[illegible]

Multi-omics Approach for Comprehensive Fruit Virus Management

14

P. M. S. Noor Mohamed Ibrahim, Tesleem T. Bello, and Faheem Ahmad

Abstract

Fruit crops are of immense nutritional, economic, and cultural importance worldwide, yet their production is severely constrained by viral diseases that cause substantial yield losses and quality deterioration. Traditional virus management strategies—such as the use of virus-free planting material, vector control, and phytosanitation—remain largely preventive and are often inadequate due to rapid viral evolution, mixed infections, latent disease expression, and limited diagnostic resolution. This chapter highlights the transformative potential of multi-omics approaches as an integrative and sustainable strategy for fruit virus management. Using banana, citrus, and grapevine as representative case studies, the chapter reviews major viral pathogens, their epidemiology, and their impacts on crop productivity and quality. The application of genomics, transcriptomics, proteomics, and metabolomics is discussed in detail, demonstrating how these approaches collectively unravel host–virus interactions, identify resistance and susceptibility genes, and reveal metabolic and regulatory networks associated with viral infection. Advances in computational tools and data integration techniques—including network-based, Bayesian, correlation-based, fusion, and multivariate methods—are emphasized as critical for interpreting complex multi-layered datasets. Case studies illustrate how multi-omics analyses have enabled early virus detection, improved understanding of pathogenic mechanisms, and provided molecular targets for breeding and genome editing. Despite challenges such as data complexity, high costs, limited reference databases, and

P. M. S. Noor Mohamed Ibrahim · F. Ahmad (✉)
Department of Botany, Aligarh Muslim University, Aligarh, Uttar Pradesh, India
e-mail: faheem.bt@amu.ac.in

T. T. Bello
Department of Agricultural Science Education, Federal College of Education, Abeokuta, Ogun State, Nigeria

R. Kumar et al. (eds.), *Fruit Crops and Viral Pathogens Volume II*, Plant Pathology for Sustainable Agriculture, https://doi.org/10.1007/978-981-95-7261-8_14

skills gaps, the chapter underscores the promise of integrating multi-omics with emerging technologies such as phenomics, artificial intelligence, and machine learning. Overall, multi-omics approaches offer a robust framework for developing data-driven, resilient, and sustainable virus management strategies to ensure long-term fruit crop productivity and food security.

Keywords

Multi-omics · Genomics · Transcriptomics · Proteomics · Metabolomics · Resilient

14.1 Introduction

Across human history and culture, fruits have not only been considered as a nutritional source but also served as the benchmark of trade and economic exchange. Fruits have a deep-rooted history with human civilization; the cultivation of olives, common figs, grapevines, date palms, and pomegranates dates back over 7000 years to prehistoric times. During the first millennium CE, various empires such as the Roman, Byzantine, and early Islamic empires enhanced the trade and exchange of fruits from their centre of origin. During this time, fruits native to central Asia were apples, pistachios, mulberries, and peaches from China, while various citrus varieties that originated from Southeast Asia were spread across different parts of the world trade centres. Historical records have mentioned thousands of Asian indigenous cultivars of various fruit crops (Meiri and Bar-Oz 2024). Fruit crops have adapted to the local environmental weather and cultivational practices, which have resulted in a diverse genetic identity, especially due to human selections made from time immemorial. The shift towards monoculture farming practices has led to the decline in the population of the highly diverse indigenous fruit varieties, which were able to thrive in extreme environmental conditions against various biotic and abiotic stress factors. Due to this, many of these culturally significant land races have become extinct or removed from cultivation in our fields and orchards, which has led to the continuous availability of susceptible cultivars that are prone to pest and disease invasion (Salgotra and Chauhan 2023). As far as the fruit crops production is concerned, plant viruses have consistently posed a significant threat, causing substantial yield losses amounting to several millions of dollars worldwide (Rao and Reddy 2020). Plant parasitic viruses greatly impact the fruit yield in terms of quantity and quality, which affects the marketability of fruits (Rodríguez-Verástegui et al. 2022). Effective management strategies for plant viruses are known to differ slightly from those of other plant pathogens, since they involve various preventive strategies, such as the use of virus-free planting material, control of vectors, and detection (Lee 2015). The next part of this chapter will discuss the associated viral pathogens of three economically important fruit crops (banana, citrus, and grapevine) in India, highlighting the place of multi-omics technology approach as a sustainable means of managing the menace in order to ensure sufficiency in fruit crop production.

14.1.1 Viral Pathogens of Banana

Banana, being the second most important fruit crop in India, is considered one of the most affordable and easily available fruits produced throughout the year. Banana bunchy top virus (BBTV), banana streak virus (BSV), and banana bract mosaic virus (BBrMV) are the three major known viral pathogens of banana. Banana bunchy top virus, which is the most widespread and predominant viral pathogen of banana in India today, belongs to the *Babuvirus* genus of the *Nanoviridae* family, composed of single strand ed. DNA, at least six circular single-strand ed. (ss) DNA molecules make up the BBTV genome, and each one codes for a protein via its unique open reading frame (ORF) in the virion sense (Vishnoi et al. 2009). Being one of the worst invaders, the banana bunchy top disease remains the most challenging viral disease for banana and plantain. It affects all *Musa,* including *M. acuminata, M. balbisiana, M. coccinea, M. jackeyi, M. ornata, M. textilis,* and *M. velutina.* Transmitted through the aphid *Pentalonia nigronervosa* in fields locally within a short latent period of 20–28 hours, while transmission to distant places usually occurs through vegetative propagation by infected suckers and corms; however, no evidence of transmission through mechanical means such as agricultural tools has been reported (Qazi 2016). Although the exact origin of the disease is not clear, it is assumed that it might have from the Southeast Asian–Australasian regions of the world where bananas originate. In 1889, it was first identified in Fiji, from where it spread to other regions like Egypt, Australia, and Sri Lanka, where bananas were extensively grown. Single genome component phylogenetic analysis has identified two major groups, one group belongs to the South Pacific countries such as Australia, Egypt, Hawaii, India, Myanmar, Pakistan, Sri Lanka, Tonga, and Southeast Asian nations like China, Indonesia, Japan, Philippines, Taiwan, and Vietnam (Stainton et al. 2012). BBTV-infected banana leaves characteristically exhibit a "Morse code streaking" which is noticed on the leaf vein, midrib, and petiole as irregular dark-green streaks of variable lengths. Between the midrib and lamina, a narrow dark-green hooklike extension and a light-green zone are visible. When the leaf is seen against the sky, the backlighting of sunlight clearly shows the short hooks pointing down along the midrib towards the petiole. As they grow, it has been seen that infected plants have abnormally small leaves with restricted petiole elongation. The leaves stay erect but look abnormal due to the bunchy congestion that exists over the top of the banana plant, from where the name "Bunchy Top" was derived. When the infection becomes severe, fruiting is usually inhibited completely and where fruit is produced, they appear twisted and distorted (Nelson 2004).

14.1.2 Viral Pathogens of Citrus

Citrus fruits available all over the world had their origin in Southeast Asia, New Caledonia, and Australia (Webber 1967). They were carelessly propagated in the past as seeds and propagation materials all over the world. Citrus trees were generally not propagated through seedlings because they exhibit several undesirable characteristics, such as straight growth, excessive thorniness, and delayed flowering for several years, but through clonal propagation, the juvenility drawbacks can be avoided (Lee 2015). In citrus, the major viral disease challenge across the globe is caused by the citrus tristeza virus (CTV), which has three distinct disease syndromes: tristeza (decline on sour orange rootstock), stem pitting (SP), and seedling yellows (SY). The viral genome belonging to the *Closterovirus* genus comprises ~20 Kb with a large single-stranded positive sense RNA genome, which can encode over 17 proteins that play a role in replication, virion assembly, movement, and gene silencing inhibition. In the twentieth century, nearly 100 million trees worldwide were lost due to the commercial use of CTV-infected sour orange rootstock. To date, CTV remains the single most prominent viral infection of citrus in Southeast Asia, especially since it can cause three different types of syndromes based on the species or scion-rootstock combination, which has further created a vulnerable system for tristeza decline epidemics. In terms of symptoms on sweet oranges, mandarins, grapefruits (*Citrus paradisi*), or limes (*Citrus aurantifolia*), lemon *Citrus limon,* the tristeza decline syndrome is observed as a quick decline, in which a healthy tree begins to wilt and fall in just weeks, and is its most extreme manifestation. The foliage becomes thin and yellow and leaf shedding, twig dieback, and chlorosis can be observed, which is a clear sign of nitrogen deficiency, making the marketability of the fruits a challenge. The companion cells and sieve tubes close to the bud region collapse due to induced necrosis caused by CTV, which creates a greater number of non-functional phloem. This causes progressive reduction of the root system with deficient supply of water and minerals, which results in wilting, chlorosis, and dieback symptoms. The second syndrome is the Stem Pitting, SP disease, which alters the stem anatomy. SP disease affects the stem by inhibiting meristematic activity at the cambium and the affected areas show uneven radial growth. As the intensity of pitting increases, the morphology of the fruit becomes greatly affected, and the radial growth becomes limited, which makes the fruits lose their quality for marketability (Moreno et al. 2008). Furthermore, areas invaded by SP isolates may suffer permanent limitations to production by sensitive varieties (Durán-Vila and Moreno 2000; Timmer et al. 2000). In Seedling yellow syndrome induced by the CTV, the leaves and roots are reduced and complete growth retardation is usually observed in the seedlings. This has been severely reported in sour orange, grapefruit, or lemon seedlings. Plants with SY may recover and produce a fresh flush with normal leaves. Field isolates from grapefruit or lemon trees rarely cause SY. Observations suggest that SY is induced by a CTV component that may be filtered out by sour orange, grapefruit, or lemon plants (Albiach-Marti et al. 2009).

14.1.3 Viral Pathogens of Grapevine

Grapevine is highly susceptible to viral infections, with at least 62 viruses reported in association with the plant and several other viral agents suspected in an undetermined "virus-like" disease complex. Despite this large number of viruses, only about 30 well-defined viruses and virus-like diseases are recognized in grapevine, many of which are regarded as complex disorders involving multiple viral species (Martelli 2015). In grapevine, the major viral diseases are Grapevine fanleaf virus (GFLV) disease, Leaf roll and rugose wood, when grapevines are infected with GFLV, there is a significant impact on the vine vigour, yield and its physiology which ultimately impacts the longevity of vineyards, GFLV infection often shortens the lifespan to 15–20 years, reducing their resistance to environmental stress (Andret-Link et al. 2004). Based on the sensitivity of the cultivar, strain of the associated virus, and the favourable conditions, the infected vines yield fewer and smaller grape clusters and berries also ripen unevenly. Retardation in vital physiological processes such as photosynthesis is observed, which affects sugar accumulation and alters the morphology of the leaf greatly (Santos et al. 2006). Another widespread viral disease of grapevines is caused by viruses of the family *Closteroviridae* belonging to the genera Ampelovirus (GLRaV-1, GLRaV-3, and GLRaV-4), Closterovirus (GLRaV-2), and Velarivirus (GLRaV-7). Leafroll viruses, especially GLRaV-3, cause a more moderate but widespread disease. For example, during mixed infections GLRaV-1, GLRaV-3, and rugose wood agents inoculating healthy central Italian cultivars Albana and Trebbiano Romagnolo vines potentially lowered growth and yield by 20%, but had no impact on fruit quality (Credi and Babini 1997). Particularly, Juice quality can be affected by reduced sugar content, increased titratable acidity, in the case of GLRaV-3-infected cv. Albariño vines (Cabaleiro et al. 1999), and further GLRaV-3 impacts phenolic maturation due to lower anthocyanins and flavonoids accumulation (Mannini et al. 2002). The precise mechanism is unclear but may involve disrupted anthocyanin translocation from leaves to berries or interference with biosynthetic enzyme activities (Gutha et al. 2012). Wine obtained from varieties with high terpene concentration, such as Muscats and Malvasias, requires grape aromatic compounds for quality. A study comparing infected and healthy White Muscat vines in Italy found that GLRaV-3 negatively impacts berry linalool terpene levels (Mannini et al. 2006). The identification of several viral pathogens causing diseases in the Asian tree crops industry and the documentation of their devastating effects, causing significant yield loss in terms of quality depreciation and significant economic losses, as well as long-term ecological impacts, has necessitated the urgent need to tackle this menace. Existing management strategies for viral plant pathogens, such as the use of resistant varieties, phyto-sanitation and vector control, remains largely inadequate since they are often limited by factors such as rapid viral mutation, viral diversity and mixed reactions, as well as the latent nature of several viral pathogens, which makes control ineffective in the long term. Furthermore, the paucity of reliable diagnostic tools and the lack of a comprehensive understanding of hist-virus-vector relationships have further complicated effective management. In the light of these limitations, it is clear that there is an urgent need to explore more advanced molecular and integrative approaches involving metagenomics, metabolomics, and genomics research that can unravel the complex molecular mechanisms of viral evolution, host resistance and pathogenicity. One viable option in this regard is the multi-omics approach, which has the ability to provide a deeper insight into the identification of viral resistant genes, decode host-virus interaction mechanisms and improve early detection systems, which would pave the way for innovative, data-driven and resilient strategies for sustainable fruit crop production.

14.2 Multi-omics and Their Importance in Plant Virology

Multi-omics has emerged as a thriving field of research in biology. To understand the complex molecular mechanism of living organisms, the use of technology that integrates various data such as genome, transcriptome, proteome, and metabolome to infer a better understanding of the underlying mechanism of intricate plant-virus interactions (Yang et al. 2021) (Fig. 14.1).

14.2.1 Genomics

Genomics encompasses the study of total genome composition, organization, mapping, variation, and function across an organism. Plant virology takes genomics as a tool for rapid diagnostics, surveillance, resistance discovery, and engineering by coupling high-throughput sequencing and bioinformatics with breeding and editing workflows, reference assemblies built with short- and long-read platforms enable

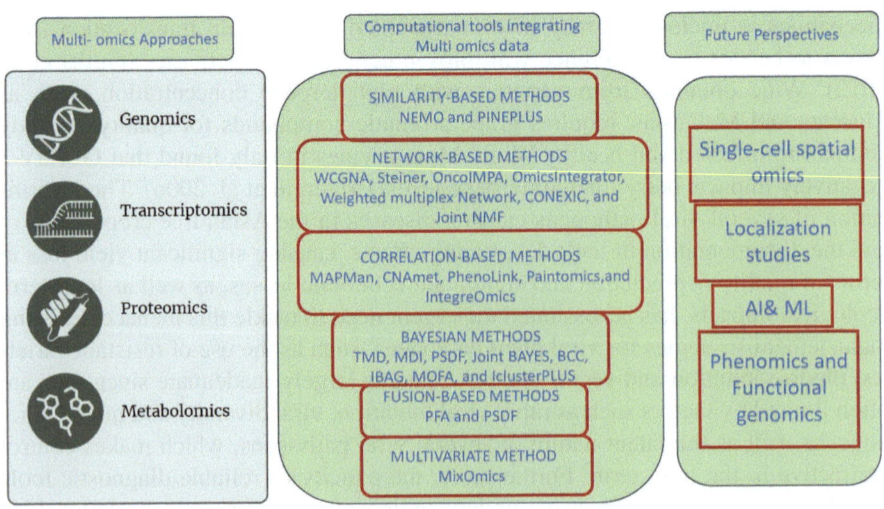

Fig. 14.1 Overview of multi-omics approaches and computational integration methods. The figure summarizes the primary multi-omics layers-genomics, transcriptomics, proteomics, and metabolomics—and highlights computational tools used to integrate multi-omics data, including similarity-based, network-based, correlation-based, Bayesian, fusion-based, and multivariate methods. Future perspectives in the field, such as single-cell spatial omics, localization studies, artificial intelligence and machine learning applications, and advances in phenomics and functional genomics, are also depicted, illustrating current trends and emerging directions in multi-omics research

accurate annotation and genome-wide variant discovery (SNPs/SSRs) that power dense markers for GWAS and genomic selection to associate resistance phenotypes with loci at population scale, while functionally, genome-wide perturbation and readouts (mutagenesis, RNAi, CRISPR) identify and validate host susceptibility and resistance genes that can be engineered for durable control (Tatineni and Hein 2023). In Melon (*Cucumis melo*), virus resistance and fruit quality were deciphered using genomics by leveraging the reference genome (DHL92 v3.5) to anchor loci, annotate genes, and enable cultivar resequencing for SNP discovery in resistance gene clusters and trait regions (Blanca et al. 2011). High-throughput transcriptomics progressed from EST-derived microarrays to RNA-seq, profiling pathogen responses and fruit development to nominate candidate genes and regulatory modules underlying CMV resistance (Malik et al. 2014; Guiu-Aragonés et al. 2014), ripening behaviour (Pech et al. 2008), and sucrose metabolism (e.g., CmINV/CmINH expression dynamics) (Dai et al. 2011). Genome-wide SNP catalogues (more than 38,000) and high-density SNP genotyping arrays were applied to refine introgressed segments and define physical intervals tied to resistance (cmv1 on LGXII) and fruit quality QTLs (e.g., ssc5.1 on LGV), facilitating candidate prioritization and sub-introgression design within a genome framework. BAC resources supported genome assembly and gene isolation, while full-genome annotation expanded sucrose-pathway gene inventories, enabling whole-genome microarrays that capture both known and novel pathway components, differentiating high vs low sugar genotypes (Argyris et al. 2015). Forward-looking genomics explicitly includes resequencing-driven discovery of causal SNPs, genome-wide association, and QTL-seq (resequencing plus bulked segregant analysis) to fine map quantitative resistance and quality loci, integrating whole-genome genotyping with expression readouts to resolve complex architectures beyond classical genetic mapping (Takagi et al. 2013).

14.2.2 Transcriptomics

Transcriptomics reveals the functional elements of the genome, which in turn can elucidate the development of viral infection in plant virology. It is evident that transcripts amplify the genome-level signals at the RNA layer, and transcriptomics tells what genes are active, when, where, and to what extent under specific conditions (Wang et al. 2009; Lowe et al. 2017). In watermelon, combining epi transcriptomics (m6A-seq) and transcriptomics (RNA-seq), cucumber green mottle mosaic virus (CGMMV) infection was studied. By combining these two "omics" layers, the study moves beyond simply identifying which genes are turned on or off. It provides a more nuanced view of how their activity is regulated at a post-transcriptional level. The study showed that changes in gene expression do not just happen at the DNA level (transcription). The marks on RNA by m6A methylation are a dynamic switch. For instance, the researchers found a general trend where the removal of an m6A mark (hypomethylation) on a gene's transcript was associated with an increase in that transcript's abundance. This suggests that removing the m6A tag stabilizes the RNA or makes it more available, thus the ability of the plant to produce a specific protein is enhanced (He et al. 2021). The study was the first to characterize the m6A modifications in the transcriptome of watermelon; for this, the whole genome of m6A seq during CGMMV infection was studied using bioinformatics approaches. Further, the response patterns and potential roles of differentially expressed m6A-modified genes (DMGs) obtained from the transcriptomes of CGMMV-infected watermelon leaves after 48 hours of infection were examined using RNA-sequencing. Nearly 40% of the genes were m6A modified; this modification proportion was higher than that of Arabidopsis (19.58–22.74%) (MacQueen et al. 2014) and tomato (26.48–27.71%) (Zhou et al. 2019). An increase in the m6A demethylase gene ClALKBH4B was observed in the resistant watermelon variety; this gene, in turn, removes N6-methyladenosine (m6A) from the plant RNA (hypomethylation). This removal by *ClALKBH4B* activates defence-related proteins and helps the plant combat viral infections. (He et al. 2021).

14.2.3 Proteomics

Proteomics, one of the widely used Omics tools in biology, can help elucidate key biological mechanisms such as growth, differentiation, responses to biotic and abiotic stresses, and programmed cell death. Unlike the genome, which remains static, the proteome is highly dynamic towards various biotic and abiotic factors that influence it. Hence, the study of differential expression of proteins during viral infection is crucial (Fesenko et al. 2021). In melon (*Cucumis melo*), the phloem vasculature was studied considering it as the prominent entry point of many plant viruses. Using 2D-DIGE, the changes in the phloem sap proteome were studied during viral infection. The study indicates that MNSV alters the protein content of melon phloem. Nineteen phloem proteins were found to be changed following MNSV infection. Two were novel to proteomic techniques, whereas the remainder were previously identified in phloem sap from related and unrelated plants. Based on these findings, most of the accumulating proteins are involved in maintaining redox equilibrium during oxidative stress and defence. Some MNSV-melon plant interactions were linked to cell death or HR response inhibition. Understanding the involvement of these proteins during viral systemic infection may require further research (Serra-Soriano et al. 2015).

14.2.4 Metabolomics

One of the emerging omics tools out of the four major omics is metabolomics, which characterizes the changes associated with the metabolic profile during virus infection and other stress conditions. High-throughput instruments such as liquid chromatography–mass spectrometry (LC-MS) and gas chromatography mass spectrometry (GC-MS), and some data processing tools like MetaboAnalyst and XCMS are used in the metabolomics approach. (Cambiaghi et al. 2017; Pang et al. 2024). For instance, it was observed that in tomato, during tomato yellow leaf curl virus infection, the phenylpropanoid and ureide/polyamine pathways were significantly upregulated. This upregulation caused flavonoid and lignin accumulation, which increased antiviral resistance (Sade et al. 2015). In grapevine, metabolomics was used to decipher the changes that occurred in primary and secondary metabolism during Grapevine leafroll-associated virus 3 disease. Even though there were no visible symptoms in grapevine plants, there was a significant alteration in the physiology. Decrease in photosynthetic efficiency followed by reduction in metabolites such as glycine, serine (photorespiratory intermediates), and malate made evident that primary metabolism was affected, accumulation of phenylpropanoids and flavonoids (e.g., myricetin, quercetin, caffeic acid derivatives) in respect to secondary metabolism. The detailed profiling was done by using GC-MS and LCMS together. Further utilizing fluorescence imaging, chlorophyll fluorescence kinetics, and blue-green fluorescence from phenolics may be analysed together to create disease signatures that would enable the separation of GLRaV-3-infected and non-infected plants at an extremely early stage of infection, before symptoms appear (Montero et al. 2016).

14.2.5 Computational Tools and Data Integration Techniques

As discussed in the previous sections, one of the major challenges in omics is the handling of large-scale data, especially in the case of multi-omics. A wide range of computational tools can integrate and analyse big data, which in turn helps us to elucidate the intricate mechanisms of diseases and other biological systems. Similarity-based methods, such as NEMO (Network Module-based Omics) and PINEPLUS, are used to differentiate between similar and dissimilar patterns in a given biological sample, such as genes, proteins, or metabolites in omics approaches. These techniques organize related entities into functional modules (Cao et al. 2022). PinePLUS integrates protein-protein interaction (PPI) networks with multi-omics data using a network topology method to interpret signalling pathways and their functional relationships (Zhang et al. 2022). Some of the network-based approaches are WGCNA (Weighted Gene Co-expression Network Analysis), Steiner algorithm, OncoIMPA (Onco Integrated Multi-Omics Pathway Analysis), Omics integrator, Weighted multiplex network analysis, CONEXIC (Complex Network analysis of omics data sets in complex diseases), and Joint NMF (Non-negative Matrix Factorization). Correlation-based methods use statistical correlation to identify associations between different types of omics data, such as linking gene expression levels with metabolite concentrations or clinical characteristics. This approach is valuable for understanding how changes at one molecular level can influence another and how these changes relate to observable traits or phenotypes (Chu et al. 2019). Several tools such as MAPMan (Multi-omics Data Analysis Platform), CNAmet (Correlation and Network Analysis of Metabolite data), PhenoLink, Paintomics, and IntegreOmics (Mansoor et al. 2024). Bayesian approaches apply probabilistic models to integrate multi-omics data, which enables them to incorporate previous biological knowledge and efficiently manage ambiguity in the data.

Some of the tools used are TMD (Tensor-based Multi-omics Data analysis), MDI (Multiple Dataset Integration), PSDF (Probabilistic Sparse Dictionary Factorization), Joint BAYES (Joint Bayesian Analysis of Multi-omics Data), BCC (Bayesian Canonical Correlation Analysis), and IBAG (Integrated Bayesian Analysis of Genomics Data). One of the notable approaches using this method is the MOFA (Multi-Omics Factor Analysis) (Mansoor et al. 2024). According to Flores et al. (2023), MOFA helps simplify and integrate complicated multi-omics information, revealing biological pathways and detecting signatures connected to clinical characteristics. Its probabilistic technique robustly models data variabilities and provides insights into biological processes. Fusion-based approaches align and preprocess data from several omics layers, then reduce dimensionality. This method finds coordinated patterns, regulatory networks, and biomolecular signatures. The objective is to build a unified picture of the data that offers insights not obvious from examining each omics layer in isolation (Baldwin et al. 2020). PFA (Pathway-based Factor Analysis) and PDF (Probabilistic Sparse Dictionary Factorization) are two different methods that implement a fusion method. PFA combines multi-omics data by leveraging various biological pathways. It depicts omics data as route activities for coordinated changes across several molecular levels within specific biological pathways (X.-T. Yu and Zeng 2018). PSDF provides interpretable representations of omics data, aiding in the discovery of biologically relevant patterns and regulatory networks (Li et al. 2014). The multivariate method analyses and integrates multi-omics data simultaneously, helping identify the complex molecular layers and their relationship. One of the prominent tools utilizing this method is MixOmics. It combines multiple data types such as gene expression, DNA methylation, miRNA expression, and metabolomics data into a unified framework for the discovery of similar patterns, biomarkers, and regulatory networks. With dimensionality reduction approaches like partial least squares (PLS) and sparse PLS, MixOmics aids in identifying hidden variables and conserving crucial information. Its visualization tools enable interactive study of interactions between variables and the influence of omics characteristics on phenotypic outcomes, while statistical analysis capabilities help hypothesis testing and biomarker identification (Jiang et al. 2018) (Fig. 14.1).

14.3 Fruit Crop Case Studies Through Multi-omics and Diagnostics

This section presents case studies on applying multi-omics and diagnostic techniques to major fruit crops affected by viral diseases. Integrative analysis of genetic, transcriptomic, protein, and metabolite data reveals host responses, susceptibility factors, and viral pathogenic mechanisms. Advanced diagnostics using molecular markers and high-throughput sequencing enhance virus detection and monitoring. The case studies demonstrate how these insights inform breeding programs, gene editing, and antiviral strategies, emphasizing the need to consider viral variability and environmental factors in developing effective virus management tailored to specific crops. Even though the use of multi-omics in fruit crop virus management is still in its infancy and the literature regarding multi-omics is scarce for important fruit crops.

14.3.1 Multi-omics and Diagnostics in Banana-Associated Virus

In banana, BBTV infection was studied through differential transcriptomic analysis of resistant wild *Musa balbisiana* and susceptible *Musa acuminata* Lakatan banana genotypes. Both genotypes were inoculated with the BBTV infection to gain insights into the resistance mechanism of the cultivars against BBTV. It was observed that both genotypes had 62 differentially expressed genes (DEGs), which were responsible for basal defence responses against BBTV. These included factors such as pathogen recognition receptors (like cysteine-rich receptor-like kinase CRK6), calcium-dependent protein kinase (CDPK), and signalling components such as mitogen-activated protein kinase (MAPK), which are generally involved in pathogen perception and defence pathways. The lakatan variety, which was the susceptible cultivar, expressed 99 unique DEGs, which were involved in virus susceptibility mechanisms. Upregulation of genes involved in the cellular machinery of the host was observed, which facilitated viral replication and maturation during infection. Whereas the resistant wild variety *M. balbisiana* expressed 151 unique DEGs involved in resistance mechanisms such as protein kinases for pathogen recognition, hypersensitive response (HR) protein that triggers localized cell death to restrict virus spread, and differential expression of genes for cell wall modification (xylan modification, lignin biosynthesis, cuticle formation) to reinforce physical barriers against infection. Further genes involved in the biosynthesis of secondary metabolites like terpenoids, carotenoids, and flavones were differently expressed in this variety. Expression of these genes is evident that they can play a role in defence and resistance against BBTV. Candidate susceptibility genes like Adagio-like protein in the susceptible Lakatan and candidate resistance genes like glucuronoxylan 4-O-methyltransferase (associated with cell wall modification) in the resistant M. balbisiana were both significantly upregulated, according to RT-qPCR validation. The differential transcriptomic analysis showed that in wild *M. balbisiana,* BBTV resistance is a complex multi-component mechanism involving enhanced pathogen recognition, signalling, hypersensitive cell death, cell wall strengthening, and secondary metabolite production, while in *Musa acuminata,* susceptibility involves viral exploitation of host translation and protein homeostasis machinery. They provide molecular targets for banana breeding and genome editing to create BBTV-resistant cultivars (Lantican et al. 2023).

14.3.2 Multi-omics and Diagnostics in Grapevine-Associated Virus

The proteome and transcriptome was profiled in grapevine fanleaf virus (GFLV), the most devastating virus of grapevine. The study was observed in *Nicotiana benthamiana* plants infected with grapevine fan leaf virus (GFLV). The emphasis was on vein-clearing symptom development in the host plant. Liquid chromatography tandem mass spectrometry (LC-MS/MS) and 3′ RNA sequencing tools were used over a time course in plants infected with wild-type and mutant GFLV strains. Changes in the host proteome and gene ontology related to immune response, gene regulation, and secondary metabolism during peak symptom expression were demonstrated. It was observed that a single amino acid in the viral RNA-dependent RNA polymerase (RdRP) directs towards distinct host proteome and transcriptome changes that underlie transient symptoms. The experimental section describes how *N. benthamiana* spreads in controlled conditions and how to use wild-type GFLV strains (GHu, symptomatic; F13, asymptomatic) and their corresponding mutants with a single residue swap in RdRP to do this. After inoculation, the apical leaf samples were collected for RNA and protein assays at 4, 7, and 12 days after infection (dpi). This time gap was chosen to observe the changes in symptoms during the arrival peak and recovery from the infection. Library preparation and sequencing were done by Illumina NextSeq500, and RNA was purified for 3′ RNA-Seq. For proteomics nano-LC – MS/MS Q Exactive system was used to extract and analyse proteins. Bioinformatics tools were used for read alignment, quality control, gene/protein abundance matrices, and gene set enrichment/network analyses, such as weighted gene co-expression network analysis (WGCNA) (Langfelder and Horvath 2008). RT-qPCR confirmed the gene expression patterns identified through sequencing. Using ELISA, systemic infection was confirmed in all test plants. The GHu strain of wild-type GFLV induced vein clearing at 7 dpi, peaking with nearly all plants affected, then faded after two weeks. Expect this strain; no other strain showed symptoms, and significant changes in both transcriptome and proteome were observed at 7 dpi in symptomatic GHu infections. With gene ontology (GO) categories enriched for nucleotide binding, RNA binding, ribosomal components, translation, and metabolism, over 2700 host genes showed varying expression patterns. Thirty host proteins, particularly those involved in translation and secondary metabolism, were expressed differentially in symptomatic plants. The lack of correlation between viral titre and symptom manifestation suggests that host regulation, rather than viral burden, drives phenotype. During peak symptoms, WGCNA displayed distinct modules with hundreds of up- or down-regulated genes in immune regulatory clusters, translation, and ribosome synthesis in symptomatic plants. Combining transcriptomics and proteomics confirmed a small set of gene/protein candidates linked to the onset of symptoms. Thus, comparing two distinct omics methods confirmed modifications in the transcriptome and proteome, with 3′RNA-Seq offering extra candidates that were less abundant than those found using LC-MS/MS. Numerous levels of transcriptional and translational control were postulated and later discovered in each dataset produced during the arms race between GFLV and its plant host. These two methods produced reliable findings that were useful for investigating interactions between GFLV and *N. benthamiana* (Roy et al. 2023).

14.4 Difficulties and Restrictions

There are many identified challenges and limitations in multi-omics research for plant virology. Key issues include handling large, diverse datasets needing standardization and integration. Complex plant genomes, especially polyploidy and repetitive elements, complicate analysis. Proteomic and metabolomic profiles vary with tissue and environment, introducing variability. Limited reference databases, cost constraints, and lack of expertise restrict wider adoption. Sample preparation biases and computational challenges add further difficulty. Addressing these requires improved bioinformatics tools, standardized protocols, collaborative data sharing, and interdisciplinary training to unlock the full potential of multi-omics for plant virus research and disease control. In the upcoming subheadings, a brief view on the constraints present in multi-omics is discussed, so that future work on plant virology can consider this point while moving towards progress in multi-omics approaches.

14.4.1 Data Standards and Complexity

The complexity and volume of data produced by genomics, transcriptomics, proteomics, and metabolomics techniques, as well as by newer omics including epigenomics, interactomics, ionomics, and phenomics, make multi-omics studies of plant–virus interactions necessary. The distinct data formats and analytical needs brought forth by each omics layer might make integration and interpretation more difficult. For example, next-generation sequencing systems such as Oxford Nanopore, PacBio, and Illumina are used in high-throughput genomics. These platforms produce huge datasets that need to be carefully normalized, annotated, and compared. Sequence assembly and subsequent analysis are complicated by the presence of repetitive DNA, heterozygosity, and polyploidy in plant genomes. Careful quality control, adapter trimming, read alignment, and quantification procedures are necessary for transcriptomic research. Proteomics and metabolomics are even more dynamic and context-dependent, with outcomes varying widely based on cultivar, tissue, developmental stage, and environmental conditions. Extraction methods greatly influence the class of molecules detected, producing systematic biases. We must integrate data from all of these omics methods (multi-omics) in order to elucidate how genotype influences phenotype. However, it also has additional issues, such as incomplete or poorly annotated reference databases for many plant species, which are the source of bioinformatic bottlenecks, which are further exacerbated by the difficulties posed by differential expression, alternative splicing, and protein post-translational modifications (Mohr et al. 2024). The complexity and high dimensionality of multi-omics datasets make it challenging to identify biologically meaningful results. Data integration requires robust experimental designs, established protocols, and advanced informatics tools, all of which are constantly changing and not always used. The growth of multi-omics-driven plant virus research is emphasized as requiring enhanced data standards, universal ontologies, and strong, accessible methodologies. According to the article's conclusion, existing concerns like batch effects, inconsistent normalization, and the absence of universal data annotation are significant challenges that must be overcome via continued methodological research and international collaboration (Luo et al. 2024).

14.4.2 Skills Gap

The implementation of multi-omics techniques for viral management involves an interdisciplinary skill set rarely present in a single laboratory or researcher. Expertise in molecular biology, advanced analytical chemistry, high-throughput instrumentation, computational biology, and biostatistics is necessary to generate, handle, and evaluate multi-layered data. However, obvious shortages persist, especially in bioinformatics, programming, and systems biology, due to traditionally narrow training in plant virology or traditional biological disciplines. Advanced integrative approaches such as network mapping, machine learning, and multi-dimensional statistical modelling are only accessible to well-trained teams or through collaborative platforms. Training programs and seminars have begun to address these deficiencies, but the rapid advancement of omics technology and analytical tools provides a changing target for skill development. Lack of suitable curriculum in many graduate programs further widens the gap between accessible technologies and active researchers' capacity to utilize them successfully. Building sustainable viral management pipelines thus demands not just technology infrastructure but focused expenditures in cross-disciplinary education, collaborative research environments, and hands-on training opportunities integrating laboratory and computational abilities. International partnerships, mobility grants, and online knowledge-sharing networks are expected to be essential to overcoming this skills shortfall in the future (Tan et al. 2009).

14.5 Conclusion and Future Prospects

Multi-omics approaches have increased the accuracy and reliability of Omics data in plant virology. Even though each omics technology produces enormous datasets on its own, this only pertains to one level of information in a biological system, which could cause some misunderstanding when attempting to comprehend and anticipate the complete process. Therefore, multi-omics investigations are necessary to prevent misunderstandings and offer thorough information on the complete flow of genetic information from genotype to phenotype. Information on the molecular-level relationships and mechanisms behind many biological processes can be obtained through the integration of complex omics datasets. As discussed in detail in the computational tools section of this chapter, a variety of computational tools have been developed, such as Bayesian methods (TMD, MDI, PSDF, Joint BAYES, BCC, IBAG, MOFA), fusion-based methods (PFA and PSDF), multivariate techniques (MixOmics), correlation-based methods (MAPMan, CNAmet, PhenoLink,

Paintomics, and IntegreOmics), similarity-based methods (NEMO and PINEPLUS), network-based approaches (WCGNA, Steiner, OmicsIntegrator, Weighted Multiplex Network, CONEXIC, and Joint NMF), and correlation-based methods (MAPMan, CNAmet, PhenoLink, PhenoLink, and IntegreOmics) (Zhang et al. 2022). Furthermore, integration of traditional omics investigations (genomics, epigenomics, transcriptomics, proteomics, and metabolomics) with phenotypic analysis allows researchers to reconstruct the whole process from genome to trait. Phenotypes include morphological, physiological, and biochemical characteristics such as plant height, root architecture, biomass, photosynthetic and respiratory efficiency, stress and disease tolerance, the composition of metabolites, hormones, and other biochemical substances that control growth and development. Emerging fields such as "phenomics," which aims at gathering non-invasive imaging (i.e., RGB, chlorophyll fluorescence, hyperspectral and thermal imaging, and spectroscopic data) and developing full tools to integrate with other omics data, can also pave new insight in plant virus management (Awada et al. 2024). Besides all the advantages, there is still a lag in the field of Omics research in reference to plant virology, hence in the future this can be compensated by combining multi-omics approaches with AI and ML, single-cell spatial omics, localization studies, and deep experimental verifications through functional analysis (Yu et al. 2023; Rodriguez-Peña et al. 2021).

Acknowledgments The facility and support provided by the Department of Botany, AMU, are gratefully acknowledged.

References

Albiach-Marti MR, Robertson C, Gowda S, Tatineni S, Belliure B, Garnsey SM, Folimonova SY, Moreno P, Dawson WO (2009) The pathogenicity determinant of Citrus tristeza virus causing the seedling yellows syndrome maps at the 3′-terminal region of the viral genome. Mol Plant Pathol 11(1):55–67. https://doi.org/10.1111/j.1364-3703.2009.00572.x

Andret-Link P, Laporte C, Valat L, Ritzenthaler C, Demangeat G, Vigne E, Laval V, Pfeiffer P, Stussi-Garaud C, Fuchs M (2004) Grapevine Fanleaf virus: still a major threat to the grapevine industry. J Plant Pathol 86(3):183–195

Argyris JM, Pujol M, Martín-Hernández AM, Garcia-Mas J (2015) Combined use of genetic and genomics resources to understand virus resistance and fruit quality traits in melon. Physiol Plant 155(1):4–11. https://doi.org/10.1111/ppl.12323

Awada L, Phillips PWB, Bodan AM (2024) The evolution of plant phenomics: global insights, trends, and collaborations (2000–2021). Front Plant Sci 15. https://doi.org/10.3389/fpls.2024.1410738

Baldwin E, Han J, Luo W, Zhou J, An L, Liu J, Zhang HH, Li H (2020) On fusion methods for knowledge discovery from multi-omics datasets. Comput Struct Biotechnol J 18:509–517. https://doi.org/10.1016/j.csbj.2020.02.011

Blanca JM, Cañizares J, Ziarsolo P, Esteras C, Mir G, Nuez F, Garcia-Mas J, Picó MB (2011) Melon transcriptome characterization: simple sequence repeats and single nucleotide polymorphisms discovery for high throughput genotyping across the species. Plant Genome 4(2):118

Cabaleiro C, Segura A, García-Berrios JJ (1999) Effects of grapevine Leafroll-associated virus 3 on the physiology and must of Vitis vinifera L. cv. Albariño following contamination in the field. Am J Enol Vitic 50(1):40–44. https://doi.org/10.5344/ajev.1999.50.1.40

Cambiaghi A, Ferrario M, Masseroli M (2017) Analysis of metabolomic data: tools, current strategies and future challenges for omics data integration. Brief Bioinform 18(3):498–510. https://doi.org/10.1093/bib/bbw031

Cao B, Li R, Xiao S, Deng S, Zhou X, Zhou L (2022) Predicting miRNA-disease association through combining miRNA function and network topological similarities based on MINE. Iscience 25(11):105299

Chu SH, Huang M, Kelly RS, Benedetti E, Siddiqui JK, Zeleznik OA, Pereira A, Herrington D, Wheelock CE, Krumsiek J, McGeachie M, Moore SC, Kraft P, Mathé E, Lasky-Su J, and on behalf of the Consortium of Metabolomics Studies Statistics Working Group (2019) Integration of Metabolomic and other omics data in population-based study designs: an epidemiological perspective. Metabolites 9(6):117. https://doi.org/10.3390/metabo9060117

Credi R, Babini AR (1997) Effects of virus and virus-like infections on growth, yield, and fruit quality of Albana and Trebbiano Romagnolo grapevines. Am J Enol Vitic 48(1):7–12. https://doi.org/10.5344/ajev.1997.48.1.7

Dai N, Cohen S, Portnoy V, Tzuri G, Harel-Beja R, Pompan-Lotan M, Carmi N, Zhang G, Diber A, Pollock S, Karchi H, Yeselson Y, Petreikov M, Shen S, Sahar U, Hovav R, Lewinsohn E, Tadmor Y, Granot D et al (2011) Metabolism of soluble sugars in developing melon fruit: a global transcriptional view of the metabolic transition to sucrose accumulation. Plant Mol Biol 76(1):1–18. https://doi.org/10.1007/s11103-011-9757-1

Durán-Vila N, Moreno P (2000) Enfermedades de los cítricos. In Electronico. Sociedad Española de Fitopatología. https://redivia.gva.es/handle/20.500.11939/7287

Fesenko I, Spechenkova N, Mamaeva A, Makhotenko AV, Love AJ, Kalinina NO, Taliansky M (2021) Role of the methionine cycle in the temperature-sensitive responses of potato plants to potato virus Y. Mol Plant Pathol 22(1):77–91

Flores RO, Lanzer JD, Dimitrov D, Velten B, Saez-Rodriguez J (2023) Multicellular factor analysis of single-cell data for a tissue-centric understanding of disease. eLife 12:e93161. https://doi.org/10.7554/eLife.93161

Guiu-Aragonés C, Monforte AJ, Saladié M, Corrêa RX, Garcia-Mas J, Martín-Hernández AM (2014) The complex resistance to cucumber mosaic cucumovirus (CMV) in the melon accession PI161375 is governed by one gene and at least two quantitative trait loci. Mol Breed 34(2):351–362. https://doi.org/10.1007/s11032-014-0038-y

Gutha LR, Alabi OJ, Naidu RA (2012) Effects of grapevine leafroll disease on photosynthesis in a red-fruited wine grape cultivar. Proceedings of the 17th congress of ICVG 7–14 October, pp 168–169

He Y, Li L, Yao Y, Li Y, Zhang H, Fan M (2021) Transcriptome-wide N6-methyladenosine (m6A) methylation in watermelon under CGMMV infection. BMC Plant Biol 21(1):516. https://doi.org/10.1186/s12870-021-03289-8

Jiang R, Zhao C, Gao B, Xu J, Song W, Shi P (2018) Mixomics analysis of breast cancer: long non-coding RNA linc01561 acts as ceRNA involved in the progression of breast cancer. Int J Biochem Cell Biol 102:1–9. https://doi.org/10.1016/j.biocel.2018.06.003

Langfelder P, Horvath S (2008) WGCNA: an R package for weighted correlation network analysis. BMC Bioinformatics 9(1):559. https://doi.org/10.1186/1471-2105-9-559

Lantican DV, Nocum JDL, Manohar ANC, Mendoza J-VS, Gardoce RR, Lachica GC, Gueco LS, Dela Cueva FM (2023) Comparative RNA-seq analysis of resistant and susceptible banana genotypes reveals molecular mechanisms in response to banana bunchy top virus (BBTV) infection. Sci Rep 13(1):18719. https://doi.org/10.1038/s41598-023-45937-z

Lee RF (2015) Chapter five—control of virus diseases of citrus. In: Loebenstein G, Katis NI (eds) Advances in virus research, vol 91. Academic Press, pp 143–173. https://doi.org/10.1016/bs.aivir.2014.10.002

Li Y, Caron RJ, Ngom A (2014) A decomposition method for large-scale sparse coding in representation learning. Int Joint Conf Neural Netw 2014:3732–3738. https://doi.org/10.1109/IJCNN.2014.6889394

Lowe R, Shirley N, Bleackley M, Dolan S, Shafee T (2017) Transcriptomics technologies. PLoS Comput Biol 13(5):e1005457. https://doi.org/10.1371/journal.pcbi.1005457

Luo Y, Zhao C, Chen F (2024) Multiomics research: principles and challenges in integrated analysis. BioDesign Res 6:0059. https://doi.org/10.34133/bdr.0059

MacQueen A, Duan H, Dore LC, Lu Z, Liu J, Chen K, Jia G, Bergelson J, He C (2014) Unique features of the m6A methylome in Arabidopsis thaliana. Nat Commun 5(1):5630. https://doi.org/10.1038/ncomms6630

Malik AA, Vashisht VK, Singh K, Sharma A, Singh DK, Singh H, Monforte AJ, McCreight JD, Dhillon NPS (2014) Diversity among melon (Cucumis melo L.) land races from the Indo-Gangetic plains of India and their genetic relationship with USA melon cultivars. Genet Resour Crop Evol 61(6):1189–1208. https://doi.org/10.1007/s10722-014-0101-x

Mannini F, Rolle L, Guidoni S (2002) Vineyard management to optimize grape quality in virus-free clones of Vitis vinifera L. VIII Int Conf Grape Genet Breed 603:121–126

Mannini F, Argamante IN, Cuozzo D, Credi R (2006) Modification in field behavior and grape quality, with focus on terpenes, after GLRaV-3 eradication in a clone of white Muscat (Vitis vinifera L.). Extended abstracts 15th meeting ICVG, Stellenbosch, South Africa, pp 3–7

Mansoor S, Hamid S, Tuan TT, Park J-E, Chung YS (2024) Advance computational tools for multiomics data learning. Biotechnol Adv 77:108447. https://doi.org/10.1016/j.biotechadv.2024.108447

Martelli GP (2015) Grapevine virology: a historical account with an eye to the studies of the last 60 years or so. Proceedings of the 18th congress of ICVG, Ankara, Turkey, 7–11 September 2015

Meiri M, Bar-Oz G (2024) Unraveling the diversity and cultural heritage of fruit crops through paleogenomics. Trends Genet 40(5):398–409. https://doi.org/10.1016/j.tig.2024.02.003

Mohr AE, Ortega-Santos CP, Whisner CM, Klein-Seetharaman J, Jasbi P (2024) Navigating challenges and opportunities in multi-omics integration for personalized healthcare. Biomedicine 12(7):1496. https://doi.org/10.3390/biomedicines12071496

Montero R, Pérez-Bueno ML, Barón M, Florez-Sarasa I, Tohge T, Fernie AR, Ouad HEA, Flexas J, Bota J (2016) Alterations in primary and secondary metabolism in Vitis vinifera 'Malvasía de Banyalbufar'upon infection with Grapevine leafroll-associated virus 3. Physiol Plant 157(4):442–452

Moreno P, Ambrós S, Albiach-Martí MR, Guerri J, Peña L (2008) Citrus tristeza virus: a pathogen that changed the course of the citrus industry. Mol Plant Pathol 9(2):251–268. https://doi.org/10.1111/j.1364-3703.2007.00455.x

Nelson SC (2004) Banana bunchy top: detailed signs and symptom. Cooperative Extension Service, College of Tropical Agriculture and Human Resources, University of Hawaii at Manoa, pp 22

Pang Z, Lu Y, Zhou G, Hui F, Xu L, Viau C, Spigelman AF, MacDonald PE, Wishart DS, Li S, Xia J (2024) MetaboAnalyst 6.0: towards a unified platform for metabolomics data processing, analysis and interpretation. Nucleic Acids Res 52(W1):W398–W406. https://doi.org/10.1093/nar/gkae253

Pech JC, Bouzayen M, Latché A (2008) Climacteric fruit ripening: ethylene-dependent and independent regulation of ripening pathways in melon fruit. Plant Sci 175(1):114–120. https://doi.org/10.1016/j.plantsci.2008.01.003

Qazi J (2016) Banana bunchy top virus and the bunchy top disease. J Gen Plant Pathol 82(1):2–11. https://doi.org/10.1007/s10327-015-0642-7

Rao GP, Reddy MG (2020) Overview of yield losses due to plant viruses. In: Applied plant virology. Elsevier, pp 531–562. https://doi.org/10.1016/B978-0-12-818654-1.00038-4

Rodriguez-Peña R, Mounadi KE, Garcia-Ruiz H (2021) Changes in subcellular localization of host proteins induced by plant viruses. Viruses 13(4):677

Rodríguez-Verástegui LL, Ramírez-Zavaleta CY, Capilla-Hernández MF, Gregorio-Jorge J (2022) Viruses infecting trees and herbs that produce edible fleshy fruits with a prominent value in the global market: an evolutionary perspective. Plants 11(2):203. https://doi.org/10.3390/plants11020203

Roy BG, DeBlasio S, Yang Y, Thannhauser T, Heck M, Fuchs M (2023) Profiling plant proteome and transcriptome changes during grapevine Fanleaf virus infection. J Proteome Res 22(6):1997–2017. https://doi.org/10.1021/acs.jproteome.3c00069

Sade D, Shriki O, Cuadros-Inostroza A, Tohge T, Semel Y, Haviv Y, Willmitzer L, Fernie AR, Czosnek H, Brotman Y (2015) Comparative metabolomics and transcriptomics of plant response to Tomato yellow leaf curl virus infection in resistant and susceptible tomato cultivars. Metabolomics 11(1):81–97. https://doi.org/10.1007/s11306-014-0670-x

Salgotra RK, Chauhan BS (2023) Genetic diversity, conservation, and utilization of plant genetic resources. Genes 14(1):174. https://doi.org/10.3390/genes14010174

Santos MT, Lurdes M, Rocha MLG, Martins JMS, Carneiro LC (2006) Morphological changes of grapevine leaves (cv. Arinto) in relation to infections by viruses and weather conditions. Extended abstracts 15th ICVG meeting, pp 248–249

Serra-Soriano M, Navarro JA, Genoves A, Pallás V (2015) Comparative proteomic analysis of melon phloem exudates in response to viral infection. J Proteome 124:11–24. https://doi.org/10.1016/j.jprot.2015.04.008

Stainton D, Kraberger S, Walters M, Wiltshire EJ, Rosario K, Halafihi M, Lolohea S, Katoa I, Faitua TH, Aholelei W, Taufa L, Thomas JE, Collings DA, Martin DP, Varsani A (2012) Evidence of inter-component recombination, intra-component recombination and reassortment in banana bunchy top virus. J Gen Virol 93(5):1103–1119. https://doi.org/10.1099/vir.0.040337-0

Takagi H, Abe A, Yoshida K, Kosugi S, Natsume S, Mitsuoka C, Uemura A, Utsushi H, Tamiru M, Takuno S (2013) QTL-seq: rapid mapping of quantitative trait loci in rice by whole genome resequencing of DNA from two bulked populations. Plant J 74(1):174–183

Tan TW, Lim SJ, Khan AM, Ranganathan S (2009) A proposed minimum skill set for university graduates to meet the informatics needs and challenges of the "-omics" era. BMC Genomics 10(3):S36. https://doi.org/10.1186/1471-2164-10-S3-S36

Tatineni S, Hein GL (2023) Plant viruses of agricultural importance: current and future perspectives of virus disease management strategies. Phytopathology 113(2):117–141. https://doi.org/10.1094/PHYTO-05-22-0167-RVW

Timmer LW, Roberts PD, Darhower HM, Bushong PM, Stover EW, Peever TL, Ibáñez AM (2000) Epidemiology and control of citrus greasy spot in different citrus-growing areas in Florida. Plant Dis 84(12):1294–1298. https://doi.org/10.1094/PDIS.2000.84.12.1294

Vishnoi R, Raj SK, Prasad V (2009) Molecular characterization of an Indian isolate of Banana bunchy top virus based on six genomic DNA components. Virus Genes 38(2):334–344. https://doi.org/10.1007/s11262-009-0331-8

Wang Z, Gerstein M, Snyder M (2009) RNA-Seq: a revolutionary tool for transcriptomics. Nat Rev Genet 10(1):57–63. https://doi.org/10.1038/nrg2484

Webber HJ (1967) History and development of the citrus industry. In: The citrus industry. University of California, pp 1–37

Yang Y, Saand MA, Huang L, Abdelaal WB, Zhang J, Wu Y, Li J, Sirohi MH, Wang F (2021) Applications of multi-omics technologies for crop improvement. Front Plant Sci 12. https://doi.org/10.3389/fpls.2021.563953

Yu X-T, Zeng T (2018) Integrative analysis of omics big data. In: Huang T (ed) Computational systems biology: methods and protocols. Springer, pp 109–135. https://doi.org/10.1007/978-1-4939-7717-8_7

Yu X, Liu Z, Sun X (2023) Single-cell and spatial multi-omics in the plant sciences: technical advances, applications, and perspectives. Plant Commun 4(3):100508

Zhang X, Zhou Z, Xu H, Liu C-T (2022) Integrative clustering methods for multi-omics data. WIREs Comput Stat 14(3):e1553. https://doi.org/10.1002/wics.1553

Zhou L, Tian S, Qin G (2019) RNA methylomes reveal the m6A-mediated regulation of DNA demethylase gene SlDML2 in tomato fruit ripening. Genome Biol 20(1):156. https://doi.org/10.1186/s13059-019-1771-7

15 Understanding Epigenetic Mechanisms in Fruit Virus Resistance

S. Kruthika, Laxmipriya Behera, Saswat Swarup Padhi, Bandana Mohapatra, Rupak Jena, Awadhesh Kumar, and Milan Kumar Lal

Abstract

Pandemics of viral diseases and significant epidemics pose a hazard to plants that provide food for people and animals, in addition to fiber or medicinal purposes. Among the food crops, fruit crops are the most susceptible to viruses that reduce yield and income for farmers. Epigenetic changes are required to transfer the stress-recovery memory to the next generations, as traditional breeding methods have failed to provide long-term remedies for the quickly spreading viral diseases. With respect to plant-pathogen interactions under unfavorable environmental conditions, epigenetic alterations are emerging as increasingly important regulatory mechanisms. Resistance to several diseases is conferred at the genetic level, preventing yield loss indefinitely. Transgenic and non-transgenic methods for creating virus-resistant crop plants have recently been made possible by RNA-based technologies, including CRISPR/Cas9, microRNA, and dsRNA- and siRNA-based RNA interference. Through epigenetic memory, the virus-infected plants can remember the previous molecular processes that enabled them to resist the virus, assisting in the achievement of sustained output, without altering the genome of the individual. When the condition recurs, they can use this knowledge to adjust and save lives. Through an explanation of several molecular breeding techniques and case studies, this chapter explores the different epigenetic mechanisms that provide resistance to fruit crops against viral diseases. Availability and high cost of the developing techniques also cause major problem for adapting these reliable and permanent solutions for improving resistance in fruit crops. A wide variety of fruit types and cultivars that may not be popular but are nonetheless nutritionally significant require more research.

S. Kruthika · L. Behera · S. S. Padhi · B. Mohapatra · R. Jena · A. Kumar · M. K. Lal (✉)
ICAR-Central Rice Research Institute, Cuttack, Odisha, India

R. Kumar et al. (eds.), *Fruit Crops and Viral Pathogens Volume II*, Plant Pathology for Sustainable Agriculture, https://doi.org/10.1007/978-981-95-7261-8_15

Integration of developing molecular breeding methods and epigenetic tools can offer virus-free fruit crops with sustainable yield and income.

Keywords

Epigenetic regulation · Fruit crops · Gene expression · Virus resistance

15.1 Introduction

The demand for fruits is increasing significantly in the exports of major tropical fruit crops, which increased by an expected 12% over prior years to a total value of approximately USD 11 billion in 2023 (FAO 2024). Fruit crops make up 17% of the total crop production in the year 2021, making up a sizeable amount of the world's agricultural output. Together, the nations of Asia and the Americas produce 76.2% of the world's tropical fruit crops (FAO 2024). Viruses infecting tropical fruits can cause yield loss of more than USD 30 billion annually, causing huge economic loss to the major fruit-producing and exporting countries (Vieira et al. 2024). The continuous population growth, particularly in poorer nations, has made tropical fruits vital to global nutritional security. These fruits are essential because they make a significant contribution to year-round food supply. Fruit trees face numerous obstacles, including viral infections, just as other crops. Because of their perennial nature and method of multiplication, fruit trees frequently act as volunteer hosts for viruses and diseases in natural conditions. Although these viruses are virulent in perennial crops, they are more severe in annual crops (Jones and Naidu 2019). Tropical tree crops are susceptible to infection during prenursery, nursery, and grafting, which frequently allows viruses to spread among trees (Maliogka et al. 2018). Viral infections are latent and exist in these immature seedlings without the grower's knowledge. In the end, a high viral incidence has been linked to a variety of yield losses in tree crops.

The following are notable examples of viruses in fruit, citrus tristeza virus in citrus (*Citrus* spp.), apple chlorotic leaf spot virus in apple (*Malus domestica*), grapevine leafroll-associated virus in grapevine (*Vitis vinifera*), and banana ringspot virus in banana (*Musa* spp.). Insect vectors or contaminated planting materials typically spread viral infections in fruit crops. For example, millions of citrus trees have already been damaged globally by CTV (Folimonova and Sun 2022). However, BBTV continues to be an important hurdle to banana production, particularly in tropical and subtropical areas. In many cases, viral infections exist in immature seedlings without the knowledge of the grower. In the end, a high viral infection occurs, which causes a variety of yield losses in tree crops (Hull 2018). Mechanisms that undergo epigenetic modifications are crucial in regulating plant-pathogen interactions during unfavorable environmental conditions. Epigenetic mechanisms enable rapid phenotypic responses to both biotic and abiotic stressors without the requirement for genetic modifications and contribute a huge role in determining how plants adapt to environmental hurdles. These changes can be reversible or permanent across generations, depending on environmental conditions, providing plants with a flexible, reliable, and robust defense system (Singh et al. 2023). There are several epigenetic mechanisms, out of which the major mechanisms that affect the response of plants to various environmental signals, including viral infections, are DNA methylation, chromatin remodeling, histone alterations, and small RNA regulation (Zhang et al. 2018).

Recently, studies have been conducted more into the area of epigenetics and its mechanisms to develop plant defense against pathogens (Sampson et al. 2024). It may help in activating or deactivating defense-promoting genes in response to the pathogen. With the growth of these techniques, it is now possible to develop a disease-resistant trait using CRISPR/Cas9, which is one of the epigenetic techniques of editing DNA sequence without altering the genome (Harris et al. 2023). With the help of these techniques, we can target only the particular gene that is responsible for susceptibility under stress conditions and can be altered easily. It is necessary to understand the epigenetic mechanisms to apply the adoptable one based on the symptoms to manage the infections. This chapter deeply explains various epigenetic mechanisms that can be used to promote resistance against viral diseases in fruit crops, followed by case studies for better understanding.

15.2 Overview of Plant-Virus Interactions

15.2.1 Common Viruses Affecting Fruit Crops

Among the horticultural crops, fruit crops are more susceptible to a broad range of viruses that affect growth, yield, and fruit taste, texture, and quality. Some of the viruses affecting major fruit crops include BBTV in banana, PRSV in papaya, CTV in citrus, GLRaV in grapevine, and ACLSV in apple and woody fruit trees (Umer et al. 2019). Some of the other fruits that are often affected by viruses are plum, sweet orange, kiwi, pear, strawberry, and cherry. These viruses belong to a diverse range of families such as *Potyviridae*, *Closteroviridae*, *Geminiviridae*, and *Betaflexiviridae*, and they exhibit different and unique genome structures and various modes of replication. Their wide host ranges and adaptability make them difficult to control and manage, particularly in mixed cropping systems and under adverse climatic conditions that influence vector dynamics (Xavier and Whitfield 2023).

15.2.2 Symptoms and Transmission of Fruit-Infesting Viruses

It has been discovered that fruit trees, namely citrus, banana, pear, grapevine, pomegranate, and stone fruits like mango, are infected by 163 viruses, which are categorized into 45 genera and 23 families. The high temperatures and tropical climate in which mangos are grown may strengthen the host's antiviral defenses and prevent virus multiplication, which may account for the limited viruses that infect them

(Szittya et al. 2003). A variety of virus species are affecting the major woody fruit trees, some of which reduce fruit production and quality, crop vigor, and the trees' resistance to frost and drought, resulting in substantial economic losses. These viruses can be able to cause observable disease symptoms in most parts of susceptible trees, including the leaves and fruits. In young leaves, the symptoms include wrinkling, folding, vein enation, crinkling, chlorosis, mottling, streaks, necrosis, and yellow-colored veins of the leaves and buds; and yellowish green-colored flecks, rings, spots, or dull translucent areas on the leaves (Jones 2021). Along with significant changes in the size, color, and flavor, the symptoms of fruit disease include the emergence of rings, spots, and uneven lines on the surface of fruits. Stem pitting, stem grooving, scaly bark, lesions on bark scales, gummosis, shorter internodes, bark splitting, and woody galls are further symptoms visible on the trunk and branches. Dwarfing or stunting, early defoliation, multiple flushing, graft incompatibility, a decline in germination and growth during the vegetative stage, fruit yield, and quality, and sudden mortality are the last systematic indications. Grafting with buds, scions, and rootstocks, and vegetative propagation in orchards are the main ways that viruses that infect woody fruit trees spread. Aphids, thrips, mealybugs, soft scales, or mites are vectors for some of them, and pollen grains and seeds can also spread them (Umer et al. 2019). For instance, BBTV causes severe stunting and a bunchy leaf look in bananas, while PRSV generates mosaic and water-soaked, greasy streaks on papaya leaves and fruits (Tripathi et al. 2008).

15.2.3 Physiology of Virus-Infected Plants

Viruses can cause irreversible impacts on the host physiology. When compared to healthy hosts, viruses change the photosynthetic ability of infected hosts. Additionally, they reduce the amount of chlorophyll in diseased leaves and prevent the growth of chloroplasts, which reduces photosynthesis. Viruses disrupt the ability of the host to metabolize carbohydrates. Virus-infected fruits are often small, less sweet, and misshapen, which ultimately paved the way for huge economic losses in fruit production (Rodríguez-Verástegui et al. 2022). They also have an impact on the host's breathing. When compared to healthy plants, the rate of respiration of the infected plants rises in the majority of virus-host pairings before rapidly declining. This may cause variation in the ripening of fruits, which may impair their shelf life and storage. The impact of viruses on host-protein metabolism is difficult to generalize. The transpiration rates and leaf water content of virus-affected plants are generally lower than those of healthy plants grown in controlled conditions. Viruses have a negative impact on the biosynthesis and metabolism of plant growth hormones, and virus-infected plants have markedly reduced auxin levels, which affect growth and development (Ertunç 2020).

15.3 Mechanisms of Epigenetic Regulation

Recent studies have focused on several aspects of plant epigenetic regulation. One of most consistent epigenetic techniques are DNA methylation, RNA-mediated silencing, histone modifications, chromatin remodeling, and gene silencing techniques, which support plants in a variety of stressful situations, including cold, drought, heat stress, nutrient deficiency, heavy metals, and pathogenic infections (Fig. 15.1). Since epigenetic mechanisms can directly suppress viral genetic material and change host defense pathways, they are essential for resisting viral infection in fruit crops (Rehman et al. 2023).

15.3.1 DNA Methylation

DNA methylation controls a variety of biological functions, including chromatin architecture, expression of genes, genome stability, and genomic imprinting, which is highly conserved. Changes in DNA methylation sequences are crucial for plants to defend against biotic and abiotic stressors, and alterations in them frequently cause serious developmental defects. Normally, there are two major methods through which DNA methylation confers antiviral defense, which are RNA-directed DNA methylation (RdDM) and cytosine methylation. Zhang et al. (2006) reported that both cytosine (5 mC) and adenine (6 mA) residues undergo methylation in the CG, CHH, and CHG contexts (where H denotes A, T, or C). This change, which ensures transcriptional silencing through transcriptional gene silencing (TGS), is particularly abundant in transposable elements (TEs). When TGS is developed at gene regulatory regions, DNA methylation greatly enhances it, and methylation of repetitive sequences and intronic TEs has been connected to altering mRNA

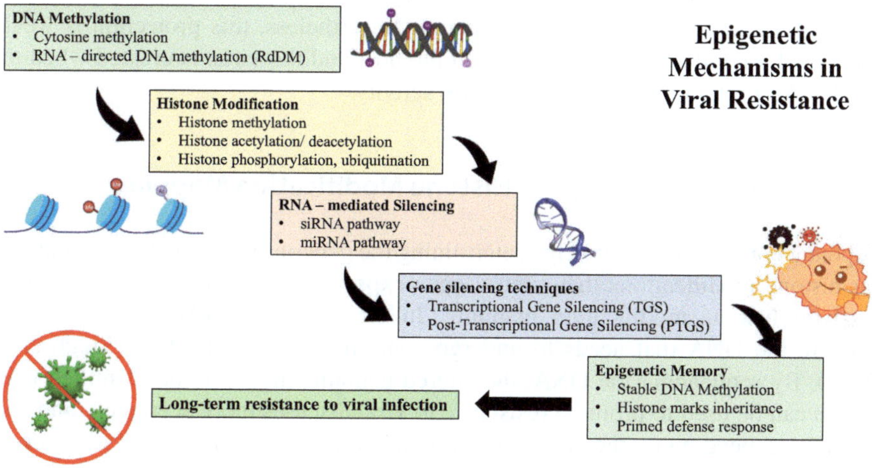

Fig. 15.1 The epigenetic mechanisms in viral resistance

splicing (Zhang et al. 2018). Thus, DNA methylation helps in silencing the viral genome and promotes resistance to viral diseases in fruits.

15.3.2 Bisulfite Sequencing for DNA Methylation Profiling

Bisulfite sequencing uses sodium bisulfite to alter the DNA sequence, which was known as a breakthrough in DNA methylation-related studies. It has numerous advantages, including being both qualitatively and quantitatively efficient in detecting 5-methylcytosine present at single-base pair precision (Frommer et al. 1992). It is the base knowledge for many other techniques that are recently being developed to better understand DNA methylation. This technique helps in identifying the differentially expressed genes responsive to viral infection, which can be downregulated using the suitable genome editing tools (Sun et al. 2019).

15.3.3 Histone Modifications

Post-transcriptional modifications (PTMs) of histone proteins include acetylation, phosphorylation, SUMOylation, ubiquitination, and methylation. Highly basic histone tails encircle the core of the nucleosome and make protein-protein and DNA-protein interactions accessible. Several enzymes known as writers and erasers catalyze the addition and deletion of histone marks, respectively (Wang et al. 2019). The proteins are known as readers because they can identify these changes or marks. The PTMs of the histone tails alter the chromatin structure and make it easier for the regulatory proteins to bind. During a viral infection, these histone modifications can shift, either boosting or suppressing the plant's defense genes as well as the virus's own gene activity. Castillo-González et al. (2015) reported that during the control of the Geminivirus infection, the plant protein SUVH4/KYP attaches to the virus's chromatin and encourages its methylation. Nevertheless, this protection is weakened, and a regulatory conflict between the host and the virus is created when the viral TrAP protein inhibits SUVH4/KYP activity.

15.3.4 ChIP Sequencing for Histone Modification Mapping

A potent epigenomic method for determining the genome-wide locations of particular histone modifications that control gene expression is ChIP sequencing (ChIP-seq) for histone modification mapping. In this method, antibodies are used to procure the DNA that needs to undergo some histone modification (Singh et al. 2024). By sequencing the DNA, the altered genomic areas can be identified later, which can be used to promote resistance to viral diseases more accurately and specifically, depending on the pathogen. This can be widely used by the researchers to identify the responsible genes and regulatory proteins responsible for conferring protection against viral infection by mapping the histone modifications like

H3K4me3 (Valensisi et al. 2015). These dynamic epigenetic alterations in response to infection are made visible and easier to understand with the help of ChIP-sequencing.

15.4 Role of RNA Interference (RNAi) and Small RNAs

Small interfering RNAs mimic and copy the complete sequences of genomes of RNA, DNA viriods, viruses, and viral satellites, and are produced using the RNA interference (RNAi) technique for conferring defense mechanisms against viruses. Thus, a detailed and well-structured virome reconstruction in a mixed infection is made possible by bioinformatics analysis and thorough sequencing of the short RNA population called small RNA-ome, in addition to the universal detection of viruses and genome remodeling (Singh et al. 2019). Like other pathogens, viral infections can also interfere with the gene silencing and RNAi mechanisms that control endogenous gene expression, suppress transposons, and endogenous viral elements integrated into the host genome, which may be liberated and cause disease (Pooggin 2018).

15.4.1 Non-coding RNAs: microRNA (miRNAs) and Small Interfering RNA (siRNAs)

Some of the noncoding RNAs (ncRNAs) that control the expression of genes at the transcriptional, post-transcriptional, and translational levels are circular RNAs (circRNAs), long noncoding RNAs (lncRNAs), microRNAs (miRNAs), and small interfering RNAs (siRNAs). ncRNAs interact with proteins, RNA, and DNA to alter their target genes (Roy et al. 2023). Evidence is increasing that ncRNAs, along with the genes coding proteins, have a significant role in determining how plants, especially fruit trees, grow and develop, as well as respond to biotic stresses like pathogenic infections and various abiotic stresses. For instance, Zhou et al. (2022) reported that the microRNAs and lncRNAs respond to low temperatures and also contribute to senescence and weakening of the plant, which could influence vulnerability to viral infections in pear.

15.5 Genome Editing Using CRISPR/Cas9 in Antiviral Defense

Clustered Regularly Interspaced Short Palindromic Repeats-CRISPR-associated protein 9 (CRISPR/Cas9) is basically a defense mechanism adapted from a genome editing mechanism used by bacteria naturally. Recently, the genomes of fungi, bacteria, yeasts, viruses, and numerous other pathogenic organisms have also using this technique to achieve successful selective mutagenesis naturally (Jiang and Doudna 2017). The primary goal of the CRISPR/Cas9 technique was to cleave DNA at various locations in vitro. The CRISPR/Cas9 technique helps in the engineering and generation of new varieties with various desired traits, including virus-resistant crop varieties. There are no barriers related to intellectual property, and the construction of CRISPR/Cas9 structures is reasonably easy and inexpensive. Tyagi et al. (2021) also reported that a single specific gene can be able to alter to get long-lasting resistance in plants using the Cas9 tool.

It uses three steps, namely, acquisition, expression, and interference, to specifically target the genetic material of the disease-causing pathogen. In the initial phase, known as acquisition, the viruses or plasmids' invasive foreign DNA spacer is broken up into little pieces, which are identified, and integrated into the locus of CRISPR (Jiang and Doudna 2017). Short RNA is produced by copying CRISPR loci, which then employs simple complementarity to guide the effector endonuclease genes to specifically target the components of the virus. A small segment of nucleotides retained that yields the DNA fragment used for identification is typically present in the protospacer adjacent motif. Using the Cas9 protein and the tracrRNA molecule, CRISPR/Cas9 expression involves efficiently converting the big pre-CRISPR RNA (pre-crRNA), which is extracted from the locus of CRISPR, into many crRNAs. Pre-crRNA processing in crRNA is facilitated by the base complementarity of tracrRNA with the repeat region of crRNA (Tyagi et al. 2021). The active crRNAs assist in locating and base-pairing to a specific target region of a foreign DNA by joining the CRISPR-associated antiviral protection complex (CASCADE). The crRNA offers protection from pathogen infections by directing and guiding the Cas9 protein towards the DNA of foreign target locations for degradation during the interference step. Gene disruption is the most popularly used technique in the CRISPR/Cas9 gene editing mechanisms, which helps in recovering the cellular Non-Homologous End Joining (NHEJ) error-prone behavior. When nucleotides are added or removed (InDel) at sgRNA-targeted sites, a frameshift mutation is created, affecting gene function. This technique is used to manufacture resistance to viruses by interfering with the function of the susceptible (S) gene or genes, which modifies the interaction between the plant and virus and lowers the fitness of the virus in the host plant system. An alternative strategy is to insert InDels in a gene's promoter region as opposed to its coding area (Fadul et al. 2023).

CRISPR-mediated promoter disruption renders the effector-binding region and the whole gene production of the vulnerable plants by controlling a virus effector from attaching to the promoter. Furthermore, using the gene clusters, it is also possible to promote virus resistance in plants using this specific method. Barman et al. (2020) reported that in order to acquire a permanent, long-lasting resistance to viral diseases, a deletion in the chromosomal region next to S gene has to be done in host plants. The application of suitable molecular methods within the Cas9 mechanisms is the primary determinant of long-lasting resistance. Additionally, the introduction of fake miRNAs or short RNA genes by CRISPR is being investigated. These constructions can target host susceptibility factors or critical virus genes. This method inherits the resistance to next generation and it is a transgene-free approach to viral resistance, which is crucial for perennial fruit crops that require everlasting, long-term protection (Borrelli et al. 2018).

15.6 Gene Silencing Techniques

One of the biggest challenges is improving crop cultivars to attain greater productivity while protecting them from pests and diseases. The primary cause of decline in fruit production is the absence of disease-resistant fruit varieties (Abas et al. 2024). The gene expression was controlled by a negative feedback mechanism to determine the fate of a cell and controls its cellular metabolism and expression over the lifetime of an organism, which is known as gene silencing. There are two silencing methods used to silence genes that are harmful to the plants.

(i) Transcriptional gene silencing (TGS)
(ii) Post-transcriptional gene silencing (PTGS)

15.6.1 Transcriptional Gene Silencing (TGS)

Chromosome remodeling and DNA methylation are important factors in TGS, which slow down the gene production to avoid consequences. In silenced transgenes, TGS shows elevated methylation in the coding areas and promoters (Vaucheret and Fagard 2001). The proteins that identify methylated cytosine by binding are facilitated by DNA methylation, which prevents the binding of transcription factors and chromatin remodeling. Histone acetylation alters the charge composition of chromatin and is one factor that may contribute to the structural instability of chromatin (Tessarz and Kouzarides 2014). Alterations in the chemicals of histone tails may send signals to the complexes of chromatin remodeling, which regulate the accessibility of the cell's transcriptional machinery to the DNA. In plants, there is a clear correlation between DNA and histone methylation, which contributes a major part in the silencing of transcriptional genes (Lippman and Martienssen 2004). This gene silence is most likely caused by antisense RNA, which was synthesized by an RNA-dependent RNA polymerase that is encoded in the host. This approach acts as a genetically inbuilt immune system that eliminates viruses naturally.

15.6.2 Post-transcriptional Gene Silencing (PTGS)

When a viral infection occurs in plants, some of the plant's genes automatically shut down to protect the plant from infection at the post-transcriptional level, a mechanism called virus-induced gene silencing (VIGS). VIGS produces a recombinant viral vector by introducing target gene segments that are capable of recognizing a function into the relevant viral genome (Zulfiqar et al. 2023). The recombinant viral vector is then used to infect the plant and inhibit the plant's endogenous gene, causing corresponding changes in the plant's phenotypic or physiological traits. By identifying and digesting the multiplication of double-stranded RNA replication in a virus, VIGS helps plants fend off viral infection (Purkayastha and Dasgupta 2009).

It helps in RNA silencing, or sequence-specific RNA breakdown. When host gene fragments are incorporated into an engineered virus, they also direct the proper host mRNA for PTGS. The viral RNA includes the breakdown of host-derived sequences (Kuo and Falk 2020). Because VIGS facilitated the targeted downregulation of particular host genes, it becomes a promising tool for functional genomics.

15.7 Epigenetic Memory in Plants

Biotic stress in plants is mostly caused by pathogens, which include fungi, bacteria, viruses, nematodes, and herbivores. These affected plants have developed an advanced immune system and are protected by physical barriers like cuticles and waxes when they are attacked by a disease. Because of their sessile nature, plants are always exposed to harmful stress circumstances, yet it has been proven that plants that have already experienced stress may react more quickly and have a higher chance of surviving than those that have not. This is a metabolic state known as priming, which may be thought of as taking place when plants are subjected to milder stressors and trigger plants' stress responses, allowing them to react more quickly and powerfully when the stress recurs (Turgut-Kara et al. 2020). Priming can be done in plants by epigenetic modifications to promote and increase resistance against biotic and abiotic stresses, which are not only activated during stress conditions but also pass on to the next generation (Hilker et al. 2016). For example, during viral infection, the affected plants can remember the molecular mechanisms that helped them to overcome the infection in the past, and implement the same mechanism when the infection recurs (Ramakrishnan et al. 2022).

15.8 Case Studies: Epigenetic Mechanisms in Specific Fruit Crops

As discussed earlier, woody fruit trees are more susceptible to virus attack, causing huge economic losses to the farmers. Using modern molecular technologies, the susceptible varieties can be made resistant. Some of the studies mentioning epigenetic mechanisms to resist viral infections in fruit crops are discussed in this section.

The banana bunchy top virus (BBTV), transmitted by aphids, is a six-component ssDNA virus that is predominant in Southeast Asia and primarily affects monocots, particularly the *Musaceae* family. Elayabalan et al. (2017) demonstrated that the hill bananas infected by BBTV recover when the BBTV exhibits an expression of the BBTV ihp rep gene temporarily. They concluded that to target DNA viruses effectively, RNAi can be designed. As this mechanism delivers temporary resistance, it can also be used to control other viral diseases of the banana. Globally, papaya production is largely affected by papaya ring spot virus (PRSV). It is now essential to develop resistant cultivars using epigenetic mechanisms, since the traditional breeding methods and strategies are not promising. Using gene silencing methods like RNAi, it can target the insect vector to restrict the spreading of infection among other trees. The new gene editing tools, like CRISPR/Cas9, provide a non-transgenic way of preventing PRSV (Jyotika et al. 2024). Jia et al. (2017) demonstrated that using the RNAi mechanism during virus replication to target the PRSV CP gene, a conserved region, can develop a broad range of resistance of isolates of PRSV collected from Hainan.

The citrus tristeza virus (CTV) seriously affects the economic standing worldwide and causes tristeza infections in a number of Rutaceae species. Because sour orange (SO) rootstock is so popular with farmers around the world, citrus varieties grafted on it suffer from degradation and death. Grafting done on rough lemon with SO plants infected with CTV showed less stunting and a slight reduction in CTV accumulation when the hypersensitive responsive genes NPR3/NPR4 associated with them were silenced. Furthermore, the CTV suppressors RNA silencing p20 and p23 inhibit the defense promoted by salicylic acid in the most virulent isolates. Long-distance and cell-to-cell CTV migration are enhanced when the genes linked to the SA and RNA silencing defense processes in SO are silenced, which ultimately flushes it away (Gomez-Munoz et al. 2017). One of the most destructive and common viral diseases affecting grapevines (*Vitis spp.*) is grapevine fanleaf disease, which is primarily caused by the grapevine fanleaf virus (GFLV). There are two major ways that GFLV can spread, one is through the vector *Xiphinema index* nematode, and another is through the propagating materials. The management and control of this vector is ineffective due to the side effects of the pesticides. The silencing of transgenes transformed in grapevines with GFLV resistance genes was demonstrated by Gambino et al. (2010) using cytosine methylation, siRNA synthesis, and variable transgene expression. Grapevine red blotch virus (GRBV), which belongs to the member of the genus *Grablovirus*, is the causative agent of red blotch disease, which affects vine growth, health, fruit quality, and yield ultimately causing a huge financial loss. The mechanism and effects of GRBV can be better understood by recent studies, however very little information is available about the molecular mechanisms governing host-virus interaction. Ault et al. (2025) studied the dynamics of sRNAs production in the leaves and berries of grapevines (*Vitis vinifera cv. Merlot*) which is infected by GRBV and they found differentially expressed viral small-interfering RNA (vsiRNAs) and miRNAs. These findings created a way to explore host-pathogen interactions and innovative defense pathways. Watermelon leaves were inoculated with cucumber green mottle mosaic virus (CGMMV) at different time intervals like 0 h, 48 h, and 25 days, and found that bZIP, HLH, MYB, WRKY, and NAC are some of the responsive transcription factors expressed during CGMMV-induced stress. It was confirmed by validating 15 differentially expressed genes with qRT-PCR analysis. The results gave information about the genes and mechanistic pathways that are responsive to CGMMV infection of watermelon leaves, which can be altered or silenced later to induce resistance in plants (Sun et al. 2019). One of the most dangerous viral diseases that affects tomato production is tomato yellow leaf curl virus (TYLCV), which is transmitted by a vector, whitefly. It causes huge yield loss to the farmers. Wang et al. (2018) studied the virus-infected tomato using lncRNAs technique, and found a sequence which is triggered by TYLCV. This epigenetic mechanism plays a major role in promoting resistance to the TYLCV by altering replication and transcription of the infecting virus. The vsiRNA of TYLCV is targeted by lincRNA (SlLNR1) to prevent the viral infection and enhance resistance to viruses in tomato, which helps in mediating the transcription processes (Yang et al. 2019). Ghorbani Faal et al. (2020) reported that tomato plants affected with TYLCV were engineered with CRISPR/Cas9, specifically targeting the regions that are encoded with the coat protein and intergenic region (IR). This technique is useful for developing resistance in tomato plants against TYLCV by cloning Cas9 with a rgsCaM promoter, which is virus-inducible to counteract the defects of Cas9. The epigenetic mechanisms adopted for viral disease resistance of major fruit crops are listed in Table 15.1.

15.9 Challenges and Future Perspectives

Advanced technologies and strict plant quarantine laws are essential to stop viral infections and disease transmission at their source level. The newly emerging techniques help in the identification of newly emerged symptoms and infection, enabling required quarantine decisions and prompt action. In addition, these developments will help molecular breeding initiatives that lay the foundations to create better horticultural crop varieties. The emerging techniques are helping researchers to better understand whether these modifications strengthen plant resistance or improve viral counter defense, which is still unclear (Gallusci et al. 2023). Epigenetic modifications, including histone modification, DNA methylation, and gene silencing performed in virus-infected fruit crops, can be used as molecular markers for screening, selection, and evaluation processes. The efficiency of the epigenetic changes on fruit crops can be predicted using epigenetic modelling, which provides information on breeding resistant crops by engineering with a suitable mechanism. The diagnosis and treatment of plant diseases can be enhanced by integrating biotechnology and an improved understanding of epigenetic mechanisms, along with ongoing research. The epigenetic responses of fruit crops other than a few major fruits have not been studied well. Sudden growth in the genome editing techniques and molecular biology studies has paved the way for a wide range of genome editing platforms available for developing disease resistance in plants. Selecting the suitable and appropriate technique for the specific species and disease is very important to get the desired changes in the traits (Zhou et al. 2024). All these new initiatives will help to create a more robust and highly sustainable fruit production system, which helps to improve resistance and meet the world's food demands (Shi et al. 2024).

15.10 Conclusion

The main epigenetic processes that underlie viral resistance in different fruit crops have been emphasized in this chapter. Since viruses cannot be managed with chemical treatments like bacterial or fungal pathogens can, managing them mostly entails managing insect vectors and creating resistant crop varieties. However, plants have

Table 15.1 Epigenetic mechanisms in virus resistance of major fruit crops

Fruit crop	Virus	Epigenetic mechanism	Key findings	Reference
Banana	Banana bunchy top virus (BBTV)	RNA interference (RNAi)	Preventing the vector from spreading the virus	Jekayinoluwa et al. (2020)
Banana	Banana bunchy top virus (BBTV)	Intron-hairpin RNA (ihpRNA) transcripts and small interfering RNA (siRNA)	Transgenic plants resistant to BBTV	Shekhawat et al. (2012)
Papaya	Papaya ringspot virus (PRSV)	MicroRNA (miRNA) modulation	Down-regulated vital infection	Patil and Tripathi (2024)
Citrus	Citrus tristeza virus (CTV)	RNA silencing	RNA silencing promoted increased suppressor activity	Gomez-Munoz et al. (2017)
Grapevine	Grapevine fanleaf virus (GFLV)	siRNAs production and cytosine methylation	GFLV infection alters small RNA profiles, influencing defense gene expression	Gambino et al. (2010)
Grapevine	Grapevine red blotch virus	miRNA and virus-derived siRNAs (vsiRNAs)	Modulated host responses and promoted resistance in leaves and berries	Ault et al. (2025)
Pear	Apple stem grooving virus (ASGV)	miRNA-mediated post-transcriptional gene regulation	Reduced ASGV titers, and up-regulated defense- and hormone-related target genes in pear shoot tips	Xie et al. (2024)
Tomato	Tomato yellow leaf curl virus (TYLCV)	Small RNA-mediated regulation (miRNAs, phasiRNAs) and DNA methylation analysis	Gene silencing and immune responses were induced, but were insufficient to block TYLCV	Romero-Rodriguez et al. (2023)

(continued)

Table 15.1 (continued)

Fruit crop	Virus	Epigenetic mechanism	Key findings	Reference
Peach	Plum pox virus (PPV)	Small RNAs (miRNA, siRNA), histone methylation	Changes in small RNA abundance and histone methylation in resistant cultivars	Corell-Sierra et al. (2024)
Watermelon	Cucumber green mottle mosaic virus (CGMMV)	Bisulfite sequencing	Found responsive TFs and sequenced for silencing	Sun et al. (2019)
Cherry	Prunus necrotic ringspot virus (PNRSV)	RNAi-mediated silencing	Transgenic rootstocks conferring viral resistance	Song et al. (2013)

their own memory of molecular defenses through various epigenetic modifications like histone alterations, chromatin remodeling, RNA modifications, and RNA-directed DNA methylation against viral infection. These molecular mechanisms analyze and decide how to react when a viral infection occurs. Most importantly, these epigenetically modified DNA patterns can be inherited by subsequent generations, thus providing a pathway for heritable resistance. Understanding and implementing the knowledge of gene editing and regulatory mechanisms will open new opportunities for new findings in biotechnology, leading to the use of better tools for virus detection. Thus, it paves the way for the most effective strategies for managing and preventing viral diseases in fruit crops.

References

Abas A, Huded S, Reddy KJ, Panigrahi CK, Padamini R, Diwan S, Tyagi V, Saiyad MM (2024) Biotechnological approaches for enhancing disease resistance and crop quality in horticultural plants. Plant Cell Biotechnol Mol Biol 25(9–10):82–104. https://doi.org/10.56557/pcbmb/2024/v25i9-108829

Ault N, Ren S, Payne D, Li Y, Srinivasan A, Zheng Y, Sunkar R, Naidu RA (2025) Dynamics of small RNAs in a red-fruited wine grape cultivar infected with Grapevine red blotch virus. BMC Genomics 26(1):417. https://doi.org/10.1186/s12864-025-11539-4

Barman A, Deb B, Chakraborty S (2020) A glance at genome editing with CRISPR-Cas9 technology. Curr Genet 66(3):447–462. https://doi.org/10.1007/s00294-019-01040-3

Borrelli VMG, Brambilla V, Rogowsky P, Marocco A, Lanubile A (2018) The enhancement of plant disease resistance using CRISPR/Cas9 technology. Front Plant Sci 9:1245. https://doi.org/10.3389/fpls.2018.01245

Castillo-González C, Liu X, Huang C, Zhao C, Ma Z, Hu T, Sun F, Zhou Y, Zhou X, Wang X-J, Zhang X (2015) Geminivirus-encoded TrAP suppressor inhibits the histone methyltransferase SUVH4/KYP to counter host defense. elife 4:e06671. https://doi.org/10.7554/eLife.06671

Corell-Sierra J, Corrêa RL, Gómez GG, Elena SF, Oliveros JC, Rodamilans B, Martínez-García PJ, Martínez-Gómez P, Rubio M (2024) Almond grafting for plum pox virus resistance triggers significant transcriptomic and epigenetic shifts in peaches. Int J Mol Sci 26(1):248. https://doi.org/10.3390/ijms26010248

Elayabalan S, Subramaniam S, Selvarajan R (2017) Construction of the BBTV rep gene RNAi vector and evaluation of the silencing mechanism through injection of Agrobacterium tumefaciens transient expression system in BBTV infected hill banana plants cv. Virupakshi (AAB). Indian J Nat Sci 7(42):12395–12403

Ertunç F (2020) Physiology of virus-infected plants. In: Applied plant virology. Elsevier, pp 199–205. https://doi.org/10.1016/B978-0-12-818654-1.00015-3

Fadul SM, Arshad A, Mehmood R (2023) CRISPR-based epigenome editing: mechanisms and applications. Epigenomics 15(21):1137–1155. https://doi.org/10.2217/epi-2023-0281

Folimonova SY, Sun Y-D (2022) Citrus tristeza virus: from pathogen to panacea. Annu Rev Virol 9(1):417–435. https://doi.org/10.1146/annurev-virology-100520-114412

Food and Agriculture Organization (FAO) (2024) Major tropical fruits market review – preliminary results 2023, Rome. https://openknowledge.fao.org/handle/20.500.14283/CD6428EN

Frommer M, McDonald LE, Millar DS, Collis CM, Watt F, Grigg GW, Molloy PL, Paul CL (1992) A genomic sequencing protocol that yields a positive display of 5-methylcytosine residues in individual DNA strands. Proc Natl Acad Sci USA 89(5):1827–1831. https://doi.org/10.1073/pnas.89.5.1827

Gallusci P, Agius DR, Moschou PN, Dobránszki J, Kaiserli E, Martinelli F (2023) Deep inside the epigenetic memories of stressed plants. Trends Plant Sci 28(2):142–153. https://doi.org/10.1016/j.tplants.2022.09.004

Gambino G, Perrone I, Carra A, Chitarra W, Boccacci P, Torello Marinoni D, Barberis M, Maghuly F, Laimer M, Gribaudo I (2010) Transgene silencing in grapevines transformed with GFLV resistance genes: analysis of variable expression of transgene, siRNAs production and cytosine methylation. Transgenic Res 19(1):17–27. https://doi.org/10.1007/s11248-009-9289-5

Ghorbani Faal P, Farsi M, Seifi A, Mirshamsi Kakhki A (2020) Virus-induced CRISPR-Cas9 system improved resistance against tomato yellow leaf curl virus. Mol Biol Rep 47(5):3369–3376. https://doi.org/10.1007/s11033-020-05409-3

Gomez-Munoz N, Velazquez K, Vives MC, Ruiz-Ruiz S, Pina JA, Flores R, Moreno P, Guerri J (2017) The resistance of sour orange to Citrus tristeza virus is mediated by both the salicylic acid and RNA silencing defence pathways. Mol Plant Pathol 18(9):1253–1266. https://doi.org/10.1111/mpp.12488

Harris CJ, Amtmann A, Ton J (2023) Epigenetic processes in plant stress priming: open questions and new approaches. Curr Opin Plant Biol 75(102432):102432. https://doi.org/10.1016/j.pbi.2023.102432

Hilker M, Schwachtje J, Baier M, Balazadeh S, Bäurle I, Geiselhardt S, Hincha DK, Kunze R, Mueller-Roeber B, Rillig MC, Rolff J, Romeis T, Schmülling T, Steppuhn A, van Dongen J, Whitcomb SJ, Wurst S, Zuther E, Kopka J (2016) Priming and memory of stress responses in organisms lacking a nervous system: priming and memory of stress responses. Biol Rev Camb Philos Soc 91(4):1118–1133. https://doi.org/10.1111/brv.12215

Hull R (2018) Viruses as pathogens: plant viruses. In: Tennant P, Fermin G, Foster JE (eds) Viruses: molecular biology, host interactions, and applications to biotechnology, p 135. https://doi.org/10.1016/B978-0-12-811257-1.00006-1

Jekayinoluwa T, Tripathi L, Tripathi JN, Ntui VO, Obiero G, Muge E, Dale J (2020) RNAi technology for management of banana bunchy top disease. Food Energy Secur 9(4):e247. https://doi.org/10.1002/fes3.247

Jia R, Zhao H, Huang J, Kong H, Zhang Y, Guo J, Huang Q, Guo Y, Wei Q, Zuo J, Zhu YJ, Peng M, Guo A (2017) Use of RNAi technology to develop a PRSV-resistant transgenic papaya. Sci Rep 7:12636. https://doi.org/10.1038/s41598-017-13049-0

Jiang F, Doudna JA (2017) CRISPR-Cas9 structures and mechanisms. Annu Rev Biophys 46(1):505–529. https://doi.org/10.1146/annurev-biophys-062215-010822

Jones RAC (2021) Global plant virus disease pandemics and epidemics. Plants 10(2):233. https://doi.org/10.3390/plants10020233

Jones RAC, Naidu RA (2019) Global dimensions of plant virus diseases: current status and future perspectives. Annu Rev Virol 6(1):387–409. https://doi.org/10.1146/annurev-virology-092818-015606

Jyotika RK, Harish S, Karthikeyan G, Kumar KK, Murugan M, Jayakanthan M, Chen T-C (2024) Molecular approaches for the management of papaya ringspot virus infecting papaya: a comprehensive review. Mol Biol Rep 51(1):981. https://doi.org/10.1007/s11033-024-09920-9

Kuo Y-W, Falk BW (2020) RNA interference approaches for plant disease control. BioTechniques 69(6):469–477. https://doi.org/10.2144/btn-2020-0098

Lippman Z, Martienssen R (2004) The role of RNA interference in heterochromatic silencing. Nature 431(7006):364–370. https://doi.org/10.1038/nature02875

Maliogka VI, Minafra A, Saldarelli P, Ruiz-García AB, Glasa M, Katis N, Olmos A (2018) Recent advances on detection and characterization of fruit tree viruses using high-throughput sequencing technologies. Viruses 10(8):436. https://doi.org/10.3390/v10080436

Patil BL, Tripathi S (2024) Differential expression of microRNAs in response to Papaya ringspot virus infection in differentially responding genotypes of papaya (Carica papaya L.) and its wild relative. Front Plant Sci 15:1398437. https://doi.org/10.3389/fpls.2024.1398437

Pooggin MM (2018) Small RNA-omics for plant virus identification, virome reconstruction, and antiviral defense characterization. Front Microbiol 9:2779. https://doi.org/10.3389/fmicb.2018.02779

Purkayastha A, Dasgupta I (2009) Virus-induced gene silencing: a versatile tool for discovery of gene functions in plants. Plant Physiol Biochem 47(11–12):967–976. https://doi.org/10.1016/j.plaphy.2009.09.001

Ramakrishnan M, Zhang Z, Mullasseri S, Kalendar R, Ahmad Z, Sharma A, Liu G, Zhou M, Wei Q (2022) Epigenetic stress memory: a new approach to study cold and heat stress responses in plants. Front Plant Sci 13:1075279. https://doi.org/10.3389/fpls.2022.1075279

Rehman UU, Ghafoor D, Ullah A, Ahmad R, Hanif S (2023) Epigenetics regulation during virus-host interaction and their effects on the virus and host cell. Microb Pathog 182(106271):106271. https://doi.org/10.1016/j.micpath.2023.106271

Rodríguez-Verástegui LL, Ramírez-Zavaleta CY, Capilla-Hernández MF, Gregorio-Jorge J (2022) Viruses infecting trees and herbs that produce edible fleshy fruits with a prominent value in the global market: an evolutionary perspective. Plants 11(2):203. https://doi.org/10.3390/plants11020203

Romero-Rodriguez B, Petek M, Jiao C, Križnik M, Zagorščak M, Fei Z, Bejarano ER, Gruden K, Castillo AG (2023) Transcriptional and epigenetic changes during tomato yellow leaf curl virus infection in tomato. Res Sq. https://doi.org/10.21203/rs.3.rs-3252600/v1

Roy L, Chatterjee O, Bose D, Roy A, Chatterjee S (2023) Noncoding RNA as an influential epigenetic modulator with promising roles in cancer therapeutics. Drug Discov Today 28(9):103690. https://doi.org/10.1016/j.drudis.2023.103690

Sampson C, Ikenwugwu TH, Okagu IU, Yahaya II, Odoh CK, Eze CN (2024) Epigenetics: toward improving crop disease resistance and agronomic characteristics. Plant Biotechnol Rep 18(1):1–20. https://doi.org/10.1007/s11816-023-00876-z

Shekhawat UKS, Ganapathi TR, Hadapad AB (2012) Transgenic banana plants expressing small interfering RNAs targeted against viral replication initiation gene display high-level resistance to banana bunchy top virus infection. J Gen Virol 93(Pt 8):1804–1813. https://doi.org/10.1099/vir.0.041871-0

Shi M, Wei Z, Zhang P, Guan C, Chachar S, Zhang J (2024) Epigenetic modifications and breeding applications in horticultural plants. Horticulturae 10(2):143. https://doi.org/10.3390/horticulturae10020143

Singh K, Dardick C, Kumar Kundu J (2019) RNAi-mediated resistance against viruses in perennial fruit plants. Plants 8(10):359. https://doi.org/10.3390/plants8100359

Singh BK, Delgado-Baquerizo M, Egidi E, Guirado E, Leach JE, Liu H, Trivedi P (2023) Climate change impacts on plant pathogens, food security and paths forward. Nature reviews. Microbiology 21(10):640–656. https://doi.org/10.1038/s41579-023-00900-7

Singh K, Balyan HS, Gupta PK (2024) Bioinformatics for analyzing the role of epigenetics in plant disease resistance. Bioinformatics Plant Res Crop Breed:351–390. https://doi.org/10.1002/9781394209965.ch15

Song G-Q, Sink KC, Walworth AE, Cook MA, Allison RF, Lang GA (2013) Engineering cherry rootstocks with resistance to Prunus necrotic ring spot virus through RNAi-mediated silencing. Plant Biotechnol J 11(6):702–708. https://doi.org/10.1111/pbi.12060

Sun Y, Fan M, He Y (2019) DNA methylation analysis of the Citrullus lanatus response to cucumber green mottle mosaic virus infection by whole-genome bisulfite sequencing. Genes (Basel) 10:344. https://doi.org/10.3390/genes10050344

Szittya G, Silhavy D, Molnár A, Havelda Z, Lovas A, Lakatos L, Bánfalvi Z, Burgyán J (2003) Low temperature inhibits RNA silencing-mediated defence by the control of siRNA generation. EMBO J 22(3):633–640. https://doi.org/10.1093/emboj/cdg74

Tessarz P, Kouzarides T (2014) Histone core modifications regulating nucleosome structure and dynamics. Nature reviews. Mol Cell Biol 15(11):703–708. https://doi.org/10.1038/nrm3890

Tripathi S, Suzuki JY, Ferreira SA, Gonsalves D (2008) Papaya ringspot virus-P: characteristics, pathogenicity, sequence variability and control. Mol Plant Pathol 9(3):269–280. https://doi.org/10.1111/j.1364-3703.2008.00467.x

Turgut-Kara N, Arikan B, Celik H (2020) Epigenetic memory and priming in plants. Genetica 148(2):47–54. https://doi.org/10.1007/s10709-020-00093-4

Tyagi S, Kumar R, Kumar V, Won SY, Shukla P (2021) Engineering disease resistant plants through CRISPR-Cas9 technology. GM Crops Food 12(1):125–144. https://doi.org/10.1080/21645698.2020.1831729

Umer M, Liu J, You H, Xu C, Dong K, Luo N, Kong L, Li X, Hong N, Wang G, Fan X, Kotta-Loizou I, Xu W (2019) Genomic, morphological, and biological traits of the viruses infecting major fruit trees. Viruses 11(6):515. https://doi.org/10.3390/v11060515

Valensisi C, Liao JL, Andrus C, Battle SL, Hawkins RD (2015) cChIP-seq: a robust small-scale method for investigation of histone modifications. BMC Genomics 16(1):1083. https://doi.org/10.1186/s12864-015-2285-7

Vieira MS, Cabral RLR, Favaratto L, Maciel LS, Xavier ADS, Zerbini FM, Fernandes PMB (2024) Tropical fruit virus resistance in the era of next-generation plant breeding. SynBio 2:267–284. https://doi.org/10.3390/synbio2030016

Wang J, Yang Y, Jin L, Ling X, Liu T, Chen T, Ji Y, Yu W, Zhang B (2018) Re-analysis of long non-coding RNAs and prediction of circRNAs reveal their novel roles in susceptible tomato following TYLCV infection. BMC Plant Biol 18(1):104. https://doi.org/10.1186/s12870-018-1332-3

Wang C, Wang C, Zou J, Yang Y, Li Z, Zhu S (2019) Epigenetics in the plant-virus interaction. Plant Cell Rep 38(9):1031–1038. https://doi.org/10.1007/s00299-019-02414-0

Xie Y-S, Zeng Q, Huang W-T, Wang J-Y, Li H-W, Yu S-Z, Liu C, Zhang X-Q, Feng C-L, Zhang W-H, Li T-Z, Cheng Y-Q (2024) A novel RAV transcription factor from pear interacts with viral RNA-silencing suppressors to inhibit viral infection. Plant J 120:1079–1093. https://doi.org/10.1111/tpj.17037

Yang Y, Liu T, Shen D, Wang J, Ling X, Hu Z, Chen T, Hu J, Huang J, Yu W, Dou D, Wang M-B, Zhang B (2019) Tomato yellow leaf curl virus intergenic siRNAs target a host long noncoding RNA to modulate disease symptoms. PLoS Pathog 15(1):e1007534. https://doi.org/10.1371/journal.ppat.1007534

Zhang X, Yazaki J, Sundaresan A, Cokus S, Chan SW-L, Chen H, Henderson IR, Shinn P, Pellegrini M, Jacobsen SE, Ecker JR (2006) Genome-wide high-resolution mapping and functional analysis of DNA methylation in arabidopsis. Cell 126(6):1189–1201. https://doi.org/10.1016/j.cell.2006.08.003

Zhang H, Lang Z, Zhu J-K (2018) Dynamics and function of DNA methylation in plants. Nature reviews. Mol Cell Biol 19(8):489–506. https://doi.org/10.1038/s41580-018-0016-z

Zhou Y, Wang X, Qi K, Bao J, Zhang S, Gu C (2022) Involvement of long non-coding RNAs in pear fruit senescence under high- and low-temperature conditions. Hortic Plant J. https://doi.org/10.1016/j.hpj.2022.08.004

Zhou L, Zeng X, Yang Y, Li R, Zhao Z (2024) Applications and prospects of CRISPR/Cas9 technology in the breeding of major tropical crops. Plants 13(23):3388. https://doi.org/10.3390/plants13233388

Zulfiqar S, Farooq MA, Zhao T, Wang P, Tabusam J, Wang Y, Xuan S, Zhao J, Chen X, Shen S, Gu A (2023) Virus-induced gene silencing (VIGS): a powerful tool for crop improvement and its advancement towards epigenetics. Int J Mol Sci 24(6). https://doi.org/10.3390/ijms24065608

RNAi and dsRNA-Based Gene Silencing for Virus Management in Fruit Crops

16

Archita Bordoloi, Adrij Pawan Neog, Anindita Saikia, Uchakankhi Kashyap, and Nima D. Namsa

Abstract

RNA interference (RNAi) and double-stranded RNA (dsRNA)-mediated gene silencing have evolvedinto novel, eco-friendly approaches for combating the viral diseases of fruit crops. Plant viruses are responsible for a great reduction in the yield andquality of fruits. Conventional methods of control, suchas insecticidal treatment and vector control, are frequently inadequate or lacking in specificity. RNAi takes advantage of the intrinsicdefense systems of the plant, in which viral RNA is targeted for destruction through sequence-specific gene silencing. Such regulation may be provided by transgenic expression ofvirus-derived dsRNA in the plant or non-transgenic applications outside the plant of synthetic or in vitro produced dsRNAs. The development of advanced dsRNA application techniques, including foliar sprays, trunk injections, as well as nanoparticle carriers, has now made the use of RNAi possible in fruit crops, which are commonly perennial having limited genetic transformation capabilities. These strategies can successfully stifle viral genes, lower the frequency of virus particles, and provide the plantwith resistance without affecting the genome. Indeed, a few reports have shown RNAi-driven immunity against fruit-infecting viruses, such as papaya ring spot virus (PRSV). The evolution of bioinformatics also contributed to the development of RNAi products, wherehighly specific dsRNAs can be designed with minimal off-target effects. RNAi and dsRNA-based gene silencing techniques provide a sustainable, targeted, and safe approachfor virus management in fruit crops that could potentially transform plant protection and strengthen worldwide fruit production systems, as discussed in this book chapter.

A. Bordoloi · A. P. Neog · A. Saikia · U. Kashyap · N. D. Namsa (✉)
Department of Molecular Biology and Biotechnology, Tezpur University, Tezpur, Assam, India
e-mail: namsa@tezu.ernet.in

R. Kumar et al. (eds.), *Fruit Crops and Viral Pathogens Volume II*, Plant Pathology for Sustainable Agriculture, https://doi.org/10.1007/978-981-95-7261-8_16

Keyword

RNAi · dsRNA-based gene silencing · Virus management · Fruit crops · Viral RNA

16.1 Introduction

16.1.1 Overview of the Interaction Between Plants and Viruses

Plants are susceptible to attack by many different viruses capable of triggering different kinds of immune responses in plants, and understanding the mechanism of interaction between these two entities is essential for developing efficient anti-viral strategies in plants. Co-evolution of viruses in plants is a noteworthy process in relation to plant-virus interactions. Here, the viruses rely on the host plant for spreading, whereas the plants develop defense mechanisms against certain specific kinds of viruses (Fontes et al. 2021).

16.1.2 Economic and Agricultural Impacts of Plant Viruses

The quality and yield of agriculturally important plants directly affect the economy. For example, one of the major viral diseases in bananas, called the banana bunchy top disease, is caused by the banana bunch top virus (BBTV), which leads to upright, short, and brittle leaves giving a bunchy appearance to the plant. The outbreak of this disease in India cost an economic loss of $50 million from 2007 to 2010 (Tatineni et al. 2023). There are also reports on viruses infecting tomato plants in Mexico, which is one of the largest producers and exporters of tomato, thereby creating difficulty in economic and agricultural yield via tomato cultivation (García-Estrada et al. 2022).

16.1.3 Need for Sustainable Virus Management Strategies

Complexity in plant viral infection arises from rapid evolution and variability of viruses that become more resistant to anti-viral responses of the host. Most viral infections are systemic in nature, causing the spread of infection to other body parts from the primary site (Ershova et al. 2022). Studies show up to 75% decrease of productivity in watermelon (*Citrullus lanatus*) upon viral infection (Kakareka et al. 2024) and up to 80% in melon (*Cucumis melo*) (Sáez et al. 2022). Chemical measures for controlling viruses have not been much of a success due to viral recombination, which generates highly aggressive strains of viruses. Therefore, the development of highly efficient virus management strategies is the need of the hour for protecting plants suffering from such attacks.

16.2 Fundamentals of RNA Interference (RNAi)

16.2.1 History of RNAi Discovery

RNA interference is a breakthrough discovery in genomics where post-transcriptional gene silencing is initiated by double-stranded RNA. Initial studies on *Caenorhabditis elegans* provided evidence of gene silencing through RNAi via dsRNA. Studies were further performed in *Drosophila melanogaster*, indicating that RNAi is a two-step process (Elbashir et al. 2001). When Guo and Kemphues accidentally inserted the anti-sense strand to stop the nematode *C. elegans* from expressing the par-1 gene, they discovered the RNAi phenomenon (Guo and Kemphues 1995). The understanding of gene silencing in plants began with a search for more purple petunia flowers. In 1990, R. Jorgensen's laboratory worked on increasing the expression of the chalcone synthase (*chsA*) gene, which is responsible for anthocyanin pigment formation. To their surprise, in some of these plants the activity of both endogenous and transgene chalcone synthase was lost, and as a result, a lot of the flowers became variegated (Napoli et al. 1990).

16.2.2 Natural Role of RNAi in Plants

There can be many different biotic stressors in plants, like bacteria, fungi, nematodes, and viruses. During plant-pathogen interactions, plants activate certain defense mechanisms that either lead to increased resistance or enhanced tolerance of the plant to fight against the progression of the pathogen. RNA silencing mechanism is an example of a natural plant defense mechanism that can block gene expression both transcriptionally (TGS) and post-transcriptionally (PTGS) (Muhammad et al. 2019). TGS results in the methylation of the promoter DNA sequence. A heterochromatin region is created in the surroundings of the gene, which ultimately causes gene transcription suppression. The phenomenon of PTGS is known as RNAi. MicroRNAs (miRNAs) and small interfering RNAs (siRNAs) are the two classes of short RNA that mediate RNAi. The origin of these small RNAs is long double-stranded RNA or the precursor of a stem-loop structure. Within the cell, siRNA and miRNA-mediated silencing processes are connected to the same metabolic pathway (Akbar et al. 2022), Fig. 16.1.

16.3 Important Plant RNAi Pathways

16.3.1 Post-Transcriptional Gene Silencing (PTGS) Pathway

In plants, we have a mechanism of gene regulation referred to as post-transcriptional gene silencing (PTGS), which allows genes to be silenced after transcription using double-stranded RNA (dsRNA) (Elbashir et al. 2001). DCL1, together with other proteins such as HYL1 and dsRBP, cleaves the pri-miRNAs into precursor miRNAs

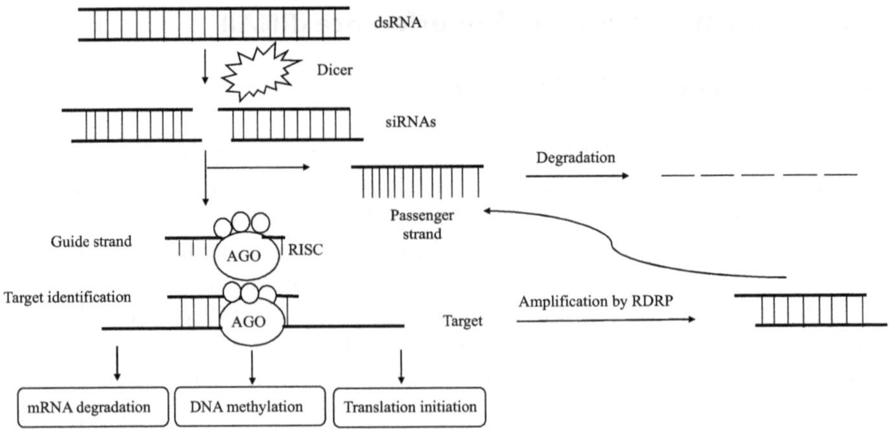

Fig. 16.1 Mechanism of the RNAi pathway in plants. dsRNA is cleaved by Dicer to produce siRNAs, which are further processed and incorporated into RISC in association with AGO protein. RISC complex then identifies the target sequence and cleaves it to suppress the expression of the desired gene

(pre-miRNAs), which are processed to yield mature miRNA duplexes (El-Sappah et al. 2021). HEN1 methylates these duplexes, and exportin-5 or its plant homolog HASTY transports them to the cytoplasm. The ribonucleoprotein complex known as the RNA-induced silencing complex (RISC) is assembled in the cytoplasm, where mature, single-stranded miRNA binds to AGO and mediates sequence-specific cleavage of target mRNAs, leading to gene repression (El-Sappah et al. 2021), Fig. 16.1.

16.3.2 Transcriptional Gene Silencing (TGS) Pathway

16.3.2.1 (RdDM) Pathway: RNA-Directed DNA Methylation

The RNA-directed DNA methylation (RdDM) pathway is a plant-specific epigenetic mechanism that guides DNA methylation, mediating transcriptional gene silencing. This process is initiated with single-stranded RNA transcription by RNA polymerase IV (Pol IV) from target regions, often transposons or repetitive sequences. This is followed by RNA conversion to dsRNA by the enzyme RNA-dependent RNA polymerase 2 (RDR2). DCL3 then processes these dsRNAs into 24-nucleotide-long siRNAs, which are stabilized by methylation and loaded onto argonaute 4 (AGO4). At the same time, non-coding scaffold RNAs are transcribed by RNA polymerase V (Pol V), providing the AGO4-siRNA complex a platform to base-pair and guide DNA methyltransferase DRM2 recruitment. Cytosine residues methylation is catalyzed by DRM2 in all sequence contexts (CG, CHG, and CHH), establishing a transcriptionally repressive chromatin state that silences gene expression at the level of transcription. Beyond the main RdDM pathway, there are proteins like RNA-dependent RNA polymerase 6 (RDR6) and the plant-specific factor needed for RDR2-independent dna methylation (NERD) that provide other routes. These factors can contribute to the exquisite specificity and efficiency of a normal RdDM response, but also build in potential for additional adaptability if it becomes necessary to silence newly active or inserted TEs (or other genomic regions) that the canonical pathway does not target. Distinct from the canonical RdDM that mainly sustains pre-existing silencing, the RDR6-dependent pathway senses active transcription of TEs to produce small RNAs, which in turn trigger and reinforce DNA methylation, adding a level of quick genome defense (Matzke and Mosher 2014).

Genomic Imprinting

Genomic imprinting in plants refers to an epigenetic phenomenon in which the expression of a gene is dependent on the parent-of-origin of the allele, thus having only one of those alleles being transcribed while silencing the other. This repression is specific and occurs mainly by means of DNA methylation and histone modifications in a parent-of-origin way, primarily restricted to the endosperm of flowering plants (Feng et al. 2010). During endosperm growth, the maternal imprinted alleles are frequently demethylated by DNA glycosylases like DEMETER and activated, while the paternal alleles remain methylated and silenced, setting up gene-specific expression profiles that are required for proper seed development and nutrient allocation(Choi et al. 2002). This parental-specific methylation pattern is shaped by the genomic proximity to transposable elements and repetitive sequences, which are loci of transcriptional suppression by DNA methylation. This transient demethylation in the central cell of the female gametophyte can activate transposons and generate siRNAs to maintain silencing in its daughter egg and sperm cells through RdDM pathways (Calarco et al. 2012). In addition, histone modification marks (e.g., H3K27me3) deposited by Polycomb Repressive Complex2 (PRC2) help maintain imprinting through repression of the hypomethylated allele, establishing stable allele-specific silencing that is important for normal plant development and genome stability (El-Sappah et al. 2021).

16.3.2.2 Paramutation

Paramutation is an epigenetic mechanism by which one allele (paramutagenic) stably silences another homologous paramutable allele (trans-silencing), leading to heritable changes in gene expression without alterations to DNA sequence. This mechanism is most well-characterized at the maize r1 locus, where the equivalent of paramutagenic alleles are molecular drivers or tandem repeats that activate an RNA-mediated process involving 25-nt small intervening RNAs (siRNAs) to trigger RNA-directed DNA methylation (RdDM) and chromatin modifications resulting in gene transcription silencing and stable heritable silencing of paramutable alleles (El-Sappah et al. 2021). In fruit-bearing crops, paramutation is correlated with the production of siRNAs and increased DNA methylation at loci like SLTAB2 in tomato, which shows transcriptional silencing and a change in leaf phenotype (chlorosis) (Z. Guo and Schnurbusch 2016). These findings expand the framework of paramutation beyond maize and illustrate its function in epigenetic control of gene expression throughout multiple plant species. The participation of sRNAs, DNAmethyltransferases, and chromatin remodelers contrasts with paramutation as a flexible mechanism that is implicated in gene expression regulation, genome stability (El-Sappah et al. 2021), and developmental processes in fruit crops (Guo and Schnurbusch 2016).

16.3.3 Major Components of the RNAi Process

16.3.3.1 DCL: Dicer-Like Enzymes

Dicer-like enzymes (DCLs) are ribonucleases required to trigger RNA interference (RNAi) via the cleavage of long double-stranded RNA (dsRNA) or precursor hairpins in small interfering RNAs (siRNAs), and microRNAs with ~20–25 nucleotides. These small RNAs are then loaded into the RNA-induced silencing complex (RISC), where they guide sequence-specific degradation or translation repression of target mRNAs, thereby regulating gene expression and protecting against viruses (Paturi and Deshmukh 2021). Organisms contain several different Dicer-like proteins, which have both separate and joint roles in producing diverse small RNA populations to ultimately modulate gene-silencing pathways (Younger et al. 2015).

16.3.3.2 AGO: Argonaute Proteins

Argonaute (AGO) proteins are indispensable factors in the RNA interference (RNAi) pathway as they constitute the catalytic component of the RISC, which is responsible for gene silencing through association with small RNAs such as siRNAs and miRNAs. Using their PAZ and PIWI domains, AGOs pair the guide RNA strand, which cleaves or inhibits complementary target mRNA to regulate expression of genes post-transcriptionally. The selective incorporation and retention of guide strands depend on thermodynamic rules and involve auxiliary proteins that assist RISC assembly and function. Mammalian AGO2 is unique among AGOs in possessing endonucleolytic "slicer" activity critical for direct mRNA cleavage, influencing development and disease. AGO proteins are evolutionarily conserved and orchestrate diverse regulatory mechanisms, from mRNA degradation to translational repression, underscoring their essential role in RNAi and gene regulation (Wu et al. 2020).

16.3.3.3 RISC: RNA-Induced Silencing Complex

RISC is a multiprotein effector assembly and is a core part of the RNAi pathway, causing gene silencing through recognition of specific sequences of target RNA. RISC assembly is initiated by the binding of small RNA species, such as siRNAs or miRNAs, which guide RISC to their complementary mRNA targets for cleavage or translational repression (Hu et al. 2006). This complex functions in both the cytoplasm, where it modulates mRNA stability and translation, and the nucleus, where it participates in transcriptional gene silencing and chromatin modification (Gaglia and Glaunsinger 2010).

16.3.3.4 RDR: RNA-Dependent RNA Polymerase

Introduction RNA-dependent RDRs are important enzymes in the RNA interference (RNAi) pathway, which amplify the signal of RNA silencing through replication of complementary strands from single-stranded RNA templates, converting them to double-stranded RNA (dsRNA) precursors for Dicer processing. RDRs are required to maintain and amplify RNAi effects in a variety of eukaryotes, particularly plants and certain fungi, as they generate secondary siRNAs that can increase gene silencing activity as well as systemic movement (Bobbin and Rossi 2016). RDRs have evolved to serve multiple functions in antiviral defense and endogenous gene regulation, with different specific RDRII isoforms functioning in specialized silencing pathways (de Farias et al. 2017). For instance, in plants, distinct RDRs are needed to boost the silencing of viral invaders and transposable element activities and preserve strong immunity genomes integrity (Sugiyama et al. 2004).

16.4 Role of dsRNA in Gene Silencing

16.4.1 dsRNA: Structure and Properties

Double-stranded RNA (dsRNA) is an important molecular architecture in RNA interference (RNAi), which comprises two interfering complementary RNA strands that can form a stable helical duplex. In-depth studies indicate that some structural elements, such as the length, thermodynamic stability, and blunt or overhanging ends of the dsRNA, affect how efficiently the dsRNA is processed and perceived in RNAi pathways (Joga et al. 2016). The features of dsRNA are crucial to its role in RNAi as an initiator and enhancer of the silencing signal. The molecular stability of dsRNA enables the molecule to avoid prompt degradation and move systematically to achieve extended silencing activity in organisms, including insects and plants (Joga et al. 2016).

16.4.2 dsRNA Origin in Plant-Virus Interactions

During plant-virus interactions, however, in plants, this viral dsRNA mainly arises as replication intermediates of the RNA viruses that infect them and acts as an essential inducer for the host RNA interference (RNAi) machinery. These dsRNA molecules can also be generated by viral transcripts that overlap, as well as structured elements of RNA, adding to the pool of substrates for RNAi in the host. Progress in exogenous or synthetic topical application of viral sequence- dsRNA replicative intermediates mimics for boosting resistance to viruses in fruit crops has shown that mimicking these replicative intermediates can stimulate RNAi pathways to provide effective protection against virus infection (Chen et al. 2025). Together, these results demonstrate the pivotal role of viral dsRNA replication and turnover in activating RNAi-based antiviral immunity in fruit crops (Pascual et al. 2019).

16.4.3 siRNA Formation via dsRNA Processing

The conversion of double-stranded RNA (dsRNA) into small interfering RNA (siRNA) is the first step of the RNA interference (RNAi) pathway and this process is carried out by the Dicer enzyme, a RNase III family protein. Dicer processes long dsRNA molecules into 21- to 23-nucleotide siRNA duplexes that bear a distinctive two-nucleotide overhang. This cleavage is essential for producing siRNAs that function as sequence-specific templates for the silencing of genes. The processing is thermodynamically and structurally specific; Dicer recognizes the termini of dsRNA, guaranteeing cuts that are exact enough to maximize target specificity and avoid off-target effects. In plants, several Dicer-like proteins generate siRNAs of specific sizes to direct different silencing pathways such as antiviral defense and transposon silencing (Das and Sherif 2020). After cleavage, the siRNA duplex is incorporated onto Argonaute proteins in the RNA-induced silencing complex (RISC) and the guide strand is retained for target recognition and the passenger strand is degraded. This strand choice is moderated by the thermodynamic stability at the duplex ends, and it is frequently dictated by accessory proteins such as R2D2 in Drosophila (Carthew and Sontheimer 2009).

16.4.4 Viral RNA Targeting via siRNA

Small interfering RNAs (siRNAs) are central to RNA interference (RNAi), where they direct the targeted destruction of viral RNA and thus function as a potent antiviral immunity in a wide range of species. Following recognition and cleavage of viral dsRNA intermediates by Dicer, the siRNAs are loaded into RISC, where the guide strand directs sequence-specific binding to complementary viral RNAs. Such interaction results in endonucleolytic cleavage of viral transcripts, impeding viral replication and spreading (Carthew and Sontheimer 2009). Therapeutically, chemically synthesized siRNAs targeting conserved viral segments have demonstrated efficacy in inhibiting viral infections by mimicking physiological RNA interference (RNAi) mechanisms, thereby emphasizing the translational potential of RNAi-based antiviral strategies (Wang et al. 2011). Nonetheless, viruses have evolved RNA silencing suppressors that either sequester siRNAs or operate to disturb the RNAi machinery in order to circumvent this defense (a dynamic molecular arms race) (Burgyán and Havelda 2011).

16.5 Virus Management in Crops via dsRNA

16.5.1 dsRNA Delivery in Plants

16.5.1.1 Genetic Transformation

Genetic transformation by RNA interference in plants. We can use RNA interference (RNAi) to get rid of those LODs directly in the plant cells themselves. It remains a principal technique of stable gene silencing despite other methods (e.g., exogenous dsRNA application and nanocarrier delivery) becoming the alternative system available for non-transgenic dsRNA uptake and systemic silencing. In this context, progress on the development of nanoparticle and protein-based delivery systems to enhance dsRNA stability and uptake in plants is expanding the potential applications of RNAi beyond traditional genetic transformation. Hence, genetic transformation with RNAi is important in the modification of gene expression for sustainable crop improvement, especially fruit crops (Pal et al. 2024).

16.5.1.2 Vectors for Transient Expression

Transient expression using vectors such as Agrobacterium is a quick and successful method for gene studies in fruit crops. Agro-injection of fruit such as tomato gives rise to high rates of transient expression level for reporter genes, allowing rapid gene regulation and silencing pathway analysis (Spolaore et al. 2001). In citrus, a highly efficient Agrobacterium-mediated transient expression system in juvenile and mature tissues, enabling rapid gene studies and facilitating transgene-free editing applications in perennial fruit crops, has been developed (Li et al. 2025).

16.5.1.3 SIGS: Spray-Induced Gene Silencing

Spray-induced gene silencing (SIGS) is a non-GMO, eco-friendly GEN with the properties that double-stranded (ds) RNA can be sprayed onto plant surfaces to silence given genes in pathogens or pests for transient protection without altering the genome of the host plant. This technology utilizes the pathogen or plant's own natural RNA interference machinery to destroy target mRNAs, mitigating disease and pest effects. SIGS has proved effective in fruit crops and cereals where fungal virulence genes have been targeted (in Fusarium or Botrytis species) and so insect pests, indicating flexibility and widespread utility (Chen et al. 2025).

16.5.2 Activation of RNAi by dsRNA

Activation of RNAi through dsRNA is a multi-step process beginning with the dicer-like enzyme forming siRNAs (20–25 nucleotides) by the cleavage of dsRNA (Mathur et al. 2025). The Argonaute component of RISC catalyzes cleavage or repression of target mRNA thereby causing gene silencing(Chen et al. 2025).

16.5.3 RISC-Mediated Cleavage of Viral RNA

RISC-mediated cleavage of viral RNA plays a vital role in RNAi-based antiviral defense in fruit crops. VsiRNAs, derived from virus mRNA and produced by plant Dicer-like enzymes following cleavage of viral double-stranded RNA, are incorporated into the Argonaute within RISC to specifically direct cleavage of viral RNAs that results in blocking replication of the viruses. RNAi pathways have been unleashed through genetic modification to potentiate defense responsiveness against viruses in perennial fruit crops like plum and tomato by stimulating "RISC"-mediated viral RNA cleavage (Ram et al., 2019).

16.5.4 Gene Silencing Amplification

Silencing signal amplification in RNAi is a major mechanism by which gene silencing efficiency has been improved in fruit crops. Secondary siRNAs are also produced in an area distal to the primary target site, transmitting the silencing effect and enhancing a local RNAi response as observed in strawberry fruit, where secondary siRNAs were detected, evidencing signal amplification (Das and Sherif 2020). These secondary siRNAs are generated via the RNA-dependent RNA polymerase and increase silencing efficiency by generating additional effective siRNAs to direct the RISC (RNA-induced silencing complex) (Sanan-Mishra et al. 2021).

16.5.5 RNAi Signal Systemic Spreading in Plants

The systemic translocation of RNAi signal was known to move small interfering RNAs (siRNAs) through plasmodesmata and the phloem in order to cause long-range gene suppression in cells that are located far away from the originally targeted ones (Zhao et al. 2024). Exogenously applied dsRNAs are taken up and transported systemically, enabling distant tissues to receive silencing signals, which allows for coordinated defense and gene regulation (Biedenkopf et al. 2020). The systemic RNAi pathway is genetically conserved and essential for processes such as antiviral defense and transgenerational epigenetic regulation (Zhao et al. 2024). Overall, systemic RNAi signaling integrates cellular responses across the plant, supporting whole-plant immunity and gene regulation (Zhao et al. 2024; Biedenkopf et al. 2020).

16.6 Approaches for dsRNA-Mediated Virus Control

16.6.1 HIGS: Host-Induced Gene Silencing

HIGS is an RNAi-based technology that enhances virus control in plants, including fruit crops, by expression of dsRNAs in the host plant that target and cause the silencing of essential viral genes. The dsRNAs are processed by plant DCLs into siRNAs, followed by transfer to the virus or pathogen where they degrade viral RNAs, effectively silencing the infection (Golubkina et al. 2018). In fruit crops such as strawberry and cucumber, HIGS has been successfully demonstrated to reduce virus symptoms by either transgenic expression of dsRNAs or exogenous application (spray-induced gene silencing, SIGS) of dsRNAs, which protect the fruits by triggering RNAi pathways against viral genomes. This method exploits the natural RNA silencing immune mechanisms of plants to systemically suppress viruses. Additionally, the mobility of siRNAs within the plant allows targeting viruses in distal tissues beyond the initial site of dsRNA production or application, making HIGS an efficient strategy for durable virus resistance in horticulture (Das and Sherif 2020).

16.6.2 SIGS: Spray-Induced Gene Silencing

SIGS is an efficient RNA interference (RNAi) technology where dsRNA targeting viral genes is sprayed onto fruit crops, triggering gene silencing without altering the plant genome. A systemic silencing signal can travel through the plant's vascular system (specifically the phloem) and initiate sequence-specific RNA degradation to protect even untreated parts of the plant. SIGS has proven to reduce viral infectiveness and disease symptoms with fruit crops, including tomato, grape (Vitis vinifera), papaya (Carica papaya), cucumber, and strawberry, providing an ecologically safe alternative approach to traditional pesticide use and development of GMOs. Obstacles for market introduction of (d)RNAi formulations are stability of dependency in the field that can, besides use of nanovesicles and nanoparticles to protect dsRNA from degradation, extend effectiveness up to several weeks (Chen et al. 2025).

16.6.3 VIGS: Virus-Induced Gene Silencing

Virus-induced gene silencing (VIGS) is an RNA interference (RNAi)-based approach that uses genetically engineered plant viruses to carry dsRNA sequences of targeted genes into plants and trigger post-transcriptional silencing of target genes in fruit crops. This strategy enables the plant's own RNAi machinery to non-specifically target and degrade both viral and host target RNAs, thereby efficiently silencing viral pathogens while also providing an immediate tool for high-throughput gene function analysis without requiring a stable transformation. Although there have also been successful applications of VIGS for silencing resistance genes in a few fruit crops like tomato, blueberry, and strawberry using vectors such as TRV to identify their function(s) (Wallace et al., unpublished data), the utility of the inhibition of viral susceptibility or ripening strategies has labeled it virus control. The dsRNA corresponding with the viral replication is subsequently cleaved by Dicer-like restriction enzymes to form small interfering RNAs (siRNAs), which, in turn, target RNA-induced silencing complex (RISC) for degradation of cognate mRNAs, thus resulting in an eventual reduction in symptoms and improvement in water mass gain. VIGS is also transmitted systemically throughout the plant through vascular tissues; this spread of gene silencing is not restricted to where the infection was initiated; this is important for efficient control of virus infection in fruit crop production. This method provides a rapid, economical, and flexible alternative for both scientific studies in all horticultural crops, as well as being used practically to manage disease (Golubkina et al. 2018).

16.6.4 Synthetic dsRNA Exogenous Application

Exogenous application of synthetic dsRNA has become a viable, nontransgenic strategy for virus control in fruit crops using RNA interference (RNAi). Recent research shows that spraying purified, high-quality dsRNA molecules specific to virus genomes can effectively reduce viral RNA levels and disease symptoms in crops such as tomato, zucchini, cucumber, and papaya. For instance, topical application of bacterially produced dsRNA provided effective protection against Tomato brown rugose fruit virus (ToBRFV) in tomato plants by reducing viral RNA independently of canonical RNA-dependent RNA polymerases and strongly suggesting a direct antiviral action for dsRNA(Weiss et al. 2025). Based on recent studies, optimizing viral target sequences and using 'chimeric dsRNAs' (designed to target two or more open reading frames, or ORFs) significantly improves antiviral activity against plant viruses, including geminiviruses (Song et al. 2013).

16.7 Case Studies and Applications

16.7.1 RNAi-Mediated Plum Pox Virus Resistance in Stone Fruits

RNAi-mediated resistance has proven highly effective for combating plum pox virus (PPV) in stone fruits, as demonstrated by the development of the "HoneySweet" plum variety. This transgenic plum expresses a hairpin RNA construct targeting the PPV coat protein gene, resulting in stable and robust virus resistance even under long-term, high infection pressure in field trials across multiple countries. Small RNA profiling confirmed that the transgene-derived siRNAs are consistently expressed at safe levels, and comprehensive animal feeding studies found no adverse health effects, supporting the safety of this approach (Callahan et al. 2019) (Table 16.1).

16.7.2 Exogenous Synthetic dsRNA Application Against Tomato Brown Rugose Fruit Virus

Recent research has shown that the application of synthetic dsRNA exogenously that targets the tomato brown rugose fruit virus (ToBRFV) can protect tomato plants

Table 16.1 Plants with successful demonstration of disease resistance using RNAi and host viral suppressor of RNAi

Fruit crop	Virus targeted	RNAi approach	Viral suppressors (VSR) involvement reported	Reference
Plum (HoneySweet)	Plum pox virus (PPV)	Hairpin RNA transgenic (HIGS, stable)	VSRs discussed in context of PPV-counterdefense	Singh et al. (2021)
Stone fruits (general; plum, etc.)	Plum pox virus (PPV)	Hairpin RNA transgenic / HIGS	Viral suppressors and counter-defence noted	Chen et al. (2025)
Cherry (Gisela6 / Gisela7 rootstocks)	Prunus necrotic ringspot virus (PNRSV)	Hairpin RNA transgenic (rootstock transformation)	VSRs considered in host–virus interactions	Song et al. (2013)
Tomato	Tomato brown rugose fruit virus (ToBRFV)	Exogenous synthetic dsRNA (SIGS/topical)	Viral suppressor activity and resistance mechanisms discussed	Weiss et al. (2026)
Tomato (other studies)	Various tomato viruses (ToMV, TSWV, TYLCV)	VIGS, HIGS, SIGS (transient and topical)	VSRs are reviewed across these cases	Ayala-Álvarez et al. (2022)

from this emerging pathogen. In a 2025 study, high-quality dsRNA molecules were produced in vivo using bacteriophage phi6 in *Pseudomonas syringae*, then purified and applied topically to tomato plants, resulting in significant reductions in both disease symptoms and the level of viral RNA(Yang et al. 2022).

16.7.3 Transgenic RNAi for Protection from Prunus Necrotic Ringspot Virus Resistance in Cherry

Transgenic RNA interference (RNAi) represents a promising new approach to develop durable resistance to prunus necrotic ringspot virus (PNRSV) in cherry rootstocks. A landmark study found that biolistic transformation of the cherry rootstocks “Gisela 6” and “Gisela 7” with a hairpin RNA construct designed against the PNRSV genome resulted in transgenic plants with minimal or no symptoms of infection and significantly reduced levels of virus, while non-transgenic controls remained susceptible to infection (or died from infection). The RNAi-mediated approach also creates an opportunity for protecting both genetically modified and conventional cherries through grafting, as the resistant rootstocks will prevent the transmission of the virus to the non-GM scion (Bachman et al. 2020).

16.7.4 RNAi-Mediated Long-Term Protection from Plum Pox Virus in "HoneySweet" Plum

RNAi-mediated long-term protection from plum pox virus (PPV) in "HoneySweet" plum has been robustly demonstrated through extensive field trials and molecular analyses. The "HoneySweet" variety, developed with a hairpin construct of the coat protein gene in PPV, remains virus-free for over a decade even when graft-challenged with infected limbs, and there is no evidence of the virus adapting to overcome this resistance. Small RNA profiling shows that transgene-derived sRNAs are stable across environments and are present at much lower levels than in symptomless, virus-infected plums, minimizing off-target risks. These findings confirm that RNAi technology in "HoneySweet" plum provides a highly effective, durable, and safe solution for managing Sharka disease in stone fruit agriculture (Callahan et al. 2019).

16.8 Challenges and Limitations

16.8.1 Stability of dsRNA in the Environment

Current research highlights that double-stranded RNA (dsRNA) is highly unstable in outdoor and biological environments due to rapid degradation by microbial nucleases, ultraviolet light, rainfall, and harsh conditions such as alkaline insect gut fluids (Palli 2023). Potential solutions for improved efficacy include the development of nanocarrier formulations and protective encapsulations to improve persistence and uptake of dsRNA. However, deployment of these technologies in a field situation is still compromised by changing environmental conditions, poor delivery to the cell, and lack of uptake of dsRNA by many resistant insect species (Ray et al. 2022).

16.8.2 Efficient Delivery Methods

RNA delivery systems for RNA interference (RNAi) have to overcome many obstacles. First, there is the rapid breakdown of RNA molecules by nucleases. Secondly, physical and biological barriers cause the inability of cells to take up the RNA molecules. Thirdly, there is limited distribution of RNA molecules throughout the target organism (Joga et al. 2016).

16.8.3 Biological Safety Concerns

The unintended silencing of genes due to sequence mismatches between the RNAi and target gene sequences, known as off-target effects, can lead to a decrease in specificity and create difficulties in predicting and controlling for off-target effects on non-target organisms as well as the persistence and potential toxic effects on the environment of RNAi molecules and also developing resistance to the effect of RNAi, thus creating a need for increased risk assessment and regulatory oversight. Furthermore, certain types of organisms may require a greater amount of effort to be successfully treated with RNAi, presenting additional obstacles regarding cytotoxicity and the effectiveness of RNAi. Ultimately, off-target effects and biosafety risks continue to create considerable barriers for the widespread acceptance and use of RNAi technology (Pugsley et al. 2021).

16.8.4 Counter-Defense Mechanism against Viruses

An impediment to RNAi technologies is represented by the presence of viral counterdefense mechanisms. Viruses produce viral suppressor(s) of RNA silencing (VSR), which act as a means to stop any of the many essential steps in the RNAi pathway, such as siRNA production, siRNA binding, and the induction of RNA silencing complexes. The suppressor proteins function by either binding to double-stranded RNA, degrading critical components of RNAi, or preventing the systemic movement of the RNAi silencing machinery. Through these mechanisms, viruses evade silencing and survive (Burgyán and Havelda 2011).

16.9 Future Perspectives

16.9.1 dsRNA Delivery via Nanoparticles

Advances in nanoparticle-based delivery systems have greatly improved RNAi technology by providing protection from environmental degradation and increasing cellular uptake in target organisms. Nanocarriers, including liposomes, chitosan, and dendrimers, offer increased stability and the ability to overcome biological barriers such as plant cell walls and insect guts. Engineered nanoparticles offer better protection against nucleases so that dsRNA can cross the endosome more easily and induce more effective gene silencing and fewer off-target effects than other methods. Additionally, polymeric nanocarriers increase the targeting ability and provide safer alternatives when used in agriculture. Ultimately, various nanocarrier strategies are addressing the challenges associated with these delivery systems, and as a result, the use of RNAi to control pests and for therapeutic intervention has become much more practical (Pugsley et al. 2021).

16.9.2 RNAi Integration with CRISPR-Cas9

By combining Genome Editing via CRISPR with Gene Silencing Using RNA interference (RNAi), it is possible to perform more precise and versatile genetic manipulation, including gene knockdowns and knockouts simultaneously. Additionally, when RNAi is used in conjunction with CRISPR, it allows for the ability to perform multiple gene-targeted perturbations with improved efficiency through the ability to modulate the activity of CRISPR guide RNA molecules using RNAi. The new combination of technologies also includes integrated delivery systems using nanovectors to co-deliver components of RNAi and CRISPR together to enhance the efficiency of editing. Additionally, the new delivery methods help to circumvent the challenge of resistance (Mujtaba et al. 2021).

16.9.3 Advances Using Synthetic Biology

Synthetic biology strategies applied to RNAi technologies offer a means of generating specific and multiple-targeted (multiplexed) gene silencing that is programmable through the use of synthetic circuitry. For instance, methods such as FAMSi enable rapid production of numerous siRNAs directed against multiple target genes in an efficient manner, improving upon silencing efficiency. Additionally, synthetic RNA-based logic circuitry offers further complexity when regulating genes within mammalian cell lines, as well as paving the path toward RNAi-based therapies for the treatment of disease(s). Many platforms, such as Syn-tasiR-VIGS, utilize synthetic designs along with viral vectors for delivering and/or expressing scalable and specific RNAi molecules within plants. Thus, applying principles of synthetic biology provides enhanced flexibility and control over RNAi, while also greatly broadening its use within medicine and agriculture (Pfeifer et al. 2023).

16.9.4 Pathways for Regulation and Commercialization

The regulatory and commercialization processes for RNA interference (RNAi) technology require careful navigation of several complex steps. These include comprehensive safety evaluations, thorough environmental impact assessments, and adherence to strict approval procedures established by regulatory bodies such as the Food and Drug Administration (FDA), the Environmental Protection Agency (EPA), and the European Medicines Agency (EMA) (De Schutter et al. 2022). Recent developments have led to the introduction of updated regulatory guidelines that specifically address RNAi therapeutics and biopesticides. These changes indicate a growing acceptance of RNAi-based products and have contributed to a more streamlined approval process (Smagghe 2016).

16.10 Conclusion

RNA interference offers a promising and sustainable method for managing viral diseases in fruit crops by enabling targeted gene silencing of viral pathogens (Chen et al. 2025). Recent advancements in delivery technologies, such as nanoparticle-mediated systems, have improved the stability and uptake of double-stranded RNA, thereby enhancing the efficiency of RNAi and expanding its practical applications

(Mathur et al. 2025). The integration of RNAi with complementary biotechnologies, including genome editing, further increases its potential for precise virus control (Touzdjian Pinheiro Kohlrausch Távora et al. 2022). Ongoing research focused on optimizing delivery methods, increasing specificity, and minimizing off-target effects will be essential to fully realize the benefits of RNAi in fruit crop virus management (Yang et al. 2022). As regulatory approvals advance and commercialization pathways develop, RNAi-based strategies are expected to become important tools for protecting fruit crops from viral threats. These approaches will contribute to global food security and promote sustainable agricultural practices (De Schutter et al. 2022).

References

Akbar S, Wei Y, Zhang M-Q (2022) RNA interference: promising approach to combat plant viruses. Int J Mol Sci 23(10):5312. https://doi.org/10.3390/ijms23105312

Ayala-Álvarez I, Valenzuela-Soto JH, De Nova-Blanca A, Déna-Félix S, García-Estrada RS, Cambrón-Crisantos JM, Ochoa-Martínez DL (2022) Viruses of Economic Impact on Tomato Crops in Mexico: From Diagnosis to Management—A Review. Viruses 14(6):1251. https://doi.org/10.3390/v14061251

Bachman P, Fischer J, Song Z, Urbanczyk-Wochniak E, Watson G (2020) Environmental fate and dissipation of applied dsRNA in soil, aquatic systems, and plants. Front Plant Sci 11. https://doi.org/10.3389/fpls.2020.00021

Biedenkopf D, Will T, Knauer T, Jelonek L, Furch ACU, Busche T, Koch A (2020) Systemic spreading of exogenous applied RNA biopesticides in the crop plant Hordeum vulgare. ExRNA 2(1). https://doi.org/10.1186/s41544-020-00052-3

Bobbin ML, Rossi JJ (2016) RNA interference (rnai)-based therapeutics: delivering on the promise? Annu Rev Pharmacol Toxicol 56(1):103–122. https://doi.org/10.1146/annurev-pharmtox-010715-103633

Burgyán J, Havelda Z (2011) Viral suppressors of RNA silencing. Trends Plant Sci 16(5):265–272. https://doi.org/10.1016/j.tplants.2011.02.010

Calarco JP, Borges F, Donoghue MTA, Van Ex F, Jullien PE, Lopes T, Gardner R, Berger F, Feijó JA, Becker JD, Martienssen RA (2012) Reprogramming of DNA methylation in pollen guides epigenetic inheritance via small RNA. Cell 151(1):194–205. https://doi.org/10.1016/j.cell.2012.09.001

Callahan AM, Dardick CD, Scorza R (2019) Multilocation comparison of fruit composition for 'HoneySweet', an RNAi based plum pox virus resistant plum. PLoS One 14(3):e0213993. https://doi.org/10.1371/journal.pone.0213993

Carthew RW, Sontheimer EJ (2009) Origins and mechanisms of miRNAs and siRNAs. Cell 136(4):642–655. https://doi.org/10.1016/j.cell.2009.01.035

Chen C, Imran M, Feng X, Shen X, Sun Z (2025) Spray-induced gene silencing for crop protection: recent advances and emerging trends. Front Plant Sci 16. https://doi.org/10.3389/fpls.2025.1527944

Choi Y, Gehring M, Johnson L, Hannon M, Harada JJ, Goldberg RB, Jacobsen SE, Fischer RL (2002) DEMETER, a DNA glycosylase domain protein, is required for endosperm gene imprinting and seed viability in arabidopsis. Cell 110(1):33–42. https://doi.org/10.1016/s0092-8674(02)00807-3

Das PR, Sherif SM (2020) Application of exogenous dsRNAs-induced RNAi in agriculture: challenges and triumphs. Front Plant Sci 11. https://doi.org/10.3389/fpls.2020.00946

de Farias ST, dos Santos Junior AP, Rêgo TG, José MV (2017) Origin and evolution of RNA-dependent RNA polymerase. Front Genet 8. https://doi.org/10.3389/fgene.2017.00125

De Schutter K, Taning CNT, Van Daele L, Van Damme EJM, Dubruel P, Smagghe G (2022) RNAi-based biocontrol products: market status, regulatory aspects, and risk assessment. Frontiers in Insect Science 1. https://doi.org/10.3389/finsc.2021.818037

Elbashir SM, Harborth J, Lendeckel W, Yalcin A, Weber K, Tuschl T (2001) Duplexes of 21-nucleotide RNAs mediate RNA interference in cultured mammalian cells. Nature 411(6836):494–498. https://doi.org/10.1038/35078107

El-Sappah AH, Yan K, Huang Q, Islam MM, Li Q, Wang Y, Khan MS, Zhao X, Mir RR, Li J, El-Tarabily KA, Abbas M (2021) Comprehensive mechanism of gene silencing and its role in plant growth and development. Front Plant Sci 12. https://doi.org/10.3389/fpls.2021.705249

Ershova N, Sheshukova E, Kamarova K, Arifulin E, Tashlitsky V, Serebryakova M, Komarova T (2022) Nicotiana benthamiana Kunitz peptidase inhibitor-like protein involved in chloroplast-to-nucleus regulatory pathway in plant-virus interaction. Front Plant Sci 13. https://doi.org/10.3389/fpls.2022.1041867

Feng S, Jacobsen SE, Reik W (2010) Epigenetic reprogramming in plant and animal development. Science 330(6004):622–627. https://doi.org/10.1126/science.1190614

Fontes EPB, Teixeira RM, Lozano-Durán R (2021) Plant virus-interactions: Unraveling novel defense mechanisms under immune-suppressing pressure. Curr Opin Biotechnol 70:108–114. https://doi.org/10.1016/j.copbio.2021.03.007

Gaglia MM, Glaunsinger BA (2010) Viruses and the cellular RNA decay machinery. WIREs RNA 1(1):47–59. https://doi.org/10.1002/wrna.3

García-Estrada RS, Diaz-Lara A, Aguilar-Molina VH, Tovar-Pedraza JM (2022) Viruses of economic impact on tomato crops in Mexico: from diagnosis to management—a review. Viruses 14(6):1251. https://doi.org/10.3390/v14061251

Golubkina N, Kekina H, Caruso G (2018) Yield, quality and antioxidant properties of Indian mustard (Brassica juncea L.) in response to foliar biofortification with selenium and iodine. Plants 7(4):80. https://doi.org/10.3390/plants7040080

Guo S, Kemphues KJ (1995) par-1, a gene required for establishing polarity in C. elegans embryos, encodes a putative Ser/Thr kinase that is asymmetrically distributed. Cell 81(4):611–620. https://doi.org/10.1016/0092-8674(95)90082-9

Guo Z, Schnurbusch T (2016) Costs and benefits of awns. J Exp Bot 67(9):2533–2535. https://doi.org/10.1093/jxb/erw140

Hu R-G, Brower CS, Wang H, Davydov IV, Sheng J, Zhou J, Kwon YT, Varshavsky A (2006) Arginyltransferase, its specificity, putative substrates, bidirectional promoter, and splicing-derived isoforms. J Biol Chem 281(43):32559–32573. https://doi.org/10.1074/jbc.m604355200

Joga MR, Zotti MJ, Smagghe G, Christiaens O (2016) RNAi efficiency, systemic properties, and novel delivery methods for pest insect control: what we know so far. Front Physiol 7. https://doi.org/10.3389/fphys.2016.00553

Kakareka NN, Volkov YG, Tolkach VF, Shchelkanov MY (2024) Diversity of phytovirus strains in the south of the Russian Far East. South Russia Ecol Develop 19(3):18–34. https://doi.org/10.18470/1992-1098-2024-3-2

Li Y, Hu W, Ganie SA, Liu Z, Cheng B, Duan H, Li Y (2025) A novel and efficient Agrobacterium-mediated transient gene expression in citrus epicotyls and mature stem tissues. Fruit Res 5(1):0–0. https://doi.org/10.48130/frures-0024-0035

Mathur S, Chaturvedi A, Ranjan R (2025) Advances in RNAi-based nanoformulations: revolutionizing crop protection and stress tolerance in agriculture. Nanoscale Adv 7(7):1768–1783. https://doi.org/10.1039/d5na00044k

Matzke MA, Mosher RA (2014) RNA-directed DNA methylation: an epigenetic pathway of increasing complexity. Nat Rev Genet 15(6):394–408. https://doi.org/10.1038/nrg3683

Muhammad T, Zhang F, Zhang Y, Liang Y (2019) RNA interference: a natural immune system of plants to counteract biotic stressors. Cells 8(1):38. https://doi.org/10.3390/cells8010038

Mujtaba M, Wang D, Carvalho LB, Oliveira JL, do Espirito Santo Pereira A, Sharif R, Jogaiah S, Paidi MK, Wang L, Ali Q, Fraceto LF (2021) Nanocarrier-mediated delivery of miRNA, RNAi, and CRISPR-Cas for plant protection: current trends and future directions. ACS Agric Sci Technol 1(5):417–435. https://doi.org/10.1021/acsagscitech.1c00146

Napoli C, Lemieux C, Jorgensen R (1990) Introduction of a chimeric Chalcone synthase gene into petunia results in reversible co-suppression of homologous genes in trans. Plant Cell:279–289. https://doi.org/10.1105/tpc.2.4.279

Pal G, Ingole KD, Yavvari PS, Verma P, Kumari A, Chauhan C, Chaudhary D, Srivastava A, Bajaj A, Vemanna RS (2024) Exogenous application of nanocarrier-mediated double-stranded RNA manipulates physiological traits and defence response against bacterial diseases. Mol Plant Pathol 25(1). https://doi.org/10.1111/mpp.13417

Palli SR (2023) RNAi turns 25:contributions and challenges in insect science. Front Insect Sci 3. https://doi.org/10.3389/finsc.2023.1209478

Pascual L, Yan J, Pujol M, Monforte AJ, Picó B, Martín-Hernández AM (2019) CmVPS41 is a general gatekeeper for resistance to cucumber mosaic virus phloem entry in melon. Front Plant Sci 10. https://doi.org/10.3389/fpls.2019.01219

Paturi S, Deshmukh MV (2021) A glimpse of "dicer biology" through the structural and functional perspective. Front Mol Biosci 8. https://doi.org/10.3389/fmolb.2021.643657

Pfeifer BA, Beitelshees M, Hill A, Bassett J, Jones CH (2023) Harnessing synthetic biology for advancing RNA therapeutics and vaccine design. Npj Syst Biol Appl 9(1). https://doi.org/10.1038/s41540-023-00323-3

Pugsley CE, Isaac RE, Warren NJ, Cayre OJ (2021) Recent advances in engineered nanoparticles for rnai-mediated crop protection against insect pests. Front Agron 3. https://doi.org/10.3389/fagro.2021.652981

Ram H, Singh BP, Katara J, Kumar P, Jaiswal R, Geetika DR, Sonah H (2019) Insertional Mutagenesis Approaches and Their Use in Rice for Functional Genomics. Plants 8(9):310. https://doi.org/10.3390/plants8090310

Ray P, Sahu D, Aminedi R, Chandran D (2022) Concepts and considerations for enhancing RNAi efficiency in phytopathogenic fungi for RNAi-based crop protection using nanocarrier-mediated dsRNA delivery systems. Front Fungal Biol 3. https://doi.org/10.3389/ffunb.2022.977502

Sáez C, Flores-León A, Montero-Pau J, Sifres A, Dhillon NPS, López C, Picó B (2022) RNA-Seq transcriptome analysis provides candidate genes for resistance to tomato leaf curl New Delhi virus in melon. Front Plant Sci 12. https://doi.org/10.3389/fpls.2021.798858

Sanan-Mishra N, Abdul Kader Jailani A, Mandal B, Mukherjee SK (2021) Secondary siRNAs in plants: biosynthesis, various functions, and applications in virology. Front Plant Sci 12. https://doi.org/10.3389/fpls.2021.610283

Singh K, Callahan AM, Smith BJ, Malinowski T, Scorza R, Jarošová J, Beoni E, Polák J, Kundu JK, Dardick C (2021) Long-Term Efficacy and Safety of RNAi-Mediated Virus Resistance in 'HoneySweet' Plum. Front Plant Sci 12(12):726881. https://doi.org/10.3389/fpls.2021.726881

Smagghe G (2016) RNAi-based biopesticides. 2016 international congress of entomology. https://doi.org/10.1603/ice.2016.93500

Song G, Sink KC, Walworth AE, Cook MA, Allison RF, Lang GA (2013) Engineering cherry rootstocks with resistance to Prunus necrotic ring spot virus through RNAi-mediated silencing. Plant Biotechnol J 11(6):702–708. https://doi.org/10.1111/pbi.12060

Spolaore S, Trainotti L, Casadoro G (2001) A simple protocol for transient gene expression in ripe fleshy fruit mediated by Agrobacterium. J Exp Bot 52(357):845–850. https://doi.org/10.1093/jexbot/52.357.845

Sugiyama T, Cam H, Verdel A, Moazed D, Grewal SIS (2004) RNA-dependent RNA polymerase is an essential component of a self-enforcing loop coupling heterochromatin assembly to siRNA production. Proc Natl Acad Sci 102(1):152–157. https://doi.org/10.1073/pnas.0407641102

Tatineni S, Hein GL (2023) Plant viruses of agricultural importance: Current and future perspectives of virus disease management strategies. Phytopathol 113(2):117–141. https://doi.org/10.1094/PHYTO-05-22-0167-RVW

Touzdjian Pinheiro Kohlrausch Távora F, de Assis dos Santos Diniz F, de Moraes Rêgo-Machado C, Chagas Freitas N, Barbosa Monteiro Arraes F, Chumbinho de Andrade E, Furtado LL, Osiro KO, Lima de Sousa N, Cardoso TB, Márcia Mertz Henning L, Abrão de Oliveira Molinari P, Feingold SE, Hunter WB, Fátima Grossi de Sá M, Kobayashi AK, Lima Nepomuceno A, Santiago TR, Correa Molinari HB (2022) CRISPR/Cas- and topical rnai-based technologies for crop management and improvement: reviewing the risk assessment and challenges towards a more sustainable agriculture. Front Bioeng Biotechnol 10. https://doi.org/10.3389/fbioe.2022.913728

Wang H, Zhang X, Liu J, Kiba T, Woo J, Ojo T, Hafner M, Tuschl T, Chua N, Wang X (2011) Deep sequencing of small RNAs specifically associated with Arabidopsis AGO1 and AGO4 uncovers new AGO functions. Plant J 67(2):292–304. https://doi.org/10.1111/j.1365-313x.2011.04594.x

Weiss D, Sede AR, Levanova AA, Leibman-Markus M, Gupta R, Mishra R, Hak H, Bar M, Poranen MM, Heinlein M, Spiegelman Z (2025) Control of tomato brown rugose fruit virus (ToBRFV) in tomato plants using in vivo-synthesized dsRNA. J Exp Bot. https://doi.org/10.1093/jxb/eraf293

Weiss D, Sede AR, Levanova AA, Leibman-Markus M, Gupta R, Mishra R, Hak H, Bar M, Poranen MM, Heinlein M, Spiegelman Z (2026) Control of tomato brown rugose fruit virus (ToBRFV) in tomato plants using in vivo synthesized dsRNA. J Exp Bot 77(3):865–879. https://doi.org/10.1093/jxb/eraf293

Wu J, Yang J, Cho WC, Zheng Y (2020) Argonaute proteins: structural features, functions and emerging roles. J Adv Res 24:317–324. https://doi.org/10.1016/j.jare.2020.04.017

Yang W, Wang B, Lei G, Chen G, Liu D (2022) Advances in nanocarriers to improve the stability of dsRNA in the environment. Front Bioeng Biotechnol 10. https://doi.org/10.3389/fbioe.2022.974646

Younger ST, Kenzelmann-Broz D, Jung H, Attardi LD, Rinn JL (2015) Integrative genomic analysis reveals widespread enhancer regulation by p53 in response to DNA damage. Nucleic Acids Res 43(9):4447–4462. https://doi.org/10.1093/nar/gkv284

Zhao J, Fu H, Wang Z, Zhang M, Liang Y, Cui X, Pan W, Ren Z, Wu Z, Zhang Y, Gui X, Huo L, Lei X, Wang C, Schnittger A, Pawlowski WP, Liu B (2024) Genetic variation in the speciesArabidopsis thalianareveals the existence of natural heat resilience factors for meiosis. Cold Spring Harbor Laboratory. https://doi.org/10.1101/2024.07.16.603532

Virus-Induced Gene Silencing (VIGS) in Functional Genomics of Fruit Crops

17

Tabasum Manzoor, Aadil Ayaz, Tabia Fayaz, Subaya Manzoor, Zakir Amin, Sajad Un Nabi, and Bushra Rasool

Abstract

Functional genomics represents the methods used in unravelling the genetic basis of complex factors such as fruit development, ripening, quality, disease resistance, and stress tolerance. Traditional genetic practices, primarily via stable transformation, are often limited by the long generation times and recalcitrant genetics of many fruit crops. A faster, more efficient reverse genetics tool known as virus-induced gene silencing was developed to target genes by transient suppression through the plant's innate RNA silencing pathways without the need for stable transgenic lines. With a focus on tobacco rattle virus (TRV), this chapter presents a detailed overview of VIGS technology and its utilization mainly in apple, strawberry, grapevine, peach, cherry, and citrus, covering the development and optimization of viral vectors. There are also some remarkable insights into important biological pathways like fruit ripening regulation, pigment biosynthesis, hormone signalling, and plant defence processes that are gained using VIGS-assisted gene silencing. In addition to recent developments improving VIGS specificity and robustness, this chapter also discusses practical issues like host specificity, off-target effects, and inconsistent silencing efficiency. Additionally, this chapter examines the new trend of combining VIGS with genome editing and innovative delivery methods, highlighting the potential of VIGS to speed up gene function analysis and molecular breeding to improve fruit crop quality and productivity.

T. Manzoor · A. Ayaz · S. Manzoor · Z. Amin · S. U. Nabi · B. Rasool (✉)
Plant Pathology Laboratory, Division of Fruit Science, ICAR-Central Institute of Temperate Horticulture, Srinagar, Jammu & Kashmir, India

T. Fayaz
Division of Plant Pathology, Faculty of Horticulture, SKUAST-Kashmir, Shalimar, Jammu & Kashmir, India

R. Kumar et al. (eds.), *Fruit Crops and Viral Pathogens Volume II*, Plant Pathology for Sustainable Agriculture, https://doi.org/10.1007/978-981-95-7261-8_17

Keywords

V*IGS* · Functional genomics · RNA silencing · Plant defence · RNA interference · Molecular breeding

17.1 Introduction

Fruit crops like apple, mango, pear, peach, citrus, etc., are an important part of human nutrition besides contributing significantly to the economy and employment generation (Nabi and Shafi 2025). Considering their importance, improving the key characteristics, viz., fruit development, ripening, quality, fruit colour, disease resistance, and stress tolerance, has become essential. Achieving these improved characteristics requires a comprehensive understanding of the genetics behind the processes. Functional genomics in fruit crops explores the underlying genes and their regulatory networks that control different biological processes, imparting different characteristics to fruit crops. However, functional genomic approaches in fruit crops are often constrained by their complex genomes, prolonged juvenile phase, and problems associated with stable genetic transformation (Wesley et al. 2001; Miki and Shimamoto 2004). To address these limitations, virus-induced gene silencing (VIGS) has emerged as an efficient approach for studying gene function. VIGS is a reverse genetics technique that utilizes the antiviral RNA silencing mechanism of plants to suppress the expression of the target gene. The approach has been successfully applied to fruit crops such as strawberry, grapevine, banana, and citrus, offering valuable insights into processes like ripening, pigment biosynthesis, hormone signalling, and defence responses (Liu et al. 2016; Fei et al. 2021). VIGS was initially observed in 1990 as a plant recovery mechanism from viral infection (Van Kammen 1997). To silence the NbPDS gene in *Nicotiana benthamiana*, the first VIGS vector (tobacco mosaic virus) was developed, which resulted in a distinct albino phenotype, paving the way for VIGS as a functional genomics tool (Ruiz et al. 1998). Since then, VIGS has become a high-throughput method that can cause heritable epigenetic changes and post-transcriptional gene silencing (PTGS), which has important implications for functional research and plant breeding (Zulfiqar et al. 2023). Ongoing developments in viral vector systems and delivery methods continue to increase the usefulness and dependability of VIGS in fruit crop research, despite certain issues with host range and silencing efficiency (Ratcliff et al. 2001).

17.1.1 Strategies for the Development of Virus-Induced Gene Silencing Technology

(a) *Criteria for Selecting Viral Vectors*

Only a small number of plant viruses have been successfully modified for virus-induced gene silencing (VIGS) in studies, despite the fact that a large variety of plant viruses are found in nature. This is mostly because it is technically challenging to transform naturally occurring plant viruses into dependable and secure gene-silencing tools. Viruses that have been reported for VIGS applications include tobacco rattle virus (TRV), tobacco mosaic virus (TMV), potato virus X (PVX), barley stripe mosaic virus (BSMV), bean pod mottle virus (BPMV), tomato golden mosaic virus (TGMV), satellite virus (SV), cabbage leaf curl virus (CbLCV), apple latent spherical virus (ALSV), prunus necrotic ringspot virus (PNRSV), cucumber mosaic virus (CMV), citrus tristeza virus (CTV), cucumber green mottle mosaic virus (CGMMV), and pea early browning virus (PEBV). While most VIGS systems were initially developed for herbaceous plants such as Nicotiana benthamiana, their application has recently been extended to woody fruit trees. The most important viruses used are TRV, ALSV, CMV, PVX, and CGMMV, which have demonstrated successful gene silencing in a variety of fruit crops (Lin et al. 2008; Cui and Wang 2017; Killiny 2020; Liu et al. 2020). For example, TRV has been used to suppress gene expression in apples (Malus × domestica) and peaches (*Prunus persica*) (Bai et al. 2016). Similarly, citrus leaf blotch virus (CLBV)-based vectors have been used in gene functional studies of Citrus species (Agüero et al. 2014). Due to its broad host range and capacity to cause gene silencing in fruit trees belonging to the *Rosaceae* family, including apples, European pears (*Pyrus communis*), and Japanese pears (*Pyrus pyrifolia*), ALSV has become one of the most promising virus-based vectors (Ito et al. 2012; Takahashi et al. 2013; Yamagishi and Yoshikawa 2009). TRV-based VIGS has been useful for researching genes associated with fruit skin colouration in Prunus species, particularly in peaches (Bai et al. 2016). This system has been used to study the functions of genes in fruit ripening, pigmentation, and drought response in sweet cherries (Li et al. 2017). A crucial prerequisite for any VIGS vector is that the host plant must not experience severe symptoms. This guarantees that observed alterations in plant physiology or development can be linked to gene silencing rather than damage caused by viruses. Since TRV usually only produces minor symptoms while maintaining a high level of gene silencing signal delivery efficiency, it effectively satisfies this requirement. TRV is still one of the most popular and trustworthy methods for analysing gene function in both herbaceous and woody plants because of its straightforward genome structure, wide host range of over 400 plant species, and capacity to enter meristematic tissues (Bachan and Dinesh-Kumar 2012).

(b) *Selection of the Reporter Gene*

In virus-induced gene silencing (VIGS), reporter genes are essential because they verify that the silencing mechanism is working and that the virus has successfully infected the plant. It's crucial to first confirm that the virus can infect, replicate, and initiate gene silencing in a new plant species before introducing a VIGS vector. The reporter gene will continue to express itself normally if the virus is unable to infect, and no alterations will be noticed. On the other hand, if the crop is a suitable host, silencing the reporter gene will result in observable alterations that are a sign of successful gene silencing. Phytoene desaturase (PDS), a crucial component in the synthesis of carotenoids, is a frequently employed reporter gene. Red fruits, like tomatoes, develop yellow patches as a result of decreased lycopene synthesis, while green tissues lose pigmentation and turn white when PDS is silenced. Because of this obvious change, PDS is a popular and trustworthy indicator for evaluating VIGS effectiveness. Similar situations have also involved the use of other pigment-related genes. For example, genes involved in the production of chlorophyll, such as Chl H, Su, and CLA1, and chalcone synthase (CHS), which aids in the biosynthesis of anthocyanins, can cause noticeable colour changes when silenced, making the effects easy to detect (Fu et al. 2006). Furthermore, under particular lighting conditions, non-pigment markers such as green fluorescent protein (GFP) provide an additional visual means of tracking gene silencing. Tomato plants have also been given genes from Antirrhinum majus, such as Delila (Del) and Roseal (Ros1), whose expression produces a purple hue that can be silenced. Additionally, ethylene signalling-related genes like SlCTR1, SlEIN2, and SlEIN3 are helpful reporter genes because their silencing can clearly alter fruit development and colouration, providing an additional means of monitoring VIGS activity.

(c) *Infection Methods*

The process of virus-induced gene silencing (VIGS) depends on infection because it allows the virus that carries the desired gene fragment to replicate efficiently and spread to the right plant tissues. Agrobacterium-mediated transformation, RNA-based delivery, and DNA-based delivery are the three main methods for introducing VIGS into plants. The most popular of these is Agrobacterium-mediated infection because of its effectiveness and ease of use. Agrobacterium infiltration has been successfully achieved using a variety of plant tissues, such as sprouts (Yan et al. 2012), roots (Ryu et al. 2004), stems (Wang et al. 2015), and leaves (Liu et al. 2012). VIGS has also targeted the fruit itself (Orzaez et al. 2006) and the carpopodium (fruit stalk) of young fruit in reproductive tissues (Fu et al. 2005). One technique is to infect the parent plant so that the virus can multiply, travel systemically to the meristematic areas, and eventually reach the fruit. This approach is simple to use, but it requires more time and frequently results in a lower virus concentration by the time it reaches the fruit, which reduces the effectiveness of gene silencing. On the other hand, infection through the fruit carpopodium enables more rapid and direct silencing in adjacent fruit tissues without resulting in physical harm. This entails injecting the base of the attached or detached fruit with an Agrobacterium suspension containing the VIGS construct (Orzaez et al. 2006). Fruit injection is typically avoided despite being used in several plant species because it can damage the fruit, cause premature abscission, or produce symptoms that make it difficult to interpret phenotypes. Another method for detached fruits is vacuum infiltration, which can also cause tissue damage. Therefore, to achieve effective silencing while preserving tissue integrity, techniques that minimize direct damage to the fruit, such as infection through sprouts (Yan et al. 2012), leaves (Liu et al. 2002), stems (Wang et al. 2015), or the fruit carpopodium (Fu et al. 2005), are typically preferred.

(d) *Verification and Analysis of Gene Silencing*

Two key factors need to be considered in order to determine whether a virus-induced gene silencing (VIGS) system has been successful. Verifying that the virus can infect the target plant species is the first step. This is typically accomplished by looking for the virus in recently developed leaves, which indicates that it has penetrated the plant, multiplied, and spread throughout the tissue. It is crucial to visually confirm virus infection by silencing a reporter gene when introducing VIGS to a species for the first time. The phytoene desaturase (PDS) gene is one commonly used reporter. A simple visual indicator of gene silencing, silencing PDS results in newly formed leaves exhibiting distinctive white or yellow bleaching patterns due to disrupted pigment formation. Real-time quantitative PCR (RT-qPCR) is used to quantify decreases in the target gene's expression in order to more accurately assess the degree of gene silencing. Silencing ripening-related genes in fruit can change the colouration pattern, frequently producing fruits with different coloured sections. A visual marker system is required to identify silenced tissue for sampling if silencing does not result in noticeable changes. Using tomato plants that have been genetically modified to express the Del and Ros1 transcription factors from Antirrhinum majus is a workable approach. It is simple to identify silenced regions when these genes are silenced alongside the target gene because the normally purple areas lose pigmentation (Orzaez et al. 2009). Transgenic tomatoes that express green fluorescent protein (GFP) are another helpful method; the lack of fluorescence under UV light indicates the presence of silenced tissues (Quadrana et al. 2011). It is crucial to use RT-qPCR to confirm the silencing at the molecular level after visually identifying the silenced tissues. To learn more about the broader biological effects of gene silencing, validated samples may then be used for more in-depth analyses like RNA sequencing, proteomic profiling, or metabolomics.

17.1.2 Molecular Mechanism of VIGS

Virus-induced gene silencing (VIGS) is a method that analyses gene function and improves fruit quality, colour, and disease resistance by taking advantage of plants' natural transcriptional (TGS) and post-transcriptional (PTGS) gene silencing defences (Rössner et al. 2022). Double-stranded RNA (dsRNA), which is present when a virus infects a plant, initiates these silencing pathways, which destroy viral RNA and stop additional infection. Viral genomes are modified to eliminate symptom-causing genes in VIGS systems mediated by *Agrobacterium tumefaciens*. This allows the creation of binary vectors containing viral cDNA sequences regulated by the CaMV35S promoter. Plant gene fragments intended for silencing can be inserted into these vectors using cloning sites (Voinnet 2001; Liu et al. 2002). These altered viruses successfully induce PTGS, which causes the corresponding plant mRNA to degrade, because they lack potent suppressors of gene silencing. A 300–500 base pair segment of the target gene is usually inserted into the viral genome to produce the VIGS construct. In order to create siRNAs that specifically silence the targeted gene without affecting other genes, this sequence was carefully selected (Xu et al. 2006). Plants can be infected with the recombinant virus through direct DNA delivery, RNA inoculation, or Agrobacterium infiltration. Following infection, viral RNA replication produces dsRNA molecules, which Dicer or Dicer-like enzymes recognize and cleave into 21–24 nucleotide-long small interfering RNAs (siRNAs). The guide strand of these siRNAs is loaded into the RNA-induced silencing complex (RISC), which includes Argonaute proteins such as AGO1. Guided by the siRNA, RISC targets complementary mRNAs for cleavage, blocking their translation and silencing the gene (Ramegowda et al. 2014). Besides *Agrobacterium*-mediated silencing via PTGS, plant viruses transmitted by vectors also use RNA virus-induced PTGS and DNA virus-induced TGS. DNA viruses replicate inside the nucleus through rolling circle replication, generating overlapping RNA transcripts that form dsRNA. These are processed by Dicer-like 3 (DCL3) into 24-nucleotide siRNAs. These siRNAs assemble with the RNA-induced transcriptional silencing (RITS) complex containing AGO4, which targets viral or host DNA for methylation, causing transcriptional gene silencing (Rössner et al. 2022). In contrast, RNA viruses replicate in the cytoplasm, producing dsRNA intermediates that are cleaved by Dicer-like enzymes DCL2 and DCL4 into 22- and 21-nucleotide siRNAs. These are methylated by HEN1 and loaded into AGO1-containing RISC complexes, which degrade viral and endogenous mRNAs, suppressing viral gene expression. This pathway involves AMP1 at the endoplasmic reticulum and allows siRNAs to move between cells through plasmodesmata, spreading the silencing effect (Rössner et al. 2022). VIGS-mediated gene silencing typically lasts around 3 weeks, after which the silencing effect fades and the plant recovers normal gene expression. However, recent techniques that adjust growth conditions to promote viral replication have extended the duration of silencing to several months. The effect is more stable and comparable to transgenic plants in certain situations, where the silencing may even be passed down to the following generation. This development greatly expands the potential applications of VIGS in crop breeding and functional genomics (Senthil-Kumar et al. 2008).

(a) *Role of RNA Interference (RNAi) in VIGS*

The primary molecular mechanism behind VIGS is RNA interference (RNAi), which breaks down particular messenger RNA (mRNA) molecules. By focusing on mRNAs that complement small RNA sequences, this conserved cellular pathway controls the expression of genes. The viral vector used in VIGS is designed to carry plant gene sequences that are meant to be silenced. Following fruit crop infection, the virus multiplies and produces intermediates of double-stranded RNA (dsRNA). DICER-like enzymes convert these dsRNAs into small interfering RNAs (siRNAs), which are subsequently incorporated into the RNA-induced silencing complex (RISC). The RISC complex effectively suppresses the expression of the targeted gene by locating and cleaving matching mRNAs in the plant cell using the siRNAs as guides. By monitoring the resultant phenotypic changes, scientists can investigate gene function through targeted gene knockdown (Li and Ding 2001).

(b) *Target Specificity: How VIGS Precisely Silences Desired Genes*

To achieve targeted gene silencing, VIGS depends on the deliberate insertion of a target gene fragment into the viral vector. Typically, a 300–500 base pair section of the target gene is cloned into the viral genome. In order to prevent inadvertently silencing non-target genes, this enables the virus to create double-stranded RNA molecules that precisely match the target gene sequence. How well the small interfering RNAs (siRNAs) produced from the dsRNA match the target mRNA sequence is a major factor in this process' accuracy. The strength of the viral promoter that drives expression, the RNA's structural arrangement, and the plant's natural RNA interference machinery are some of the variables that can impact the overall silencing efficiency. Furthermore, the specificity and effectiveness of gene silencing are greatly influenced by both the viral vector system selection and the insertion site of the gene fragment within the vector (Xu et al. 2006).

(c) *Key Factors Influencing VIGS Specificity and Effectiveness*

Many factors influence the efficacy and specificity of virus-induced gene silencing. The length and sequence of the target gene fragment inserted into the viral genome are important factors; fragments between 300 and 500 base pairs typically produce the best siRNA. The choice of viral vector is also crucial; to facilitate effective post-transcriptional gene silencing, viruses with weak or nonexistent gene silencing suppressors are favoured. Processing dsRNA and degrading complementary mRNAs requires the plant's own RNA interference machinery, specifically DICER-like enzymes and the RNA-induced silencing complex (RISC). Additionally, the target mRNA's secondary structure and conserved nucleotide motifs influence siRNA binding and, in turn, the effectiveness of silencing. Viral replication rates and the length of gene silencing have been demonstrated to be influenced by environmental factors, such as temperature and light exposure, with some growth conditions prolonging the silencing effect (Senthil-Kumar and Mysore 2011).

17.2 VIGS Applications in Functional Genomic Approaches of Fruit Crops

In many fruit crops, virus-induced gene silencing (VIGS) is used for rapid gene function analysis (Table 17.1). Tobacco rattle virus (TRV) is the preferred viral vector, and tomatoes continue to be the most widely used fruit model. Additionally, VIGS has been successfully applied to fruits where stable genetic transformation is frequently challenging, such as apples, peaches, pears, litchi, and mangos. The VIGS vector is usually injected into detached fruits in these studies, but there is little proof that these species can successfully infect entire plants with the virus. It is therefore advised to verify the effectiveness of VIGS in these fruit crops. VIGS is widely used to study genes associated with pigment biosynthesis and fruit ripening. Affected fruit sections may exhibit phenotypes such as irregular ripening or altered

Table 17.1 Fruit crop functional genomics using virus-induced gene silencing as a tool

S.no.	Fruit crop	Virus	Challenge	Target gene	Reference
01.	Citrus	Citrus leaf blotch virus	Long juvenile period	CiFT	Velázquez et al. (2016)
02.	Apple	Apple latent spherical virus	Long juvenile period	MdTFL1	Yamagishi et al. (2016)
03	Apple	Tobacco rattle virus	Disease resistance (*Botryosphaeria dothidea*)	MdCNGC2	Zhou et al. (2020)
04	Apple	Tobacco rattle virus	Biosynthesis of anthocyanin	MdHB1	Jiang et al. (2017)
05	Strawberry	Tobacco rattle virus	Fruit ripening	FvTCP9	Xie et al. (2020)
06	Peach	Tobacco rattle virus	Coloration	PpGST1	Zhao et al. (2019)
07	Mango	Tobacco rattle virus	Fruit ripening and coloration	RCCR	Liu et al. (2018)
08	Litchi	Tobacco rattle virus	Development of seed	LcCWIN2, LcCWIN5	Zhang et al. (2018)
09	Pear	Tobacco rattle virus	Anthocyanin accumulation	PpBBX16	Bai et al. (2019)
10	Mulberry	Tobacco rattle virus	Anthocyanin accumulation	MmCHI2	Chao et al. (2021)
11	Kiwi	Tobacco rattle virus	Anthocyanin accumulation	AcMYB10	Yu et al. (2019)
12	Loquat	Tobacco rattle virus	Fruit coloration and carotenoid accumulation	CYC-B	Hong et al. (2019)
13	Grape berry	Tobacco rattle virus	Biosynthesis of anthocyanin	VvMYBA1	Zhang et al. (2020)
14	Citrus	Tobacco rattle virus	Biosynthesis of limonoids	CiOSC	Wang et al. (2017)
15	Sweet cherry	Tobacco rattle virus	Fruit ripening and softening	PaPME1, PaPME2	Qi et al. (2020)
16	Sweet cherry	Tobacco rattle virus	Flavonoid biosynthesis	ANS	Qi et al. (2018)

colouration in comparison to untreated sections when a target gene is silenced. Silenced regions may mature before normal tissue if the gene functions as a negative ripening regulator. On the other hand, inhibiting positive regulators frequently causes incomplete or delayed ripening, impacting processes like the breakdown of chlorophyll and the build-up of pigments like anthocyanins and lycopene. These alterations offer a practical way to verify the functions of genes. Gene families linked to fruit development are also screened using this technology. SlNOR-like1 is a positive regulator of ripening, according to Gao et al. (2018), who used VIGS to evaluate 34 NAC transcription factors in tomatoes. RT-qPCR verified that the uneven ripening patterns caused by this gene's silencing were correlated with lower expression levels. Similarly, Meng et al. (2018) screened seven HD-Zip genes and showed that SlBEL11 inhibits the synthesis of chlorophyll in tomato fruit, with silencing increasing the accumulation of chlorophyll in specific regions. In another instance, tomato fruit ripening was postponed by using PVX vectors to silence the HD-Zip homeobox gene LeHB-1 (Lin et al. 2008). VIGS has been useful in characterizing ripening-related proteins in addition to transcription factors. Wang et al. (2014) found that two ubiquitin-conjugating enzymes, SlUBC32 and SlUBC41, are involved in the control of ripening. Patchy ripening was observed in fruits silenced for these genes, with delayed yellowing regions exhibiting notable downregulation of the target genes. Zhu et al. (2015) found that two long non-coding RNAs, lncRNA1459 and lncRNA1840, affect the timing of ripening; fruit maturation was significantly delayed when these RNAs were silenced. The high-throughput potential of VIGS is one of its main benefits. More than 100 candidate genes can be effectively screened in tomato fruits using sprout vacuum infiltration (Yan et al. 2012). For this, 300–500 bp fragments of candidate genes are cloned into TRV vectors, transformed into *Agrobacterium tumefaciens* GV3101, and then inoculated into groups of Micro-Tom tomato sprouts. Following vacuum infiltration, seedlings are cultivated in controlled environments until they produce fruit. Fruits with uneven ripening reveal potential genes that influence ripening and can be chosen for in-depth research. Each experimental batch's infection efficiency is monitored with the aid of control plants infected with TRV that carries the PDS gene.

17.2.1 Application of VIGS in Disease Resistance

(a) *R-Gene-Mediated Resistance Through VIGS*

A wide range of pathogens, such as bacteria, fungi, and viruses, are consistently found in plants. They depend on both highly specialized defence mechanisms regulated by resistance (R) genes and general immune responses to survive. A potent technique for examining these genes' roles in plant immunity is virus-induced gene silencing (VIGS) (Senthil-Kumar et al. 2008). Researchers can use VIGS to silence particular genes and track changes in disease resistance. The function of the N gene in protecting *Nicotiana benthamiana* from the tobacco mosaic virus (TMV) has been confirmed using this method. Numerous genes involved in this defence have been found through functional studies using VIGS, including transcription factors like WRKY and MYB, hormone-signalling genes like COI1 and CTR1-like, RanGAP2, RAR1, EDS1, NPR1, MEK1, and NTF6. VIGS has helped identify gene functions involved in immunity to a variety of fungal pathogens in addition to viral resistance. It has assisted in identifying important genes for resistance to powdery mildew in both monocots and dicots, as well as for leaf rust (wheat), leaf mould (tomato), blue mould (tobacco), and late blight (potato). Specifically, VIGS has been used to functionally validate genes associated with Cf-mediated resistance to *Cladosporium fulvum*, including ACIK1, HSP90, nuclear GTPase, L19 ribosomal protein, and NB-LRR proteins. Additionally, VIGS has shown efficacy in the search for tomato Pto-mediated resistance, which offers defence against *Pseudomonas syringae* pv. *tomato*. It has been demonstrated that genes like MEK1, MEK2, and NTF6 are important in this defence mechanism (Uppalapati et al. 2007). Similarly, this method has been used to analyse RPS2-dependent responses in *Arabidopsis thaliana*. In addition to identifying resistance genes, VIGS has demonstrated the role of salicylic acid in defence signalling and clarified the role of pathogen-derived virulence factors like coronatine (COR). This technique has also been used to study the hypersensitive response (HR), a form of programmed cell death that aids in limiting the spread of pathogens and sheds light on the genes involved. VIGS remains an essential method in plant pathology and immunity research because of its capacity to quickly silence genes and reveal their role in defence signalling (Senthil-Kumar et al. 2008).

(b) *Non-host Resistance and Its Genetic Insights Through VIGS*

The most prevalent and reliable type of plant defence is non-host resistance, which offers defence against a variety of pathogens that are unable to infect a specific plant species (Mysore and Ryu 2004). Many of the intricate signalling networks and molecular interactions involved in this type of immunity have been revealed through the use of virus-induced gene silencing (VIGS). The SGT1 protein is linked to ubiquitin ligase complexes, has been shown by VIGS to be an important component of both specific R-gene defence and more general non-host resistance mechanisms (Peart et al. 2002). Researchers have used VIGS to silence a number of defence-related genes and investigate their roles in plant-pathogen interactions using *Nicotiana benthamiana* as a model. As a model of their functional significance in preserving non-host resistance, silencing NbHSP90c-1 and NbHSP70c-1, two members of the heat shock protein family, reduced resistance against *Pseudomonas cichorii*. Similarly, plants exhibited reduced defence and a compromised hypersensitive response (HR) to the same pathogen when the WIPK and SIPK genes encoding stress-related MAP kinases were silenced separately or jointly (Sharma et al. 2003). Via VIGS, another gene, NbCD1, which codes for an ethylene-responsive element-binding protein, was demonstrated to control cell death and play a major role in non-host resistance (Nasir et al. 2005). This demonstrates how immune regulation is influenced by hormone signalling and transcriptional control.

In addition to plant genes, VIGS has made it easier to study pathogen effectors. Through a mechanism dependent on NbSGT1, the type III effector protein DspA/E from *Erwinia amylovora* has been found to induce cell death, offering insight into how pathogens interact with and try to suppress plant immunity. VIGS is still being used in ongoing studies to find more genes involved in non-host resistance, particularly those that might provide long-term defence against bacterial diseases that are economically significant. According to Senthil-Kumar et al. (2008), these findings open the door for crop improvement techniques based on boosting natural immune pathways.

17.3 Advantages of Virus-Induced Gene Silencing (VIGS)

Virus-induced gene silencing (VIGS) is an effective approach that enables researchers to rapidly investigate the functions of particular genes in plants. VIGS uses viruses to quickly "turn off" genes, in contrast to traditional genetic modification, which can take a long time to create plants with permanent changes in their DNA. In a much shorter amount of time, researchers can observe the effects of gene suppression thanks to this temporary silencing. VIGS's ability to function even with only partial gene fragments is one of its main advantages; it does not require the complete gene sequence. When researching recently identified or poorly characterized genes, this adaptability is particularly useful (Dagnachew et al. 2019). Furthermore, VIGS can target a single gene or several related genes simultaneously, which helps comprehend how plant gene groups interact. Because it can infect a wide variety of plants, the tobacco rattle virus (TRV) vector is frequently used in this technique, which uses viral vectors to deliver the gene-silencing signals. Many species, including significant research models like *Nicotiana benthamiana* and *Arabidopsis thaliana*, have successfully used TRV-based VIGS. Because of its wide host range, VIGS can be modified to be used in a variety of plant species, including both crops and wild ones. Because some plants, like *Nicotiana benthamiana*, are more vulnerable to viral infections, VIGS works particularly well in these species. Because of their vulnerability, the virus can effectively spread throughout the plant, suppressing the target genes in a powerful and long-lasting way. Because of this, VIGS is an effective tool for rapidly identifying gene functions and researching plant biology, which eventually helps scientists enhance crop characteristics and gain a deeper understanding of how plants react to their surroundings (Dagnachew et al. 2019).

17.4 Limitations of Virus-Induced Gene Silencing (VIGS)

VIGS has a number of intrinsic drawbacks despite being a potent tool for gene function research. Its reliance on a successful viral infection to cause gene silencing is one major problem. According to Dagnachew et al. (2019), a number of crop species, such as beans, cucumbers, peas, peppers, potatoes, and tomatoes, have genetic resistance to specific viruses that prevents infection and lessens the efficacy of VIGS in these plants. Safety of the environment is another factor. Careful biosafety procedures are required to stop VIGS vectors from spreading accidentally. While it is theoretically possible for small RNA molecules that mediate gene silencing to cross between organisms, there is no proof that this happens in typical VIGS conditions. Without experimental evidence, the possibility of silencing signals being transferred to nearby plants or soil microbes remains a theoretical risk. The lack of viral vectors that are appropriate for all plant species is another drawback. Gene sequences from related species that react favourably to VIGS can be used to solve this issue, or new viral vectors tailored to the target plants can be created. Potential remedies include viruses that can infect a wide variety of plants, such as Apple latent spherical virus (ALSV) and tobacco rattle virus (TRV). Finally, it is still difficult to effectively deliver the viral vectors into plants. Although Agrobacterium-mediated delivery is frequently employed and successful in many dicotyledonous plants, it is not always applicable, and more effective delivery techniques are required to increase the utility of VIGS (Dagnachew et al. 2019).

17.5 Conclusion

In the field of functional genomics, virus-induced gene silencing (VIGS) has become a potent and cutting-edge tool, especially for fruit crops, whose complex genomes, lengthy life cycles, and resistance to transformation frequently make traditional genetic manipulation difficult. VIGS speeds up the study of gene function directly in fruit tissues by enabling quick and targeted downregulation of particular genes without the need for stable genetic modification. This enables researchers to understand the molecular underpinnings of critical characteristics like fruit development, ripening, pigmentation, hormone signalling, stress response, and disease resistance. The effective use of VIGS in a variety of fruit species, such as tomato, grape, banana, citrus, and strawberry, demonstrates its wide host range and adaptability. This quick gene silencing technology offers an economical and effective substitute for conventional techniques, enabling high-throughput screening of potential genes and making it easier to identify important regulators that affect fruit productivity and quality. Although VIGS has many benefits, issues like host specificity, inconsistent silencing effectiveness, and possible off-target effects still need to be addressed. Overcoming these constraints could be greatly aided by ongoing developments in vector design, delivery strategies, and integration with new genome editing technologies. These advancements will increase the accuracy and reliability of gene silencing while also expanding the use of VIGS to a greater range of fruit crops and characteristics. In the future, VIGS will be a key technology in the molecular breeding era, facilitating quicker functional validation of genes and hastening the creation of improved fruit varieties with better agronomic and quality traits. In the end, it supports initiatives to increase crop resilience, yield, and nutritional value by helping to understand intricate genetic networks in fruit biology.

References

Agüero J, Vives MC, Velázquez K, Pina JA, Navarro L, Moreno P, Guerri J (2014) The effectiveness of gene silencing induced by viral vectors based on citrus leaf blotch virus is different in Nicotiana benthamiana and citrus plants. Virology 460–461:154–164

Bachan S, Dinesh-Kumar SP (2012) Tobacco rattle virus (TRV)-based virus-induced gene silencing. Methods Mol Biol 894:83–92

Bai SL, Tuan PA, Tatsuki M et al (2016) Knockdown of carotenoid cleavage dioxygenase 4 (CCD4) via virus-induced gene silencing confers yellow coloration in peach fruit: evaluation of gene function related to fruit traits. Plant Mol Biol Rep 34(1):257–264

Bai SL, Tao RY, Tang YX et al (2019) BBX16, a B-box protein, positively regulates light-induced anthocyanin accumulation by activating MYB10 in red pear. Plant Biotechnol J 17(10):1985–1997

Chao N, Wang RF, Hou C et al (2021) Functional characterization of two chalcone isomerase (CHI) revealing their responsibility for anthocyanins accumulation in mulberry. Plant Physiol Biochem 161:65–73

Cui HG, Wang AM (2017) An efficient viral vector for functional genomic studies of Prunus fruit trees and its induced resistance to Plum pox virus via silencing of a host factor gene. Plant Biotechnol J 15(3):344–356

Dagnachew B, Kassahun T, Asnake F (2019) Applications of virus induced gene silencing (VIGS) in plant functional genomics studies. J Plant Biochem Physiol 7:229. https://doi.org/10.4172/2329-9029.1000229

Fei Y, Nyiko T, Molnar A (2021) Non-perfectly matching small RNAs can induce stable and heritable epigenetic modifications and can be used as molecular markers to trace the origin and fate of silencing RNAs. Nucleic Acids Res 49(4):1900–1913

Fu DQ, Zhu BZ, Zhu HL et al (2005) Virus-induced gene silencing in tomato fruit. Plant J 43(2):299–308

Fu DQ, Zhu BZ, Zhu HL et al (2006) Enhancement of virus-induced gene silencing in tomato by low temperature and low humidity. Mol Cells 21(1):153–160

Gao Y, Wei W, Zhao XD et al (2018) A NAC transcription factor, NORlike1, is a new positive regulator of tomato fruit ripening. Hortic Res 5:75

Hong M, Chi ZH, Wang YQ et al (2019) Expression of a chromoplast-specific lycopene beta-cyclase gene (CYC-B) is implicated in carotenoid accumulation and coloration in the loquat. Biomolecules 9(12):874

Ito H, Ochiai M, Kato H, Shiratake K, Takemoto D, Otagaki S, Matsumoto S (2012) Rose phytoene desaturase gene silencing by apple latent spherical virus vectors. HortScience 47:1278–1282

Jiang YH, Liu CH, Yan D et al (2017) MdHB1 down-regulation activates anthocyanin biosynthesis in the white-fleshed apple cultivar 'granny Smith'. J Exp Bot 68(5):1055–1069

Killiny N (2020) The efficacy of citrus tristeza virus as a vector for virus induced gene silencing in Huanglongbing-affected citrus. Trop Plant Pathol 45(3):327–333

Li H, Ding SW (2001) Viral suppressors of RNA silencing. Curr Opin Biotechnol 12(2):150–154

Li JJ, Fang L, Qian M et al (2017) Characteristics and regulatory pathway of the PrupeSEP1 SEPALLATA gene during ripening and softening in peach fruits. Plant Sci 257:63–73

Lin ZF, Hong YG, Yin MG et al (2008) A tomato HD-zip homeobox protein, LeHB-1, plays an important role in floral organogenesis and ripening. Plant J 55(2):301–310

Liu YL, Schiff M, Dinesh-Kumar SP (2002) Virus-induced gene silencing tomato. Plant J 31(6):777–786

Liu HP, Fu DQ, Zhu BZ et al (2012) Virus-induced gene silencing in eggplant (Solanum melongena). J Integr Plant Biol 54(6):422–429

Liu N, Xie K, Jia Q, Zhao J, Chen T, Li H, Wei X, Diao X, Hong Y, Liu Y (2016) Foxtail mosaic virus-induced gene silencing in monocot plants. Plant Physiol 171(3):1801–1807

Liu KL, Zhao ZC, Gao AP et al (2018) Cloning RCCR gene from mango (Mangifera indica) and the construction of gene silencing vector. Mol Plant Breed 16(10):3174–3181. [In Chinese with English abstract]

Liu M, Liang ZL, Aranda MA et al (2020) A cucumber green mottle mosaic virus vector for virus-induced gene silencing in cucurbit plants. Plant Methods 16(1):9

Meng LH, Fan ZQ, Zhang Q et al (2018) BEL1-LIKE HOMEODOMAIN 11 regulates chloroplast development and chlorophyll synthesis in tomato fruit. Plant J 94(6):1126–1140

Miki D, Shimamoto K (2004) Simple RNAi vectors for stable and transient suppression of gene function in rice. Plant Cell Physiol 45(4):490–495

Mysore KS, Ryu CM (2004) Nonhost resistance: how much do we know? Trends Plant Sci 9(2):97–104

Nabi S, Shafi SM (eds) (2025) Conventional to emerging advances in plant disease diagnosis and pathogen detection. CRC Press

Nasir KHB, Takahashi Y, Ito A, Saitoh H, Matsumura H, Kanzaki H et al (2005) High-throughput in planta expression screening identifies a class II ethylene-responsive element binding factor-like protein that regulates plant cell death and non-host resistance. Plant J 43(3):491–505

Orzaez D, Mirabel S, Wieland WH et al (2006) Agroinjection of tomato fruits: a tool for rapid functional analysis of transgenes directly in fruit. Plant Physiol 140(1):3–11

Orzaez D, Medina A, Torre S et al (2009) A visual reporter system for virus-induced gene silencing in tomato fruit based on anthocyanin accumulation. Plant Physiol 150(3):1122–1134

Peart JR, Lu R, Sadanandom A, Malcuit I, Moffett P, Brice DC et al (2002) Ubiquitin ligase-associated protein SGT1 is required for host and nonhost disease resistance in plants. Proc Natl Acad Sci USA 99(16):10865–10869

Qi XL, Li M, Liu CL et al (2018) Construction of TRV-mediated virus induced gene silencing (VIGS) system in sweet cherry fruit. J Fruit Sci 35(11):1309–1315

Qi XL, Li M, Liu CL et al (2020) Functional characterization of sweet cherry PaPME1 and PaPME2 during fruit ripening and softening. J Fruit Sci 37(10):1455–1463

Quadrana L, Rodriguez MC, López M et al (2011) Coupling virus-induced gene silencing to exogenous green fluorescence protein expression provides a highly efficient system for functional genomics in Arabidopsis and across all stages of tomato fruit development. Plant Physiol 156(3):1278–1291

Ramegowda V, Mysore KS, Senthil-Kumar M (2014) Virus-induced gene silencing is a versatile tool for unraveling the functional relevance of multiple abiotic-stress-responsive genes in crop plants. Front Plant Sci 5:323

Ratcliff F, Martin-Hernandez AM, Baulcombe DC (2001) Tobacco rattle virus as a vector for analysis of gene function by silencing. Plant J 25(2):237–245

Rössner C, Lotz D, Becke A (2022) VIGS goes viral: how VIGS transforms our understanding of plant science. Annu Rev Plant Biol 73:703–722

Ruiz MT, Voinnet O, Baulcombe DC (1998) Initiation and maintenance of virus-induced gene silencing. Plant Cell 10(6):937–946

Ryu CM, Anand A, Kang L et al (2004) Agrodrench: a novel and effective agroinoculation method for virus-induced gene silencing in roots and diverse Solanaceous species. Plant J 40(2):322–331

Senthil-Kumar M, Anand A, Uppalapati SR, Mysore KS (2008) Virus-induced gene silencing and its applications. CABI Rev 3:11

Senthil-Kumar M, Mysore KS (2011) Virus-induced gene silencing can persist for more than 2 years and also be transmitted to progeny seedlings in Nicotiana benthamiana and tomato. Plant Biotechnol J 9(7):797–806

Sharma PC, Ito A, Shimizu T, Terauchi R, Kamoun S, Saitoh H (2003) Virus-induced silencing of WIPK and SIPK genes reduces resistance to a bacterial pathogen but has no effect on the INF1-induced hypersensitive response (HR) in Nicotiana benthamiana. Mol Gen Genomics 269(5):583–591

Takahashi R, Yamagishi N, Yoshikawa N (2013) A MYB transcription factor controls flower color in soybean. J Hered 104:149–153

Uppalapati SR, Ishiga Y, Wangdi T, Kunkel BN, Anand A, Mysore KS et al (2007) The phytotoxin coronatine contributes to pathogen fitness and is required for suppression of salicylic acid accumulation in tomato inoculated with Pseudomonas syringae pv. tomato DC3000. Mol Plant-Microbe Interact 20(8):955–965

Van Kammen AJ (1997) Virus-induced gene silencing in infected and transgenic plants. Trends Plant Sci 2(10):409–411

Velázquez K, Agüero J, Vives MC, Aleza P, Pina JA, Moreno P, Navarro L, Guerri J (2016) Precocious flowering of juvenile citrus induced by a viral vector based on citrus leaf blotch virus: a new tool for genetics and breeding. Plant Biotechnol J 14(10):1976–1985

Voinnet O (2001) RNA silencing as a plant immune system against viruses. Trends Genet 17(8):449–459

Wang YY, Wang WH, Cai JH et al (2014) Tomato nuclear proteome reveals the involvement of specific E2 ubiquitin-conjugating enzymes in fruit ripening. Genome Biol 15(12):548

Wang T, Wen LW, Zhu HL (2015) Effectively organ-specific virus induced gene silencing in tomato plant. J Nat Sci 1:e34

Wang FS, Wang M, Liu XN et al (2017) Identification of putative genes involved in limonoids biosynthesis in citrus by comparative transcriptomics analysis. Front Plant Sci 8:782

Wesley SV, Helliwell CA, Smith NA, Wang M, Rouse DT, Liu Q et al (2001) Construct design for efficient, effective, and high-throughput gene silencing in plants. Plant J 27(6):581–590

Xie YG, Ma YY, Bi PP et al (2020) Transcription factor FvTCP9 promotes strawberry fruit ripening by regulating the biosynthesis of abscisic acid and anthocyanins. Plant Physiol Biochem 146:374–383

Xu P, Zhang Y, Kang L, Roossinck MJ, Mysore KS (2006) Computational estimation and experimental verification of off-target silencing during posttranscriptional gene silencing in plants. Plant Physiol 142(2):429–440

Yamagishi N, Yoshikawa N (2009) Virus-induced gene silencing in soybean seeds and the emergence stage of soybean plants with apple latent spherical virus vectors. Plant Mol Biol 71:15–24

Yamagishi N, Li C, Yoshikawa N (2016) Promotion of flowering by apple latent spherical virus vector and virus elimination at high temperature allow accelerated breeding of apple and pear. Front Plant Sci 7:171

Yan HX, Fu DQ, Zhu BZ et al (2012) Sprout vacuum-infiltration: a simple and efficient agroinoculation method for virus-induced gene silencing in diverse solanaceous species. Plant Cell Rep 31(9):1713–1722

Yu M, Man YP, Wang YC (2019) Light- and temperature-induced expression of an R2R3-MYB gene regulates anthocyanin biosynthesis in red-fleshed kiwifruit. Int J Mol Sci 20(20):5228

Zhang JX, Wang FR, Zhang CY et al (2018) A novel VIGS method by agroinoculation of cotton seeds and application for elucidating functions of GhBI-1 in salt-stress response. Plant Cell Rep 37(8):1091–1100

Zhang PF, Dong YM, Wen HY et al (2020) Knockdown of VvMYBA1 via virus-induced gene silencing decreases anthocyanin biosynthesis in grape berries. Can J Plant Sci 100(2):175–184

Zhao Y, Dong WQ, Zhu YC et al (2019) PpGST1, an anthocyanin-related glutathione S-transferase gene, is essential for fruit coloration in peach. Plant Biotechnol J 18(5):1284–1295

Zhou HJ, Bai SH, Wang N et al (2020) CRISPR/Cas9-mediated mutagenesis of MdCNGC2 in apple callus and VIGS-mediated silencing of MdCNGC2 in fruits improve resistance to Botryosphaeria dothidea. Front Plant Sci 11:575477

Zhu BZ, Yang YF, Li R et al (2015) RNA sequencing and functional analysis implicate the regulatory role of long non-coding RNAs in tomato fruit ripening. J Exp Bot 66(15):4483–4495

Zulfiqar S, Farooq MA, Zhao T, Wang P, Tabusam J, Wang Y et al (2023) Virus-induced gene silencing (VIGS): a powerful tool for crop improvement and its advancement towards epigenetics. Int J Mol Sci 24(6):5608

Cross-Protection Strategies Using Mild Viral Strains in Fruit Crops

18

Lishini Arucho, Mohor Roy, Roshani Singh, Manoj K. Singh, and Vanita Chandel

Abstract

Management of fruit viruses is nearly impossible once plants are infected by the viruses. But there is an decades old technique used for the management of plant viruses known as cross-protection. As the name suggests, mild strains of the virus prevent the infection by the severe strain of the same virus. This technique has been successfully used for the management of many important fruit viruses such as citrus tristeza virus, tobacco mosaic virus, papaya ringspot virus, and apple mosaic virus. The success of the technique depends upon the selection of a suitable mild strain. The mechanism behind cross-protection is explained by the classical view, superinfection exclusion, interference, systemic acquired resistance, and RNA interference. The cross-protection has significance in the management of viruses, along with a few limitations, which can be overcome. Where genetically modified plants come with a baggage of societal, ethical, and environmental issues, cross-protection is the safest technique with minimum ethical clearance requirements. Fruits where this technique has helped in the management of virus infection are apple, citrus, papaya, and tomato.

Keywords

Infection · Exclusion · Interference · Systemic acquired resistance · RNA interference · Mild strain

L. Arucho · M. Roy · R. Singh · V. Chandel (✉)
Amity Institute of Virology & Immunology, Amity University Uttar Pradesh, Noida, Uttar Pradesh, India
e-mail: vchandel@amity.edu

M. K. Singh
National Academy of Agriculture Sciences, NASC Complex, Pusa, New Delhi, India

R. Kumar et al. (eds.), *Fruit Crops and Viral Pathogens Volume II*, Plant Pathology for Sustainable Agriculture, https://doi.org/10.1007/978-981-95-7261-8_18

18.1 Introduction

Cross-protection is a strategy for plant disease management that involves infecting a plant with a mild strain of a virus to protect it from severe strain infections. The concept is similar to mammalian vaccination and is thus also called "pre-immunization" (Pechinger et al. 2019; Folimonova et al. 2020). The concept was first demonstrated by McKinney in 1929, when he had infected healthy tobacco plants with tobacco mosaic virus (TMV), a rampant disease-causing agent in tobacco plants. His observation formed the basis of understanding cross-protection, as the results demonstrated the absence of infection when a tobacco plant, previously infected with a mild strain of TMV, was challenged with a more severe form of the same virus (Hull 2002). A precedent report by Wingard in 1928 explained recovery of tobacco plants from ringspot disease, highlighting that the plants remained infectious to naïve plants but never gave rise to ringspot symptoms after reinoculation (Hernando and Aranda 2023). Since then, cross-protection has been experimented on and applied to various commercially important fruit crops to ensure survival and thus, enhance yield. Cross-protection is currently a need of the era, with an increasing global population leading to higher yield requirements, and a shift towards a more sustainable, efficient, and reliable approach to fruit crop cultivation. Despite viruses being the main pathogens causing emerging infections across many species of fruits, methods to combat them are limited. Most techniques rely on the use of chemicals, which harm not only the soil, water, and land but also the associated biodiversity and us humans, as we end up consuming the chemicals with fruits (Pechinger et al. 2019). Thus, a shift to cross-protection works in favour of the crops, the environment, and the consumers of the current and future generations. This strategy has had a history spanning almost a century. After the initial experiment by McKinney, Salaman in 1933 extended the findings through the first practical demonstration of cross-protection in tobacco plants through infection of a mild isolate of potato virus X, which resisted infection by the aggressive isolate. Further experimentation in the 1940s and 1950s established cross-protection as a viable method of protection. For fruit crops, citrus was experimented on the most, followed by papaya, with more fruits following in the suite, albeit some disappointing results (Folimonova et al. 2020; Hernando and Aranda 2023). Cross-protection was slightly sidelined as the years went by, but has been successfully brought back recently due to the pressing challenges of virus management. This chapter will explore cross-protection and its mechanisms in detail, shedding light on its significance and addressing the limitations faced by scientists in its application.

18.2 Cross-Protection

Cross-protection can be defined as the use of a mild strain of virus to infect plants that are subsequently protected against economic damage, which can be caused by a more severe strain of the same virus (Desbiez and Lecoq 2021). Cross-protection has been applied in research to study relationships between viruses and to control viral diseases in several crops, which include citrus, tomato, zucchini, and papaya, against diseases caused by severe viral strains (Folimonova 2013; Hernando and Aranda 2023; Ziebell and Carr 2010). The most successful example of cross-protection in fruit production is the control of citrus tristeza virus (CTV) in citrus orchards. For instance, in Brazil, over 80 million Pera sweet orange trees were protected using mild CTV isolates, resulting in increased crop production and reduced economic losses (Zhu et al. 2023). Several mechanisms contribute to cross-protection. The primary mechanism is RNA-mediated gene silencing, a plant defence mechanism in which double-stranded RNA (dsRNA) is cleaved by an RNA-induced silencing complex (RISC) into small interfering RNA (siRNA), which guides the RISC complex to degrade complementary viral RNAs, inducing systemic antiviral response. The specificity of this process supports why cross-protection primarily works between closely related viruses (Poogin 2018; Pechinger et al. 2019). Another important mechanism is superinfection exclusion (SIE), where infection with a mild virus isolate prevents secondary infection with more severe isolates of the same virus (Zhu et al. 2023; Hernando and Aranda 2023). These mechanisms will be discussed in detail in the upcoming sections.

18.2.1 Mild Strain Cross-Protection (MSCP)

Mild strain cross-protection, as the name suggests, is where we need to identify a mild strain and use it to inoculate a plant to provide resistance against the virulent strain of a related virus (Ziebell and Carr 2010; Pechinger et al. 2019). All available methods to manage plant viruses are either gene editing or transgenic approaches, whereas MSCP is still considered the safest approach to manage plant viruses (Pechinger et al. 2019). Major fruit crops where this technique has been used successfully are citrus, papaya, and tomato (Table 18.1). This approach was chemical and gene modification free, so it was widely accepted by people (Foley et al. 2011; Van Bueren et al. 2018; Pechinger et al. 2019). Though finding a mild strain remains challenging in this approach (Ziebell and MacDiarmid 2017). Sometimes when isolates are selected for cross-protection, some isolates protect only one or a few plant species, which means it is not effective against all species or varieties of a plant. For instance, in the case of CTV in citrus isolate was able to protect one cultivar but not another cultivar (Müller and Rezende 2004; Pechinger et al. 2019). Thus, it becomes important to check this aspect when we are selecting isolates for cross-protection (Ziebell and MacDiarmid 2017).

18.2.1.1 Classical Method of Identification of Mild Strains

An ancient method to identify a mild strain was very simple; carefully observe the symptoms in the field and categorize plants into symptomatic, asymptomatic, but infected and with fewer symptoms. After basic diagnosis using ELISA and PCR to confirm the virus, the virus is transmitted to host plants and observed for the display of symptoms. So we can separate mild strains and virulent strains on the basis of the symptoms observed on the host plants (McKinney,1929). For instance, mild strains

Table 18.1 Viruses for which mild strain cross-protection (MSCP) has been used successfully in the fruit crops

Severe (challenging) isolate	Mild (protecting) isolate	Host plant	References
Citrus tristeza virus (CTV) severe isolate	CTV, "IAC" isolate	*Citrus sinensis* ("Pera" sweet orange)	Müller and Rezende (2004) and Pechinger et al. (2019)
Apple mosaic virus (ApMV), severe strain "A"	ApMV, mild strain "M" and very mild strain "Y"	*Malus domestica* (Apple)	Chamberlain et al. (1964)
Pepino mosaic virus (PepMV) CH2 strains	PepMV, CH2 mild isolate SP13	*Solanum lycopersicum L.* (Tomato)	Gomez et al. (2012), Agüero et al. (2018), and Hernando and Aranda (2023)
Papaya ringspot virus type-P (PRSV-P)	PRSV-P mild strain	*Carica papaya L.* (Papaya)	Tripathi et al. (2008), Desbiez and Lecoq (2020), and Medina-Salguero et al. (2021)
Zucchini yellow mosaic virus (ZYMV), ZYMV-CA strain	ZYMV-WK mild strain	*Cucurbita pepo* L. (Zucchini)	Wang et al. (1991)

of tobacco mosaic virus (TMV) were initially identified as they showed fewer symptoms in tobacco plants, whereas other strains caused severe mosaic symptoms in tobacco plants. So it's important to check symptoms on the plants in the field and also to compare with symptoms on the inoculated plants (Jridi et al. 2006). Following identification, the selected mild isolates are then inoculated onto the host plants, which are subsequently infected by challenging isolates to determine the effectiveness of mild strain cross-protection. Plants are challenged with severe strains through mechanical, graft, vector inoculation, or, in some cases, plants are exposed to natural infection under field conditions (Pechinger et al. 2019). It is noteworthy that Chamberlain et al. (1964) demonstrated the use of the leaf infection index to measure disease severity (chlorosis) in apple mosaic virus (AMV). In doing so, they could distinguish between mild and severe isolates of apple mosaic virus (AMV). It was also evident that trees infected with a mild strain of AMV were protected against infection by severe strains of the same virus. This cross-protection mechanism provides a useful technique for detecting the presence of mild or symptomless strains of AMV (Chamberlain et al. 1964).

18.2.1.2 Mild Strains Selection

1. Alternative methods to identify mild strains are: Alternative Host: This technique promotes the selection of mild strains by passing viruses through different hosts. Here alternate host is used for identifying mild strains by infecting an alternate host with the virus and checking symptoms (Yeh and Cheng 1989).

2. Heat Treatment: Heat therapy is used to reduce virulence of viruses before inoculation (Desjardins et al. 1959; Pechinger et al. 2019).
3. Use of Chemicals: Certain chemicals can be used to create mutations and reduce the virulence.
4. Genome Sequence: Genome of plant viruses can be studied using sequencing methods (Pechinger et al. 2019).

In order to select mild strains, some basic criteria were mentioned by Fulton (1986) and Lecoq (1998) (Hernando and Aranda 2023). Only related plant species should be used; yield should not be affected by mild strain and should be easy to grow on alternative host plants. Also use of genetically modified viral mild strain is not recommended (e.g., Goh et al. 2023; Sehgal and Krause 1968; Hernando and Aranda 2023).

18.3 Applications of Mild-Strain Cross-Protection in Fruit Crops

18.3.1 Citrus

Citrus tristeza virus is considered one of the largest and most complex viruses in the *Closteroviridae* family. This family includes plant viruses that are known to cause significant economic losses in several crops, including vegetables, grains, grapes, and fruit trees (Bar-Joseph et al. 1979; Dolja et al. 1994, 2006; Agranovsky 1996; Karasev 2000, as cited in Folimonova 2013). Citrus tristeza virus is known to cause two major diseases that have significantly affected citrus production worldwide. The first disease is known as quick decline, which primarily affects trees grafted onto sour orange (*Citrus aurantium* L.) rootstock. This condition arises from a virus-induced graft incompatibility between the scion and rootstock (Moreno et al. 2008). The second disease is called stem pitting, which is caused by some CTV strains, and it affects citrus trees regardless of the rootstock used. This condition arises when the virus replication interferes with the normal cambial development, leading to the formation of pits on the stem (Brlansky et al. 2002; Tatineni and Dawson 2012; Folimonova 2013). Among the two diseases caused by CTV, stem pitting is considered to be more difficult to manage as it affects both the scion and the rootstock; therefore, replacing the rootstock is not effective. The only reliable method to protect citrus varieties from severe stem pitting is via cross-protection using mild isolates. This strategy has been most widely used in Brazil, where over 80 million Pera sweet orange trees are protected. It has also been successfully used in other countries, including Australia, South Africa, Peru, Argentina, and Japan, making it possible to cultivate citrus varieties in regions where aggressive or severe CTV strains of the virus are present (da Graça and van Vuuren 2010; Roistacher et al. 2010; Folimonova 2013).

18.3.2 Papaya

India is one of the biggest producers of Papaya, thus contributing to the economy of India. But the incidence of pathogens in the fruit has led to the economic losses both quantitative and qualitative, particularly papaya ringspot virus (PRSV) (*Potyviridae*), which is impacting the crop throughout the papaya growing regions in the world (Medina-Salguero et al. 2021). Major symptoms caused by PRSV are mosaic, ringspots, and most importantly, fruits with deformations. This pathogen was not managed using cultural methods such as managing vectors and removing diseased plants. MSCP was used as an alternate strategy to manage PRSV. In the literature, there are examples where researchers have managed PRSV in papaya using MSCP (Medina-Salguero et al. 2021). Though in this study it was mentioned that cross-protection was not long term, a delay in symptoms or less severe symptoms can definitely reduce the economic losses to the growers (Medina-Salguero et al. 2021).

18.3.3 Tomato

Tomato is considered a vegetable and a fruit. Botanically, it is a fruit as it develops from the flower ovary and has seeds. Production of tomato is hampered by many viruses, and pepino mosaic virus (PepMV) (*Flexiviridae*) has caused significant losses in its production (Hernando and Aranda 2023). Plants showed remarkable symptoms such as yellowing and mosaic in the leaves and chlorosis in the veins. Hernando and Aranda (2023) used mild strain isolate (Sp13 or PS5) and a combination of the two strains (Sp13+PS5) against PepMV infecting tomatoes. Those tomato plants that were infected by mild strains did not show any symptoms. In contrast, plants pre-inoculated with a combination of Sp13+PS5 showed no symptoms (Agüero et al. 2018; Hernando and Aranda 2023). Their research describes that the effectiveness of cross-protection depends on strain compatibility, and protection is strain-specific. In addition, strict hygiene practices, seed certification, and regular molecular monitoring are also essential. So, when these approaches are applied correctly, cross-protection can improve the yield and quality of fruit for an entire season (Hernando and Aranda 2023). Commercial MSCP products such as "V10", which combines two mild isolates, have been licensed in European countries, protecting thousands of hectares of tomato crops annually (Pechinger et al. 2019).

18.3.4 Apple

In pome fruits apple is the most cherished fruit throughout the world. The incidence of viruses has resulted in losses in the apple fruit in places where it's grown. Viral management has always been tricky to deal with, as once in the field, it cannot be managed. A research group of Posnette and Cropley reported the incidence of three strains of apple mosaic virus from England. To manage the virus, they used MSCP technology (1952) and mentioned that mild and moderate strains of the virus were able to provide protection to the trees against the virus (Chamberlain et al. 1964). Later, Chamberlain and his coworkers in 1964 evaluated different plants of the Golden Delicious variety of apple to identify a mild strain of AMV to protect plants against virulent strains of AMV. After carefully observing the trees, they found that mild strains were able to provide protection to apple trees. All the trees were examined, and results showed that both mild and very mild strains gave protection in healthy plants when challenged with severe strains. On the other hand, when trees already infected with the severe strains were inoculated with mild M and very mild Y strains, the protection was not very effective. Only the very mild Y strain gave moderate protection, indicating a slight reduction in symptoms (Chamberlain et al. 1964). This study provided one of the earliest reports of cross-protection in perennial fruit crops, wherein a mild or very mild strain could successfully confer protection against severe strains when introduced first.

18.4 Cross-Protection Mechanisms

Cross-protection harnesses the power of a mild "protector" virus to set up a shield in the plant against an aggressive "challenger" virus. Resistance against secondary infection is generally seen with viruses that share high genomic similarity to the protector strain, with the exception of SAR. This phenomenon is not controlled by a single, isolated mechanism. Instead, it is an intricate interplay of different mechanisms that overlap to form a robust, multilayered barrier. This section will explore the different mechanisms at an individual level to understand the underlying molecular mechanisms that confer the final resistance to the plants.

18.4.1 Classical View

The concept of pathogen-derived resistance (PDR) from gene products was first demonstrated by Sanford and Johnston in 1985 using bacteriophage Qβ. The hypothesis stated that four Qβ gene constructs (coat protein gene, modified replicase gene, gene encoding for replicase binding site, and maturation protein gene, along with antisense RNA) can be inserted into *E.coli*, causing transformation and providing protection to the bacteria from the virus (Sanford and Johnston 1985) They suggested the broader scope of PDR for other organisms, which was first experimentally demonstrated by Abel et al. in 1986 through the creation of a transgenic tobacco plant (*Nicotiana tabacum* cv. Xanthi and cv. Samsun). This plant was modified to express the tobacco mosaic virus (TMV) coat protein (CP) gene and challenged with virulent strains of TMV. Transgenic plants showed greater resistance than non-transgenic plants, indicating successful PDR (Abel et al. 1986). Subsequent studies were performed to determine the exact mechanism behind this protection, leading to the formulation of classical theories of PDR, which can be extended to fruit crops as well. Abel et al.'s breakthrough experiment, where the CP of TMV provided PDR, provided a base for the classical theory. It was observed that some transgenic plants showed no symptoms, whereas others developed disease symptoms within 2 weeks of inoculation. The resistance efficacy was also correlated to the level of CP expression, which meant increased doses could confer greater resistance (Abel et al. 1986). Subsequent experiments also demonstrated the specificity of resistance wherein superinfection with Tobamovirus related to TMV showed greater resistance than distantly related viruses (Nejidat 1990). CP-mediated PDR towards TMV is thought of as the CP interfering in virus disassembly through subunit-subunit interaction, therefore preventing infection by the challenger virus (Register and Beachy 1988). Through a series of experiments conducted from late 1980s to early 1990s, it became increasingly evident that any genetic element of the virus can confer resistance, for example, genes encoding for: complete or partial replicase proteins from RNA dependent RNA polymerase readthrough domain (Golemboski et al. 1990), movement proteins (Malyshenko et al. 1993), different RNA sequences (Harrison et al. 1987; Kollàr et al. 1993), and even non-coding regions (Nelson et al. 1993; Zaccomer et al. 1993). This paved the way for the theory to be substantiated with the idea that the challenger virus replication can be interfered with at multiple points of the life cycle, depending on which protein is made to express in the transgenic plants (like CP for virus disassembly, movement proteins for intracellular trafficking, etc.) (Gottula and Fuchs 2009; Lindbo and Falk 2017). Over the ensuing years, scientists tested this theory in many transgenic plants, where different data were generated regarding its efficacy, especially with CP-mediated PDR. The conclusion derived was that different plant virus systems can express this PDR in differing capacities, thus requiring optimization of transgenic protocols. Practical application of this theory was seen through the development of commercial transgenic fruit crops against viruses like plum pox virus, papaya ringspot virus, zucchini yellow mosaic virus, and watermelon mosaic virus, which affects plum, papaya, squash and other cucurbits, and watermelon. (Lindbo and Falk 2017).

18.4.2 Superinfection Exclusion

Superinfection exclusion (SIE) is a phenomenon in which a cell with primary virus infection prevents a secondary infection with the same or closely related viruses. The concept of SIE was first discovered by H.H. McKinney in 1929 during his work with tobacco mosaic virus (TMV). He observed that on inoculating a tobacco plant with a green mosaic virus (mild strain), the plant did not show symptoms when inoculated with yellow mosaic virus (severe strain) (Hull 2002). In its earlier years, SIE was used by biologists to determine whether two virus isolates are of the same strain. Subsequently, it was incorporated in agriculture as a method of cross-protection where infection of a plant with mild strains would protect the plant from a severe challenger virus. The underlying molecular mechanism handling SIE in plants has been studied intricately in recent years. RNA silencing by the protector virus that leads to sequence-specific degradation of the superinfector has been widely considered as the key mechanism. However, certain studies argue that RNA silencing-based SIE failed even when the protector and superinfector had shared identical sequences (Folimonova et al. 2010; Zhang and Qu 2016). Other studies have established a protein-based mechanism wherein specific viral proteins (like p33 of citrus tristeza virus) involved in pathogenicity or viral life cycle can block the same protein upon superinfection (Folimonova 2012; Bergua et al. 2016; Zhang et al. 2018). SIE is believed to operate at two levels—cellular and systemic, which confers resistance to the whole plant despite inoculation in a specific part of the plant, as demonstrated with CTV (Bergua et al. 2014; Folimonova et al. 2020; Folimonova 2013). Some notable examples of fruit crop viruses against which resistance has been conferred through SIE are citrus tristeza virus (citrus plants), papaya ringspot virus (papaya), Tobamovirus (tomato), and zucchini yellow mosaic virus (squash, watermelon, and melon) (Folimonova et al. 2020; Pechinger et al. 2019; Hernando and Aranda 2023). The limitations associated with SIE are high strain specificity requirements and possible incomplete protection at higher doses of superinfector (Bergua et al. 2016). Despite these challenges, SIE stands out as a potential mechanism for cross-protection.

18.4.3 Interference

Interference, as defined by Bennette in 1951, is a phenomenon in which a plant or a part of a plant, when invaded by a virus, becomes immune to infection by a second virus or strain of challenging virus (Bennett 1951). Interference is a broad mechanism, explainable by experimental evidence and inferences. Some of the proposals are: (1) inhibition of challenger virus due to high multiplication of the protector virus which occupies the host receptor sites; (2) unavailability of resources for challenger to complete life cycle due to protector virus monopolizing limiting resources (Ziebell et al. 2007; Sánchez-Tovar et al. 2025); (3) triggering antiviral defence systems like pattern-triggered immunity (PTI) and effector-triggered immunity (ETI), translational repression by activation of nuclear shuttle protein (NSP)-interacting kinase 1 (NIK1) by protector virus which leads to phosphorylation of translation factors, making the host machinery unavailable for challenger virus, and ubiquitination of viral components followed by recognition and degradation by proteasome as a part of Ubiquitin-Proteasome System (Calil and Fontes 2016); (4) interference through contribution of defective proteins to the replication complex when variants of same species is involved (Whitaker-Dowling and Youngner 1987; Sánchez-Tovar et al. 2025); and (5) preventing the uncapping and replication of challenger virus by protector virus (Singhal et al. 2021; Sánchez-Tovar et al. 2025). Essentially, interference is an antagonistic phenomenon as there exists a rivalry between the protector and challenger virus for resources and host machinery. The protector eventually triumphs over the challenger through the mechanisms described above (Sánchez-Tovar et al. 2025). For fruit crops, this mechanism has been applied against citrus tristeza virus (CTV), papaya ringspot virus (PRSV), plum pox virus (PPV), and grapevine fanleaf virus (GFLV) as mentioned across literature

(Folimonova et al. 2020; Komar et al. 2008; Silva-Rosales et al. 2019; Ravelonandro et al. 2008).

18.4.4 Systemic Acquired Resistance

Systemic acquired resistance (SAR), also called whole plant resistance, is an inducible, broad-spectrum, long-lasting immunity against reinfection by diverse pathogens like viruses. It is a form of induced resistance and can be triggered via pathogens or chemical elicitors (Fig. 18.1). The protection will last against a broad range of pathogens (virus, bacteria, and fungi) even if the protector strain was a viral strain (Reimer-Michalski and Conrath 2016; Kamatham et al. 2017; Mandadi and Scholthof 2013). A major characteristic of SAR is its ability to confer distal immunity. Local activation of pattern-triggered immunity (PTI) and effector-triggered immunity (ETI) primes immunity in distal plant tissues through long-distance signalling (Ito et al. 2024). The onset of SAR begins with pathogen invasion detected by PRRs, which induces the production of salicylic acid (SA) primarily by isochorismate synthase 1 (ICS1) in the ICS pathway. Accumulation of SA affects cellular redox and triggers a cascade reaction, which induces gene expression of nonexpressor-of-PR-genes1 (NPR1) to produce antimicrobial PR proteins like PR1, PR2, and PR5 locally for protection. Parallel to this reaction, mobile signals are also produced, including *N*-hydroxypipecolic acid (NHP) as the major molecule, azelaic acid (AzA), glycerol-3-phosphate (G3P), dehydroabietinal amines (DA), and methyl salicylic acid (MeSA). NHP is produced from its precursor pipecolic acid by

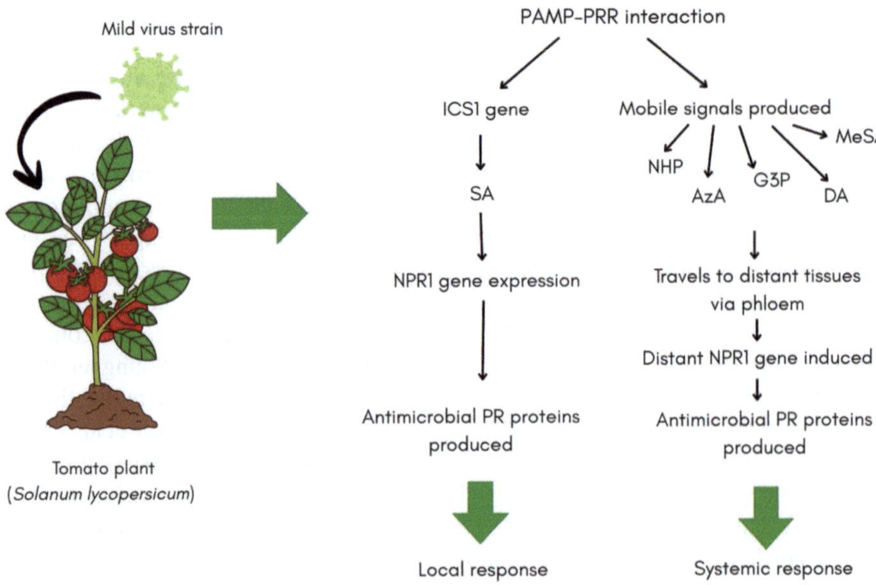

Fig. 18.1 Schematic diagram representing SAR in the tomato plant

flavin-dependent monooxygenase 1 (FMO1). These molecules travel to distant uninoculated tissues via phloem to induce antimicrobial PR genes, priming the whole plant against secondary infection. Thus, the name systemic acquired resistance is justified. Deficiency of any of the key mediators (SA or NHP) due to genetic mutations in the synthesis pathway can lead to SAR deficiency, leaving the plant susceptible to disease (Ito et al. 2024). It must be noted that the SAR primed state is associated with acetylation and methylation of SAR-promoted gene promoters, which leads to eventual priming of plant defence genes, allowing SAR to be passed onto the future generations (Ito et al. 2024; Ahmad Rather et al. 2018).

SAR has been studied as an effective mechanism of cross-protection in fruit plants like mulberry (Ahmad Rather et al. 2018). The broad-spectrum action has also been demonstrated by Shopova et al. where SAR induced by compatible and incompatible tomato mosaic virus (ToMV) was effective against *Pseudomonas syringae* pv. *tomato* and *Xanthomonas vesicatoria* in the tomato plant (Shopova et al. 2020). Multiple studies are conducted nowadays to explore the use of synthetic chemicals to induce SAR, including acibenzolar-*S*-methyl against tomato yellow leaf curl disease and other SA analogs (Monci et al. 2019), and β-aminobutyric acid (BABA) for many plants (Zhou and Wang 2018).

18.4.5 RNA Interference

RNA interference (RNAi) is a type of post-transcriptional gene silencing (PTGS) mechanism used by plants to defend themselves against viral infections. Viral infections are one of the major causal agents of plant virus epidemics that lead to hefty crop and economic losses yearly. To combat this, plants can be given cross-protection against certain viral strains via the RNAi pathway. The silencing process is initiated after mild virus inoculation by the recognition and cleaving of dsRNA produced during viral replication. It leads to the formation of small fragments of 21–25 nucleotides referred to as small interfering RNA (siRNA). There exists high sequence fidelity for this process of recognition. The enzyme responsible for the cleavage is Dicer-like (DCL) proteins, a ribonuclease (RNase III). Next, these siRNAs are unwound by helicase and loaded onto the RiSC complex (RNA-induced Silencing Complex) containing the Argonaute (AGO) protein. The 5′ end of the less stable strand is selected as the "guide" strand, and the second strand is the "passenger strand", which is later discarded. RiSC has endonuclease activity, which mediates the degradation or repression of the target RNA (Fig. 18.2; Akbar et al. 2022; Venu et al. 2024). The mechanism's continuity is maintained by RNA-dependent RNA polymerase (RdRp), which amplifies the small RNA molecules produced in the first cleaving cycle till all viral RNA is degraded. There are different types of RNAi: virus-induced gene silencing (VIGS), wherein the viral genome is modified by removing virulent genes and cloned into a binary vector. The vector enters the host plant via Agrobacterium-mediated transformation or through the gene gun. It undergoes replication (in RNA virus) or transcription (in DNA virus), both involving RdRp, which creates the dsRNA to initiate RNAi. The other, advanced technique is

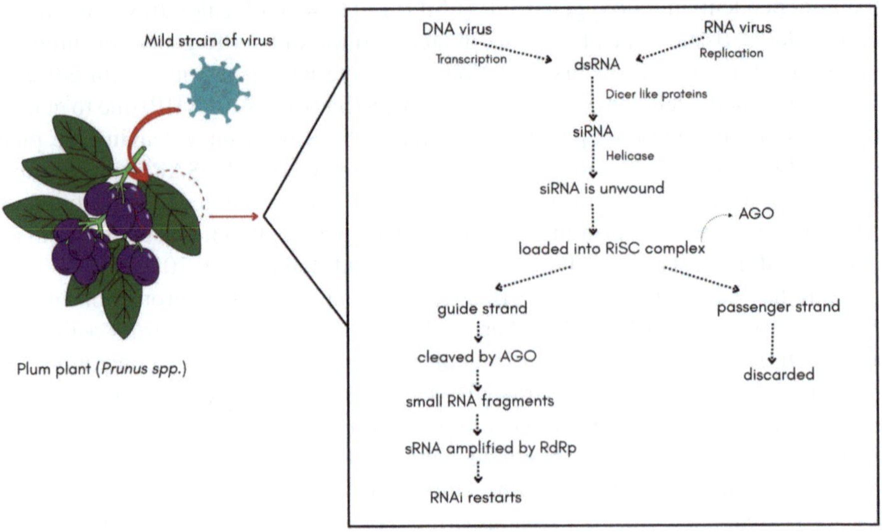

Fig. 18.2 Schematic overview of RNAi in the plum plant

spray-induced gene silencing (SIGS), where dsRNA is externally brushed or syringe-infiltrated into leaves (Akbar et al. 2022).

Many economically significant plant viruses have been targeted through RNAi, such as papaya ringspot virus (PRSV) (Jia et al. 2017; Bau et al. 2003), citrus tristeza virus (CTV) (Soler et al. 2011; Borah et al. 2024), Plum pox virus (PPV) (Ravelonandro et al. 2019), etc. However, multiple limitations with this mechanism have been noted, including off-target gene silencing, high sequence fidelity to the first RNA that triggered protection, which allows genetically divergent viruses to evade RiSC, unavailability of a detailed mechanism of highly localized RNAi, and possible competition between protective and challenge strains for resources, along with high cost, large timeframe, and regulation issues. Current research aims to overcome these challenges and establish RNAi as a viable cross-protection mechanism (Akbar et al. 2022; Pechinger et al. 2019).

18.5 Significance of Cross-Protection

In the current global scenario, the population has come to stand at 8+ billion. There has been an extremely high jump in population within the last century, which has prompted the need for increased food production more than ever. The land mass is shrinking as more people require housing and infrastructure to maintain the quality of life. This leads to agriculture having a very limited area for the cultivation of food and fruit crops. On the other hand, Earth's climate has taken a hit. We are witnessing extreme climactic conditions and soaring temperatures of the planet due to greenhouse gas emissions. In the context of agriculture, these emissions are generated from industries producing chemicals for pesticide to be used in the disease management, as fertilizers, etc., and from the heavy use of these in farming. There is also a need to be able to cultivate crops in harsh climates without compromising on the yield and quality. These problems have elicited the need to adopt sustainable farming methods which reduce dependence on chemicals for horticulture crop cultivation, and decrease damage to habitat and people. It aims to adopt biological methods to support agriculture and focuses on saving the environment from the utter disaster conditions we humans have created on our planet (Pechinger et al. 2019; Nishiguchi and Kobayashi 2011).

Fruit crops are an essential component of agriculture, serving as the nutritional pillars that provide us with fibre, essential vitamins, and minerals which strengthen our immune system. Beyond nourishment, they are also environmental stewards: strengthening the soil through their deep root systems, providing nectar for pollinators, and serving as habitats for insects (Eeraerts et al. 2021). A major hurdle faced by these crops is the threat of emerging and re-emerging virus infections, which leads to devastating losses both economically and societally. Despite known epidemiology, the strategies to control viral infections are low. Examples of some fruit crops which have faced high virus infection rates include citrus like oranges, lemons, and mandarin (by citrus tristeza virus, citrus leprosis virus, citrus variegation virus, etc.), pome fruits like apples (by apple chlorotic leaf spot virus, and apple mosaic virus), and stone fruits like plum, prunes, and cherries (by plum pox virus, prunus necrotic ringspot virus, prune dwarf virus and little cherry virus 1 and 2) (Umer et al. 2019). Due to their importance, fruit crops have been heavily experimented upon to establish pathogen-resistant varieties. Cross-protection harnesses the natural interaction between viruses to establish a shield within the plant, protecting it from possible future infections by more aggressive strains. MSCP is relatively cost-effective and a sustainable strategy for protection and increasing the yield of fruits while reducing dependence on unsustainable resources like agricultural chemicals (Pechinger et al. 2019). It has also been noted that through cross-protection, the plant is capable of tolerating certain abiotic stresses, too, like ageing and water deficit (Hernando and Aranda 2023). Many of the fruits mentioned above and other fruits like cucumber (infected by melon yellow spot virus, zucchini yellow mosaic virus, etc.) have been cross-protected from multiple viruses through different mechanisms as discussed in the previous section (Tomitaka et al. 2024).

18.6 Limitations Associated with Cross-Protection

The main goal of MSCP is to provide protection to crops (in this context, fruit crops) against viruses that have lethal consequences, often leading to disastrous levels of crop and economic losses. Over the years, it has given us advantages that have saved fruit crops across multiple locations globally. However, even a technique like MSCP is laced with certain inherent limitations. As stated in literature, the key limitations are: (1) risk of mild strain mutating into an aggressive strain, which undermines protection and leads to high rate of infection and death, (2) genetic recombination between different viruses during mixed infection, (3) infection by a strain which is not genetically similar to protector virus prevents cross-protection action, (4) low-dose inoculation leading to incomplete protection, (5) synergism between mild strain with other strains or unrelated virus during mixed infection which could amplify disease symptoms, (6) heteroencapsidation (genetic material of one virus is encapsulated within the capsid of another virus) or heteroassistance (one virus provides functional proteins to other virus) in mixed infection altering specificity, efficiency and transmission, (7) breakdown of cross-protection, and (8) possible spread of protective strain to other fruit plants where it may exhibit severe infection (Ziebell and Carr 2010; Tomitaka et al. 2024; Abbas et al. 2005; Hernando and Aranda 2023). An example: Hanssen et al. in 2010 conducted a greenhouse trial wherein they infected tomato plants with a mild LP isolate of *pepino mosaic virus* and later challenged it with an aggressive CH2 isolate. Analysis of several assays conducted showed symptom severity instead of cross-protection, which was likely due to synergism (Hanssen et al. 2010). Many of these limitations can be avoided through good inoculation practices and performing genetic modification to stabilize the protective virus. Another limitation that has been pointed out is the hesitation of farmers to inoculate their crops with a virus (Ziebell and Carr 2010). The reason is understandable: it is widely known that viruses are the main culprits behind losses in many fruit crops. This can be easily solved through creating awareness amongst the growers, ensuring them that the inoculation is of a mild strain, which will protect the crop instead of destroying it. Ultimately, these limitations are not a roadblock that cannot be overcome. With precise strain selection and modifications, along with stringent real-time monitoring, better MSCP can be achieved.

References

Abbas M, Khan MM, Mughal SM, Khan IA (2005) Prospects of classical cross protection technique against citrus tristeza closterovirus in Pakistan: a review. Hortic Sci 32(2):74–83. https://doi.org/10.17221/3769-hortsci

Abel PP, Nelson RS, De B, Hoffmann N, Rogers SG, Fraley RT, Beachy RN (1986) Delay of disease development in transgenic plants that express the tobacco mosaic virus coat protein gene. Science 232(4751):738–743. https://doi.org/10.1126/science.3457472

Agranovsky AA (1996) Principles of molecular organization, expression, and evolution of clostero viruses: over the barriers. Adv Virus Res 47:119–158

Agüero J, Gomez-Aix C, Sempere RN, García-Villalba J, García Núñez J, Hernando Y, Aranda MA (2018) Stable and broad spectrum cross-protection against Pepino mosaic virus attained by mixed infection. Front Plant Sci 9:1810

Ahmad Rather D-R, Younus Wani M, Rani S, Mir M, Kamili AS, Younus Wani CM, Mehraj S, Rather R, Hajam O, Ganie N, Baqual M (2018) SAR: a novel strategy for plant protection with reference to mulberry. Int J Chem Stud 6(2). https://www.researchgate.net/publication/326741076

Akbar S, Wei Y, Zhang M-Q (2022) RNA interference: promising approach to combat plant viruses. Int J Mol Sci 23(10):5312. https://doi.org/10.3390/ijms23105312

Bar-Joseph M, Garnsey SM, Gonsalves D (1979) The clos teroviruses: a distinct group of elon gatedplantviruses. AdvVirus Res 25:93–168

Bau H-J, Cheng Y-H, Yu T-A, Yang J-S, Yeh S-D (2003) Broad-Spectrum resistance to different geographic strains of papaya ringspot virus in coat protein gene transgenic papaya. Phytopathology 93(1):112–120. https://doi.org/10.1094/PHYTO.2003.93.1.112

Bennett CW (1951) Interference phenomena between plant viruses. Ann Rev Microbiol 5(1):295–308. https://doi.org/10.1146/annurev.mi.05.100151.001455

Bergua M, Zwart MP, El-Mohtar C, Shilts T, Elena SF, Folimonova SY (2014) A viral protein mediates superinfection exclusion at the whole-organism level but is not required for exclusion at the cellular level. J Virol 88(19):11327–11338. https://doi.org/10.1128/JVI.01612-14

Bergua M, Kang S-H, Folimonova SY (2016) Understanding superinfection exclusion by complex populations of citrus tristeza virus. Virology 499:331–339. https://doi.org/10.1016/j.virol.2016.10.001

Borah M, Nath PD, Chaudhury SP, Biswas KK, Patil BL, Voloudakis A (2024) Topical application of dsRNAs targeting citrus tristeza virus (CTV) reduces its titer in the CTV infected sweet orange (Citrus sinensis). Eur J Plant Pathol 168(2):273–278. https://doi.org/10.1007/s10658-023-02752-6

Brlansky RH, Howd DS, Broadbent P, Damsteegt VD (2002) Histology of sweet orange stem pit ting caused by an Australian isolate of citrus tristeza virus. Plant Dis 86:1169–1174

Calil IP, Fontes EPB (2016) Plant immunity against viruses: antiviral immune receptors in focus. Ann Bot mcw200. https://doi.org/10.1093/aob/mcw200

Chamberlain EE, Atkinson JD, Hunter JA (1964) Cross-protection between strains of apple mosaic virus. N Z J Agric Res 7(4):480–490. https://doi.org/10.1080/00288233.1964.10416375

da Graça JV, van Vuuren SP (2010) Managing citrus tristeza virus losses using cross protection. In: Karasev AV, Hilf ME (eds) Citrus Tristeza virus complex and Tristeza diseases. APS Press, Eagan, pp 247–260

Desbiez C, Lecoq H (2020) Papaya ringspot virus (Potyviridae). Reference module in life. Elsevier, Sciences

Desbiez C, Lecoq H (2021) Papaya ringspot virus (Potyviridae). In: Bamford DH, Zuckerman M (eds) Encyclopedia of virology, 4th edn. Academic Press, pp 671–678. https://doi.org/10.1016/B978-0-12-809633-8.21241-7

Desjardins PR, Wallace JM, Wollman ESH, Drake RJ (1959) A separation of virus strains from a tristeza-seedling-yellows complex by heat treatment of infected lime seedlings. Citrus Virus Dis 91–95

Dolja VV, Kreuze JF, Valkonen JPT (2006) Comparative and functional genomics of closteroviruses. Virus Res 117:38–51. https://doi.org/10.1016/j.virusres.2006.02.002

Dolja VV, Karasev AV, Koonin EV (1994) Molecular biology and evolution of closteroviruses: sophisticated build-up of large RNA genomes. Annu Rev Phytopathol 32:261–285

Eeraerts M, Van Den Berge S, Proesmans W, Verheyen K, Smagghe G, Meeus I (2021) Fruit orchards and woody semi-natural habitat provide complementary resources for pollinators in agricultural landscapes. Landsc Ecol 36(5):1377–1390. https://doi.org/10.1007/s10980-021-01220-y

Foley JA, Ramankutty N, Brauman KA, Cassidy ES, Gerber JS, Johnston M, Mueller ND, O'Connell C, Ray DK, West PC et al (2011) Solutions for a cultivated planet. Nature 478:337–342

Folimonova SY (2012) Superinfection exclusion is an active virus-controlled function that requires a specific viral protein. J Virol 86(10):5554–5561. https://doi.org/10.1128/JVI.00310-12

Folimonova SY (2013) Developing an understanding of cross-protection by citrus tristeza virus. Front Microbiol 4. https://doi.org/10.3389/fmicb.2013.00076

Folimonova SY, Robertson CJ, Shilts T, Folimonov AS, Hilf ME, Garnsey SM, Dawson WO (2010) Infection with strains of citrus Tristeza virus does not exclude superinfection by other strains of the virus. J Virol 84(3):1314–1325. https://doi.org/10.1128/JVI.02075-09

Folimonova SY, Achor D, Bar-Joseph M (2020) Walking together: cross-protection, genome conservation, and the replication machinery of citrus tristeza virus. Viruses 12(12):1353. https://doi.org/10.3390/v12121353

Fulton RW (1986) Practices and precautions in the use of cross protection for plant virus disease control. Annu Rev Phytopathol 24:67–81
Goh RP, Xie XY, Lin YC, Cheng HW, Raja JAJ, Yeh SD (2023) Rapid selection of potyviral cross-protection effective mutants from the local lesion host after nitrous acid mutagenesis. Mol Plant Pathol 24:973–988
Golemboski DB, Lomonossoff GP, Zaitlin M (1990) Plants transformed with a tobacco mosaic virus nonstructural gene sequence are resistant to the virus. Proc Natl Acad Sci 87(16):6311–6315. https://doi.org/10.1073/pnas.87.16.6311
Gomez P, Sempere RN, Aranda MA (2012) Pepino mosaic virus and tomato torrado virus. Two emerging viruses affecting tomato crops in the Mediterranean basin. Adv Virus Res 84:505–532
Gottula J, Fuchs M (2009) Toward a quarter century of pathogen-derived resistance and practical approaches to plant virus disease control. Adv Virus Res 75(09). Elsevier. https://doi.org/10.1016/s0065-3527(09)07505-8
Hanssen IM, Gutiérrez-Aguirre I, Paeleman A, Goen K, Wittemans L, Lievens B, Vanachter ACRC, Ravnikar M, Thomma BPHJ (2010) Cross-protection or enhanced symptom display in greenhouse tomato co-infected with different Pepino mosaic virus isolates. Plant Pathol 59(1):13–21. https://doi.org/10.1111/j.1365-3059.2009.02190.x
Harrison BD, Mayo MA, Baulcombe DC (1987) Virus resistance in transgenic plants that express cucumber mosaic virus satellite RNA. Nature 328(6133):799–802. https://doi.org/10.1038/328799a0
Hernando Y, Aranda MA (2023) Cross-protection against Pepino mosaic virus, more than a decade of efficient disease control. Ann Appl Biol 184(2):174–182. https://doi.org/10.1111/aab.12884
Hull R (2002) Induction of disease 2: virus–plant interactions. In: Matthews' plant virology. Elsevier, pp 437–484. https://doi.org/10.1016/B978-012361160-4/50061-X
Ito S, Sakugawa K, Novianti F, Arie T, Komatsu K (2024) Local application of Acibenzolar-S-methyl treatment induces antiviral responses in distal leaves of Arabidopsis thaliana. Int J Mol Sci 25(3):1808. https://doi.org/10.3390/ijms25031808
Jia R, Zhao H, Huang J, Kong H, Zhang Y, Guo J, Huang Q, Guo Y, Wei Q, Zuo J, Zhu YJ, Peng M, Guo A (2017) Use of RNAi technology to develop a PRSV-resistant transgenic papaya. Sci Rep 7(1):12636. https://doi.org/10.1038/s41598-017-13049-0
Jridi C, Martin J-F, Marie-Jeanne V, Labonne G, Blanc S (2006) Distinct viral populations differentiate and evolve independently in a single perennial host plant. J Virol 80:2349–2357. https://doi.org/10.1128/jvi.80.5.2349-2357.2006
Kamatham S, Pallu R, Pasupulati AK, Singh SS, Gudipalli P (2017) Benzoylsalicylic acid derivatives as defense activators in tobacco and Arabidopsis. Phytochemistry 143:160–169. https://doi.org/10.1016/j.phytochem.2017.07.014
Karasev AV (2000) Genetic diver sity and evolution of clostero viruses. Annu Rev Phytopathol 38:293–324
Kollàr À, Dalmay T, Burgyàn J (1993) Defective interfering RNA-mediated resistance against cymbidium ringspot Tombusvirus in transgenic plants. Virology 193(1):313–318. https://doi.org/10.1006/viro.1993.1127
Komar V, Vigne E, Demangeat G, Lemaire O, Fuchs M (2008) Cross-protection as control strategy against grapevine fanleaf virus in naturally infected vineyards. Plant Dis 92(12):1689–1694. https://doi.org/10.1094/PDIS-92-12-1689
Lecoq H (1998) Control of plant virus diseases by cross protection. In: Hadidi A, Khetarpal RK, Koganezawa H (eds) Plant virus disease control. American Phytopathological Society Press, pp 33–40
Lindbo JA, Falk BW (2017) The impact of "coat protein-mediated virus resistance" in applied plant pathology and basic research. Phytopathology 107(6):624–634. https://doi.org/10.1094/PHYTO-12-16-0442-RVW
Malyshenko SI, Kondakova OA, Nazarova JV, Kaplan IB, Taliansky ME, Atabekov JG (1993) Reduction of tobacco mosaic virus accumulation in transgenic plants producing non-functional viral transport proteins. J Gen Virol 74(6):1149–1156. https://doi.org/10.1099/0022-1317-74-6-1149

Mandadi KK, Scholthof K-BG (2013) Plant immune responses against viruses: how does a virus cause disease? Plant Cell 25(5):1489–1505. https://doi.org/10.1105/tpc.113.111658

Medina-Salguero AX, Cornejo-Franco JF, Grinstead S, Mowery J, Mollov D, Quito-Avila DF (2021) Genetic characterization of a mild isolate of papaya ringspot virus type-P (PRSV-P) and assessment of its cross-protection potential under greenhouse and field conditions. PLoS One 16(2 February). https://doi.org/10.1371/journal.pone.0241652

Monci F, García-Andrés S, Sánchez-Campos S, Fernández-Muñoz R, Díaz-Pendón JA, Moriones E (2019) Use of systemic acquired resistance and whitefly optical barriers to reduce tomato yellow leaf curl disease damage to tomato crops. Plant Dis 103(6):1181–1188. https://doi.org/10.1094/PDIS-06-18-1069-RE

Moreno P, Ambros S, Albiach-Martí MR, Guerri J, Peña L (2008) Citrus tristeza virus: a pathogen that changed the course of the citrus industry. Mol Plant Pathol 9:251–268

Müller GW, Rezende JA (2004) Preimmunization: applications and perspectives in virus disease control. In: Diseases of fruits and vegetables volume I: diagnosis and management, vol I. Springer Netherlands, Dordrecht, pp 361–395

Nejidat A (1990) Transgenic tobacco plants expressing a coat protein gene of tobacco mosaic virus are resistant to some other Tobamoviruses. Mol Plant-Microbe Interact 3(4):247. https://doi.org/10.1094/MPMI-3-247

Nelson A, Roth DA, Johnson JD (1993) Tobacco mosaic virus infection of transgenic Nicotiana tabacum plants is inhibited by antisense constructs directed at the 5′ region of viral RNA. Gene 127(2):227–232. https://doi.org/10.1016/0378-1119(93)90724-H

Nishiguchi M, Kobayashi K (2011) Attenuated plant viruses: preventing virus diseases and understanding the molecular mechanism. J Gen Plant Pathol 77(4):221–229. https://doi.org/10.1007/s10327-011-0318-x

Pechinger K, Chooi KM, MacDiarmid RM, Harper SJ, Ziebell H (2019) A new era for mild strain cross-protection. Viruses 11(7):670. https://doi.org/10.3390/v11070670

Poogin MM (2018) Small RNA-omics for plant virus identification, virome reconstruction, and antiviral defense characterization. Front Microbiol 2018(9):2779. https://doi.org/10.3389/fmicb.2018.02779

Ravelonandro M, Briard P, Glasa M, Adam S (2008) The ability of a mild isolate of plum pox virus to cross-protect against sharka virus. Acta Hortic 781:281–286. https://doi.org/10.17660/ActaHortic.2008.781.42

Ravelonandro M, Scorza R, Briard P (2019) Innovative RNAi strategies and tactics to tackle plum pox virus (PPV) genome in Prunus domestica-plum. Plants 8(12):565. https://doi.org/10.3390/plants8120565

Register JC, Beachy RN (1988) Resistance to TMV in transgenic plants results from interference with an early event in infection. Virology 166(2):524–532. https://doi.org/10.1016/0042-6822(88)90523-5

Reimer-Michalski E-M, Conrath U (2016) Innate immune memory in plants. Semin Immunol 28(4):319–327. https://doi.org/10.1016/j.smim.2016.05.006

Roistacher CN, da Graça JV, Müller GW (2010) Cross protection against citrus tristeza virus–a review. In: Hilf ME, Timmer LW, Milne RG, da Graça JV (eds) Proceeding conference of the International Organization of Cit Rus virology, vol 17. IOCV, Riverside, pp 1–27

Sánchez-Tovar MR, Rivera-Bustamante RF, Saavedra-Trejo DL, Guevara-González RG, Torres-Pacheco I (2025) Mixed plant viral infections: complementation, interference and their effects, a review. Agronomy 15(3):620. https://doi.org/10.3390/agronomy15030620

Sanford JC, Johnston SA (1985) The concept of parasite-derived resistance—deriving resistance genes from the parasite's own genome. J Theor Biol 113(2):395–405. https://doi.org/10.1016/S0022-5193(85)80234-4

Sehgal OP, Krause GF (1968) Efficiency of nitrous acid as an inactivating and mutagenic agent of intact tobacco mosaic virus and its isolated nucleic acid. J Virol 2:966–971

Shopova E, Mihailova B, Todorova D, Sergiev I, Stoimenova E (2020) Systemic acquired resistance induced by compatible and incompatible tomato mosaic viruses effectively controls bacterial spot and speck diseases in tomato. Agriculture 10(7):302. https://doi.org/10.3390/agriculture10070302

Silva-Rosales L, Vargas P, Alcalá-Briseño R, Casarrubias K (2019) Revisiting cross-protection in the form of antagonism: the case of papaya mosaic virus protecting papaya plants against papaya ringspot virus. Acta Hortic 1250:55–62. https://doi.org/10.17660/ActaHortic.2019.1250.9

Singhal P, Nabi SU, Yadav MK, Dubey A (2021) Mixed infection of plant viruses: diagnostics, interactions and impact on host. J Plant Dis Prot 128(2):353–368. https://doi.org/10.1007/s41348-020-00384-0

Soler N, Fagoaga C, Chiibi S, López C, Moreno P, Navarro L, Flores R, Peña L (2011) RNAi-mediated protection against citrus Tristeza virus in transgenic citrus plants. In: Non coding RNAs in plants. Springer Berlin Heidelberg, Berlin, Heidelberg, pp 447–460. https://doi.org/10.1007/978-3-642-19454-2_27

Tatineni S, Dawson WO (2012) Enhancement or attention of dis ease by deletion of genes from citrus tristeza virus. J Virol 86:7850–7857

Tomitaka Y, Shimomoto Y, Ryang B-S, Hayashi K, Oki T, Matsuyama M, Sekine K-T (2024) Development and application of attenuated plant viruses as biological control agents in Japan. Viruses 16(4):517. https://doi.org/10.3390/v16040517

Tripathi S, Suzuki JY, Ferreira SA, Gonsalves D (2008) Papaya ringspot virus-P: characteristics, pathogenicity, sequence variability and control. Mol Plant Pathol 9(3):269–280. https://doi.org/10.1111/j.1364-3703.2008.00467.x

Umer M, Liu J, You H, Xu C, Dong K, Luo N, Kong L, Li X, Hong N, Wang G, Fan X, Kotta-Loizou I, Xu W (2019) Genomic, morphological and biological traits of the viruses infecting major fruit trees. Viruses 11(6):515. https://doi.org/10.3390/v11060515

Van Bueren ETL, Struik PC, van Eekeren N, Nuijten E (2018) Towards resilience through systems-based plant breeding. A review. Agron Sustain Dev 38:42

Venu E, Ramya A, Babu PL, Srinivas B, Kumar S, Reddy NK, Babu YM, Majumdar A, Manik S (2024) Exogenous dsRNA-mediated RNAi: mechanisms, applications, delivery methods and challenges in the induction of viral disease resistance in plants. Viruses 17(1):49. https://doi.org/10.3390/v17010049

Wang HL, Gonsalves D, Provvidenti R, Lecoq HL (1991) E ectiveness of cross-protection by a mild strain of zucchini yellow mosaic-virus in cucumber, melon, and squash. Plant Dis 75:203–207. [CrossRef]

Whitaker-Dowling P, Youngner JS (1987) Viral interference-dominance of mutant viruses over wild-type virus in mixed infections. Microbiol Rev 51(2):179–191. https://doi.org/10.1128/mr.51.2.179-191.1987

Yeh SD, Cheng YH (1989) Use of resistant Cucumis metuliferus for selection of nitrous-acid induced attenuated strains of papaya ringspot virus. Phytopathology 79:1257–1261

Zaccomer B, Cellier F, Boyer J-C, Haenni A-L, Tepfer M (1993) Transgenic plants that express genes including the 3′ untranslated region of the turnip yellow mosaic virus (TYMV) genome are partially protected against TYMV infection. Gene 136(1–2):87–94. https://doi.org/10.1016/0378-1119(93)90451-8

Zhang X-F, Qu F (2016) Cross protection of plant viruses: recent developments and mechanistic implications. In: Current research topics in plant virology. Springer International Publishing, pp 241–250. https://doi.org/10.1007/978-3-319-32919-2_10

Zhang X-F, Zhang S, Guo Q, Sun R, Wei T, Qu F (2018) A new mechanistic model for viral cross protection and superinfection exclusion. Front Plant Sci 9. https://doi.org/10.3389/fpls.2018.00040

Zhou M, Wang W (2018) Recent advances in synthetic chemical inducers of plant immunity. Front Plant Sci 9:1613. https://doi.org/10.3389/fpls.2018.01613

Zhu M, Liu S, Wang Z, Yu C, Yuan X (2023) A new strategy of cross-protection based on attenuated vaccines: RNA viruses are used as vectors to control DNA viruses. Agronomy 13(9). https://doi.org/10.3390/agronomy13092334

Ziebell H, Carr JP (2010) Cross-protection: a century of mystery. In: Advances in virus research, vol 76. Academic Press, pp 211–264. https://doi.org/10.1016/S0065-3527(10)76006-1
Ziebell H, MacDiarmid R (2017) Prospects for engineering and improvement of cross-protective virus strains. Curr Opin Virol 26:8–14. https://doi.org/10.1016/j.coviro.2017.06.010
Ziebell H, Payne T, Berry JO, Walsh JA, Carr JP (2007) A cucumber mosaic virus mutant lacking the 2b counter-defence protein gene provides protection against wild-type strains. J Gen Virol 88(10):2862–2871. https://doi.org/10.1099/vir.0.83138-0

Nanotechnology in Managing Viral Diseases in Fruit Crops

19

Anindita Saikia, Darshana Goswami, Archita Bordoloi, Adrij Pawan Neog, and Nima D. Namsa

Abstract

Plant viral diseases, particularly those caused by tobamoviruses, continue to challenge global agriculture despite more than a century of scientific advancements. This chapter outlined the interplay between traditional and advanced strategies for managing plant viral diseases, with particular emphasis on tobamoviruses. Conventional methods such as chemical control, resistant varieties, and vector management often fall short due to viral diversity, rapid mutation rates, and environmental concerns. Nanotechnology introduces a novel paradigm by enabling precise delivery systems, early diagnostics, and enhanced plant protection strategies. Nanoparticles (NPs), including silver, gold, silica, and polymer-based types, have demonstrated strong antiviral properties either by directly inactivating viruses or by boosting plant defense mechanisms. One of the most promising applications is the use of nanoparticles for the targeted delivery of antiviral agents or double-stranded RNA (dsRNA), to induce RNA interference (RNAi) in plants. For instance, chitosan and liposome-based nanocarriers have been used to deliver dsRNA effectively in fruit crops like papaya and citrus to combat papaya ringspot virus (PRSV) and citrus tristeza virus (CTV), respectively. Nanobiosensors enable rapid and sensitive detection of early infection, thereby facilitating timely decisions for management of disease. The integration of nanotechnology with precision agriculture tools further enhances disease monitoring and management. The safety, scalability, and regulatory aspects of nano-based interventions are under active research. Nanotechnology holds immense potential to revolutionize viral disease management in fruit crops by

A. Saikia · D. Goswami · A. Bordoloi · A. P. Neog · N. D. Namsa (✉)
Department of Molecular Biology and Biotechnology, Tezpur University, Tezpur, Assam, India
e-mail: namsa@tezu.ernet.in

R. Kumar et al. (eds.), *Fruit Crops and Viral Pathogens Volume II*, Plant Pathology for Sustainable Agriculture, https://doi.org/10.1007/978-981-95-7261-8_19

offering targeted, eco-friendly, and highly effective solutions for sustainable agriculture, as discussed in this chapter.

Keywords

Nanotechnology · Viral diseases · Fruit crops · Nanoparticles · RNA interference

19.1 Introduction

Agriculture is a mainstream science that focuses on the sustainable cultivation of crops to help in fighting the problem of hunger globally, and with the world expected to reach almost 10 billion people by 2050, securing food security has become a global agenda (Gill and Garg 2014). However, physical stress factors such as biotic and abiotic stresses severely affect crop production, with bacteria, fungi, and viruses being the most devastating. Therefore, to curb their effects, measures such as quarantine and trade limitations have been practiced in the past, and farmers still rely on chemical, biological, and integrated pest management (IPM) methods (Almeida 2018). Notwithstanding these, 20–40% of agricultural produce is lost to pests per year worldwide, with 16% of the loss caused by plant pathogens (Gullino et al. 2021). Crop losses due to diseases, pests, and weeds stand at 15–25% each year in India (Ficke et al. 2018). Plant viruses are the greatest menace among pathogens, contributing about 47% of global losses associated with plant diseases (Kumar et al. 2021). Viral infections distort the physiology of plants, which includes stunting, chlorosis, distortion of leaves, mottling of fruits, and failure to flower or set fruit; these eventually lead to low yields and reduced marketability. Remarkable examples include sunflower necrosis disease caused by tobacco streak virus (TSV) in India and geminivirus-induced pod issues in cotton, legumes, tomato, okra, and chili (Reddy and Zehr 2004). Insect vectors such as whiteflies, aphids, thrips, and leafhoppers play a major role in dispersing most plant viruses, so vector control is a major topic in disease management. However, overuse of chemical pesticides has caused pesticide resistance, loss of biodiversity, and environmental contamination, creating a pressing need to adopt alternative and sustainable methods (Khurana and Marwal 2016). Nanotechnology opens a pathway in the management of plant diseases. The unique features of nanoparticles (NPs), including their very small size (below 100 nm), large surface-to-volume ratio, and high reactivity, render them highly suitable for agriculture (Jeevanandam et al. 2018). In recent research, NPs have been investigated extensively for biomolecular detection, advanced diagnostics, targeted antimicrobial delivery, and crop protection techniques over the last 20 years (Dutta 2012). Nano-encapsulated pesticides contain fewer active ingredients than conventional formulations, release their contents in a controlled manner, and enhance crop protection—lowering environmental exposure risks. Moreover, NPs can be used as vectors, increasing bioavailability and target specificity of active compounds, while reducing harm to beneficial organisms. In addition to crop protection, NPs enhance plant immunity, nutrient uptake, and tolerance to environmental stress. Notably, metallic nanoparticles, including silver (Ag), zinc oxide (ZnO), and titanium dioxide (TiO_2), exhibit antiviral effects through the inactivation of viral particles or induction of systemic resistance. This indicates that nanotechnology may soon integrate disease diagnosis, pathogen control, and plant health management into a coherent and environmentally friendly system for sustainable crop production (Mahmood et al. 2017). Historically, tobamoviruses have played a pivotal role in plant virology research. Tobacco mosaic virus (TMV), the first virus ever studied in detail, together with related tobamoviruses such as tomato mosaic virus (ToMV) and those infecting pepper (*Capsicum* spp.), caused serious crop losses for decades. By the late twentieth century, the incorporation of resistance genes into commercial tomato and pepper cultivars provided effective control, although tobamoviruses continued to serve as important models in molecular biology, biotechnology, and bio-nanotechnology. In recent years, however, tobamoviruses have re-emerged as serious crop pathogens with the appearance of tomato brown rugose fruit virus (ToBRFV)—capable of overcoming TMV and ToMV resistance—and the worldwide spread of cucumber green mottle mosaic virus, which threatens cucurbit crops (Muronga et al. 2021). To combat these viruses, a variety of molecular approaches are being explored, including cross-protection with mild strains, transgenic expression of viral genes to disrupt infection, RNA silencing strategies (via engineered or externally applied double-stranded RNA), CRISPR-mediated editing of host susceptibility factors, and the transfer or optimization of natural resistance genes (Muronga et al. 2021). One of the most concerning instances of viral persistence is observed with ToBRFV, which is seed-borne, soil-borne, and mechanically transmitted, making it very hard to eliminate. ToBRFV-infected crops act as reservoirs in monoculture systems, and soils contaminated with ToBRFV infect new seedlings, causing severe yield loss. In current research, wild tobacco species were inoculated with ToBRFV, and the survival of the virus in soil for more than 5 months was observed. High ToBRFV titers were found in soil samples through high-throughput sequencing (HTS), showing 99.92% sequence similarity to the wild-type (WT) strain. This isolate, called ToBRFV-A134T, harbored a non-synonymous mutation in the movement protein (A134T substitution). Remarkably, the A134T variant of ToBRFV was maintained in Tm-22-resistant tomatoes (e.g., cv. Ikram), and co-inoculation with ToBRFV-WT confirmed the co-occurrence of the two isolates. Comparative symptom analysis showed that WT infections produced mottling and mosaic patterns with infrequent shoestring leaves, whereas A134T infections resulted in more severe symptoms, such as strong shoestring-like, felt-like leaves and fern-like leaflets. Gene expression profiling compared to RT-qPCR in resistant plants showed that RNA-dependent RNA polymerase 6b, Dicer-like 4, and Dicer-like 2b were altered by ToBRFV-A134T, indicating that ToBRFV-A134T induces stronger mild disease via molecular pathways distinct from the WT strain (Molad et al. 2025).

19.2 Plant Viral Disease Management: Nanotechnology Measures

Nanotechnology is the manipulation, design, and use of materials at the molecular, atomic, or supramolecular scale. Due to their large surface-area-to-volume ratio, nanoparticles exhibit unusual physicochemical characteristics such as higher thermal conductivity, chemical stability, catalytic activity, and biological functionality compared to their bulk forms. These properties make nanoparticles extremely useful in sensitive biomolecular detection, disease diagnostics, antimicrobial development, and therapeutic applications. The combination of nanotechnology and biological sciences has given rise to biotechnology, a fast-developing discipline with enormous potential. Bio-nanotechnology focuses on combining biotechnology and nanoscience to establish environmentally friendly and biologically motivated processes for the production of nanomaterials, providing viable solutions to contemporary scientific and medical challenges (Sobha and Surendranath 2010). Nanomaterials have been important in plant protection—either as direct protectants or as carriers that deliver pesticides with great accuracy through adsorption, encapsulation, or conjugation. They exhibit high antifungal, antiviral, antipyretic, and antibacterial activity and are used as foliar sprays, seed treatments, or root applications. Commonly studied nanoparticles include silver, copper, zinc oxide, gold, titanium oxide, and iron, which are effective against a wide variety of plant pathogens such as *Rhizoctonia solani*, *Fusarium oxysporum*, *Sclerotinia sclerotiorum*, *Colletotrichum* spp., *Xanthomonas oryzae*, and plant viruses such as BYMV, CMV, and PVY (Khan at el. 2019). Their antimicrobial effects include interference with cell membranes, inhibition of RNA and protein synthesis, disruption of nutrient flow, and suppression of toxin production. Moreover, nano-fertilizers improve plant health and resistance and have been successfully used to control diseases, including Fusarium wilt, Verticillium wilt, and root rots in various crops (Elmer et al. 2018).

19.3 Nanophytovirology: Plant Virus Management

Nanophytovirology is a technology that uses nanotechnology in the detection, diagnosis, and treatment of plant viruses and their arthropod vectors at early stages to protect crops against economically important viral diseases. Viral plant diseases are particularly difficult to contain because they propagate rapidly through vectors, posing severe challenges to global agriculture. Conventional management techniques, usually founded on virus ecology, have not achieved significant success in reducing crop losses. Effective control requires a deeper understanding of plant–virus–vector interactions. Nanotechnology in diagnostics and management has become a trend in recent years due to its cost-effectiveness, simplicity, and efficiency. Nano-formulations have benefits such as reduced toxicity, longer shelf life, and enhanced solubility of poorly soluble pesticides, making them environmentally friendly. Nanoparticles may be used in two primary applications to manage plant viruses: (i) nanoparticles themselves as protectants or (ii) nanoparticles as nanocarriers of pesticides and RNA interference molecules. Additionally, nanoparticles have potential as bio-stimulants and nano-biosensors for early and precise virus detection (Marwal and Gaur 2019).

19.4 Nanoparticles Used for Plant Virus Control

Nanoparticles (NPs) exhibit strong virucidal activity against a broad spectrum of viruses and can reduce viral infectivity. Nanomaterials—including non-metals, metalloids, metallic oxides, and carbon-based nanomaterials—are widely used in the prevention of viral diseases in plants. The earliest published case of nanoparticle use in virus detection was gold nanoparticles (AuNPs) conjugated with silver staining for detecting cervical cancer cells containing human papillomavirus. Among the widely studied nanoparticles are iron, gold, titanium oxide, copper, zinc oxide, and especially silver (Wang and Xia 2004). Recent studies have explored sustainable nanoparticle production, such as using rice straw to create bio-nanocomposites consisting of nanoparticle-coated biopolymer matrices. These green bio-nanocomposites offer improved surface area, strength, barrier properties, and aspect ratio compared to biopolymers alone, making them promising eco-friendly materials. Similarly, wheat straw has gained attention as a viable filler for polymer matrices and a rich biomass source for synthesizing stable nanoparticles. Nanoparticles are thought to interfere mainly with early stages of viral replication, although more research is needed to fully elucidate their effects on viral mobility, replication, encapsidation, and transmission. Their antiviral mechanisms may include the release of metal ions, generation of oxidative and non-oxidative stresses, and the inactivation or denaturation of viral capsid proteins, nucleic acids (RNA/DNA), and other essential proteins. Additionally, NPs may inhibit virion binding, fusion, and infectivity by producing reactive oxygen species (ROS) and oxidizing viral proteins. By interacting with virus surface glycoproteins, NPs can also block host cell recognition (Warghane et al. 2024). In agriculture, the activity of different types of synthesized NPs has been assessed against numerous phytopathogenic viral pathogens, and results show strong inhibitory effects (Table 19.1). ZnO NPs and Fe_3O_4 NPs inhibit tobacco mosaic virus (TMV) in vitro. Schiff-base nanosilver NPs reduce TMV infection. Engineered nanomaterials (ENMs), both metal-based (Fe_2O_3, TiO_2) and carbon-based (MWCNTs, C_{60}), applied as daily foliar sprays, decreased Turnip mosaic virus (TuMV) coat protein levels by 15–60% in *Nicotiana benthamiana* while enhancing biomass and phytohormone levels, suggesting a dual role in disease suppression and plant growth promotion. AgNPs inhibit local lesions, suppress tomato spotted wilt virus (TSWV), and induce resistance to potato virus Y (PVY). Graphene oxide-silver NPs (GO-AgNPs) reduce tomato bushy stunt virus (TBSV) infection and disease severity. AuNPs dissociate barley yellow mosaic virus (BaYMV) particles in vitro. SiO_2 NPs reduce the severity of tomato yellow leaf curl virus (TYLCV). Carbon nanotubes disrupt TMV movement and replication while inducing resistance. TiO_2 NPs lower disease severity caused by broad bean stain virus (BBSV). NiO NPs decrease both the severity and concentration of cucumber

Table 19.1 Antiviral properties of potential nanoparticles

Nanoparticle	Size	Target virus	Host plant	Effect
Zinc oxide NPs (ZnONPs)	18 nm	Tobacco mosaic virus (TMV)	*Nicotiana benthamiana*	Suppression of the speed of TMV invasion
Silicon dioxide NPs (SiO_2NPs)	100 nm	Tomato yellow leaf curl virus (TYLCV)	*Solanum lycopersicum*	Reduction in disease severity and TYLCV concentration
Carbon nanotubes (CNTs)	–	TMV	*N. benthamiana*	TMV movement and replication inhibition and induction of resistance
Graphene oxide-silver NPs (GO-AgNPs)	3050 nm	Tomato bushy stunt virus (TBSV)	*Lactuca sativa*	Decrease in virus concentration, infection percentage, and disease severity
Nickel oxide NPs (NiONPs)	15–20 nm	Cucumber mosaic virus (CMV)	*Cucumis sativus*	Reduction in disease severity and CMV concentration
Silver nanoparticles (AgNPs)	12.6 ± 5 nm	Tomato spotted wilt virus (TSWV)	*Solanum tuberosum*	Inhibition of local lesions and reduction in TSWV infection
Gold NPs (AuNPs)	–	Barley yellow mosaic virus (BaYMV)	*H. vulgare*	Dissociation of viral particles in vitro
Titanium dioxide NPs (TiO_2NPs)	~3–5 µm	Broad bean stain virus (BBSV)	*Vicia faba L.* (Fabaceae)	Reduction in disease severity
Schiff base nanosilver NPs	–	TMV	*Nicotiana tabacum*	Reduction in infection caused by TMV
Iron oxide NPs (Fe_3O_4)	0.19 nm	TMV	*N. benthamiana*	Induction of plant resistance and activation of antioxidants against TMV by upregulation of SA genes
AgNPs	12 nm	Potato virus Y (PVY)	*S. tuberosum*	Induction of resistance in host plant against virus
Cerium oxide NPs (CeO_2NPs)	–	TMV	*Datura stramonium* and *Nicotiana tabacum*	Suppression of viral symptoms

mosaic virus (CMV) in cucumber plants. CeO_2 NPs have been reported to suppress TMV symptoms (Dutta et al. 2022).

Beyond management, NPs can be conjugated with nucleic acids, antibodies, or aptamers to develop nano-biosensors for virus detection or to identify virus-carrying vectors. Future research should focus on the combined use of metallic nanoparticles and organic polymer-based nanocomposites to develop effective, field-ready strategies for plant viral disease control (Dutta et al. 2022).

19.5 Mechanism of Action of Nanomaterials Against Plant Viruses

Direct Method Nanoparticles directly deal with the viruses that interfere with the replication of the viral genome, which hampers protein biosynthesis of viruses, specifically the viral coat protein. NPs alter the packaging of viral genomes and the degradation of the capsid protein that is required for the packaging of viral particles. They disturb the electron transport system of plant cells infected with viruses and might be responsible for increasing the level of cellular barriers like cell walls or lignin deposition.

Indirect Method The NP(s) do not directly inactivate the viruses. They activate the defense system of plants that is involved in eliminating viral infection. NPs in plants during viral infection lead to increased antioxidant enzyme (peroxidase, polyphenol oxidase, phenylalanine ammonia-lyase, and superoxide dismutase) activity.

19.6 Nanoparticles in Pathogen Suppression

Pathogen-induced crop losses are a major concern for global agriculture, and though conventional management strategies exist, none provide absolute protection (Hak and Spiegelman 2021). This highlights the necessity of integrating innovative methods with traditional practices to improve disease control. Among such innovations, nanotechnology has emerged as a promising tool. Nanoforms of metals and other compounds have shown considerable potential in managing plant diseases effectively. For instance, studies have demonstrated that nanoforms of micronutrients such as $CuSO_4$ and $Na_2B_4O_7$ are particularly effective in controlling rust disease in field peas, while manganese and zinc nanoparticles have suppressed damping-off and charcoal-rot in sunflowers (Singh et al. 2013). It is therefore possible to find several protective measures against plant pathogens using metallic nanoparticles (Gupta et al. 2018). Among the innovations, there is the application of nanoparticles as nanocarriers to deliver double-stranded RNA (dsRNA) as a protective measure against viral infections by RNA interference (RNAi). RNAi operates as a component of the plant immune system where mRNA stability was suppressed by the injection of a specific pathogenic nucleic acid, showed that the layered double hydroxide (LDH) nanosheets were able to deliver the RNAi into plants with a pathogenic nucleic acid, which resulted in viral immunity against pepper mild mottle virus (PMMoV) and cucumber mosaic virus (CMV) (Harrison and Wilson 1999). Notably, the nanoparticles were carriers, which disintegrated naturally as time went by, making the system sustainable. Nanoparticles also have a direct antimicrobial effect. This is because their size allows them to work effectively due to their nanoscale size and their large surface-to-volume ratio, increasing the ability to interact with microbial membranes more than just by releasing metal ions. Currently, as an illustration, nanoparticles of ZnO and gold have demonstrated high levels of antibacterial action against E. coli and Staphylococcus aureus, whereas silver (AgNPs) and copper nanoparticles have been effective on S. aureus, Pseudomonas aeruginosa, and Salmonella enterica (Ghazy et al. 2021). Nevertheless, the ecological effects of nanoparticles, including AgNPs in soil ecosystems, should be taken into account cautiously, because they affect both the pernicious and advantageous communities of bacteria. Silver nanoparticles have been extensively researched on the basis of antimicrobial activity. There are several mechanisms put forward to describe their mode of action, although the actual mechanism is controversial (Feng et al. 2008). AgNPs have the ability to adsorb to and traumatize bacterial cell walls, changing the membrane permeability, and eventually leading to cell death. They can also produce free radicals that weaken the integrity of the membranes as reported previously (Durán et al. 2010). The other hypothesis is the discharge of silver ions, which interact with thiol (−SH) functional groups of essential enzymes and interfere with their functionality. These ions are also able to block respiratory enzymes and result in the generation of reactive oxygen species (ROS) that cause destruction of the cell formation. Also, AgNPs can participate in the interaction with DNA, binding sulfur and phosphorus elements, thus impeding the process of replication and causing the death of the cell. AgNPs can also disrupt bacterial signaling pathways. They have been demonstrated to modify phospho-tyrosine levels in bacterial peptides, preventing signal transduction that is required to induce growth. In addition, silver ions react with -SH groups within the proteins, which also causes further inactivation of bacteria. Together, the antimicrobial activity of AgNPs has been described to occur via five broad categories: (i) release of toxic ions binding to sulfur proteins, (ii) genotoxic effects resulting in DNA damage, (iii) electron transport disruption, (iv) membrane potential collapse, and (v) oxidative stress through ROS (Harrison and Wilson 1999). Nanotechnology has a massive potential in managing plant diseases. In addition to having a high antimicrobial effect against fungal and bacterial pathogens, nanoparticles are effective carriers of the dsRNA-based viral resistance. Silver nanoparticles, specifically, exhibit multi-dimensional effects of pathogen inhibition, but more studies are needed to clarify their activities completely and assess their effect on the ecological situation. A combined approach of these nano-enabled traditional practices improves crop resistance to pathogens.

19.7 Manufactured Resistance to Tobamoviruses

Plant pathology took a new turn at the end of the nineteenth century with the identification of tobacco mosaic virus (TMV), the first virus to be known, the predecessor of the more general idea of viruses as infectious agents independent of other microorganisms, and the birth of the discipline of virology (Kenyon et al. 2014). TMV was discovered due to the dire necessity to know and control mosaic disease in tobacco (*Nicotiana tabacum* L.), which was leading to massive economic losses by the time of its discovery. A series of other related tobamoviruses were identified afterwards, such as tomato mosaic virus (ToMV) and pepper mild mottle virus

(PMMoV). These viruses were some of the most devastating agents of economic crops like tomato (*Solanum lycopersicum* L.) and pepper (*Capsicum annuum* and related species), causing severe losses in yields and economic costs over decades (Scholthof 2004). Tobamoviruses are also known to be extremely hard to control because of their very high infectivity and transmissibility. Transmission is mostly mechanical, also known as contact transmission, in which the virus in the inoculum infects the plant tissue by entering through wounds or abrasions when in contact with such tissue. They are extraordinarily stable with their encapsidated single-stranded, positive-sense RNA genomes, which are in the form of a rigid rod shape (Ishibashi et al. 2023). This stability makes tools, equipment, and facilities harder to disinfect and decontaminate, as tobamoviruses can live not only in live and dead plant material, but also in soil, irrigation water, frozen or liquid storage, tobacco products, and even worker clothing. The strength of the tobacco mosaic virus particles has therefore been of primary importance in their persistence and inability to be eliminated. TMV research and similar viruses helped to achieve breakthroughs in the study of the structure of viruses, their genome structure, modes of replication, interactions with the host, and evolution of viruses (Symonds et al. 2018). Moreover, unexpected applications of biotechnology have been made with the use of tobamoviruses. In the early twenty-first century, TMV was repurposed as a platform for drug delivery systems and as a building block in bio-based nanotechnology. Similarly, PMMoV has recently been applied as a reliable tracer for detecting sewage contamination in water systems (Hanssen et al. 2010). Despite their historical significance, tobamoviruses began to decline in importance as threats to tomato cultivation by the late twentieth and early twenty-first centuries, particularly in developed countries. During this period, authoritative reviews of tomato viral diseases barely mentioned tobamoviruses, instead highlighting other pathogens such as cucumber mosaic virus (CMV), tomato spotted wilt virus (TSWV), pepino mosaic virus (PepMV), tomato yellow leaf curl virus (TYLCV), and members of the *Torradovirus and Crinivirus* genera as the primary threats (Gilbertson et al. 2015). This decline occurred in two major stages. The first was the introduction of cross-protection, a strategy whereby plants were deliberately inoculated with mild strains of TMV or ToMV to protect against more virulent strains. While initially effective, cross-protection lost prominence with the development of tomato varieties carrying resistance genes derived from wild relatives. The Tm-1 gene, introgressed from *Solanum habrochaites* (syn. *Lycopersicon hirsutum*), and its allelic variants Tm-2 and Tm-22, derived from *S. peruvianum* (syn. *L. peruvianum*), became central to breeding programs (Pfitzner 2006). The Tm-1 and tm-1 alleles encode proteins of 754 amino acids that differ at 25 residues. The Tm-1 protein, but not the tm-1 variant, binds to the RNA helicase domain of the viral 126/183 kDa replicase protein, thereby blocking the formation of the replication complex. Consequently, Tm-1 confers resistance to TMV and ToMV, whereas tm-1 does not. However, both alleles inhibit tobamoviruses such as tobacco mild green mosaic virus (TMGMV) and PMMoV, which explains differences in tobamovirus host range. Unfortunately, TMV and ToMV frequently evolved resistance-breaking mutations, limiting the durability of Tm-1 alone. To counter this, breeders developed tomato lines stacking Tm-1 with Tm-2 or Tm-22, creating multi-layered resistance that imposed a high evolutionary cost on viral mutants and remained effective for decades (Dunigan et al. 1987). Hypersensitive response (HR)-type resistance is provided by the Tm-2 and Tm-22 genes, in which viral infection is limited to hundreds of cells at the site of inoculation. This localized defense can be seen in the form of necrotic lesions brought about by programmed cell death (Saito et al. 1987). These two genes encode NB-LRR (nucleotide-binding, leucine-rich repeat) immune receptors, which identify the viral movement protein as an elicitor. Upon recognition, the receptors trigger HR and activate downstream defense signaling pathways to prevent systemic viral spread. The durability of Tm-2 and Tm-22 conferred strong protection in tomato for nearly 60 years, until the emergence of tomato brown rugose fruit virus (ToBRFV) in 2014, which overcame this resistance. Nevertheless, in many low- and middle-income countries, where older tomato varieties are still cultivated, TMV and ToMV continue to pose serious threats, particularly for smallholder farmers. In tobacco, tobamovirus resistance has largely been mediated by the N and N′ genes, originally identified in *Nicotiana glutinosa* and *N. sylvestris*, respectively, and introgressed into cultivated tobacco (*N. tabacum*). Like the tomato Tm genes, N and N′ encode NB-LRR immune receptors that confer HR-type resistance. The N protein recognizes an elicitor sequence in tobamovirus replication proteins, while N′ detects specific amino acids in the viral coat protein (CP) (Ojinaga et al. 2022). The N gene confers broad-spectrum resistance against most tobamoviruses, but the N′ gene is more limited, protecting against ToMV but not the common TMV-U1 strain (Poulicard et al. 2024). Although it is effective, N-gene resistance was not used throughout the breeding of tobacco, which contributes to a continued pool of TMV inoculum that can infect other crops (Poulicard et al. 2024). In pepper production, tobamoviruses are also still of importance to the agriculture sector. Such viruses are a frequent occurrence in isolation or with other pathogens, making them have a destructive effect. Breeders have used various alleles of L genes in commercial pepper varieties. Similar to the tomato and tobacco resistance genes, NB-LRR proteins are encoded by the L alleles here; however, the elicitors are amino acid sequences within the viral CP orthologs. Pepper lines with L genes are very prone to systemic infection with a wide array of tobamoviruses, which include TMV, ToMV, TMGMV, PMMoV, and paprika mild mottle virus. The L gene family (L1–L4) has different resistance patterns, forming a hierarchical arrangement of the pathotypes of the tobamoviruses. TMV is suppressed by L1, but in some PMMoV strains, the most troublesome tobamoviruses in pepper, even L4, the most extensive resistance allele (Salem et al. 2016). Other management practices, in conjunction with genetic resistance, such as strict phytosanitary practices like disinfection of equipment, handwashing, and protective clothing, have contributed to the reduction of mechanical transmission. It has also been used in the cultivation of plants under glasshouses or tunnels, which has lowered the cases of tobamovirus infection due to low contact with outside sources of the inoculum. Such combined strategies made sure that by the beginning of the twenty-first century, the challenge of tobamoviruses in crops like tomato and pepper could no longer be considered the primary problem in dealing with viruses, but it still posed a serious threat when conditions were favorable. The history of the tobamoviruses, therefore, presents the dynamic nature of the interaction between pathogens, host resistance, and agricultural practices. Since the nineteenth century, when they were first identified as pathogenic to tobacco, through their use in molecular virology, and now in biotechnology, tobamoviruses have kept the limelight of plant virology studies. Even though their importance as crop pathogens has declined in most developed agricultural systems through the introduction of resistance genes and better agricultural practices, they are still a problem in resource-constrained environments, and they also evolve, as seen with the development of ToBRFV. Their biology, stability, and interactions with their hosts are also of vital importance in not only the protection of plants but also the exploitation of their potential in biotechnology and in environmental monitoring (Luria et al. 2017).

19.8 Insights on ToBRFV and CGMMV Challenges in Global Crop Production

The latest renewed interest in the control of the tobamoviruses is due to the 2014–2015 discovery in Israel and Jordan of tomato brown rugose fruit virus (ToBRFV), a novel tobamovirus that was able to overcome the tomato resistance gene Tm-2 and Tm-22, and was not affected by Tm-1 or tm-1 products (Jones 2021; Zinger et al. 2021). ToBRFV, with a wide host range (including tomato, pepper, tobacco, and various non-solanaceous plants), has resulted in the most losses in tomato, with up to 70% yield loss because of leaf deformation, fruit rugosity, discoloration, and uneven ripening (Salem et al. 2023; Fox et al. 2020; Hak and Spiegelman 2021). As phylogenetically related to TMV than ToMV, ToBRFV has a different movement protein (MP) sequence, enabling it to avoid Tm-22-mediated resistance, albeit with a lower efficiency of intercellular transport than TMV/ToMV MPs (Hak et al. 2023; Obregón et al. 2023). This fitness cost notwithstanding, the ability of ToBRFV to circumvent resistance has seen it triumph over tomato populations immunized against TMV and ToMV, and spread around the world in trade routes, with Australia being currently an exception (Murphy et al. 2023). N and N′ genes provide resistance in tobacco, and peppers with L alleles display temporary resistance, i.e., L-based resistance is of short value. Symptomatic and asymptomatic peppers play a role in inoculum reservoirs, sabotaging tomato production. This is similar to the previous explanation given by Holmes of TMV resistance breeding in tobacco, which was based on the reduction of sources of inoculum and not the crop itself (Lindbo and Dougherty 2005). Cucumber green mottle mosaic virus (CGMMV) is another issue that has appeared along with ToBRFV. It can be transmitted by seeds and pollination by hosts, making it a troublesome host-restricted disease to the cucurbits, including cucumber, melon, and watermelon (Lius et al. 1997). Even the possible Antarctic associations are reported. Combining ToBRFV and CGMMV, classical resistance barriers begin to be eroded, and the importance of pre-emptive research into biotechnological interventions becomes increasingly important as a valid approach to effective and lasting control of tobamoviruses compared to broad-spectrum induced resistance (Lius et al. 1997).

19.9 Antiviral Strategy Using RNA Silencing

RNA silencing or RNA interference (RNAi) is one of the best innate silencing mechanisms. Such a process becomes activated in the event the plant identifies the presence of RNA molecules that are made in a double-stranded form (dsRNA), when the virus is replicating, and cleaves them into small interfering RNAs (siRNAs). The siRNAs are integrated into RNA-induced silencing complexes (RISCs) that direct the cleavage and translational silencing of complementary viral RNA, limiting viral proliferation (Gonsalves et al. 1998). Initial research in Arabidopsis thaliana showed the pivotal position of RNA silencing in antiviral immunity, when antiviral immunity mutants, defective in major RNAi pathway components like Dicer-like (DCL) or Argonaute (AGO) proteins, were found to be more vulnerable to tobamoviruses (Lopez-Gomollon and Baulcombe 2022). Equivalent results in other flora species proved that RNA silencing is a universal immune response, resulting in the conservation of the mechanism against RNA viruses. The small RNA sequencing has, in most instances, discovered the accumulation of virus-derived siRNAs (vsiRNAs) in case of infection by *tobamovirus*, additionally depicting the significance of this route in inhibiting viral multiplication (Lin et al. 2007). Nevertheless, the antiviral effects of RNA silencing are met with significant mutation and the evolution of viral suppressors of RNA silencing (VSRs). Tobamoviruses have proteins, including the 126 kDa replicase subunit, that can inhibit siRNA biogenesis or activity so as to suppress host defense. Such suppression leads to a new endless molecular arms race, in which plants constantly improve their silencing processes as viruses counter them. The degradation of resistance to ToBRFV tomato is an example of the efficacy of silencing repression, which has the potential to facilitate the development and transmission of viruses (Gaafar and Ziebell 2020). RNA silencing, in spite of these difficulties, is found to be one of the strong bases of biotechnological strategies geared towards improving antiviral resistance. Transgenic strategies that overexpress hairpin RNAs or small RNA precursors have already managed to transfer resistance to several tobamoviruses in model systems. In addition, RNA silencing can be used in conjunction with other methods of resistance, such as R gene deployment or genome editing, promising the potential of long-term protection.

19.10 Artificial microRNAs (amiRNAs)

Artificial microRNAs (amiRNAs) are an advanced use of the RNA silencing cascade in developing resistance to viruses. However, in contrast to traditional hairpin RNA biosensors that have the capacity to produce heterogeneous groups of minute interfering RNAs (siRNAs), amiRNAs are engineered to appear as natural microRNAs (miRNAs), but they are reconfigured to attack particular viral sequences with specificity (Niehl et al. 2018). In this approach, the miRNA biogenesis pathway of the plant is used, which causes the generation of one or a few fixed 21–22 nucleotide small RNAs that mediate sequence-specific cleavage of viral RNA (Uslu and Wassenegger 2020). AmiRNA design generally entails the substitution of the miRNA/miRNA* sequence in a precursor hairpin by viral gene-of-interest complementary sequences. In plants, this precursor is cleaved by the Dicer-like 1 (DCL1) into the mature amiRNA, which is inserted into Argonaute 1 (AGO1) into the RNA-induced silencing complex (RISC) (Pugsley et al. 2024). AmiRNAs can be used to potently inhibit the replication and dissemination of viruses by greatly narrowing down on the conserved areas of the viral genomes, e.g., the movement protein (MP) or replicase genes. One of the major pros of amiRNAs is its specificity. AmiRNAs can be computationally used to contain a minimum sequence that overlaps with the host genome without altering the titer (Mitter et al. 2017a, b). Multiplexed amiRNA builds multiple essential genes or conserved motifs that offer resistance in case of sequence variability. The use of amiRNAs to confer resistance to plant viruses has been proven to be successful in several studies. As an illustration, tomato and Nicotiana benthamiana plants that had amiRNAs against ToMV and TMV exhibited a high level of resistance, a low level of symptoms, and a low level of viruses (Jain et al. 2022). Although a particular large-scale application against ToBRFV remains in its initial phases, the methodology is viewed as a high-potential one due to the ability of the virus to defeat traditional resistance through the utilization of the R-genes. In the future, amiRNA technology can be integrated with CRISPR/Cas-based editing of host susceptibility genes or exogenous delivery of dsRNAs to create a multilayered and combined protection against tobamoviruses. Furthermore, synthetic biology approaches may allow the development of inducible amiRNA systems that activate only under infection conditions, reducing metabolic costs to the host while preserving antiviral efficacy.

19.11 Topical dsRNA and Nanocarrier Delivery

The application of exogenous double-stranded RNA (dsRNA) to plants has emerged as a practical and non-transgenic approach to induce RNA silencing and achieve antiviral resistance. The idea behind this approach is that externally delivered dsRNAs, after being internalized by the plant, can be processed into small interfering RNAs (siRNAs), which in turn lead to the degradation of their corresponding complementary viral RNAs (Uslu et al. 2020). In the case of the tobamoviruses, like ToBRFV and CGMMV, which are severe threats to resistance breeding, topical application of the dsRNA treatment has a very promising future and can be applied within a very short period, without the necessity of genetic modification. Among the first examples of such a strategy was the extrinsic expression of TMV coat protein (CP) and motion protein (MP) genes as a result of the exogenous expression of the TMV dsRNAs. It was shown that plants that were treated with such molecules accumulated fewer viruses and developed fewer symptoms, proving this method as a proof-of-concept of the concept of protection by the use of dsRNA (Carr 2024). Even more recent efforts have extended the approach to ToBRFV and CGMMV, with the results demonstrating that the use of dsRNAs against viral replication-related proteins can greatly reduce the viral loads in tomato and cucumber, respectively (Scholthof 2017). Topical application of dsRNA has issues to do with stability and delivery, although it promises to be a promising technique. The naked dsRNA molecules are easily degraded in the environment by nucleases, and their survival on the plant surface is short-lived, reducing their efficacy in the field. To overcome this, scientists have resorted to nanotechnology-based delivery systems, such as clay nanosheets, carbon nanotubes, and liposomal carriers, which have the ability to protect the molecules of transported dsRNA against degradation and release it on demand (Dunigan et al. 1987). Such nanocarriers increase the absorption of dsRNA on the leaf surfaces or root systems, which increases the duration of protection and enhances resistance. In the case of tobamoviruses, the nanocarrier-assisted delivery is particularly appealing as the practical option, due to the ability to move protective organ oleic dsRNAs both locally and systemically in the plant. Recent studies with layered double hydroxide (LDH) nanosheets, or BioClay, have demonstrated that up to 20 days of protection of loaded plants with dsRNAs can be achieved when nanosheets are utilized (Palukaitis and Yoon 2024). This sort of durability is essential in managing the rapid spread of tobamoviruses in the business place. The other benefit of exogenous delivery of dsRNA is that it is regulatory-flexible as compared to transgenic methods. Since plants are not modifiable, there might be fewer regulatory challenges in most countries with the adoption of dsRNA sprays, which will lead to faster adoption by growers (Palukaitis and Yoon 2024). Still, there is a technical issue in the cost-effective, large-scale production of high-quality dsRNAs and also in the consistency of performance in varying environmental conditions. As a perspective, the combination of precision-targeted dsRNA with nanocarrier technology is one of the areas of the future of managing tobacco mosaic. It could be feasible to create an environmentally friendly and field-qualified platform to control tobamoviruses by developing a system that targets conserved viral targets (with potentially multiplexed combinations) and uses them in stable nanocarrier formulations. Additionally, the possibility of cross-protection approaches, i.e., with a set of dsRNAs against a variety of tobamoviruses being targeted, the resulting broad-spectrum protection deserves further investigation.

19.12 Overcoming Resistance Breakdown in Tomato, Pepper, and Tobacco by R-Gene Transfer and Resistance

The introduction of ToBRFV, which overcomes the popular resistance gene Tm-22, is one of the most important recent problems of tomato virology (Whitham et al. 1994). In contrast to TMV and ToMV, which are susceptible to Tm-22, ToBRFV is also resistant to this defense mechanism because it harbors mutations in its movement protein (MP). Though this MP variant is not as efficient in transporting between the cells as the TMV or ToMV variants, the resistance to Tm-22 has provided ToBRFV with a highly competitive edge in resistant tomato populations. This has seen the virus spread fast among tomato-growing areas in the world (Liu et al. 2002). To acquire long-term resistance in tomato, scientists are incorporating measures like pyramiding of R genes, finding new allelic relatives of Tm-2, and transferring resistance in wild Solanum kinsfolk. Developments in high-throughput sequencing and other technologies aimed at gene-editing, especially the allele modification using CRISPR/Cas, provide opportunities to introduce new resistance phenotypes into commercial cultivars (Pelletier and Moffett 2022). One of the recent challenges in tomato virology is the introduction of ToBRFV which has the ability to circumvent the Tm-22 resistance gene that is highly deployed. ToBRFV is not vulnerable to Tm-22, as TMV and ToMV are, but carries mutations in its movement protein (MP), which allows it to evade this defense mechanism. This type of MP is not as effective in intercellular movement as the TMV or ToMV forms, but a Tm-22 evasion capacity has provided ToBRFV with a potent competitive edge in populations with tomato resistance. As a result, the virus has spread rapidly across global tomato-growing regions (Ishikawa et al. 2022). To restore durable resistance in tomato, researchers are exploring strategies such as stacking multiple resistance (R) genes, identifying novel allelic variants of Tm-2, and transferring resistance from wild Solanum relatives. Advances in high-throughput sequencing and gene-editing technologies, particularly CRISPR/Cas-mediated allele modification, offer promising tools for introducing new resistance traits into commercial cultivars (Kravchik et al. 2022).

19.13 Genome Editing and RNAi-Based Approaches: CRISPR/Cas-Mediated Editing of Susceptibility Genes

One of the most powerful new tools for engineering tobamovirus resistance is CRISPR/Cas gene editing, which allows precise modification of plant genes required for virus replication or movement. Unlike R-gene strategies, which rely on recognition of viral effectors, susceptibility (S) gene editing aims to remove essential host factors exploited by the virus. A landmark example involves TOM1 (TOBAMOVIRUS MULTIPLICATION 1), a seven-pass transmembrane protein that anchors tobamovirus replication complexes to intracellular membranes. CRISPR/Cas9-mediated knockout of all four TOM1 homologs in tomato conferred complete immunity to TMV, ToMV, and YoMV, with no apparent adverse effects on plant growth or fertility (Chandrasekaran et al. 2016). Similarly, combined knockouts of TOM1a and TOM3, another replication cofactor, rendered tomato plants resistant to ToBRFV, whereas single knockouts had only partial effects (Nishikiori et al. 2011). These studies demonstrate that targeting host factors through genome editing can produce broad-spectrum and durable resistance, less likely to be overcome by viral mutation. Importantly, TOM1 orthologs are conserved across a wide range of plant taxa, raising the possibility of applying this strategy in other solanaceous and cucurbit crops. Beyond TOM1 and TOM3, multiple host proteins contribute to different stages of tobamovirus infection (Taylor and Carr 2000). Candidate factors include TOM2A, which supports tobamovirus replication as a cofactor, translation factors (eEF1A, eIF3-associated proteins such as GCD10) that facilitate viral protein synthesis, auxin/indole-3-acetic acid proteins, which regulate localization of viral proteins, plasmodesmata-associated proteins such as pectin methylesterase and synaptotagmin A, which interact with tobamovirus movement proteins, and phloem-associated proteins (glycine-rich proteins, callose-binding proteins) that regulate systemic spread (Padmanabhan et al. 2008; Tran et al. 2014). Editing or silencing these factors could provide new routes to resistance. However, because many of these proteins also serve essential cellular functions, careful functional studies and selective editing will be required to avoid negative trade-offs in plant growth and development. RNA interference (RNAi) remains one of the most extensively tested biotechnological strategies for plant virus resistance. Early approaches involved hairpin RNA constructs, designed to fold into double-stranded RNA structures that are processed into siRNAs by plant Dicer-like enzymes. These siRNAs guide the cleavage of complementary viral RNAs, providing targeted resistance (Carr 2024). Examples of success include papaya ringspot virus (PRSV) resistance in papaya through coat protein-derived hairpins and TMV and PMMoV resistance in *N. benthamiana* and tobacco via stable and transient hairpin constructs. Although these strategies are effective, constitutive expression of RNAi constructs in transgenic crops raises regulatory, biosafety, and consumer acceptance concerns. Moreover, generating transgenic lines is often time-consuming and may be limited to transformable species. An emerging alternative to stable transformation is the exogenous application of synthetic dsRNAs. These can be delivered by spraying, infiltration, or root uptake and are processed by the plant's silencing machinery into siRNAs that target viral genomes. The double-stranded RNA molecules (dsRNAs) can be synthesized quickly and deployed rapidly in response to emerging viral threats, and are effective in crops that are recalcitrant to transformation. However, naked dsRNA is unstable, being highly susceptible to nucleases and UV degradation. To overcome this, researchers have developed nanoparticle carriers to enhance stability and uptake. Examples include chitosan nanoparticles, which encapsulate viral dsRNAs and may also induce host resistance responses; cationic polymers, proposed for controlled release of dsRNAs under field conditions; and layered double hydroxide (LDH) nanoparticles, also known as "BioClay," which extend dsRNA protection duration from 5 to 20 days and enable systemic resistance in new plant tissues. Among these, LDH systems appear most promising for tobamoviruses, having shown strong protection against PMMoV and potential applicability for ToBRFV. Importantly, mixtures of dsRNAs targeting multiple viral genes could provide broad-spectrum and resistance-breaking protection. Future resistance strategies will likely combine RNAi-based approaches with CRISPR/Cas editing. For example, edited plants lacking susceptibility factors (e.g., TOM1 knockouts) could be supplemented with topical dsRNA applications to reduce viral pressure and minimize the risk of resistance-breaking mutations. Similarly, amiRNAs designed against conserved tobamovirus sequences could complement host gene edits, creating multi-layered and durable resistance.

19.14 Mutation-Driven Symptom Variability in ToBRFV: Insights for Nanotechnology-Based Management in Tomato

Tomato brown rugose fruit virus (ToBRFV), a member of the genus *Tobamovirus* (family *Virgaviridae*), is one of the most destructive pathogens of tomato (*Solanum lycopersicum*) and pepper (*Capsicum annuum*). It is very challenging to eliminate it in agricultural systems due to its stability, seed-borne character, and effective mechanical transfer. The virus is also soil-borne, and it remains in the contaminated soils and crop debris for up to months, thereby becoming a source of new infections during the next planting seasons. ToBRFV has been especially problematic in resistant tomato plants because it can break Tm-22 resistance, which breeders have traditionally used to safeguard against the tobamoviruses, including tobacco mosaic virus (TMV) and tomato mosaic virus (ToMV). One rather alarming finding has been that isolates of ToBRFV have been found with nonsynonymous mutations in the movement protein (MP), including the A134T mutation, which increases the severity of symptoms and alters the host–pathogen interactions. The molecular dynamics of ToBRFV-A134T is therefore critical to formulate integrated management strategies and new technologies, including the RNA delivery system by nanotechnology.

19.15 ToBRFV Transmission and Soil Persistence

ToBRFV is effectively transmitted through seeds, soil, irrigation water, and mechanical contact with contaminated surfaces. The soils in monoculture tomato production are the viral reservoirs. In a recent study, wild tomato species infected with ToBRFV grew for 5 months, and then viral RNA was retrieved by high-throughput sequencing (HTS) of soil samples. The isolate showed a high similarity of 99.92 to the wild-type strain, but possessed one important MP mutation (A134T). ToBRFV persistence in soil not only underscores its epidemiological risk but also proves the ability of the virus to persist in the environment over time and select favorable variants, which will enable it to overcome plant resistance and cause new outbreaks.

19.16 Molecular Basis of the A134T Mutation

Tobamoviruses have a movement protein (MP) that is essential in cell-to-cell transport via plasmodesmata. The interactions between the amino acid residues within MP and host proteins, such as pectin methylesterase (PME), control the porosity of the cell wall. The A134T replacement is found in the amino acid domain (130–185) of MP, which is involved in binding PME and spreading viruses. It is interesting to note that although previous researchers found the amino acids 60–186 to be necessary to overcome Tm-22 resistance, A134 does not directly interact with the Tm-22 protein. Rather, structural modeling indicates that A134T alters MP conformational change, promoting PME contact and allowing systemic dissemination to be more effective. This structural change is probably the reason why A134T infections are severe in terms of symptomatology, with shoestring-like and fern-like leaflets, which are not commonly observed in WT ToBRFV infections.

19.17 Symptom Severity in A134T Infections

Resistant tomato cv. Ikram was comparatively inoculated with ToBRFV-WT and ToBRFV-A134T, with results showing different symptoms. WT infections led to mottling and mosaic appearance, with slight cases of shoestring leaves, whereas A134T infections resulted in more pronounced shoestring and fern leaflets. The symptoms on fruit were also severe, with both isolates causing the rugose phenotypes that were yellow-brown; A134T infections were associated with the additional deformed leaf phenotype. The A134T phenomenon persisted throughout several generations of plants, which validated the genetic stability of the phenomenon and made A134T a serious new threat (Adams et al. 2025).

19.18 Gene Expression and RNA Silencing Pathways

Plants defend against viruses using RNA silencing pathways, including RNA-dependent RNA polymerase 6 (RDR6), Dicer-like 4 (DCL4), and Dicer-like 2b (DCL2b). RDR6 and DCL4 were downregulated, and auxin response factors (ARFs), ARF3 and ARF4, were upregulated in Ikram plants with WT infections, which are associated with shoestring leaf symptoms. Conversely, infection by A134T elicited concomitant increases in RDR6, DCL4, and ARF3/ARF4, which indicates that the mutant circumvents or gets co-opted into the regular regulatory pathways. Notably, DCL2b, which is reported to suppress the shoestring symptoms during TMV infections, was not altered in A134T infections, which may be the reason behind the increased severity of leaf deformities (Adams et al. 2025). This deviation suggests that ToBRFV-A134T is regulated by different mechanisms than WT and can therefore have a fitness benefit by simultaneously affecting multiple RNA silencing pathways.

19.19 Stability and Coexistence of ToBRFV-A134T and WT

Sequential and mixed inoculation studies showed that the A134T variant is stable across plant generations and does not revert to the wild-type (WT) virus. Interestingly, in co-inoculations, the two isolates had the ability to coexist in the same host, although mostly the WT dominated systemically. Nonetheless, the plants that were originally inoculated with A134T still held the mutant, which validated the mutant as being stable under field-like conditions. This coexistence suggests a risk of mixed-variant infections in commercial fields, where A134T may increase disease severity even in the presence of the WT strain.

19.20 Implications for Resistance Breeding

The Tm-22 allele was previously known to provide long-lasting resistance to tobamoviruses, but ToBRFV has become a resistance-breaking virus (Hak and Spiegelman 2021). Other mutations, such as A134T, have complicated breeding approaches, as they increase the virulence of resistant cultivars. Although A134 does not directly bind to Tm-22, its proximity to Tm-22-binding domains indicates that loss of resistance can be due to synergistic interaction between MP conformational changes and PME interactions. Therefore, resistance plans should take into account the adaptive evolution of ToBRFV and its capacity to acquire mutations, which can refine interactions with hosts.

19.21 Role of Nanotechnology in Managing ToBRFV Variants

This is because ToBRFV is highly persistent and adaptable, thereby requiring control strategies beyond traditional methods. Nanotechnology has provided novel approaches, particularly the introduction of double-stranded RNA (dsRNA) molecules that induce RNA interference (RNAi) against viral genes. The use of layered double hydroxide (LDH) clay nanosheets has been identified as an effective carrier for the delivery of dsRNA, where RNA is not degraded and the nanosheets can be absorbed by the plant cells. It is possible to design such systems so that they focus on conserved MP regions, which would silence WT and A134T alike. Nanoparticles are biodegradable, hence making the environment sustainable and improving the stability and efficacy of the treatment using dsRNA. Besides that, metallic nanoparticles like silver (AgNPs) and copper (CuNPs) exhibit antiviral effects and are able to form reactive oxygen species as well as destabilize viral proteins. They might be combined with the delivery systems of the dsRNAs, which would build synergistic plant protection methods.

19.22 Future Prospects

The management of tobamoviruses in the future depends on the clever combination of various methods, which involve traditional breeding, sophisticated molecular technologies, and the latest nanotechnology. Some of the main directions that can be identified as potential directions of research and implementation are pointed out below.

19.23 Delivery Systems Based on Nanotechnology

The future era of antiviral measures will focus on nanoparticles. One of the most promising developments is the creation of biodegradable nanocarriers to deliver dsRNA, including layered double hydroxide (LDH) nanosheets. These systems prevent the degradation of dsRNA, increase the duration of protection, and enable silencing of viral genes specifically. Future work should aim at optimizing the use of these platforms in large field areas, making them cost-effective, and reducing environmental hazards. Combining nanoparticles (e.g., silver or copper) with RNA-based strategies could further increase antiviral potential while decreasing the use of chemical pesticides.

19.24 Genome Editing: CRISPR/Cas-Mediated

CRISPR/Cas technology is the most precise in engineering virus resistance through gene targeting of susceptibility (S) genes. *TOM1* homolog knockout in tomato has already shown long-term resistance to numerous tobamoviruses. It will be necessary to expand this to other crops and host factors. Nonetheless, genome editing has to balance viral resistance and plant fitness cautiously, as many S genes are involved in standard cellular functions. New base-editing and prime-editing tools could potentially assist in revising these changes with low off-target impact.

19.25 Artificial microRNA (amiRNAs) and RNA Silencing

RNA interference will remain a potent antiviral platform. Artificially produced microRNAs (amiRNAs) with viral sequence targets can be designed to provide specific and long-lasting resistance and are less likely to cause off-target effects than more traditional RNAi vectors. Simultaneously targeting several viral genes with multiplexed amiRNAs can stop the appearance of resistance-breaking strains such as ToBRFV-A134T. Besides, inducible expression systems that activate antiviral responses only under infection would reduce the energy expenditure on the plant.

19.26 Exogenous Non-transgenic Resistance: dsRNA Sprays

Application of topical dsRNA will bring a revolution in crop protection because it will offer resistance without transgenes. In combination with nanocarriers, dsRNA sprays could be rapidly designed against new variants of viruses, making them responsive to changing agricultural threats. Technical bottlenecks in scaling up dsRNA production and in-field stability still exist, but recent progress in microbial fermentation and synthetic biology could soon address these challenges.

19.27 Breeding of Resistance and Wild Relatives

Classical resistance breeding, especially introgression of Tm and L alleles of wild tomatoes and pepper relatives, has been effective against tobamoviruses. However, the disintegration of resistance by ToBRFV highlights the importance of ongoing study of genetic diversity. Modern genomics can reveal novel resistance alleles already present in wild germplasm, and gene pyramiding and genomic selection can be used to make them more useful in breeding programs.

19.28 Nanotechnology and Integrated Disease Management (IDM)

In the future, nanotechnology-based interventions will likely be used together with traditional practices of sanitation, crop rotation, and resistant cultivars. For example, nano-enabled disinfectants might enhance surface sanitation, whereas nano-fertilizers could improve plant strength and health. These combined strategies will not only enhance effectiveness but also decrease the use of chemical pesticides, aligning with global sustainability targets.

19.29 Farmer Adoption, Biosafety, and Regulation

With nanotechnology and genetic engineering methods moving out of the laboratory and into the field, biosafety and regulatory frameworks will become very important. Comprehensive risk assessments are needed to determine nanoparticle persistence in soils, effects on beneficial microbes, and non-target impacts. Likewise, public acceptance of genome-edited crops and nano-enabled products will rely on transparent communication, participatory decision-making, and demonstrating material benefits to farmers and consumers. Global cooperation will be critical for policies that promote access and affordability for smallholders.

19.30 Predictive Modeling and Systems Biology

Omics technologies and computational modeling, combined with systems biology, will enable recognition of plant–virus–nanoparticle interactions as a whole. These tools can foresee virus evolution, discover novel resistance points, and optimize nanoformulations for efficiency and safety. Models based on artificial intelligence (AI) can facilitate real-time decision-making in disease management, correlating diagnostics with customized intervention strategies.

19.31 Conclusion

Although scientific achievements have been made for more than a century, plant viral diseases still pose a problem to world agriculture, especially those caused by tobacco mosaic virus (TMV), tomato mosaic virus (ToMV), and tomato brown rugose fruit virus (ToBRFV), among others. Their refractory nature due to innate characteristics of stability, mechanical transmission, and large host range makes eradication very hard. As a result of the relentless evolutionary arms race between viruses and their hosts, the development of strains such as ToBRFV, which overcomes a long-established genetic defense like Tm-22, demonstrates the evolution of virulent strains. In this regard, nanotechnology has emerged as a radical horizon in plant pathology, providing new strategies for diagnosis, treatment, and sustainable management. This chapter described the interaction of conventional and new approaches to control plant viral diseases, specifically the tobamoviruses. Traditional interventions, including quarantine, sanitation, pesticide use, and R-gene application, have played a significant role in curbing virus spread. Their success is, however, becoming compromised by viral evolution, pesticide resistance, and ecological issues. These shortcomings demonstrate the importance of innovative remedies that not only increase disease resistance but also protect the environment.

Nanotechnology provides a unique collection of tools to deal with these challenges. Silver, zinc oxide, and titanium dioxide are metallic nanoparticles that demonstrate high antiviral activity, either by destroying viral particles or by inducing plant defense. Carbon-based nanomaterials and nanocomposites have also demonstrated potential in disrupting viral replication and movement. Notably, nanocarriers allow precise delivery of biomolecules, like double-stranded RNA (dsRNA), making RNA interference (RNAi) strategies feasible in contexts not reachable using transgenic methods. These innovations align with the need for field-ready, non-transgenic, and environmentally friendly procedures for plant protection. Nanotechnology in plant virology is also likely to transform diagnostics. Quick, sensitive, and portable detection of viruses can be achieved using nano-biosensors functionalized with antibodies, aptamers, or nucleic acids, allowing suitable management decisions. These tools, combined with high-throughput sequencing and bioinformatics, can accelerate the detection of new viral variants and aid in monitoring the epidemiology of emerging threats such as ToBRFV-A134T. On the molecular level, examination of host–virus interactions has provided insights into resistance and resistance breakdown. Host factors, including TOM1, TOM2A, and PME, which enable viral replication and movement, have paved the way for CRISPR/Cas-mediated editing to eliminate susceptibility genes. Similarly, RNA-silencing approaches such as artificial microRNAs and exogenous dsRNAs show potential for harnessing innate plant immunity for sustainable viral defense. In short, nanotechnology, molecular biology, and improved breeding strategies are converging, transforming the management of plant viral diseases. Although challenges persist—especially regarding biosafety, cost-effectiveness, and ecological impact—the accelerated pace of innovation indicates that integrated nano-enabled platforms are poised to become pillars of sustainable agriculture.

References

Adams MJ, Buzkan N, Ghorbani M, Salem N, Kamenidou S, Winter S et al (2025) Tomato brown rugose fruit virus isolate containing A134T amino acid substitution in the movement protein enhanced symptom severity in tomato plants. Plant Dis 109(9):1901–1912

Almeida RPP (2018) Emerging plant disease epidemics: biological research is key but not enough. PLoS Biol 16:e2007020. https://doi.org/10.1371/journal.pbio.2007020

Carr JP (2024) Engineered resistance to tobamoviruses. Viruses 16(7):1007. https://doi.org/10.3390/v16071007

Chandrasekaran J, Brumin M, Wolf D, Leibman D, Klap C, Pearlsman M et al (2016) Development of broad virus resistance in non-transgenic cucumber using CRISPR/Cas9 technology. Mol Plant Pathol 17:1140–1153. https://doi.org/10.1111/mpp.12375

Dunigan DD, Golemboski DB, Zaitlin M (1987) Analysis of the N gene of *Nicotiana*. In: Evered D, Harnett S (eds) *Plant resistance to viruses*, Ciba Foundation Symposium 133. Wiley, Chichester, pp 120–135

Durán N, Marcato PD, Conti RD, Alves OL, Costa F, Brocchi M (2010) Potential use of silver nanoparticles on pathogenic bacteria, their toxicity and possible mechanisms of action. J Braz Chem Soc 21:949–959

Dutta A (2012) How helpful is nanotechnology in agriculture? Adv Nat Sci Nanosci Nanotechnol 3:033002. https://doi.org/10.1088/2043-6262/3/3/033002

Dutta P, Kumari A, Mahanta M et al (2022) Advances in nanotechnology as a potential alternative for plant viral disease management. Front Microbiol 13:935193. https://doi.org/10.3389/fmicb.2022.935193

Elmer W, Ma C, White J (2018) Nanoparticles for plant disease management. Curr Opin Environ Sci Health 6:66–70. https://doi.org/10.1016/j.coesh.2018.08.002

Feng QL, Wu J, Chen GQ, Cui FZ, Kim TN, Kim JO (2008) A mechanistic study of the antibacterial effect of silver ions on *Escherichia coli* and *Staphylococcus aureus*. J Biomed Mater Res 52(4):662–668

Ficke A, Cowger C, Bergstrom G, Brodal G (2018) Understanding yield loss and pathogen biology to improve disease management: Septoria nodorum blotch—a case study in wheat. Plant Dis 102:696–707. https://doi.org/10.1094/PDIS-09-17-1375-FE

Fox A, Anderson H, Everett M (2020) Plant pest factsheet: Tomato Brown rugose fruit virus. DEFRA, London

Gaafar YZA, Ziebell H (2020) Novel targets for engineering *Physostegia chlorotic mottle* and tomato brown rugose fruit virus-resistant tomatoes: in silico prediction of tomato microRNA targets. PeerJ 8:e10096. https://doi.org/10.7717/peerj.10096

Ghazy NA, El-Hafez A, Omnia A, El-Bakery AM, El-Geddawy DI (2021) Impact of silver nanoparticles and two biological treatments to control soft rot disease in sugar beet (*Beta vulgaris* L). Egypt J Biol Pest Control 31(1):1–12

Gilbertson RL, Batuman O, Webster CG, Adkins S (2015) Role of the insect supervectors *Bemisia tabaci* and *Frankliniella occidentalis* in the emergence and global spread of plant viruses. Annu Rev Virol 2:67–93. https://doi.org/10.1146/annurev-virology-031413-085410

Gill HK, Garg H (2014) Pesticide: environmental impacts and management strategies. In: Soloneski S (ed) Pesticides—toxic aspects. IntechOpen, London. https://doi.org/10.5772/57399

Gonsalves C, Cai W, Tennant P, Gonsalves D (1998) Effective development of papaya ringspot virus resistant papaya with untranslatable coat protein gene using a modified microprojectile transformation method. Acta Hortic 461:311–314. https://doi.org/10.17660/ActaHortic.1998.461.34

Gullino ML, Albajes R, Al-Jboory I, Angelotti F, Chakraborty S, Garrett KA et al (2021) Scientific review of the impact of climate change on plant pests. Presented at Secretariat of the International Plant Protection Convention. FAO

Gupta N, Upadhyaya CP, Singh A, Abd-Elsalam KA, Prasad R (2018) Applications of silver nanoparticles in plant protection. In: Nanobiotechnology applications in plant protection. Springer, Cham, pp 247–265

Hak H, Spiegelman Z (2021) The tomato brown rugose fruit virus movement protein overcomes *Tm-22* resistance in tomato while attenuating viral transport. Mol Plant-Microbe Interact 34:1024–1032. https://doi.org/10.1094/MPMI-01-21-0023-R

Hak H, Raanan H, Schwarz S, Sherman Y, Dinesh-Kumar SP, Spiegelman Z (2023) Activation of *Tm-22* resistance is mediated by a conserved cysteine essential for tobacco mosaic virus movement. Mol Plant Pathol 24:838–848. https://doi.org/10.1111/mpp.13318

Hanssen IM, Lapidot M, Thomma BP (2010) Emerging viral diseases of tomato crops. Mol Plant-Microbe Interact 23:539–548. https://doi.org/10.1094/MPMI-23-5-0539

Harrison BD, Wilson TMA (1999) Milestones in the research on tobacco mosaic virus. Philos Trans R Soc Lond Ser B Biol Sci 354:521–529. https://doi.org/10.1098/rstb.1999.0403

Ishibashi K, Kubota K, Kano A, Ishikawa M (2023) Tobamoviruses: old and new threats to tomato cultivation. J Gen Plant Pathol 89:305–321. https://doi.org/10.1007/s10327-023-01141-5

Ishikawa M, Yoshida T, Matsuyama M, Kouzai Y, Kano A, Ishibashi K (2022) Tomato brown rugose fruit virus resistance generated by quadruple knockout of homologs of TOBAMOVIRUS MULTIPLICATION1 in tomato. Plant Physiol 189:679–686. https://doi.org/10.1093/plphys/kiac103

Jain RG, Fletcher SJ, Manzie N, Robinson KE, Li P, Lu E et al (2022) Foliar application of clay-delivered RNA interference for whitefly control. Nat Plants 8:535–548. https://doi.org/10.1038/s41477-022-01152-8

Jeevanandam J, Barhoum A, Chan YS, Dufresne A, Danquah MK (2018) Review on nanoparticles and nanostructured materials: history, sources, toxicity and regulations. Beilstein J Nanotechnol 9:1050–1074. https://doi.org/10.3762/bjnano.9.98

Jones RAC (2021) Global plant virus disease pandemics and epidemics. Plants. 10:233. https://doi.org/10.3390/plants10020233

Kenyon L, Kumar S, Tsai WS, Hughes JD (2014) Virus diseases of peppers (*Capsicum* spp.) and their control. Adv Virus Res 90:297–354. https://doi.org/10.1016/B978-0-12-801246-8.00006-8

Khan MR, Ahamad F, Rizvi TF (2019) Effect of nanoparticles on plant pathogens. In: Advances in phytonanotechnology. Academic, pp 215–240

Khurana SMP, Marwal A (2016) Recent developments towards detection and diagnosis for management of plant viruses. Indian Phytopathol 69:30–34

Kravchik M, Shnaider Y, Abebie B, Shtarkman M, Kumari R, Kumar S et al (2022) Knockout of *SlTOM1* and *SlTOM3* results in differential resistance to tobamovirus in tomato. Mol Plant Pathol 23:1278–1289. https://doi.org/10.1111/mpp.13227

Kumar J, Murali-Baskaran RK, Jain SK, Sivalingam PN, Dixit A, Mallikarjuna J et al (2021) Biotic stresses of agricultural crops in India: re-visiting national status and mitigation strategies. Curr Sci 120:264–265

Lin SS, Henriques R, Wu HW, Liu QW, Yeh SD, Chua NH (2007) Strategies and mechanisms of plant virus resistance. Plant Biotechnol Rep 1:125–134. https://doi.org/10.1007/s11816-007-0021-8

Lindbo JA, Dougherty WG (2005) Plant pathology and RNAi: a brief history. Annu Rev Phytopathol 43:191–204. https://doi.org/10.1146/annurev.phyto.43.040204.140228

Liu Y, Schiff M, Dinesh-Kumar SP (2002) Virus-induced gene silencing in tomato. Plant J 31:777–786. https://doi.org/10.1046/j.1365-313x.2002.01394.x

Lius S, Manshardt RM, Fitch MM, Slightom JL, Sanford JC, Gonsalves D (1997) Pathogen-derived resistance provides papaya with effective protection against papaya ringspot virus. Mol Breed 3:161–168. https://doi.org/10.1023/A:1009614508659

Lopez-Gomollon S, Baulcombe DC (2022) Roles of RNA silencing in viral and non-viral plant immunity and in the crosstalk between disease resistance systems. Nat Rev Mol Cell Biol 23:645–662. https://doi.org/10.1038/s41580-022-00496-5

Luria N, Smith E, Reingold V, Bekelman I, Lapidot M, Levin I et al (2017) A new Israeli Tobamovirus isolate infects tomato plants harboring *Tm-22* resistance genes. PLoS One 12:e0170429. https://doi.org/10.1371/journal.pone.0170429

Mahmood S, Lakra N, Marwal A, Sudheep NM, Anwar K (2017) Crop genetic engineering: an approach to improve fungal resistance in plant system. In: Singh DP (ed) Plant-microbe interactions in agro-ecological perspectives. Springer, Singapore, pp 581–591. https://doi.org/10.1007/978-981-10-6593-4_23

Marwal A, Gaur R (2019) Nanophytovirology: an emerging field for disease management. In: Plant diseases—current threats and management trends. IntechOpen, London. https://doi.org/10.5772/intechopen.86653

Mitter N, Worrall EA, Robinson KE, Li P, Jain RG, Taochy C et al (2017a) Clay nanosheets for topical delivery of RNAi for sustained protection against plant viruses. Nat Plants 3:16207. https://doi.org/10.1038/nplants.2016.207

Mitter N, Worrall EA, Robinson KE, Xu ZP, Carroll BJ (2017b) Induction of virus resistance by exogenous application of double-stranded RNA. Curr Opin Virol 26:49–55

Molad O, Leibman D, Luria N, Sela N, Lachman O, Smith E, Reches M, Dombrovsky A (2025) Tomato brown rugose fruit virus isolate containing A134T amino acid substitution in the movement protein enhanced symptom severity in tomato plants. Plant Dis. 109(9):1901–1912. https://doi.org/10.1094/PDIS-07-24-1533-RE

Muronga M, Quispe C, Tshikhudo PP, Msagati TAM, Mudau FN, Martorell M et al (2021) Three selected edible crops of the genus *Momordica* as potential sources of phytochemicals: biochemical, nutritional, and medicinal values. Front Pharmacol 12:625546. https://doi.org/10.3389/fphar.2021.625546

Murphy AM, Jiang S, Elderfield JAD, Pate AE, Halliwell C, Glover BJ et al (2023) Biased pollen transfer between virus-infected and non-infected plants by bumblebees favors the paternity of infected plants in cross-pollination. *iScience* 26:106116. https://doi.org/10.1016/j.isci.2023.106116

Niehl A, Soininen M, Poranen MM, Heinlein M (2018) Synthetic biology approach for plant protection using dsRNA. Plant Biotechnol J 16:1679–1687. https://doi.org/10.1111/pbi.12904

Nishikiori M, Mori M, Dohi K, Okamura H, Katoh E, Naito S et al (2011) A host small GTP-binding protein ARL8 plays crucial roles in tobamovirus RNA replication. PLoS Pathog 7:e1002409. https://doi.org/10.1371/journal.ppat.1002409

Obregón VG, Ibañez JM, Lattar TE, Juszczak S, Groth-Helms D (2023) First report of Tomato brown rugose fruit virus in tomato in Argentina. New Dis Rep 48:e12203. https://doi.org/10.1002/ndr2.12203

Ojinaga M, Guirao P, Larregla S (2022) A survey of main pepper crop viruses in different cultivation systems for the selection of the most appropriate resistance genes in sensitive local cultivars in northern Spain. Plants 11:719. https://doi.org/10.3390/plants11060719

Padmanabhan MS, Kramer SR, Wang X, Culver JN (2008) Tobacco mosaic virus replicase–auxin/indole acetic acid protein interactions: reprogramming the auxin response pathway to enhance virus infection. J Virol 82:2477–2485. https://doi.org/10.1128/JVI.01865-07

Palukaitis P, Yoon JY (2024) Defense signaling pathways in resistance to plant viruses: crosstalk and finger pointing. Adv Virus Res 118:77–212. https://doi.org/10.1016/bs.aivir.2024.01.002

Pelletier A, Moffett P (2022) N and N'-mediated recognition confers resistance to tomato brown rugose fruit virus. microPubl Biol. https://doi.org/micropub.biology.000660

Pfitzner AJP (2006) Resistance to tobacco mosaic virus and Tomato mosaic virus in tomato. In: Loebenstein G, Carr JP (eds) Natural resistance mechanisms of plants to viruses. Springer, Dordrecht, pp 399–413

Poulicard N, Pagán I, González-Jara P, Mora MÁ, Hily JM, Fraile A et al (2024) Repeated loss of the ability of a wild pepper disease resistance gene to function at high temperatures suggests that thermoresistance is a costly trait. New Phytol 241:845–860. https://doi.org/10.1111/nph.19371

Pugsley CE, Isaac RE, Warren NJ, Stacey M, Ferguson CTJ, Cappelle K et al (2024) Effective delivery and selective insecticidal activity of double-stranded RNA via complexation with diblock copolymer varies with polymer block composition. Pest Manag Sci 80:669–677. https://doi.org/10.1002/ps.7793

Reddy KVS, Zehr UB (2004) Novel strategies for overcoming pests and diseases in India. In: New directions for a diverse planet. Proceedings of 4th International Crop Science Congress; Brisbane, Australia

Saito T, Meshi T, Takamatsu N, Okada Y (1987) Coat protein gene sequence of tobacco mosaic virus encodes a host response determinant. Proc Natl Acad Sci USA 84:6074–6077. https://doi.org/10.1073/pnas.84.17.6074

Salem N, Mansour A, Ciuffo M, Falk BW, Turina M (2016) A new tobamovirus infecting tomato crops in Jordan. Arch Virol 161:503–506. https://doi.org/10.1007/s00705-015-2677-7

Salem NM, Jewehan A, Aranda MA, Fox A (2023) Tomato brown rugose fruit virus pandemic. Annu Rev Phytopathol 61:137–164. https://doi.org/10.1146/annurev-phyto-021622-120703

Scholthof KB (2004) Tobacco mosaic virus: a model system for plant biology. Annu Rev Phytopathol 42:13–34. https://doi.org/10.1146/annurev.phyto.42.040803.140322

Scholthof KB (2017) Spicing up the N gene: F.O. Holmes and Tobacco mosaic virus resistance in *Capsicum* and *Nicotiana* plants. Phytopathology 107:148–157. https://doi.org/10.1094/PHYTO-07-16-0264-RVW

Singh D, Kumar A, Singh AK, Tripathi HS (2013) Induction of resistance in field pea against rust disease through various chemicals/micronutrients and their impact on growth and yield. Plant Pathol J 12:36–49

Sobha K, Surendranath K (2010) Emerging trends in nanobiotechnology. Biotechnol Mol Biol Rev 5:1–12

Symonds EM, Nguyen KH, Harwood VJ, Breitbart M (2018) Pepper mild mottle virus: a plant pathogen with a greater purpose in (waste) water treatment development and public health management. Water Res 144:1–12. https://doi.org/10.1016/j.watres.2018.06.066

Tran PT, Choi H, Kim SB, Lee HA, Choi D, Kim KH et al (2014) A simple method for screening of plant NBS-LRR genes that confer a hypersensitive response to plant viruses and its application for screening candidate pepper genes against Pepper mottle virus. J Virol Methods 201:57–64. https://doi.org/10.1016/j.jviromet.2014.02.003

Taylor DN, Carr JP (2000) The GCD10 subunit of yeast eIF-3 binds the methyltransferase-like domain of the 126 and 183 kDa replicase proteins of Tobacco mosaic virus in the yeast two-hybrid system. J Gen Virol 81:1587–1591. https://doi.org/10.1099/0022-1317-81-6-1587

Uslu VV, Wassenegger M (2020) Critical view on RNA silencing-mediated virus resistance using exogenously applied RNA. Curr Opin Virol 42:18–24. https://doi.org/10.1016/j.coviro.2020.03.004

Uslu VV, Bassler A, Krczal G, Wassenegger M (2020) High-pressure-sprayed double-stranded RNA does not induce RNA interference of a reporter gene. Front Plant Sci 11:534391. https://doi.org/10.3389/fpls.2020.534391

Wang Y, Xia Y (2004) Bottom-up and top-down approaches to the synthesis of monodispersed spherical colloids of low melting-point metals. Nano Lett 4:2047–2050. https://doi.org/10.1021/nl048689j

Warghane A, Saini R, Shri M, Andankar I, Ghosh DK, Chopade BA (2024) Application of nanoparticles for management of plant viral pathogen: current status and future prospects. Virology 592:109998. https://doi.org/10.1016/j.virol.2024.109998

Whitham S, Dinesh-Kumar SP, Choi D, Hehl R, Corr C, Baker B (1994) The product of the tobacco mosaic virus resistance gene N: similarity to toll and the interleukin-1 receptor. Cell 78:1101–1115. https://doi.org/10.1016/0092-8674(94)90283-6

Zinger A, Lapidot M, Harel A, Doron-Faigenboim A, Gelbart D, Levin I (2021) Identification and mapping of tomato genome loci controlling tolerance and resistance to Tomato brown rugose fruit virus. Plants. 10:179. https://doi.org/10.3390/plants10010179

Transforming Post-harvest Viral Disease Management: A Twenty-First-Century Perspective on Fruit Crops

20

Bernadetta Rina Hastilestari, Purnama Isti Khaerani, Muhammad Yusril Hardiansyah, Azmi Nur Karimah Amas, and Vietgar Membalik

Abstract

Post-harvest handling often includes primary processing for all commodities from harvest until they can be consumed "fresh" or in preparation for further processing. This primary processing usually does not change the appearance and some marketing criteria like nutritional values. Fruits usually have high water content, which accelerates the respiration and transpiration process and favors disease development during harvest and consumption periods. Post-harvest diseases are usually caused by microbial pathogens, such as viruses. Viral infections are a significant problem in post-harvest fruit handling, potentially endangering consumers' health. These diseases, often latent during harvest, can also manifest during storage and distribution, affecting significant losses in product quality and economic losses. This chapter describes mitigating strategies for post-harvest viral infections, emphasizing integrated traditional ways and twenty-first-century innovations, such as remote sensing for early detection and Internet of Things (IoT) for control. Moreover, advances in molecular diagnostics enable the accurate detection of viral pathogens and give new possibilities for breeding fruits that are resistant to biotic agents such as viruses. Furthermore, the chapter

B. R. Hastilestari (✉) · P. I. Khaerani
Research Center for Genetic Engineering, National Research and Innovation Agency (BRIN), Jakarta, Indonesia
e-mail: bern003@brin.go.id

M. Y. Hardiansyah
Institute of Biotechnology, College of Bioresources and Agriculture, National Taiwan University, Taipei, Taiwan

A. N. K. Amas
Hasanuddin University, Makassar University, Makassar, Indonesia

V. Membalik
Wageningen University and Research, Wageningen, The Netherlands

R. Kumar et al. (eds.), *Fruit Crops and Viral Pathogens Volume II*, Plant Pathology for Sustainable Agriculture, https://doi.org/10.1007/978-981-95-7261-8_20

explains the comprehensive approach for reducing post-harvest viral diseases. These strategies will play a crucial role in increasing crop value and also resilience in food sustainability.

Keywords

Biotechnology · Food sustainability · Post-harvest fruits · Viral diseases

20.1 Transformation of Post-harvest Viral Disease Management for Fruit Crops

Traditional paradigms in post-harvest disease management are always considered to be a major challenge. However, recent advances in plant virology, biotechnology, and post-harvest technology have been widely reported in evaluating this paradigm. The use of resistant varieties, bio-fungicides, and appropriate technology can increase plant resistance to disease while maintaining post-harvest product quality. This is certainly related to food security and the sustainability of agricultural products. This sub-chapter will comprehensively discuss the development of post-harvest viral diseases, from technical issues and challenges to innovative approaches and strategies grounded in the twenty-first century for sustainable fruit disease management.

20.1.1 Historical Study of Post-harvest Viral Diseases

Historically, viral infections were thought to be resolved during the pre-harvest phase, with the assumption that post-harvest plant metabolic activity slowed, thus inhibiting pathogen development (Sharma 2014). However, recent research has shown that viruses can continue to influence metabolic and physiological processes post-harvest and affect shelf life (Arah et al. 2016). Consequently, in the context of the global fresh fruit trade, post-harvest viral diseases have become a significant concern (Hounmanou et al. 2018). The understanding of post-harvest viral diseases in horticultural fruit crops has evolved over the decades, with various areas of focus reflecting the dynamic nature of the efforts to address post-harvest viral problems. This section will present a historical timeline and the phases of post-harvest viral diseases in fruit crops. The 1910s marked the initial identification of plum pox virus (PPV), also known as Sharka. This virus is one of the most devastating viral diseases affecting stone fruits, particularly plums, peaches, and apricots. In 1929, papaya ringspot virus (PRSV) was identified, causing yellowing and deformation of papaya leaves and fruit (Jyotika et al. 2024). In the 1940s, two additional viruses were identified in the Caribbean: cocoa light mosaic virus and cocoa yellow vein virus. These viruses significantly impacted cocoa plantations (Ullah et al. 2021). In 1950–1951, the discovery of potato virus Y (PVY) led to a significant expansion in the cultivation of economically important crops, such as peppers and tomatoes (Torrance and Talianksy 2020). In the 1960s, research shifted to a new challenge, identified as citrus tristeza virus (CTV), which rapidly impacted global citrus production. This led to a sharp decline in citrus orchards (Moreno et al. 2008). These mid-twentieth-century discoveries laid the foundation for subsequent advances in virus research. The 1970s saw the advent of a new era characterized by the development of molecular tools, leading to a paradigm shift in virus characterization. In the 1980s, there was a paradigm shift in research focus toward the study of insect-borne viruses, including cucumber mosaic virus (CMV), PRSV, and tomato yellow leaf curl virus (TYLCV). These viruses, transmitted by aphids and whiteflies, have a significant destructive impact on fruit and vegetable crops (Naveed et al. 2023). The 1980s and 1990s marked a period of research on the identification of symptoms and morphological characterization of pathogens at the molecular level. The early 2000s witnessed significant transformations with the introduction of molecular techniques, including real-time PCR and RT-PCR, and the identification of damaging viruses. In particular, tobacco mosaic virus (TMV), which was extensively isolated in the early twentieth century, laid the groundwork for the development of advanced molecular diagnostic techniques, such as real-time multiplex RT-PCR for the simultaneous detection of five major grapevine viruses (López-Fabuel et al. 2013). Multiplex real-time RT-PCR for the simultaneous detection of five major grapevine viruses. Studies focused on the genetic characterization and alteration of viral genes have highlighted the importance of genetic variation for virus management (Kumar et al. 2017). These developments marked a new era in which post-harvest viruses were no longer considered passive entities, but rather as active agents causing fruit damage. In the 2010s, next-generation sequencing (NGS) and metagenomic analysis technologies presented new perspectives. Post-harvest viruses were discovered not as single entities, but rather as part of complex viral communities, often interacting with storage environmental factors and the fruit microbiota (Prabha et al. 2013). For example, grapevine leafroll-associated virus (GLRaV) in grapes, which contains multiple strains, and papaya meleira virus (PMeV) in papaya, which specifically damages fruit tissue (Kumar et al. 2024). Studies in this century have examined viruses' interaction with plant cellular components, the defense mechanisms activated by plants, and factors influencing viral virulence (García-Ruíz 2018; Fontes et al. 2021). These discoveries have led to the development of more virus-resistant crop varieties through genetic engineering and gene editing (Niraula and Fondong 2021). Recent developments in the 2020s have highlighted the significant challenges that post-harvest viruses pose to the global food system. Well-known horticultural viruses, such as citrus tristeza virus (CTV) in citrus, papaya ringspot virus (PRSV) in papaya, and banana bunchy top virus (BBTV) in banana, continue to hamper tropical fruit exports to global markets (Kumar et al. 2015). Recent research aims to correlate the economic impact of viral infections with agricultural policies to develop more sustainable and effective management strategies. Advances in diagnostic and management technologies are crucial at this stage, with a focus on rapid and accurate detection and the implementation of monitoring systems to alert farmers to potential virus outbreaks (Maina et al. 2024). Furthermore, a recent study also focused on the impact of climate change on virus infection dynamics (Adeosun and Loots 2024). Therefore, the evolution of post-harvest viral disease management in horticultural fruits reflects a shift from descriptive to transdisciplinary science. The world now needs solutions to ensure food security, ensure supply chain operations, and enhance the competitiveness of horticultural products in the market.

20.1.2 Limitations of Twentieth-Century Approaches in Addressing Contemporary Viral Challenges

The development of methodologies for virus discovery and management in the twentieth century has contributed significantly to the fundamental knowledge of virology. However, these methods have various limitations when faced with the more complex challenges of the contemporary era (Table 20.1).

The earliest approach to this was visual symptom observation. However, this method lacks specificity, as similar symptoms can also occur due to nutritional deficiencies or other pathogens (Khakimov et al. 2022). One study improved the precision of observation using electron microscopy, which facilitates direct visualization of virus particles. Although EM provides strong empirical evidence, it requires high costs, specialized equipment, and trained personnel for rapid diagnostics in horticultural post-harvest systems (Richert-Pöggeler et al. 2019; Vac et al. 2023). Significant progress has also been made through serological techniques, particularly

Table 20.1 Limitations of twentieth-century approaches in addressing contemporary viral challenges

Twentieth-century approach	Main	Contemporary	References
Visual observation of symptoms	Morphological and organoleptic identification of fruit	Nonspecific; difficult to differentiate viral infection from physiological disorders or multiple infections.	Dhiman et al. (2023) and Umer et al. (2019)
Electron microscopy (EM)	Direct detection of virus particles in plant tissue	Expensive, requires high expertise, not practical for rapid post-harvest diagnosis	Kaur et al. (2020)
Serology (ELISA)	Antigen–antibody detection, widely used for PRSV, CTV	Limited sensitivity to new strains; results are affected by tissue condition	Cassedy et al. (2021) and Shang et al. (2024)
Mechanical isolation and inoculation on indicator plants	Use of test plants to show characteristic symptoms	Slow process, impractical for international trade; does not detect latent infections	Jiménez et al. (2024)
Tissue culture in vitro	Use of in vitro cultures to test virus reactions	Not sensitive to multi-virus detection; limited to specialized laboratories	Steinberger and Voytas (2025) and Pamies et al. (2022)
Conventional sanitation and quarantine	Eradication of diseased plants and transport restrictions	Insufficient for global trade; does not address viruses carried by post-harvest fruit	Budhathoki et al. (2022) and Pandya et al. (2023)

the enzyme-linked immunosorbent assay (ELISA). This method can detect new strain variations is limited and is affected by the physiological state of the plant tissue (Boonham et al. 2014). Furthermore, the mechanical isolation and inoculation methods used on indicator plants in fundamental research are inefficient and impractical (Bukhamsin et al. 2025). Based on a management perspective, the twentieth century was marked by the dominance of traditional practices, including sanitation, eradication, and isolation. While these methods were effective in local contexts with limited trade, they proved inadequate to address the rapid movement of horticultural products across borders (Gergerich et al. 2015). Faced with contemporary virus challenges, twentieth-century methodologies have proven inadequate in providing practical solutions. It is important to have a new paradigm combining molecular technologies, bioinformatics, and a more adaptive, systemic approach to address the more complex food security and competitiveness of horticulture in the future.

20.2 Biological Dynamics of Post-harvest Viral Infections on Fruit Crops

Viral infections typically occur early in the plant life cycle, for example, during vegetative growth or flowering in fruit plants (Nicaise 2014; Zaffaroni et al. 2021; Manzoor et al. 2025). After entering the host, the viral genome is released into the cytoplasm, where it can exploit the host's molecular machinery for replication or persist in an episomal state (Harper, 2002). Certain plant viruses encode repressor proteins that interfere with RNA silencing, one of the most potent plant antiviral defense mechanisms (Leonetti et al. 2021). Virulence of viruses may regulate their multiplication and transmission within the host to maintain viral titers below a damaging threshold (Burgyán 2008). Alternatively, viruses manipulate hormonal signaling pathways in the body, such as salicylic acid, jasmonic acid, and ethylene, thereby reducing the induced systemic defense response (Zhang et al. 2020; Müllender et al. 2021). During the latent phase, the viral genome remains within the host cell, but its replication is severely restricted by antiviral protection mechanisms such as RNA silencing (Folimonova et al. 2008), resulting in fruit that exhibits no visible disease symptoms (Roossinck 2010). The asymptomatic active phase occurs when viral replication begins to increase moderately, primarily due to viral suppressor proteins that counteract the host silencing response (Burgyán 2008). At this stage, the virus is metabolically active, but symptoms are either absent or too subtle to be detected by routine visual inspection. Finally, in the symptomatic active phase, viral replication increases, suppressor proteins effectively enhance host protection, and the viral load accumulates to very high levels (Ando et al. 2019).

20.2.1 The Role of Fruit Physiology and Severe Post-harvest Infection

Once fruits are separated from their parent stock, they rely solely on internal reserves to grow and carry out other physiological processes. Without nutrients, fruits are more susceptible to stress, which accelerates aging and quality decline (Ufitinema et al. 2024). This damage can be divided into physiological or mechanical post-harvest losses and losses due to biotic stress from plant pathogens. After harvest, fruits continue to respire and undergo related metabolic processes (Giovannoni 2004; Pott et al. 2020). Recovery, dehydration, and microbial activity are the three most relevant factors that influence and accelerate fruit senescence (Yang et al. 2014). Beyond physiological and mechanical stressors, biotic stresses, including pests and diseases, represent another major source of post-harvest losses. Microbial infections can establish either before or after harvest, typically gaining entry through natural openings or wounds in the produce. Post-harvest decay is caused mainly by fungi and bacteria, whereas viruses are confined to the field and rarely contribute to deterioration after harvest. The resulting diseases may cause two forms of losses: quantitative, when extensive decay renders the product unmarketable, and qualitative, when superficial blemishes lower its commercial value. Infections may persist without symptoms in the field and merely emerge post-harvest, especially during storage and ripening. Pathogens are disseminated through several routes, such as contaminated soil, handling equipment, containers, wash water, or direct contact with infected produce. Apart from microbial diseases, insect pests are a major cause of post-harvest losses in grains and legumes, yet they are generally a minor problem in fresh produce, except during local outbreaks such as those caused by the potato moth (*Phthorimaea operculella*) (Kroschel et al. 2016). Although losses from insects and vertebrates tend to be smaller after harvest compared to microbial diseases, pests, including fruit flies, weevils, rodents, and other animals, can still inflict significant localized damage during storage and distribution.

20.2.2 Case Studies of Active Viral Diseases After Harvest

One of the major viral infections in citrus cultivation is caused by citrus tristeza virus (CTV), invading the phloem tissue and is transmitted primarily via grafting. This virus disseminates systemically within the phloem of citrus plants and can remain infectious for more than 2 weeks in insect vectors such as the Asian citrus psyllid (*Diaphorina citri*), with transmission potential persisting up to 24 days (Kang et al. 2021; Wang et al. 2023). This persistence enables efficient spread to healthy hosts. Although CTV is not transmitted through harvested fruit, it survives in vegetative planting material, including rootstocks and scion wood, making grafting and nursery production high-risk stages for dissemination (FAO 2013). In severe cases, CTV infection causes tree decline or death and shortens orchard lifespan. Therefore, strict sanitation practices and monitoring are crucial in both orchards and nurseries, as virus persistence in propagation material significantly increases management costs (FAO 2013; Kang et al. 2021; Wang et al. 2023). Furthermore, tomato brown rugose fruit virus (ToBRFV) can infect almost the entire tomato crop, causing 25–70% yield losses due to unmarketable fruit, reduced plant vigor, and shortened production cycles (DEFRA 2025). ToBRFV threatens the Australian vegetable industry, estimated at AUD $5.8 billion, with losses of up to 75%, forcing growers to replant entire crops and comply with costly quarantine protocols (PIRSA 2024). The epidemiological hallmark of ToBRFV is its remarkable stability, with the virus persisting for months on fruit, packaging, equipment, and various surfaces that are vulnerable to post-harvest contamination (DEFRA 2025; PIRSA 2024). Although ToBRFV does not pose a risk to human food safety, its presence creates a bottleneck in the global horticultural food chain (DEFRA 2025). Banana bunchy top virus (BBTV) represents another important plant viral disease in banana cultivation. It threatens global food security, especially in climate-vulnerable regions (Retkute et al. 2025). Southeast Asia is recognized as the center of BBTV diversity (Dale et al. 2017; Stainton et al. 2015), and it is believed that BBTV originated in Southeast Asia and spread sporadically to other banana-growing regions, such as Africa and North and South America, over the past millennium. As transmission occurs primarily through infected plant material or banana aphids (*Pentalonia nigronervosa*), managing and restricting the disease are both complex and resource-intensive, particularly for subsistence farmers in tropical and subtropical regions (Kumar et al. 2015); Table 20.2. It is reported that BBTV poses a major threat to global food availability and millions of smallholder farmers in Sub-Saharan Africa, which is considered a hotspot for epidemics (Moffat et al. 2021). Furthermore, papaya ringspot virus (PRSV) is characterized by leaf chlorosis, fruit deformities, dwarfing of infected plants, and yield reductions ranging from 85–100% in severely affected areas (Tripathi et al. 2023). PRSV poses a serious threat to both commercial and subsistence farming in tropical and subtropical papaya-growing regions (Yeh et al. 1999).

20.2.3 Conventional Detection Systems Against Latent Infections

Traditional serological tests for detecting latent virus infections in plants have several shortcomings, particularly at low viral loads. Various plant viruses occur at very low concentrations in infected tissues, and this may complicate assay accuracy, such as ELISA to detect viral antigens, frequently yielding false-negative results (Matthews 1967). Viral identification may be impeded by phenolic compounds and other substances in plant tissues that interfere with antigen extraction or precipitate proteins, reducing the availability of antigenically active virus particles (Matthews 1967). In addition, genetic variability between viral strains also plays a role in reducing detection accuracy by impeding antibody binding when antibodies do not recognize all existing variants (Regassa 2022). The performance of serological assays also depends heavily on the quality and specificity of the antibodies; poor-quality antibodies may lead to cross-reactivity or reduced sensitivity (Matthews 1967; Kanapiya et al. 2024). Molecular techniques are common techniques using

Table 20.2 Characteristics of selected plant viruses, highlighting latency periods, transmission factors, and associated economic impacts

Plant virus	Latency period	Transmission factors/triggers	Economic impact	References
Citrus tristeza virus (CTV)	Several years of latency reported	Aphid transmission: Rootstock interactions trigger tristeza and stem pitting	Devastated citrus industry; ~100 million trees lost; tree death, stunting, poor fruit quality; decline on sensitive rootstocks	EPPO (2024), FAO (1989, 2013), and IPPC 2023
Tomato brown rugose fruit virus (ToBRFV)	Latency is not precisely defined; symptoms appear within weeks	Mechanical inoculation; no resistant tomato varieties; sap is infective for months	Severe losses in greenhouse tomatoes; no cure; eradication costly; fruit deformation, necrosis, yield, and market loss	Salem et al. (2016), Yan et al. (2019), and EPPO (2024)
Banana bunchy top virus (BBTV)	An extended latency of months to years is occasionally reported	Circulative transmission by banana aphids (*Pentalonia* spp.)	Most destructive banana virus; up to 100% yield loss (Uganda); major income loss for smallholders; costly but essential control	Qazi (2016), Ocimati et al. (2024), and Hooks et al. (2008)
Papaya ringspot virus (PRSV)	Latency in fruits is poorly documented; orchard infections may establish within 3–7 months.	Aphid and mechanical transmission; symptoms influenced by environmental factors	Severe yield loss (70–100%); fruit quality is unmarketable; rapid spread threatens orchards and livelihoods	Tripathi et al. (2023) and Premchand et al. (2023)

PCR. However, field-applicable tests such as lateral flow assays (LFA), loop-mediated isothermal amplification (LAMP), and recombinase polymerase amplification (RPA) provide faster results and high scalability (Silva et al. 2018). In addition, reliable diagnosis requires equipment such as thermal cyclers or fluorescence readers, which are rarely accessible in distribution facilities (Table 20.3).

20.3 Innovations in Technology and Smart Management Approaches for Fruit Cultivation

The development of digital technology is currently having a significant impact on modern fruit cultivation systems. Challenges such as climate change, soil degradation, and attacks by plant pests are driving the rapid adoption of advanced technology. Consumer demands for food quality and safety are driving the need for more adaptive and precise innovations in post-harvest fruit management. This section will discuss innovations in technology and smart management approaches for fruit cultivation.

Table 20.3 Comparison of serological and molecular detection methods for latent plant viral infections across diagnostic performance criteria

Comparison criteria	Traditional serological detection (e.g., ELISA)	Molecular detection (e.g., RT-PCR, qPCR)
Sample preparation	Plant sap or tissue extract preparation	Nucleic acid (DNA/RNA) extraction from plant tissue
Target detected	Viral proteins (antigens, e.g., coat proteins)	Viral nucleic acid sequences (DNA or RNA)
Detection principle	Antigen–antibody binding with enzyme-linked color change	Amplification with specific primers (PCR/qPCR)
Sensitivity	Moderate; may miss low viral loads in latent infections	High; detects low viral loads in early or latent infections
Specificity	Good; risk of cross-reactivity	Very high with well-designed primers/probes
Quantification	Semi-quantitative; relative antigen levels only	Quantitative; real-time PCR/qPCR allows viral load estimation
Time	3–6 h (including incubation)	1–3 h (faster than serology)
Equipment	ELISA reader, microplate incubator	Thermal cycler, real-time PCR machine
Field use	Limited field use (lateral flow assays)	Mostly lab-based; portable qPCR platforms are emerging
Detection of new variants	Restricted to known viral proteins	Detects novel viruses with sequence data
False negatives in latency	Higher risk due to low antigen expression	Lower risk due to nucleic acid amplification

20.3.1 Revolution of Detection Frameworks: PCR, ELISA, Biosensor Methods, and Cutting-Edge Nanotechnological Progress

Several strategies, notably polymerase chain reaction (PCR) and enzyme-linked immunosorbent assay (ELISA), establish credible avenues for identifying plant pathogens esteemed for their reliability. PCR diagnostics, most notably quantitative real-time PCR, facilitate the sensitive identification of pathogens at subclinical levels, thereby empowering agrarians to execute preventive strategies. In the same way, ELISA is regarded for its affordable pricing, diverse applications, and compatibility with real agricultural settings, particularly in places with few resources (Mahaffee et al. 2024; Venbrux et al. 2023). Nanotechnology's integration has elevated sensitivity and diminished detection limits. Biosensors are now utilizing nanoparticles more than ever to boost signal amplification and enable ultra-sensitive detection of pathogens. Besides, nanomaterials make it possible to detect several pathogens simultaneously while offering various functionalities. These developments indicate a movement away from laboratory-centered practices toward versatile and finely tuned systems necessary for the effective farming process (Rahman et al. 2025; Vandana et al. 2024; Karaagac and Öçsoy 2023).

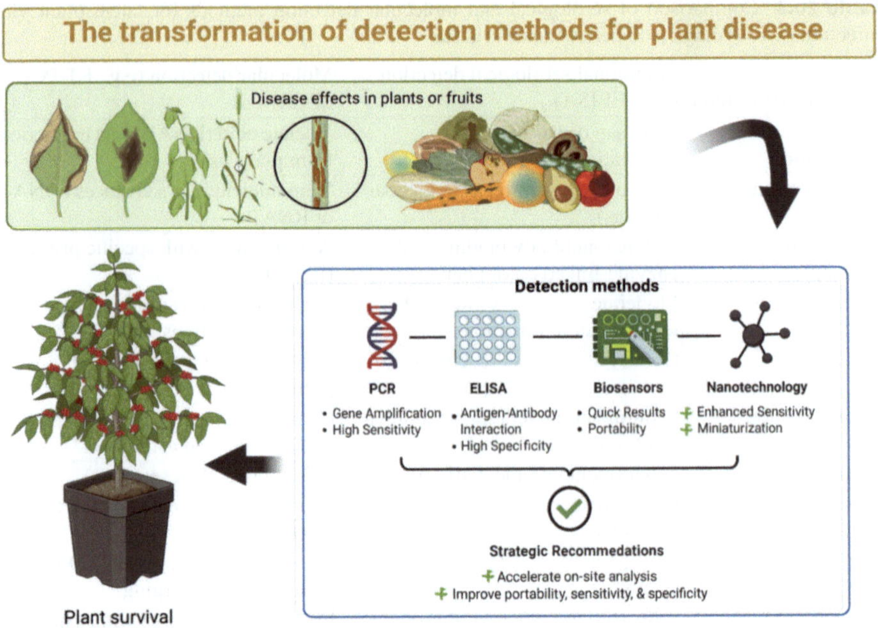

Fig. 20.1 Transformation of diagnostic techniques employed for the identification of phytopathologies

Figure 20.1 outlines the changes in detection systems for phytopathogens and fruit pathogens, showcasing their heightened sensitivity, specificity, and practicality. The original approach using polymerase chain reaction (PCR) for exact genetic identification eventually transitions into enzyme-linked immunosorbent assay (ELISA), enhancing specificity through interactions between antigens and antibodies. The rise of biosensors denotes an essential evolution by offering speedy and handheld diagnostic tools, alongside nanotechnology's contribution to bettering detection efficacy and shrinking device dimensions. All of these strategies illustrate the movement from diagnostics centered in laboratories to portable tools, promoting faster, more trustworthy, and efficient handling of plant diseases.

20.3.2 Application of IoT, Blockchain Technology, and AI Approaches in Tracking and Overseeing Diseases

Digital agricultural technologies are augmenting the monitoring of crops by integrating sophisticated systems into the framework of agricultural management. The Internet of Things (IoT) empowers instant surveillance of ecological and agricultural health attributes via coordinated sensor networks. Apparatus based on IoT innovations gauge soil moisture, humidity, canopy temperature, and disease vectors, presenting uninterrupted data that fosters early identification and preemptive

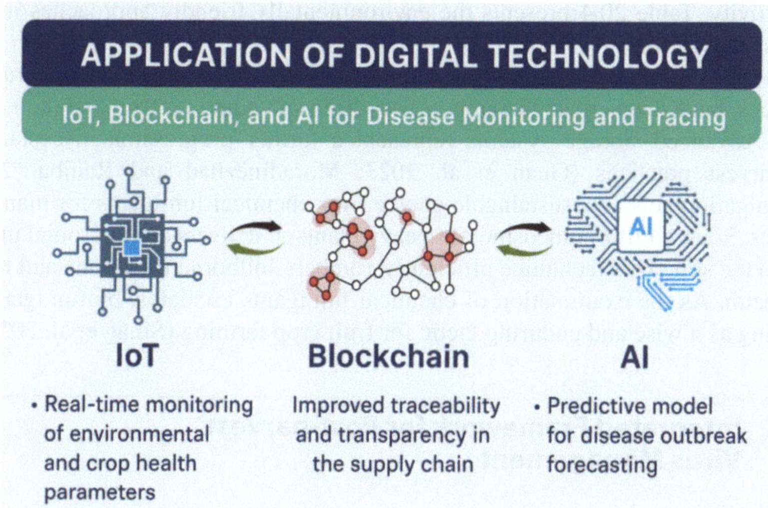

Fig. 20.2 Application of digital technologies (IoT, blockchain, AI) in disease monitoring and tracing within fruit agriculture

strategies. These enhancements are significant in converting ordinary orchards into intelligent farming businesses (Nie et al. 2021; Surakusumah et al. 2025).

Figure 20.2 highlights how IoT sensors gather ongoing agricultural data, which AI uses to analyze and recognize early disease symptoms. The implementation of blockchain technology assures the protection and documentation of the tracking process, nurturing transparency and dependability in the supply chain from producers to global arenas. AI and machine learning significantly contribute to IoT systems by changing unrefined data into practical insights. Computer vision powered by AI is widely utilized for identifying diseases in fruits, including grapes, apples, and citrus, successfully recognizing powdery mildew, rust, or viral infections with high accuracy. In addition, AI-driven forecasting models identify pest and disease surges by correlating climatic conditions with previous disease occurrences, promoting a management style that is anticipatory rather than just responsive. This predictive capability strengthens resilience in fruit cropping systems (Palani et al. 2023; Srinivasa et al. 2023). With blockchain innovations, tracking improves significantly and trust is heightened in fruit supply chains.

20.3.3 Sustainable Methods: Antiviral Surfaces, Intelligent Storage Solutions, and Biological Fumigation

The ambition for sustainable disease oversight in fruit cultivation resonates with global strategies to lessen reliance on synthetic inputs while safeguarding productivity. The ambition for sustainable disease oversight in fruit cultivation resonates with global strategies to lessen reliance on synthetic inputs while safeguarding productivity. Table 20.4 presents the environmentally friendly approaches in fruit crop management.

Coatings designed to be antiviral, using natural extracts alongside biodegradable resources, have been created to stop viruses from spreading on fruit and their packaging. Advanced storage systems represent a further progression in sustainable post-harvest practices (Guan et al. 2023; Moradinezhad and Ranjbar 2023). Biofumigation acts as a sustainable choice over chemical fumigants for managing diseases. Sourced from glucosinolate-heavy plants, it delivers antimicrobial materials into the soil. This technique efficiently combats soilborne infections and boosts soil health. As the examination of chemical fumigants escalates, biofumigation is surfacing as a wise and enduring tactic for fruit crop farming (Sihag et al. 2022).

20.4 Integrated Framework for Post-harvest Virus Management

Post-harvest virus management is central to food security and international trade. Proper management preserves product quality and economic value, protecting supply chains and market access. A sustainable approach integrates phytosanitary regulations with international standards, institutional capacity building, and research on virus prediction and system resilience. These pillars collectively drive sustainable agriculture and global food security. According to the FAO (2022), the goal of sustainable agricultural development and global food security is realized through the vision of “a sustainable world and safe food for all,” which shows the direction of the sustainable transformation of the global food system.

20.4.1 The Importance of Post-harvest Phytosanitary Regulations in International Standards

Post-harvest phytosanitary regulations are essential for the safety of agricultural products in global trade. They are key to biosecurity, competitiveness, and facilitating safe international commerce. The global movement of fresh products increases pest risks, potentially harming ecosystems and agriculture. Regulations based on international standards systematically mitigate these threats (IPPC 2023). Post-harvest phytosanitary regulations act as a last checkpoint against the spread of viruses in international trade. The International Plant Protection Convention (IPPC) and its International Standards for Phytosanitary Measures (ISPM), such as ISPM 14 and ISPM 31, offer clear global benchmarks (IPPC 2021). The WTO Sanitary and Phytosanitary Measures (SPS) Agreement ensures scientific grounding and prevents unnecessary trade barriers (Gupta et al. 2019; WTO 2019). The establishment of the World Trade Organization (WTO) and the entry into force of the Sanitary and Phytosanitary (SPS) Agreement have significantly impacted international trade in livestock and livestock products. Sanitary and Phytosanitary policies are guided by international standards, such as those recommended by the Office International des

Table 20.4 Environmentally friendly approaches in fruit crop management

Approach	Description	Application in fruit crops	Benefits	References
Antiviral coatings	Use of natural or synthetic compounds (e.g., chitosan, essential oils, nanoparticles) as edible coatings with antiviral/ antimicrobial properties	Applied to fruits such as apples, citrus, berries, and tomatoes to prevent post-harvest viral and fungal contamination	Extends shelf life, reduces viral/fungal infection, minimizes synthetic pesticide use	Valdés et al. (2017) and Priya et al. (2023)
Nanotechnology-based coatings	Incorporation of silver, zinc oxide, or silica nanoparticles into coatings or packaging	Used on strawberries, grapes, and bananas for controlled release of antimicrobial agents	Improves barrier properties, prevents microbial spoilage, enhances storage quality	Priya et al. (2023)
Smart storage systems	Use of IoT sensors, controlled atmosphere (CA), and modified atmosphere packaging (MAP) to monitor and control storage conditions	Employed for apples, kiwifruit, citrus, and mangoes to regulate humidity, ethylene, and temperature	Reduces post-harvest losses, improves quality retention, and allows real-time monitoring	Mowla et al. (2023) and Zhou et al. (2019)
Biofumigation (using brassica crops or natural volatiles)	Use of glucosinolate-rich plant residues (mustard, cabbage, radish) or essential oils for natural fumigation	Applied in storage rooms or soil treatment to suppress fungal, bacterial, and viral pathogens in fruit crops	Eco-friendly pest and pathogen control reduces synthetic fumigant use, improves soil health	Bonanomi et al. (2018)
Essential oil vapors	Application of plant-derived volatiles such as thymol, carvacrol, and cinnamaldehyde	Applied to citrus, berries, and grapes in storage or packaging	Antifungal and antiviral activity, natural preservative, residue-free	Valdés et al. (2017)
Biopolymer films (chitosan, alginate, starch-based)	Development of biodegradable films with antimicrobial activity	Used for strawberries, papaya, and tomatoes to improve coating effectiveness	Reduces environmental pollution, enhances fruit firmness, and delays ripening	Valdés et al. (2017)

Epizooties (Abdisa et al. 2023). The Sanitary and Phytosanitary Agreement allows countries to take legitimate measures to protect the life and health of consumers, animals, and plants, provided that these measures are scientifically justified and do not unnecessarily hinder trade (Eshetu and Abraham 2016; Abdisa et al. 2023). The international standards framework, which forms the core of regulation, is established by global institutions such as the International Plant Protection Convention (IPPC) and the World Trade Organization (WTO) through the Sanitary and Phytosanitary (SPS) agreement. These international standards are then formalized in technical documents, such as the International Standards for Phytosanitary Measures (ISPM), which serve as the scientific and legal framework for implementing global phytosanitary regulations. Harmonizing regulations within these international standards aims to ensure that quarantine and surveillance measures do not become discriminatory technical barriers but instead strengthen the resilience of the international food system. Furthermore, the national regulatory section emphasizes the importance of implementing international standards into domestic or national policies. In Indonesia, this regulation is implemented by the Indonesian Quarantine Agency through several regulations, such as Law No. 16 of 1992 and Minister of Agriculture Regulations No. 42/2012 and No. 43/2012, which govern quarantine measures for fresh plants, tubers, fruits, and vegetables. This regulation covers inspection systems, export-import certification, and management of carriers of plant pests and diseases. Nationally, Indonesia has implemented Law Number 21 of 2019 concerning Animal, Fish, and Plant Quarantine, which strengthens the role of the Agricultural Quarantine Agency in preventing the entry and spread of quarantined plant pests and diseases (Agricultural Quarantine Agency 2023). Plant quarantine helps facilitate safe global trade in agriculture by assisting producers and exporters in providing a reliable phytosanitary certification system to meet trading partner requirements (Gupta et al. 2019).

20.4.2 Strengthening Institutional Capacities Across the Value Chain

Institutional capacity is crucial for preventing the spread of viruses in the post-harvest phase, including enhancing technical competence, providing diagnostic facilities, and implementing quality management practices. Laboratories involved in virus detection must be accredited according to ISO/IEC 17025 to ensure reliable test results (ILAC 2017). Regular proficiency testing is also crucial to maintain consistent laboratory performance and ensure methods are in line with scientific advances (OIE 2020). Good Agricultural Practices (GAP) and Good Handling Practices (GHP) are crucial for ensuring food safety meets international and national standards. Optimizing post-harvest management supports the efficiency of the agricultural supply chain and ensures the availability of safe, nutritious, and high-quality food for the public (Alegbeleye and Sant'Ana 2021). At the business level, the implementation of Good Handling Practices (GHP) and Hazard Analysis and Critical Control Points (HACCP) is crucial for preventing viral contamination during product handling, storage, and distribution. One factor in successful post-harvest management is worker education and training on hygiene, equipment sanitation, and early detection of viral infection symptoms (Singh et al. 2023). Synergy between government, academia, industry, and farmers is crucial for building a resilient system. Case studies in Indonesia demonstrate that post-harvest biological control requires a strong institutional framework. High-value horticultural commodities, such as fruits and vegetables, contribute significantly to global trade flows, yet they are vulnerable to viral pathogens that trigger import rejections. Therefore, national agricultural institutions are required to have molecular detection capabilities, sanitation certification, and rapid response to changes in global phytosanitary regulations, meeting IPPC and SPS-WTO standards. Research by Dwiastuti et al. (2021) demonstrated the effectiveness of *Candida* spp., *Bacillus* spp., and *Trichoderma* spp. as biological agents useful for controlling post-harvest diseases in citrus, apples, grapes, and strawberry crops in Indonesia. However, this control is also successful because it is supported by the capabilities of research institutions and extension services that can systematically produce, distribute, and train users of this technology.

20.4.3 Future Research Directions for Virus Infection

Post-harvest virus management depends on the ability of agricultural systems to predict, detect, and respond to viral infections and build product resilience to post-harvest degradation. Food safety is a crucial global challenge, with post-harvest losses estimated at 20–50% due to microbial spoilage, fungal and viral infections, and physiological damage threatening food security and economic stability (Rhouma 2025; Taha et al. 2025). Future research is directed at early prediction of virus spread and improving detection systems during the post-harvest phase. One key research direction is the development of big data and artificial intelligence (AI)-based prediction systems that utilize information from field sensors, remote sensing, and genomic data. Advances in digital technologies such as machine learning and remote sensing have opened up new opportunities for predicting the risk of viral infections before and after harvest (Upasani and Goswami 2025). Predictive models based on weather data, vector distribution patterns, and environmental parameters can help determine timely and effective intervention strategies (Zhang et al. 2020). Furthermore, the use of biosensor-based rapid detection technologies and CRISPR-Cas has shown potential to identify viral infections with high sensitivity within minutes. One of the latest technologies relevant to the smallholder farmer sector is DragonFruitQualityNet, a lightweight CNN system capable of performing real-time dragon fruit quality inspections via smartphone with 93.98% accuracy (Haquea et al. 2025). This technology enables farmers to identify virus-free fruit before packaging and shipping. Future integration of big data, edge computing, Internet of Things (IoT) sensors, and blockchain within agricultural logistics systems is necessary to establish secure and transparent tracking mechanisms. Such integration can enhance post-harvest resilience and enable both government and private sector stakeholders to respond rapidly to potential virus outbreaks at distribution or storage points. The integration of Internet of Things (IoT) and blockchain technologies into storage systems offers an opportunity to increase transparency in post-harvest handling, from harvest to distribution, enabling accurate tracking of virus transmission (Kollia et al. 2021). This technology can provide targeted, data-driven solutions to improve agricultural efficiency, productivity, and sustainability. Optimizing agricultural yields, reducing the risk of losses, and improving food security can be achieved through big data analysis, making it easier for farmers (Morchid et al. 2024). Molecular methods such as polymerase chain reaction, loop-mediated isothermal amplification, and CRISPR-based assays can detect pathogens quickly and precisely at the genetic level. Meanwhile, non-destructive technologies such as near-infrared spectroscopy and hyperspectral imaging can capture biochemical and morphological changes in produce, enabling real-time monitoring.

References

Abdisa T, Getu M, Etana M (2023) Review on sanitary and phytosanitary agreement. J Vet Health Sci 4(2):53–63

Adeosun WB, Loots DT (2024) Medicinal plants against viral infections: a review of metabolomics evidence for the antiviral properties and potentials in plant sources. Viruses 16(2):218. https://doi.org/10.3390/v16020218

Agricultural Quarantine Agency (2023) Rencana Strategis 2024. Kementerian Pertanian Republik Indonesia

Alegbeleye OO, Sant'Ana AS (2021) Risks associated with the consumption of irrigation water contaminated produce: on the role of quantitative microbial risk assessment. Curr Opin Food Sci 41:88–98. https://doi.org/10.1016/j.cofs.2021.03.013

Ando S, Miyashita S, Takahashi H (2019) Plant defense systems against cucumber mosaic virus: lessons learned from CMV–Arabidopsis interactions. J Gen Plant Pathol 85(3):174–181. https://doi.org/10.1007/s10327-019-00845-x

Arah IK, Ahorbo GK, Anku EK, Kumah EK, Amaglo H (2016) Postharvest handling practices and treatment methods for tomato handlers in developing countries: a mini review. Adv Agric 2016(1):6436945. https://doi.org/10.1155/2016/6436945

Bonanomi G, Lorito M, Vinale F, Woo SL (2018) Biofumigation: a comprehensive review of its agronomic, biochemical and ecological aspects. Crop Prot 104:65–76. https://doi.org/10.1146/annurev-phyto-080615-100046

Boonham N, Kreuze J, Winter S, van der Vlugt R, Bergervoet J, Tomlinson J, Mumford R (2014) Methods in virus diagnostics: from ELISA to next generation sequencing. Virus Res 186:20–31. https://doi.org/10.1016/j.virusres.2013.12.007

Budhathoki P, kumar Pandey S, Magar PD (2022) Review on post-harvest handling of fruits and vegetables to minimize loss. Food Agri Econ Rev (FAER) 2(1):37–40. https://doi.org/10.26480/faer.01.2022.37.40

Bukhamsin A, Kosel J, Mccabe MF, Blilou I, Salama KN (2025) Early and high-throughput plant diagnostics: strategies for disease detection. Trends Plant Sci 30(3):324–337. https://doi.org/10.1016/j.tplants.2024.10.003

Burgyán J (2008) Role of silencing suppressor proteins. In: Plant virology protocols: from viral sequence to protein function, pp 69–79. https://doi.org/10.1007/978-1-59745-102-4_5

Cassedy A, Parle-McDermott A, O'Kennedy R (2021) Virus detection: a review of the current and emerging molecular and immunological methods. Front Mol Biosci 8:637559. https://doi.org/10.3389/fmolb.2021.637559

Dale J, James A, Paul JY, Khanna H, Smith M, Peraza-Echeverria S, Garcia-Bastidas F, Kema G, Waterhouse P, Mengersen K, Harding R (2017) The global distribution of banana bunchy top virus reveals little evidence for frequent recent introductions to new regions. PLoS Pathog 13(1):e1006205. https://doi.org/10.1038/s41467-017-01670-6

DEFRA-Department for Environment, Food and Rural Affairs (2025) Tomato brown rugose fruit virus (ToBRFV) policy review. DEFRA. https://planthealthportal.defra.gov.uk/assets/ToBRFV-policy-review.pdf

Dhiman P, Kaur A, Kukreja V (2023) Citrus fruit disease detection techniques: a survey and comparative analysis of relevant approaches. Int J Comput Digit Syst 14(1):10127–10148. https://doi.org/10.12785/ijcds/140187

Dwiastuti ME, Soesanto L, Aji TG, Devy NF, Hardiyanto (2021) Biological control strategy for postharvest diseases of citrus, apples, grapes, and strawberries fruits and application in Indonesia. Egypt J Biol Pest Control 31:141. https://doi.org/10.1186/s41938-021-00488-1

EPPO-European and Mediterranean Plant Protection Organization (2024) Updated version of the pest risk analysis for Tobamovirus fructirugosum (tomato brown rugose fruit virus) (revised 2024). EPPO. https://pra.eppo.int/pra/f8bed8a7-392b-41fe-8737-0a05b31e8c00

Eshetu E, Abraham Z (2016) Review on live animal and meat export marketing system in Ethiopia: challenges and opportunities. J Sci Innov Res 5:59–64. https://doi.org/10.31254/jsir.2016.5206

FAO (2022) Sustainable food systems: concept and framework. Food and Agriculture Organization of the United Nations

Folimonova SY, Folimonov AS, Tatineni S, Dawson WO (2008) Citrus tristeza virus: survival at the edge of the movement continuum. J Virol 82(13):6546–6556. https://doi.org/10.1128/JVI.00515-08

Fontes EPB, Teixeira RM, Lozano-Durán R (2021) Plant virus-interactions: unraveling novel defense mechanisms under immune-suppressing pressure. Curr Opin Biotechnol 70:108–114. https://doi.org/10.1016/j.copbio.2021.03.007

Food and Agriculture Organization of the United Nations (1989) Prevention of post-harvest food losses: fruits, vegetables and root crops (FAO Training Series No. 17/2). FAO. https://www.fao.org/3/t0073e/t0073e00.htm

Food and Agriculture Organization of the United Nations (FAO) (2013) Save and grow: cassava—a guide to sustainable production intensification. FAO. http://www.fao.org/docrep/018/i3278e/i3278e.pdf

García-Ruíz H (2018) Susceptibility genes to plant viruses. Viruses 10(9):484. https://doi.org/10.3390/v10090484

Gergerich RC, Welliver RA, Gettys S, Osterbauer NK, Kamenidou S, Martin RR et al (2015) Safeguarding fruit crops in the age of agricultural globalization. Plant Dis 99(2):176–187. https://doi.org/10.1094/pdis-07-14-0762-fe

Giovannoni J (2004) Genetic regulation of fruit development and ripening. Plant Cell 16:170–181. https://doi.org/10.1105/tpc.019158

Guan J, Zeng K, Chen Z (2023) Editorial: postharvest disease management in fruits and vegetables: recent advances and mechanisms. Front Microbiol 14. https://doi.org/10.3389/fmicb.2023.1203010

Gupta S, Verma S, Thakur R (2019) Phytosanitary requirement for import of horticulture crops. Int J Curr Microbiol App Sci 8(2):2871–2886. https://doi.org/10.20546/ijcmas.2019.802.337

Harper JW, Burton JL, Solomon MJ (2002) The anaphase-promoting complex: it's not just for mitosis any more. Genes Dev 16(17):2179–206. https://doi.org/10.1101/gad.1013102

Haquea MZ, Sarker Y, Mahi MFS, Jaman SJ, Islam MR (2025) DragonFruitQualityNet: a lighweight convolutional neural network for real-time dragon fruit quality inspection on mobile devices. arXiv preprint, 2508.07306. https://doi.org/10.48550/arXiv.2508.07306

Hooks CRR, Wright MG, Kabasawa DS, Manandhar R, Almeida RPP (2008) Effect of banana bunchy top virus infection on morphology and growth characteristics of banana. Ann Appl Biol 153(1):1–9. https://doi.org/10.1111/j.1744-7348.2008.00233.x

Hounmanou YMG, Mdegela RH, Dougnon TV, Achoh ME, Mhongole OJ, Agadjihouèdé H et al (2018) Tilapia lake virus threatens tilapiines farming and food security: socio-economic challenges and preventive measures in Sub-Saharan Africa. Aquaculture 493:123–129. https://doi.org/10.1016/j.aquaculture.2018.05.001

ILAC (2017) ISO/IEC 17025: general requirements for the competence of testing and calibration laboratories

IPPC (2021) International Standards for Phytosanitary Measures (ISPM), vol No. 14 & No. 31. International Plant Protection Convention

IPPC (2023) ISPM 15: regulation of wood packaging material in international trade. International Plant Protection Convention

Jiménez J, Sadras VO, Espaillat N, Moreno A, Fereres A (2024) Interplay between drought and plant viruses co-infecting melon plants. Sci Rep 14(1):15833. https://doi.org/10.1038/s41598-024-66344-y

Jyotika RK, Harish S, Karthikeyan G, Kumar KK, Murugan M, Jayakanthan M, Chen TC (2024) Molecular approaches for the management of papaya ringspot virus infecting papaya: a comprehensive review. Mol Biol Rep 51(1):981. https://doi.org/10.1007/s11033-024-09920-9

Kanapiya A, Amanbayeva U, Tulegenova Z, Abash A, Zhangazin S, Dyussembayev K, Mukiyanova G (2024) Recent advances and challenges in plant viral diagnostics. Front Plant Sci 15. https://doi.org/10.3389/fpls.2024.1451790

Kang B, Liu H, Liu Y, Zhang Y, Zhou C (2021) Transmission characteristics and epidemiology of citrus tristeza virus in grafted citrus plants. Plant Pathol J 37(3):307–318. https://doi.org/10.5423/PPJ.OA.12.2020.0210

Karaagac Z, Öçsoy (2023) Nanomaterials for the rapid identification of agriculturally important plant pathogens, pp 179–198. https://doi.org/10.1007/978-981-99-5787-3_10

Kaur SI, Kashyap PL, Kang SS, Sharma A (2020) Detection and diagnosis of seed-borne viruses and virus-like pathogens. In: Seed-borne diseases of agricultural crops: detection, diagnosis & management. Springer, Singapore, pp 169–199

Khakimov A, Salakhutdinov I, Omolikov A, Utaganov S (2022) Traditional and current-prospective methods of agricultural plant diseases detection: a review. In: IOP conference series: earth and environmental science. (vol. 951, no. 1, p. 012002). IOP Publishing. https://doi.org/10.1088/1755-1315/951/1/012002

Kollia I, Stevenson J, Kollias S (2021) AI-enabled efficient and safe food supply chain. Comput Sci arXiv:2105.00333. https://doi.org/10.48550/arXiv.2105.00333

Kroschel J, Sporleder M, Carhuapoma P (2016) Potato tuber moth, Phthorimaea operculella (Zeller 1873). In: Kroschel J, Mujica N, Carhuapoma P, Sporleder M (eds) Pest distribution and risk atlas for Africa: potential global and regional distribution and abundance of agricultural and horticultural pests and associated biocontrol agents under current and future climates. International Potato Center (CIP), pp 7–23. https://doi.org/10.4160/9789290604761-1

Kumar PL, Selvarajan R, Iskra-Caruana ML, Chabannes M, Hanna R (2015) Biology, etiology, and control of virus diseases of banana and plantain. Adv Virus Res 91:229–269. https://doi.org/10.1016/bs.aivir.2014.10.006

Kumar R, Jeevalatha A, Baswaraj R, Sharma S, Nagesh M (2017) A multiplex RT-PCR assay for simultaneous detection of five viruses in potato. J Plant Pathol:37–45. https://doi.org/10.4454/jpp.v99i1.3824

Kumar M, Gupta PK, Kumar Y (2024) Chapter-15 Papaya Mosaic Virus. In: Viral diseases of vegetable & fruit crops, vol 33, p 207

Leonetti P, Stuttmann J, Pantaleo V (2021) Regulation of plant antiviral defense genes via host RNA-silencing mechanisms. Virol J 18(1):194. https://doi.org/10.1186/s12985-021-01664-3

López-Fabuel I, Wetzel T, Bertolini E, Bassler A, Vidal E, Torres LB et al (2013) Real-time multiplex RT-PCR for the simultaneous detection of the five main grapevine viruses. J Virol Methods 188(1–2):21–24. https://doi.org/10.1016/j.jviromet.2012.11.034

Mahaffee WF, Thiessen LD, King KM, Choudhury RA (2024) Detection and monitoring of plant pathogens and pests. PhytoFrontiers™. https://doi.org/10.1094/phytofr-02-24-0005-fi

Maina S, Donovan NJ, Plett K, Bogema D, Rodoni BC (2024) High-throughput sequencing for plant virology diagnostics and its potential in plant health certification. Front Hortic 3:1388028. https://doi.org/10.3389/fhort.2024.1388028

Manzoor S, Nabi SU, Ayaz A, Rasool B, Sharma SK, Chesti MH et al (2025) Unveiling viral threats to temperate pome fruits: characterization, transmission, and sustainable management strategies. Front Microbiol 16:1560720. https://doi.org/10.3389/fmicb.2025.1560720

Matthews R (1967) Serological techniques for plant viruses. In: Methods in virology, pp 199–241. https://doi.org/10.1016/b978-1-4832-3219-5.50011-1

Moffat J, Harper G, Oduor G, Alicai T, Kumar PL, Mahuku G (2021) Banana bunchy top disease in Africa—predicting continent-wide spread and impact. Plant Pathol 70(6):1217–1229. https://doi.org/10.1111/ppa.13764

Moradinezhad F, Ranjbar A (2023) Advances in postharvest diseases management of fruits and vegetables: a review. Horticulturae. https://doi.org/10.3390/horticulturae9101099

Morchid A, Alami RE, Raezah AA, Sabbar Y (2024) Applications of internet of things (IoT) and sensors technology to increase food security and agricultural sustainability: benefits and challenges. Ain Shams Eng J 15(102509):1–15. https://doi.org/10.1016/j.asej.2023.102509

Moreno P, Ambrós S, Albiach-Martí MR, Guerri J, Pena L (2008) Citrus tristeza virus: a pathogen that changed the course of the citrus industry. Mol Plant Pathol 9(2):251–268. https://doi.org/10.1111/j.1364-3703.2007.00455.x

Mowla MN, Mowla N, Shah AS, Rabie KM, Shongwe T (2023) Internet of things and wireless sensor networks for smart agriculture applications: a survey. IEEe Access 11:145813–145852. https://doi.org/10.1109/ACCESS.2023.3346299

Müllender M, Varrelmann M, Savenkov EI, Liebe S (2021) Manipulation of auxin signalling by plant viruses. Mol Plant Pathol 22(11):1449–1458. https://doi.org/10.1111/mpp.13122

Naveed H, Islam W, Jafir M, Andoh V, Chen L, Chen K (2023) A review of interactions between plants and whitefly-transmitted Begomoviruses. Plants 12(21):3677. https://doi.org/10.3390/plants12213677

Nicaise V (2014) Crop immunity against viruses: outcomes and future challenges. Front Plant Sci 5:660. https://doi.org/10.3389/fpls.2014.00660

Nie P, Zhang Q, He Y (2021) IoT management of field crops and orchards. Springer, Cham, pp 291–303. https://doi.org/10.1007/978-3-030-65702-4_10

Niraula PM, Fondong VN (2021) Development and adoption of genetically engineered plants for virus resistance: advances, opportunities and challenges. Plants 10(11):2339. https://doi.org/10.3390/plants10112339

Ocimati W, Ogwal G, Tazuba AF, Kubiriba J, Tugume J, Erima R, Okurut W, Mahuku G, Kutunga D, Blomme G (2024) Mapping the vulnerability of banana production landscapes in Uganda to banana bunchy top disease. Front Agron 6:1401478. https://doi.org/10.3389/fagro.2024.1401478

OIE (2020) Proficiency testing in laboratories: guidelines and best practices

Palani HK, Ilangovan S, Senthilvel PG, Thirupurasundari DR, Rajesh Kumar K (2023) AI-powered predictive analysis for pest and disease forecasting in crops:950–954. https://doi.org/10.1109/iccsai59793.2023.10421237

Pamies D, Leist M, Coecke S, Bowe G, Allen D, Gstraunthaler G et al (2022) Guidance document on good cell and tissue culture practice 2.0 (GCCP 2.0). ALTEX. https://doi.org/10.14573/altex.2111011

Pandya Y, Sharma A, Bakshi M (2023) Edible coatings in fruits: effectiveness and applicability: a review. Food Sci Indian J Res Food Sci Nutr 10:1–10. https://doi.org/10.15613/fijrfn/2022/v10i1/220470

PIRSA-Primary Industries and Regions South Australia (2024) Tomato brown rugose fruit virus (ToBRFV): biosecurity advice for growers. PIRSA. https://www.pir.sa.gov.au/biosecurity/plant_health/emergency_and_significant_plant_pests/tomato_brown_rugose_fruit_virus

Pott DM, Vallarino JG, Osorio S (2020) Metabolite changes during postharvest storage: effects on fruit quality traits. Metabolites. 10:187. https://doi.org/10.3390/metabo10050187

Prabha K, Baranwal VK, Jain RK (2013) Applications of next generation high throughput sequencing technologies in characterization, discovery and molecular interaction of plant viruses. Indian J Virol 24(2):157–165. https://doi.org/10.1007/s13337-013-0133-4

Premchand U, Mesta RK, Devappa V, Basavarajappa MP, Venkataravanappa V, Narasimha Reddy LRC, Shankarappa KS (2023) Survey, detection, characterization of papaya ringspot virus from southern India and management of papaya ringspot disease. Pathogens 12(6):824. https://doi.org/10.3390/pathogens12060824

Priya K, Thirunavookarasu N, Chidanand DV (2023) Recent advances in edible coating of food products and its legislations: a review. J Agric Food Res 12:100623. https://doi.org/10.1016/j.jafr.2023.100623

Qazi J (2016) Banana bunchy top virus and the bunchy top disease. J Gen Plant Pathol 82(1):2–11. https://doi.org/10.1007/s10327-015-0642-7

Rahman MA, Karthikeyan M, Johnson I, Raja K, Sekar C, Mary XA, Basha J (2025) Ensuring food security through rapid and in-field detection of diseases in food crops using real time and portable sensors. Anal Biochem:115925. https://doi.org/10.1016/j.ab.2025.115925

Regassa B (2022) Advanced and commonly used serological techniques for detection and diagnosis of plant virus: review. Adv Life Sci Technol. https://doi.org/10.7176/alst/91-02

Retkute R, Gilligan CA (2025) Developing a spatio-temporal model for banana bunchy top disease: leveraging remote sensing and survey data. Front Plant Sci 16:1521620. https://doi.org/10.3389/fpls.2025.1521620

Rhouma A (2025) Innovations in post-harvest disease detection: from molecular diagnostics to AI-based imaging. Mor J Agric Sci 6(2):78–84. https://doi.org/10.5281/zenodo.15433446

Richert-Pöggeler KR, Franzke K, Hipp K, Kleespies RG (2019) Electron microscopy methods for virus diagnosis and high resolution analysis of viruses. Front Microbiol 9:3255. https://doi.org/10.3389/fmicb.2018.03255

Roossinck MJ (2010) Plant virus evolution: what we know and why it matters. Annu Rev Phytopathol 48:27–52. https://doi.org/10.1007/978-3-540-75763-4

Salem N, Mansour A, Ciuffo M, Falk BW, Turina M (2016) A new tobamovirus infecting tomato crops in Jordan. Arch Virol 161: 503–506. https://doi.org/10.1007/s00705-015-2677-7

Shang P, Xu L, Cheng T (2024) Serological and molecular detection of citrus tristeza virus: a review. Microorganisms 12(8):1539. https://doi.org/10.3390/microorganisms12081539

Sharma N (2014) Biological controls for preventing food deterioration: strategies for pre-and post-harvest management. Wiley. https://doi.org/10.1002/9781118533024.ch1

Sihag M, Kumar V, Rana M, Srivastava S, Singh SR (2022) Biofumigation: prospects for control of soil borne plant diseases. J Biopest 15(2):136–149. https://doi.org/10.57182/jbiopestic.15.2.136-149

Silva G, Oyekanmi J, Nkere CK, Bömer M, Kumar PL, Seal SE (2018) Rapid detection of potyviruses from crude plant extracts. Anal Biochem 546:17–22. https://doi.org/10.1016/j.ab.2018.01.019

Singh KJ, Kapoor DS, Sharma A, Thakur K, Bajaj T, Tomar A, Mittal S, Singh B, Agarwal R (2023) Internet of things in agriculture industry: implementation, applications, challenges and potential. Lect Notes Electr Eng 976. https://doi.org/10.1007/978-981-99-0412-9_29

Srinivasa A, Krishna R, Dheepthi R, Sri T, Reddy SS (2023) Machine learning-based citrus plant disease detection and management system using computer vision. IEEE, pp 1137–1145. https://doi.org/10.1109/icesc57686.2023.10193336

Stainton D, Martin DP, Muhire B, Lolohea S, Halafihi M, Lepoint P, Bellaoui M, Sharman M, Harding R (2015) The global distribution of Banana bunchy top virus reveals little evidence for frequent recent introductions to new regions. Virus Res 195:70–77. https://doi.org/10.1016/j.virusres.2014.09.01

Steinberger AR, Voytas DF (2025) Virus-induced gene editing free from tissue culture. Nat Plants:1–11. https://www.nature.com/articles/s41477-025-02025-6

Surakusumah RF, Yusuf M, Tarif EAK (2025) IoT-based environmental health monitoring system in agriculture: a case study on watermelon cultivation to support sustainable community well-being. Int J Sci Res 4(2):62–67. https://doi.org/10.25299/ijsr.2024.23686

Taha MF, Mao H, Zhang Z, Elmasry G, Awad MA, Abdalla A, Mousa S, Elwakeel AE, Elsherbiny O (2025) Emerging technologies for precision crop management towards agriculture 5.0: A comprehensive overview. Agriculture 15(6):582. https://doi.org/10.3390/agriculture15060582

Torrance L, Talianksy ME (2020) Potato virus Y emergence and evolution from the Andes of South America to become a major destructive pathogen of potato and other solanaceous crops worldwide. Viruses 12(12):1430. https://doi.org/10.3390/v12121430

Tripathi S, Suzuki JY, Ferreira SA, Gonsalves D, Manshardt RM (2023) Survey, detection, and characterization of papaya ringspot virus: global implications for disease management. Front Plant Sci 14:10301999. https://doi.org/10.3389/fpls.2023.10301999

Ufitinema B, De Dieu Niyomugabo J, Hakizima M (2024) Relationship between damages and other postharvest stress factors affecting quality of fruit and vegetables. J Food Technol Nutr Sci:1–5. https://doi.org/10.47363/jftns/2024(6)178

Ullah I, Daymond AJ, Hadley P, End MJ, Umaharan P, Dunwell JM (2021) Identification of cacao mild mosaic virus (CaMMV) and cacao yellow vein-banding virus (CYVBV) in cocoa (Theobroma cacao) germplasm. Viruses 13(11):2152. https://doi.org/10.3390/v13112152

Umer M, Liu J, You H, Xu C, Dong K, Luo N et al (2019) Genomic, morphological and biological traits of the viruses infecting major fruit trees. Viruses 11(6):515. https://doi.org/10.3390/v11060515

Upasani N, Goswami C (2025) Plant disease detection using machine learning techniques: a comprehensive approach. Int J Environ Sci 11(12):1268–1480. https://doi.org/10.64252/qm5z0g23

Vac CS, Andreica I, Roman IA (2023) Prospects for research, development, innovation and technology transfer in Romanian horticulture. Sustainability 2023(15):10215. https://doi.org/10.3390/su151310215

Valdés A, Ramos M, Beltrán A, Jiménez A, Garrigós MC (2017) State of the art of antimicrobial edible coatings for food packaging applications. Coatings 7(4):56. https://doi.org/10.3390/coatings7040056

Vandana JN, Sharavanan PT, Johnson I, Anandham R, Kalimuthu R, Karthikeyan G, Kathithachalam A, Vaithiyalingan M (2024) Nanobiosensors for early detection of plant pathogens. Plant Sci Today 12:1. https://doi.org/10.14719/pst.4285

Venbrux MMJ, Crauwels S, Rediers H (2023) Current and emerging trends in techniques for plant pathogen detection. Front Plant Sci 14. https://doi.org/10.3389/fpls.2023.1120968

Wang N, Trivedi P, Hu J, Levy A (2023) Emerging insights into citrus tristeza virus: ecology, epidemiology, and management strategies. Front Plant Sci 14:1178342. https://doi.org/10.3389/fpls.2023.1178342

WTO (2019) The WTO agreement on the application of sanitary and phytosanitary measures

Yan ZY, Ma HY, Han SL, Geng C, Tian YP, Li XD (2019) First report of tomato brown rugose fruit virus infecting tomato in China. Plant Disease 103(11):2973. https://doi.org/10.1094/PDIS-05-19-1045-PDN

Yang Z, Cao S, Su X, Jiang Y (2014) Respiratory activity and mitochondrial membrane associated with fruit senescence in postharvest peaches in response to UV-C treatment. Food Chem 161:16–21. https://doi.org/10.1016/j.foodchem.2014.03.120

Yeh SD, Jan FJ, Chiang CH, Doong TJ, Chen MC, Chung PH, Bau HJ (1999) Complete nucleotide sequence and genetic organization of papaya ringspot virus RNA. J Gen Virol 80(7):1787–1794. https://doi.org/10.1099/0022-1317-80-7-1787

Zaffaroni M, Rimbaud L, Mailleret L, Cunniffe NJ, Bevacqua D (2021) Modelling interference between vectors of non-persistently transmitted plant viruses to identify effective control strategies. PLoS Comput Biol 17(12):e1009727. https://doi.org/10.1371/journal.pcbi.1009727

Zhang H, Li L, He Y, Qin Q, Chen C, Wei Z et al (2020) Distinct modes of manipulation of rice auxin response factor OsARF17 by different plant RNA viruses for infection. Proc Natl Acad Sci 117(16):9112–9121. https://doi.org/10.1073/pnas.1918254117

Zhou J, Chen S, Wang Z (2019) Intelligent agriculture with deep learning applications. IEEE Access 7:181906–181918. https://doi.org/10.1109/ICSSE.2018.8520209

21 Transgenic Approaches for Virus Resistance in Fruit Crops: Progress and Policy

Prashant S. Waghrulkar, Nagendran Krishnan, V. Balasubramanian, and R. Selvarajan

Abstract

Viral diseases cause severe yield losses in fruit crops worldwide, estimated at over USD 30 billion annually. Conventional breeding for resistance is limited by long timelines and the scarcity of natural resistance genes, highlighting the need for modern biotechnological solutions. This chapter reviews progress in developing virus-resistant fruit crops through transgenic and genome-editing approaches. Pathogen-derived resistance (e.g., coat protein-mediated and RNAi-based strategies) has produced successful examples such as papaya ring spot virus-resistant papaya and plum pox virus-resistant plum. Emerging tools, including artificial microRNAs and CRISPR-Cas9, offer precise, sequence-specific protection. While scientific advances demonstrate strong potential, adoption is shaped by diverse global regulatory policies and public acceptance. Balanced, evidence-based frameworks are needed to realize role of biotechnology in ensuring sustainable fruit production and food security. This chapter focuses on the scientific progress, regulatory challenges, and future prospects of transgenic and gene-edited fruit crops for virus resistance.

Keywords

Virus resistance · Transgenic crops · CRISPR · RNAi · Biosafety regulations · Food security · Fruit pathogens

P. S. Waghrulkar · N. Krishnan · V. Balasubramanian · R. Selvarajan (✉)
Molecular Virology Lab., ICAR-National Research Center for Banana, Tiruchirappalli, Tamil Nadu, India

R. Kumar et al. (eds.), *Fruit Crops and Viral Pathogens Volume II*, Plant Pathology for Sustainable Agriculture, https://doi.org/10.1007/978-981-95-7261-8_21

21.1 Introduction

In 2022, the world produced nearly 900 million tons of fruits, accounting for about 9.4% of total global food production (FAO 2023). The fresh fruit market was valued at USD 594.3 billion in 2023 and is expected to grow to USD 831.4 billion by 2032, at an annual growth rate of 3.8% (Skyquest 2024). China, India, and Brazil together produce about 30% of the world's fruits, though most of this is consumed domestically, limiting their share in global trade (FAO 2025). Agriculture employs nearly 857 million people worldwide, making fruit production critical for both food security and livelihoods. Plant viruses are one of the major threats to fruit production, causing yield losses of over USD 30 billion annually (Sastry and Zitter 2014). To manage these viral diseases in fruit crops, different strategies are available, including conventional breeding, marker-assisted selection (MAS), transgenic approaches, and modern genome editing techniques such as CRISPR-Cas. Conventional breeding for virus resistance faces serious challenges. It often requires several generations of backcrossing and selection, which can take many years or even decades. A major obstacle is the lack of natural resistance genes in elite cultivars. For example, in bananas, wild relatives such as *Musa balbisiana* show resistance to banana bunchy top virus (BBTV), but they are difficult to cross with commercial varieties (Dela Cueva et al. 2023). Most cultivated banana hybrids also produce sterile progeny, making traditional breeding impractical (Reddy et al. 2009). Even when resistance genes are available, their transfer into cultivated varieties is not easy. The process of introgression is often hampered by interspecific barriers and linkage drag, where undesirable traits from wild relatives (such as poor fruit quality) are inherited along with the resistance gene (Cabrera et al. 2024). Such constraints highlight the importance of modern biotechnological approaches, including genetic engineering and genome editing, to accelerate the development of virus-resistant fruit crops (Manzoor et al. 2024). Traditional genetic engineering usually involves inserting foreign DNA from unrelated species into the plant genome, which can sometimes disrupt native genes or regulatory regions. In contrast, site-directed nuclease (SDN)-based genome editing methods, such as CRISPR-Cas9, TALENs, or ZFNs, allow precise modifications at specific locations in the genome, often without adding foreign DNA. SDN-1 and SDN-2 methods usually produce plants indistinguishable from those with natural mutations, making them more acceptable in some regulatory systems. For example, CRISPR-edited high-oleic soybeans were not classified as transgenic in certain countries, which speed up their adoption (Tome 2022). On the other hand, SDN-3 approaches involve inserting large DNA fragments, which results in genetically modified (GM) crops similar to those produced by traditional transgenesis (Ahmad et al. 2023). Given the limitations of conventional breeding and the precision offered by modern biotechnological tools, transgenic approaches and genome editing hold great promise for developing virus-resistant fruit crops. This chapter explores the progress made so far, highlights successful case studies, and discusses the policy and regulatory frameworks that shape the future of these technologies in ensuring sustainable fruit production and global food security.

21.1.1 Genetic Engineering in Virus Resistance for Fruit Crops

Genetic engineering has played a central role in developing virus-resistant fruit crops by enabling targeted interventions that are not possible through conventional breeding. Successful examples illustrate its potential to provide rapid, durable, and broad-spectrum resistance. A landmark case is the development of papaya resistant to papaya ringspot virus (PRSV), where the viral coat protein (CP) gene was introduced *via* biolistic particle bombardment. Resistant plants were generated within a short period and later commercialized in Hawaii, saving the papaya industry from collapse (Gonsalves et al. 2004a, b). Beyond single-virus resistance, genetic engineering also enables the transfer of resistance genes across species barriers. For instance, squash engineered with CP genes from zucchini yellow mosaic virus (ZYMV) and watermelon mosaic virus 2 (WMV2) exhibited resistance against multiple viruses simultaneously (Tricoli et al. 1995). Precision targeting of viral mechanisms is another major advantage. RNA interference (RNAi) constructs and CRISPR/Cas9-based editing can silence or disrupt viral genes, blocking replication and systemic spread. A notable example is the CRISPR-mediated knockout of the tomato *SlPelo* gene, which encodes the ribosome-recycling factor Pelota (PELO). Edited plant prevented tomato yellow leaf curl virus (TYLCV) from hijacking host translation machinery, conferring strong resistance (Pramanik et al. 2021). These approaches are particularly valuable in crops with limited natural resistance sources. In banana, CRISPR was used to inactivate integrated banana streak viruses (BSV's) sequences in the genome of plantain (*Musa* AAB). Edited plants remained largely asymptomatic under stress conditions, demonstrating the feasibility of safely using B-genome bananas in breeding programs (Tripathi et al. 2019). Similarly, RNAi-based strategies have been highly effective against banana bunchy top virus (BBTV). Transgenic banana plants expressing intron-hairpin RNA (ihpRNA) constructs targeting the viral Rep gene showed complete resistance to BBTV infection (Shekhawat et al. 2012). Durable resistance has also been demonstrated in tree fruits. The transgenic plum clone C5 (cv. HoneySweet), engineered to express sequences of plum pox virus (PPV), showed stable RNAi-mediated resistance for more than 12 years under natural infection pressure, even when exposed to mixed infections with other viruses such as prune dwarf virus (PDV) and apple chlorotic leaf spot virus (ACLSV) (Polák et al. 2017). Gene stacking strategies have further expanded resistance to multiple pathogens, as seen in transgenic squash engineered with viral CP genes from ZYMV, WMV2, and cucumber mosaic virus (CMV), which provided protection against all three viruses (Tricoli et al. 1995). Hence, genetic engineering enables the development of virus-resistant fruit crops that are not only rapid to generate but also durable and adaptable to diverse viral threats. The major strategies and case examples of transgenic approaches for virus resistance in fruit crops are summarized in Table 21.1.

Table 21.1 Virus resistance in fruit crops is achieved through genetic engineering

Fruit crop	Target virus	Viral gene target	Reference
Apple			
Fuji	Cucumber mosaic virus (CMV)	CMV-CP	Lee et al. (2003)
Banana			
Dwarf Brazilian	Banana bunchy top virus (BBTV)	BBTV *rep*	Borth et al. (2011)
Virupakshi	BBTV	BBTV *rep*	Elayabalan et al. (2012)
Rasthali	BBTV	BBTV Rep ihpRNA	Shekhawat et al. (2012)
Citrus			
Citrus sinensis cv. Valencia/ Hamlin	Citrus tristeza virus (CTV)	CTV-CP	Sun et al. (2019)
C. sinensis cv. Itaborai	CTV	CTV-392/393	
C. aurantifolia (Christm.) Swing	CTV	3DF1scFv	
C. aurantifolia (Christm.) Swing.; *C. paradisi* Macf.; *C. sinensis* (L.) Osbeck	CTV	p20/23/25	
C. sinensis (L.) Osbeck	Citrus psorosis virus (CPsV)	CPsV-CP ihpCP	
Grapes			
Rootstock 110 Richter (*Vitis Berlandieri* x *V. rupestris*)	Grapevine chrome mosaic virus (GCMV)	GCMV-CP	Le Gall et al. (1994)
V. vinifera (Nebbiolo, Lumassina, and Blaufrankisch)	Grapevine fanleaf virus (GFLV)	GFLV-CP	Gambino et al. (2005)
Papaya			
Sunset	Papaya ringspot potyvirus (PRSV)	PRSV-CP	Gonsalves et al. (1998)
Yuanyou No.1	PRSV	PRSV *rep*	Tuo et al. (2023)
Plum			
Bluebyrd	Plum pox virus (PPV)	PPV-CP	Ravelonandro et al. (1997) and GC (2023)

21.2 Transgenic Approaches for Virus Resistance

21.2.1 Pathogen-Derived Resistance (PDR)

Pathogen-derived resistance (PDR) was the first transgenic strategy that convincingly demonstrated the potential of genetic engineering for virus management in plants. The concept, first proposed by Sanford and Johnston (1985), suggested that parts of a viral genome could be introduced into the host plant to block infection by the same or related viruses. This approach soon proved effective in several crops, including fruit crops, and provided some of the earliest examples of practical virus resistance through genetic engineering. The principle behind PDR is that viral genes or gene fragments, when expressed in plants, can interfere with the replication, movement, or accumulation of the invading virus. Early efforts focused on the use of coat protein (CP) genes, which in many cases reduced viral infection by disrupting processes such as uncoating, replication, or assembly. Similar strategies were later applied using movement protein (MP) genes, which are required for virus trafficking through plasmodesmata (Lapidot et al. 1993), and replicase genes, which encode the RNA-dependent RNA polymerase essential for viral genome replication (Guo and García 1997). In each case, the transgene could disrupt critical steps in the infection cycle, thereby reducing the ability of the virus to spread and cause disease. Further research revealed that much of the resistance conferred by PDR was not due to the proteins themselves but rather to the RNA transcripts produced by the transgenes. This discovery led to the recognition of RNA-mediated resistance (RMR), also known as post-transcriptional gene silencing (PTGS) or RNA interference (RNAi). In this process, double-stranded or hairpin-like RNA molecules derived from viral gene fragments are processed into small interfering RNAs (siRNAs), which guide the degradation of complementary viral RNAs (Waterhouse et al. 1998; Baulcombe 1999). This mechanism provides a highly specific antiviral defense system that is effective against a broad range of plant viruses, including both RNA and DNA viruses. Several fruit crops provide striking examples of successful PDR. In papaya, the introduction of the PRSV coat protein gene produced cultivars such as "SunUp" and "Rainbow," which displayed strong resistance and rescued the Hawaiian papaya industry from devastation in the 1990s (Fitch et al. 1992; Gonsalves et al. 2004a, b; Jia et al. 2017). In citrus, transgenic Mexican lime plants expressing the *Citrus tristeza virus* coat protein showed variable but significant resistance, with some lines exhibiting symptom attenuation and others demonstrating immunity to certain viral strains (Domínguez et al. 2002). In plum, the cultivar "HoneySweet" (also known as C5) was engineered with a PPV coat protein gene. Resistance in this case was later shown to be RNAi-mediated, and the trees have remained stably resistant for more than a decade under natural infection pressure in both Europe and the United States (Polák et al. 2017). These achievements established PDR as the foundation of transgenic virus resistance in fruit crops. The success of papaya, citrus, and plum illustrates not only the durability and effectiveness of PDR but also its pivotal role in shaping the development of next-generation approaches, including targeted genome editing with CRISPR.

21.2.2 Non-pathogen-Derived Strategies

Non-pathogen-derived strategies for developing virus resistance represent a newer generation of transgenic and genome-editing approaches that go beyond the direct use of viral genes. While pathogen-derived resistance has been highly successful, especially in papaya and plum, its reliance on viral sequences sometimes raises concerns about durability, strain specificity, and biosafety. To address these challenges, researchers have developed strategies that exploit host-based mechanisms or artificial constructs to achieve targeted and often broader resistance. One such approach involves the use of artificial microRNAs (amiRNAs), which are engineered small RNAs designed to mimic endogenous plant microRNAs while specifically targeting viral genomes. AmiRNAs offer high precision because a single base mismatch can abolish their activity, allowing them to be tailored for strain-specific or broad-spectrum resistance. They have been successfully applied against several plant viruses, including those infecting fruit crops, by silencing essential viral genes (Niu et al. 2006; Qu et al. 2007). Compared to conventional RNA interference, amiRNAs require shorter sequences, minimize the risk of off-target effects, and provide a more stable form of silencing, making them an attractive alternative for transgenic virus resistance. Another promising method is host-induced gene silencing (HIGS), which takes advantage of the natural cross-kingdom movement of small RNAs. In this strategy, plants are engineered to produce double-stranded RNAs or small RNAs that target viral genes. During infection, these molecules are transferred into the virus or its replication intermediates, where they trigger RNA degradation. HIGS has been widely applied in fungal and insect pathogen systems, and in the case of plant viruses, it represents a non-pathogen-derived way to achieve durable resistance without necessarily inserting viral sequences into the plant genome. Recent advances in genome editing have further expanded non-pathogen-derived resistance strategies. The CRISPR/Cas system allows precise targeting of either viral genomes or host susceptibility genes. One application is the inactivation of integrated viral sequences, such as endogenous banana streak viruses (BSV's) elements found in the B genome of *Musa* species. These sequences can become activated under stress, leading to spontaneous infections. Using CRISPR/Cas9, Tripathi et al. (2019) successfully edited and disrupted endogenous BSV sequences in the banana genome containing the B genome, producing plants that remained symptom-free under conditions that would normally trigger viral activation. This achievement demonstrated the potential of genome editing to eliminate virus reservoirs from host genomes, which was not possible with conventional breeding or earlier transgenic methods. Another CRISPR-based approach involves knocking out host genes that viruses exploit during infection. For many RNA viruses, host translation initiation factors, especially *eIF4E* and its isoforms, are critical for viral replication and systemic spread. Mutations in these factors have been associated with natural resistance in several crops. Genome editing now allows targeted disruption of these susceptibility genes in elite cultivars without introducing foreign DNA. For example, editing of the *eIF4E* gene has conferred resistance to potyviruses in solanaceous crops (Chandrasekaran et al. 2016), and similar strategies are being explored for fruit crops where natural resistance alleles are absent or difficult to introgress. Together, these non-pathogen-derived strategies highlight the progression of virus resistance research from transgene-based expression of viral genes to precise manipulation of host and viral targets. These innovations not only overcome some of the biological and regulatory limitations of pathogen-derived resistance but also pave the way for next-generation fruit crops with improved viral resilience and wider acceptance.

21.3 Safety Assessment of Transgenic Fruits

The commercialization of transgenic fruits requires a comprehensive safety assessment to ensure that they are as safe for human consumption and the environment as their conventional counterparts. This process generally involves three major dimensions: molecular and genetic characterization of the transgene, food and feed safety evaluation, and environmental risk assessment, all of which are guided by international biosafety frameworks such as those of the Codex Alimentarius and the Cartagena Protocol on Biosafety (CAC 2003; CBD 2000). At the molecular level, characterization focuses on identifying the nature of the inserted DNA, its copy number, genomic location, and stability across generations. Techniques such as Southern blotting, PCR, and next-generation sequencing are commonly used to confirm that the transgene is stably inherited and does not disrupt essential plant genes (Chowdhury et al. 2009). This provides a baseline assurance of genetic integrity before food safety and environmental studies are conducted. Food and feed safety assessments include tests for potential toxicity, allergenicity, and nutritional equivalence. Transgenic fruits are compared with their non-transgenic counterparts to ensure that they do not contain unintended toxins or allergenic proteins and that their macronutrient and micronutrient profiles remain within natural variation. For example, studies with transgenic plum (C5) and papaya (*SunUp* and *Rainbow*) cultivars confirmed no differences in allergenicity, toxicity, or nutritional content compared with conventional crops, supporting their safety for human consumption (APHIS 2007; Fermín et al. 2011). Similar compositional analyses have been conducted for other engineered fruits to demonstrate equivalence with their non-GM counterparts (Su et al. 2020). Environmental biosafety is another critical component, particularly for perennial fruit crops that may persist for decades in orchards and vineyards. Risk assessments evaluate whether transgenic fruits could affect non-target organisms, promote weediness, or facilitate gene flow to wild relatives or conventional cultivars. For viral resistance traits, particular attention is given to the possibility of viral recombination or the emergence of novel viral strains. However, long-term monitoring of virus-resistant papaya and plum has not provided evidence of such risks, and transgenic plants often reduce viral inoculum pressure in the environment by limiting host susceptibility (Gonsalves 2006; Polák et al. 2017). The study found that transgenic grapevines and plums expressing viral coat protein genes had no detectable effect on the emergence of recombinant virus species or on non-target arthropod populations over multi-year field trials (Vigne et al. 2004; Fuchs et al. 2007). No translocation of transgene-derived products to non-transgenic scions occurred, and engineered resistance remained stable even in co-infected plants. Another innovative example is the use of trans-grafting in stone fruits, where a transgenic rootstock expressing RNAi constructs confers resistance to *Plum pox virus* (PPV) in a non-transgenic scion. This approach reduces regulatory concerns while still providing the benefits of engineered resistance (Scorza et al. 2013). Taken together, the accumulated evidence from molecular, nutritional, and ecological studies strongly supports the safety of virus-resistant transgenic fruits. Where deployed, they have not posed risks to consumers or the environment, and in many cases, they have contributed positively to disease management by reducing reliance on chemical inputs and limiting the spread of destructive viruses (Tome et al. 2024).

21.4 Regulatory Landscapes and Policy Perspectives

21.4.1 Defining GMOs and Gene-Edited Organisms

The regulation of transgenic and gene-edited fruits remains a critical factor influencing their development, commercialization, and public acceptance. A central challenge in regulation is the definition of what constitutes a genetically modified organism (GMO) and how to classify products derived from new breeding technologies such as genome editing. Traditional GMOs are typically defined as organisms containing foreign DNA introduced by recombinant DNA techniques. In contrast, gene-edited organisms produced using site-directed nucleases such as CRISPR/Cas9 may not contain foreign sequences, particularly in the case of SDN-1 and SDN-2 modifications, which introduce small deletions or substitutions indistinguishable from natural mutations (Ahmad et al. 2023). This distinction has created regulatory divergence, as some regions regulate based on the process used and others on the product developed.

21.4.2 Regional Regulatory Frameworks

The cultivation of GM crops has been authorized in over 30 countries as of October 2024 (Tome et al. 2024). The European Union follows a restrictive, process-based approach, classifying all genome-edited plants as GMOs under the 2018 Court of Justice ruling (Court of Justice of the EU 2018), though reforms are under discussion (European Commission 2021). North America applies product-based systems: the US exempts some gene-edited crops without foreign DNA from GMO regulation (Waltz 2016), while Canada regulates all "plants with novel traits" regardless of the method (CFIA 2025). Latin America adopts case-by-case frameworks, with Argentina, Brazil, and Chile exempting genome-edited crops from GMO oversight if no foreign DNA remains (Candia et al. 2020). Africa has emerging frameworks: South Africa regulates GM crops, while Nigeria and Kenya recently advanced biosafety approvals; policies for gene editing remain under development (Schmidt et al. 2020). In the Asia-Pacific, approaches are diverse: China is streamlining approvals for gene-edited crops (Zhu 2022; GLP 2025), Japan exempts certain CRISPR plants without foreign DNA (Tsuda et al. 2019), and Australia and the Philippines apply pragmatic, science-based models (AG 2020). An overview of regional regulatory frameworks for GM and gene-edited fruits is presented in Table 21.2, highlighting the contrasting process-based and product-based approaches.

Table 21.2 Global regulatory approaches regarding GM crops. (Adopted from Schmidt et al. 2020)

Country/countries/ region	Stance	Remark
Canada	Not regulated unless trait identified as novel	Genome-edited crops are not regulated as GMOs
USA	Most non-transgenic plants not regulated	
Colombia, Brazil, Argentina, Chile	Case-by-case: If no foreign DNA, then not regulated as GMO	
Israel	If no foreign DNA, then not regulated as GMO	
Australia	If no foreign DNA, then not regulated as GMO	
Japan	If no foreign DNA, then not regulated as GMO	
Norway	Proposal: If no foreign DNA, then not regulated as GMO	Discussion is ongoing
UK	Discussion ongoing for post-Brexit	
Nigeria, Kenya	Likely case-by-case: If no foreign DNA, then not regulated as GMO	
Paraguay, Uruguay	Likely case-by-case: If no foreign DNA, then not regulated as GMO	
India, Bangladesh	GMO definition encompasses genome editing. Discussion is ongoing	
Philippines, Indonesia	Discussion is ongoing	
EU	Genome-edited crops are regulated as GMOs	Genome-edited crops are regulated as GMOs
New Zealand	Genome-edited crops are regulated as GMOs	

21.4.3 Need for Harmonization and Science-Based Regulation

Despite this diversity, there is broad recognition of the need for harmonization and science-based regulation. Divergent rules across regions complicate international trade, delay deployment of virus-resistant fruit crops, and create uncertainty for both developers and farmers. Science-based risk assessment frameworks emphasize evaluating the product rather than the process, focusing on measurable risks such as allergenicity, toxicity, and environmental impact (NASEM 2016). Greater international alignment would reduce regulatory asymmetry, support innovation, and enable the safe and responsible use of both transgenic and gene-edited fruits to address challenges of food security, climate resilience, and viral disease management. Harmonized policies would reduce uncertainty, facilitate global acceptance, and accelerate the safe adoption of both transgenic and gene-edited fruits.

21.5 Broader Implications

21.5.1 Role in Reducing Food Losses and Waste

Virus-resistant transgenic and gene-edited fruits have the potential to significantly reduce food losses, which account for nearly one-third of global fruit production. Viral diseases such as PRSV, BBTV, and PPV can devastate orchards and plantations, leading to massive yield losses and shortened shelf life. By providing durable resistance, biotechnological solutions not only safeguard yields but also stabilize fruit supply chains, reduce waste at both farm and market levels, and enhance farmer income.

21.5.2 Impacts on Agrobiodiversity and Genetic Erosion

Concerns are often raised that widespread adoption of transgenic cultivars could narrow genetic diversity, increasing vulnerability to new pathogens. However, in practice, virus-resistant fruits can contribute to agrobiodiversity conservation. For instance, transgenic papaya preserved Hawaii's papaya industry and prevented genetic erosion by maintaining traditional cultivars like "Kapoho" in production. Similarly, resistant lines can complement, rather than replace, existing germplasm, serving as a tool to preserve heritage varieties by integrating resistance traits into them through trans-grafting or targeted editing. Balanced deployment strategies are therefore essential to enhance, rather than reduce, genetic diversity.

21.5.3 Influence of Advocacy Groups

Advocacy groups play an important role in shaping public perception and regulatory outcomes. Supportive organizations emphasize the contributions of GM crops to food security, reduced pesticide use, and resilience against climate change. In contrast, opposing groups raise concerns about biosafety, corporate dominance, and consumer choice. These debates often extend beyond scientific evidence, highlighting issues of trust, transparency, and the role of multinational companies. Effective science communication, stakeholder engagement, and participatory decision-making are essential to address polarized narratives and ensure that farmers and consumers are well-informed. Several organizations have played influential roles in shaping global debates on food security, future foods, and the use of GM crops, either in support or in opposition (Table 21.3).

21.5.4 Ethical and Socio-economic Considerations

Beyond science and regulation, ethical and socio-economic dimensions strongly influence the adoption of transgenic and gene-edited fruits. Equity and access are

Table 21.3 List of some organizations raising food security concerns, promoting future foods, and advocating for or against GM crops

Organizations raising food security issues			
Organization/Body/ Institution	Head office	Concern	
Action Against Hunger	New York, USA	Food security	
FIAN International	Heidelberg, Germany		
Food and Agriculture Organization (FAO)	Rome, Italy		
Food Security Coordination Office (Ethiopia)	Addis Ababa, Ethiopia		
Global Agriculture and Food Security Program (GAFSP)	Washington D.C., USA		
Integrated Food Security Phase Classification (IPC)	Rome, Italy		
International Land Coalition	Rome, Italy		
La Via Campesina	Jakarta Selatan, Indonesia		
Office of Global Food Security (U.S. State Department)	Washington, D.C., USA		
World Food Programme (WFP)	Rome, Italy		
Organizations promoting future foods			
Organization	*Head office*	*Key focus*	*Type*
Air Protein	USA	Carbon-negative protein from CO2, backed by NASA.	Lab-Grown Protein
Aleph Farms	Rehovot, Israel	Premium cultivated beef/ lamb, sustainability-focused.	Lab-Grown Meat
BlueNalu	San Diego, USA	Cell-based tuna, mahi-mahi, addressing overfishing.	Lab-Grown Seafood
ByFlow	Eindhoven, Netherlands	Focus on sustainable food shapes using fresh/discarded ingredients.	3D Food Printing
Digital Food Processing Initiative (DFPI)	Leuven, Belgium	Advances 3D-printed food for personalized nutrition.	Consortium
EIT Food	Leuven, Belgium	Funds projects on alternative proteins, sustainability, and future foods.	EU Innovation Network

(continued)

Table 21.3 (continued)

Organizations raising food security issues			
Organization/Body/ Institution	Head office	Concern	
Food 4 Future Expo	Bilbao, Spain	Awards innovations in FoodTech, including lab-grown and insect-based foods.	Industry Platform
Food Ink.	Netherlands	First 3D-printing restaurant with edible sculptures.	3D-Printed Dining
Good Food Institute (GFI)	Washington D.C., USA	Promotes lab-grown meat and alternative proteins.	Advocacy & Research
Gourmet Grubb	Cape Town, South Africa	Produces Entomilk (insect-based milk) and ice cream from sustainable bug protein.	Food Startup
Institute for Stem Cell Biology and Regenerative Medicine (InSTEM)	Bangalore, India	Pioneered research on cockroach milk protein crystals as a complete food supplement.	Research Institution
Institute of Food Technologists (IFT)	Chicago, USA	Advances food science, including novel superfoods and tech-driven solutions.	Professional Society
Meatable	Delft, Netherlands	Cultivated pork & beef using opti-ox tech, scaling in Europe.	Lab-Grown Meat
Mosa Meat	Maastricht, Netherlands	Cultivated beef production, cost reduction via plant-based growth media.	Lab-Grown Meat
Natural Machines (Foodini)	Barcelona, Spain	Creates 3D printers for customized plant-based and insect-enriched foods.	3D Food Printing
NOVAMEAT	Barcelona, Spain	Plant-based steak with fibrous texture.	3D-Printed Plant-Based Meat
NuGluCol	Spain	Develops cholesterol-lowering seaweed-based food tech.	Health Food Startup
Procusini	Freising, Germany	Chocolate, pasta, and minced meat 3D printers.	3D Food Printing
Revo Foods	Vienna, Austria	Vegan salmon & tuna via 3D printing.	3D-Printed Plant-Based Seafood
Savor	California, USA	Creating alternative fats without using animals, plants, or farmland. Backed by Bill Gates.	Food Tech Startup

(continued)

Table 21.3 (continued)

Organizations raising food security issues			
Organization/Body/ Institution	Head office	Concern	
Steakholder Foods	Ness Ziona, Israel	Bioprinted plant-based & cultivated meats.	3D-Printed & Lab-Grown Meat
The Good Food Institute (GFI)	Washington D.C., USA	Promotes cultivated meat and alternative proteins, including policy support.	Advocacy & Research
UPSIDE Foods	Berkeley, USA	FDA-approved cultivated chicken, US market leader.	Lab-Grown Meat
UWE Bristol Research Team	Bristol, UK	Develops 3D-printed meals for dysphagia patients.	Academic/ Research
Vow	Sydney, Australia	Cultivated kangaroo, alpaca, zebra for gourmet markets.	Lab-Grown Exotic Meat
Wageningen University & Research	Wageningen, Netherlands	Studies insects as food, including cockroach milk scalability challenges.	Academic/ Research
Organizations advocating for GM crops			
Organization/Body/ Institution	*Head office*	*Key advocacy/Position*	
Academy of Science of South Africa	Pretoria, South Africa	Advocates for GM crops to address African agricultural challenges.	
American Association for the Advancement of Science (AAAS)	Washington, D.C., USA	Supports GM crop safety, opposes mandatory labeling.	
Biotechnology Innovation Organization (BIO)	Washington, D.C., USA	Advocates for biotech policies, supports GM crop innovation.	
Brazilian Academy of Sciences	Rio de Janeiro, Brazil	Supports GM crop adoption for food security.	
European Academies Science Advisory Council (EASAC)	Halle (Saale), Germany	Represents 29 EU science academies supporting GM crop research.	
Indian National Academy of Agricultural Sciences	New Delhi, India	Advocates for GM crops to improve nutrition and yield.	
International Service for the Acquisition of Agri-biotech Applications (ISAAA)	Los Baños, Laguna, Philippines	Promotes biotech solutions for food security, publishes GM crop adoption data.	
National Academy of Sciences (NAS)	Washington, D.C., USA	Endorses GM crops as safe and beneficial.	
Pontifical Academy of Sciences (Vatican)	Vatican City, Vatican City	Endorses GM crops for global food security.	
Royal Society (UK)	London, United Kingdom	Publishes reports supporting GM crop safety and benefits.	
Organizations against GM crops			
Organization/Body/ Institution	*Head office*	*Key opposition stance*	

(continued)

Table 21.3 (continued)

Organizations raising food security issues		
Organization/Body/ Institution	Head office	Concern
Center for Food Safety (CFS)	Washington, D.C., USA	Legal advocacy against GMOs, promotes organic-only agriculture, opposes GMO labeling exemptions.
Environmental Working Group (EWG)	Washington, D.C., USA	Advocates against GMOs, promotes organic farming, publishes the "Dirty Dozen" list of pesticide-laden produce.
Greenpeace International	Amsterdam, Netherlands	Actively campaigns against GM crops, including protests and field destruction; opposes Golden Rice.
Green America	Washington, D.C., USA	Advocates for regenerative organic agriculture, opposes GMOs via the "GMO Inside" campaign.
Just Label It (Housed at EWG)	Washington, D.C., USA	Campaigns for mandatory GMO labeling, funded by organic industry groups.
Consumers Union (Consumer Reports)	Yonkers, New York, USA	Criticizes GM food safety testing, supports mandatory labeling, and promotes organic alternatives.
Food Democracy Now	Clear Lake, Iowa, USA	Opposes corporate control of agriculture, advocates for non-GMO food systems.
Non-GMO Project	Bellingham, Washington, USA	Certifies non-GMO products, promotes GMO transparency and labeling.
Soil Association	Bristol, UK	Campaigns against GM crops, supports organic farming, opposes gene-editing deregulation.
GM-Free Australia Alliance	Australia	Advocates for GM-free food policies, publishes shopping guides for non-GMO products.
Centre for Environment (CZŽS)	Czech Republic	Campaigns against GM crops, promotes organic farming and biodiversity.
Australian Greens	Australia	Political party advocating for a moratorium on GMOs, strict labeling, and bans on gene patents.

central issues: while large corporations dominate initial technology development, public-sector initiatives and partnerships have been crucial in ensuring that smallholder farmers also benefit, as seen in publicly funded papaya transformation programs in Hawaii and China. Concerns about corporate control, intellectual property, and seed sovereignty remain, particularly in developing regions where farmers rely on saving and sharing planting materials. To maximize benefits, policies must encourage open access, public–private collaboration, and region-specific solutions that align with local agricultural systems.

21.6 Conclusions and Future Perspectives

The development of virus-resistant fruit crops through transgenic and genome-editing technologies has transformed our ability to manage some of the most destructive plant diseases. From the success of transgenic papaya in Hawaii to RNAi-mediated resistance in plum and CRISPR-based inactivation of BSVs, these advances demonstrate that modern biotechnology can provide durable, precise, and sustainable solutions where conventional breeding has failed. The capacity to engineer resistance directly into elite cultivars or to use trans-grafting strategies offers new opportunities to protect fruit production without compromising quality or consumer preference. Safety assessments conducted over the past three decades consistently show that virus-resistant transgenic fruits are as safe as their conventional counterparts in terms of nutrition, allergenicity, and environmental impact. Nevertheless, regulatory diversity remains a major barrier to global deployment. The coexistence of restrictive, process-based systems in regions such as the European Union with more pragmatic, product-based approaches in North and South America has created uncertainty and slowed adoption. Harmonization of science-based regulatory frameworks, focusing on product characteristics rather than breeding method, will be essential for realizing the full potential of these technologies. Looking forward, genome editing offers new prospects for precision and acceptability. Site-directed nuclease (SDN)-based approaches, particularly SDN-1 and SDN-2, create mutations indistinguishable from natural variation, offering a regulatory and social advantage. Combining gene editing with RNAi, gene stacking, and host factor knockouts will enable broader and more durable resistance to multiple viruses. At the same time, public-sector initiatives and international collaborations must ensure equitable access, particularly for smallholder farmers in developing countries where viral diseases cause the most severe losses. Ethical, socio-economic, and advocacy dimensions will continue to shape the trajectory of these crops. Ensuring transparency, farmer participation, and consumer trust is as important as technological progress. Addressing concerns over corporate control, genetic erosion, and long-term ecological impacts will be necessary to secure broad acceptance. With climate change expected to exacerbate viral disease pressures, biotechnological tools will play an increasingly critical role in safeguarding global fruit production. In conclusion, transgenic and gene-edited approaches offer powerful, science-driven solutions for managing viral diseases in fruit crops. Their responsible development, combined with balanced regulation, inclusive policies, and effective communication, can help secure resilient fruit systems, reduce food losses, and support global food and nutritional security in the decades to come.

References

AG (2020) Gene Technology Amendment (2019 Measures No. 1) Regulations 2019. Federal Register of Legislation. Australian Government. Retrieved November 25, 2025, from https://classic.austlii.edu.au/au/legis/cth/num_reg_es/gta2019mn1r2019201900573487.html

Ahmad A, Jamil A, Munawar N (2023) GMOs or non-GMOs? The CRISPR conundrum. Front Plant Sci 14. https://doi.org/10.3389/fpls.2023.1232938

APHIS (2007) Approval of USDA-ARS request (04-264-01P) seeking a determination of non-regulated status for C5 plum resistant to plum pox virus. Animal and Plant Health Inspection Service/U.S. Department of Agriculture. Retrieved September 18, 2025, from https://www.aphis.usda.gov/sites/default/files/04_26401p_com.pdf

Baulcombe DC (1999) Fast forward genetics based on virus-induced gene silencing. Curr Opin Plant Biol 2(2):109–113. https://doi.org/10.1016/s1369-5266(99)80022-3

Borth W, Perez E, Cheah K, Chen Y, Xie W, Gaskill D, Khalil S, Sether D, Melzer M, Wang M, Manshardt R, Gonsalves D, Hu J (2011) Transgenic banana plants resistant to banana bunchy top virus infection. Acta Hortic 897:449–457. https://doi.org/10.17660/actahortic.2011.897.61

Cabrera JÁ, Carbonell P, Salinas JF, Grau A et al (2024) Effect of introgression of Ty-1 and ty-5 genes on productivity, quality, and antioxidant compounds in De la Pera Tomato breeding lines. Agriculture 14(7):1192. https://doi.org/10.3390/agriculture14071192

CAC (2003) Guideline for the conduct of food safety assessment of foods derived from recombinant-DNA plants, CAC/GL 45-2003. Codex Alimentarius Commission, Geneva. Retrieved September 17, 2025, from https://www.fao.org/fao-who-codexalimentarius/sh-proxy/en/?lnk=1&url=https%253A%252F%252Fworkspace.fao.org%252Fsites%252Fcodex%252FStandards%252FCXG%2B45-2003%252FCXG_045e.pdf

Candia NB, Ríos DF, Vicién C (2020) Paraguay's path toward the simplification of procedures in the approval of GE crops. Front Bioeng Biotechnol 8. https://doi.org/10.3389/fbioe.2020.01023

CBD (2000) Cartagena protocol on biosafety to the convention on biological diversity: text and annexes. Secretariat of the Convention on Biological Diversity, Montreal. https://www.cbd.int/doc/legal/cartagena-protocol-en.pdf

CFIA (2025) Plants with novel traits. Canadian Food Inspection Agency. Government of Canada. Retrieved August 06, 2025, from https://inspection.canada.ca/en/plant-health/plant-varieties/novel-traits

Chandrasekaran J, Brumin M, Wolf D, Leibman D et al (2016) Development of broad virus resistance in non-transgenic cucumber using CRISPR/Cas9 technology. Mol Plant Pathol 17(7):1140–1153. https://doi.org/10.1111/mpp.12375

Chowdhury MK, Chowdhury AC, Hoque MI & Sonnino A. (Eds.). (2009). Biosafety of genetically modified organisms: basic concepts, methods and issues (By Bangladesh Agricultural Research Council (BARC) & Food and Agriculture Organization of the United Nations). Food and Agriculture Organization of the United Nations. https://www.fao.org/4/i1252e/i1252e.pdf

Dela Cueva FM, Perez NAM, Benjamin AFA, Yanos LA, Gueco LS, Thomas JE (2023) Resistance of *Musa balbisiana* accessions of The Philippines to *Banana bunchy top virus*. Plant Dis 107(7):1973–1978. https://doi.org/10.1094/pdis-10-22-2427-sc

Domínguez A, de Mendoza AH, Guerri J, Cambra M et al (2002) Pathogen-derived resistance to *Citrus tristeza virus* (CTV) in transgenic Mexican lime (Citrus aurantifolia (Christ.) Swing.) plants expressing its p25 coat protein gene. Mol Breed 10:1–10. https://doi.org/10.1023/a:1020347415333

Elayabalan S, Kalaiponmani K, Subramaniam S, Selvarajan R et al (2012) Development of Agrobacterium-mediated transformation of highly valued hill banana cultivar Virupakshi (AAB) for resistance to BBTV disease. World J Microbiol Biotechnol 29(4):589–596. https://doi.org/10.1007/s11274-012-1214-z

European Commission (2021) EC study on new genomic techniques. European Commission. Retrieved August 06, 2025, from https://ec.europa.eu/food/plant/gmo/modern_biotech/new-genomic-techniques_en

FAO (2023) Agricultural production statistics 2000–2022, FAOSTAT Analytical Briefs, No. 79. Food and Agriculture Organization of the United Nations, Rome. https://doi.org/10.4060/cc9205en

FAO (2025) Chapter 1. Fruits and vegetables: an overview on socio-economical and technical issues. Food and Agriculture Organization of the United Nations, Rome. Retrieved June 15, 2025, from https://www.fao.org/4/y4358e/y4358e04.htm

Fermín G, Keith RC, Suzuki JY, Ferreira SA et al (2011) Allergenicity assessment of the papaya ringspot virus coat protein expressed in transgenic rainbow papaya. J Agric Food Chem 59(18):10006–10012. https://doi.org/10.1021/jf201194r

Fitch MMM, Manshardt RM, Gonsalves D, Slightom JL, Sanford JC (1992) Virus resistant papaya plants derived from tissues bombarded with the coat protein gene of papaya ringspot virus. Nat Biotechnol 10(11):1466–1472. https://doi.org/10.1038/nbt1192-1466

Fuchs M, Cambra M, Capote N, Jelkmann W et al (2007) Safety assessment of transgenic plums and grapevines expressing viral coat protein genes: new insights into real environmental impact of perennial plants engineered for virus resistance. J Plant Pathol 89(1):5–12. http://www.jstor.org/stable/41998352

Gambino G, Gribaudo I, Leopold S, Schartl A, Laimer M (2005) Molecular characterization of grapevine plants transformed with GFLV resistance genes: I. Plant Cell Rep 24(11):655–662. https://doi.org/10.1007/s00299-005-0006-4

GC (2023) Novel food information: virus resistant C5 plum. Government of Canada. Retrieved September 16, 2025, from https://www.canada.ca/en/health-canada/services/food-nutrition/genetically-modified-foods-other-novel-foods/approved-products/virus-resistant-plum/document.html

GLP (2025) China: crops/food. Global Gene Editing Regulation Tracker Genetic Literacy Project. Retrieved September 18, 2025, from https://crispr-gene-editing-regs-tracker.geneticliteracyproject.org/china-crops-food/

Gonsalves D (2006) Transgenic papaya: development, release, impact and challenges. Adv Virus Res 67:317–354. https://doi.org/10.1016/S0065-3527(06)67009-7

Gonsalves D, Ferreira S, Manshardt R, Fitch M, Slightom J (1998) Transgenic virus resistant papaya: new hope for control of Papaya ringspot virus in Hawaii. Online. APSnet Feature, American Phytopathological Society. https://www.apsnet.org/edcenter/apsnetfeatures/Pages/Papaya.aspx

Gonsalves C, Lee DR, Gonsalves D (2004a) Transgenic virus-resistant papaya: the Hawaiian 'Rainbow' was rapidly adopted by farmers and is of major importance in Hawaii today. American Phytopathological Society (APS). https://doi.org/10.1094/APSnetFeature-2004-0804

Gonsalves D, Gonsalves C, Ferreira S, Pitz K, Fitch M, Manshardt R, Slightom J (2004b) Transgenic virus resistant papaya: From hope to reality for controlling papaya ringspot virus in Hawaii. American Phytopathological Society (APS). https://doi.org/10.1094/APSnetFeature-2004-0704

Guo HS, García JA (1997) Delayed resistance to plum Pox potyvirus mediated by a mutated RNA replicase gene: involvement of a Gene-Silencing mechanism. Mol Plant-Microbe Interact 10(2):160–170. https://doi.org/10.1094/mpmi.1997.10.2.160

Jia R, Zhao H, Huang J, Kong H, Zhang Y et al (2017) Use of RNAi technology to develop a PRSV-resistant transgenic papaya. Sci Rep 7(1). https://doi.org/10.1038/s41598-017-13049-0

Lapidot M, Gafny R, Ding B, Wolf S et al (1993) A dysfunctional movement protein of tobacco mosaic virus that partially modifies the plasmodesmata and limits virus spread in transgenic plants. Plant J 4(6):959–970. https://doi.org/10.1046/j.1365-313x.1993.04060959.x

Le Gall O, Torregrosa L, Danglot Y, Candresse T, Bouquet A (1994) Agrobacterium-mediated genetic transformation of grapevine somatic embryos and regeneration of transgenic plants expressing the coat protein of grapevine chrome mosaic nepovirus (GCMV). Plant Sci 102(2):161–170. https://doi.org/10.1016/0168-9452(94)90034-5

Lee CH, Hyung NI, Lee GP, Choi JY et al (2003) Transformation of fuji apple plant harbouring the coat protein gene of cucumber mosaic virus. Plant Pathol J 19(3):162–165. https://doi.org/10.5423/ppj.2003.19.3.162

Manzoor S, Nabi SU, Rather TR et al (2024) Advancing crop disease resistance through genome editing: a promising approach for enhancing agricultural production. Front Genome Ed 6. https://doi.org/10.3389/fgeed.2024.1399051

NASEM (2016) Genetically engineered crops: experiences and prospects. The National Academies Press, Washington, DC. https://doi.org/10.17226/23395

Niu Q-W, Lin S-S, Reyes JL, Chen K et al (2006) Expression of artificial micro RNAs in transgenic Arabidopsis thaliana confers virus resistance. Nat Biotechnol 24(11):1420–1428. https://doi.org/10.1038/nbt1255

Polák J, Kundu JK, Krška B, Beoni E et al (2017) Transgenic plum *Prunus domestica* L., clone C5 (*cv.* HoneySweet) for protection against sharka disease. J Integr Agric 16(3):516–522. https://doi.org/10.1016/s2095-3119(16)61491-0

Pramanik D, Shelake RM, Park J, Kim MJ, Hwang I, Park Y, Kim J (2021) CRISPR/Cas9-Mediated Generation of Pathogen-Resistant Tomato against *Tomato Yellow Leaf Curl Virus* and Powdery Mildew. Int J Mol Sci 22(4):1878. https://doi.org/10.3390/ijms22041878

Qu J, Ye J, Fang R (2007) Artificial MicroRNA-mediated virus resistance in plants. J Virol 81(12):6690–6699. https://doi.org/10.1128/jvi.02457-06

Ravelonandro M, Scorza R, Bachelier JC, Labonne G, Levy L, Damsteegt V, Callahan AM, Dunez J (1997) Resistance of transgenic prunus domestica to plum pox virus infection. Plant Dis 81(11):1231–1235. https://doi.org/10.1094/pdis.1997.81.11.1231

Reddy DVR, Sudarshana MR, Fuchs M, Rao NC, Thottappilly G (2009) Genetically engineered virus-resistant plants in developing countries: current status and future prospects. Adv Virus Res:185–220. https://doi.org/10.1016/s0065-3527(09)07506-x

Sanford JC, Johnston SA (1985) The concept of parasite-derived resistance—deriving resistance genes from the parasite's own genome. J Theor Biol 113(2):395–405. https://doi.org/10.1016/s0022-5193(85)80234-4

Sastry SK, Zitter TA (2014) Management of virus and viroid diseases of crops in the tropics. In: Plant virus and viroid diseases in the tropics, vol. 2, epidemiology and management. Springer, Dordrecht, pp 149–480. https://doi.org/10.1007/978-94-007-7820-7_2

Schmidt SM, Belisle M, Frommer WB (2020) The evolving landscape around genome editing in agriculture. EMBO Rep 21(6). https://doi.org/10.15252/embr.202050680

Scorza R, Callahan A, Dardick C, Ravelonandro M et al (2013) Genetic engineering of Plum pox virus resistance: 'HoneySweet' plum—from concept to product. Plant Cell Tissue Organ Cult (PCTOC) 115(1):1–12. https://doi.org/10.1007/s11240-013-0339-6

Shekhawat UKS, Ganapathi TR, Hadapad AB (2012) Transgenic banana plants expressing small interfering RNAs targeted against viral replication initiation gene display high-level resistance to banana bunchy top virus infection. J Gen Virol 93(8):1804–1813. https://doi.org/10.1099/vir.0.041871-0

Skyquest (2024) Fresh fruits market size, share, and growth analysis. Retrieved June 15, 2025, from https://www.skyquestt.com/report/fresh-fruits-market#:~:text=Global%20Fresh%20Fruits%20Market%20size,is%20moving%20the%20market%20forward

Su S, Ezhuthachan ID, Ponda P (2020) Genetically modified foods and food allergy. J Food Allergy 2(1):111–114. https://doi.org/10.2500/jfa.2020.2.200012

Sun L, Nasrullah N, Ke F, Nie Z, Wang P, Xu J (2019) Citrus genetic engineering for disease resistance: past, present and future. Int J Mol Sci 20(21):5256. https://doi.org/10.3390/ijms20215256

Tome KGN (2022) What is the difference between genetic engineering and gene editing? In: Science speaks. International Service for the Acquisition of Agri-biotech Applications (ISAAA). https://www.isaaa.org/blog/entry/default.asp?BlogDate=5/4/2022

Tome KG, Dionglay C, Escasura JC (2024) Countries approving GM crop cultivation. Science speaks. International Service for the Acquisition of Agri-biotech Applications (ISAAA). Retrieved September 18, 2025, from https://www.isaaa.org/blog/entry/default.asp?BlogDate=10/31/2024

Tricoli DM, Carney KJ, Russell PF, McMaster JR et al (1995) Field evaluation of transgenic squash containing single or multiple virus coat protein gene constructs for resistance to cucumber mosaic virus, watermelon mosaic virus 2 and zucchini yellow mosaic virus. Nat Biotechnol 13(12):1458–1465. https://doi.org/10.1038/nbt1295-1458

Tripathi JN, Ntui VO, Ron M, Muiruri SK et al (2019) CRISPR/Cas9 editing of endogenous banana streak virus in the B genome of Musa spp. overcomes a major challenge in banana breeding. Commun Biol 2(1). https://doi.org/10.1038/s42003-019-0288-7

Tsuda M, Watanabe KN, Ohsawa R (2019) Regulatory status of genome-edited organisms under the Japanese Cartagena Act. Front Bioeng Biotechnol 7. https://doi.org/10.3389/fbioe.2019.00387

Tuo D, Ma C, Yan P, Kong H, Zhou P, Guo A, Zhao H, Shen W (2023) Genetic transformation and gene delivery strategies in *Carica papaya* L. Trop Plants 2(1):0. https://doi.org/10.48130/tp-2023-0005

Vigne E, Komar V, Fuchs M (2004) Field safety assessment of recombination in transgenic grapevines expressing the coat protein gene of Grapevine fanleaf virus. Transgenic Res 13(2):165–179. https://doi.org/10.1023/b:trag.0000026075.79097.c9

Waltz E (2016) Gene-edited CRISPR mushroom escapes US regulation. Nature 532(7599):293. https://doi.org/10.1038/nature.2016.19754

Waterhouse PM, Graham MW, Wang M (1998) Virus resistance and gene silencing in plants can be induced by simultaneous expression of sense and antisense RNA. Proc Natl Acad Sci 95(23):13959–13964. https://doi.org/10.1073/pnas.95.23.13959

Zhu J (2022) The future of gene-edited crops in China. Natl Sci Rev 9(4). https://doi.org/10.1093/nsr/nwac063

Tsuda M, Watanabe KN, Ohsawa R (2019) Regulatory status of genome-edited organisms under the Japanese Cartagena Act. Front Bioeng Biotechnol 7. https://doi.org/10.3389/fbioe.2019.00387

[illegible]

[illegible]

[illegible]

[illegible]

Current and Emerging Biotech Solutions for Virus Resistance in Perennial Fruit Crops

22

Marcella S. Vieira, Giovanni Ferreira Montovaneli, Marlonni Maurastoni, Silas Pessini Rodrigues, and Patricia M. B. Fernandes

Abstract

Biotechnological advances have provided promising solutions to manage viral diseases affecting perennial fruit crops, a strategic sector of global agriculture. Innovative strategies, such as RNA interference (RNAi), CRISPR-based gene editing, gene pyramiding, and manipulation of the host microbiome, have emerged as key tools in the development of more resilient cultivars. Successful examples, such as *Carica papaya* L. resistant to papaya ringspot virus (PRSV) and *Prunus domestica* L. resistant to plum pox virus (PPV), demonstrate the potential of these cutting-edge technologies to produce high-quality, virus-resistant plants. These approaches are particularly relevant for perennial species, which present distinct challenges due to their long life cycles and the need for multi-year evaluations, factors that complicate both the development and regulatory approval of genetically modified (GM) cultivars. Furthermore, genetic variability and spontaneous mutations can drive the emergence of viral variants capable of overcoming resistance mechanisms developed through synthetic biology, raising concerns about the durability of these biotechnological solutions. Nevertheless, the integration of advanced biotechnologies remains essential for promoting a more sustainable and resilient perennial fruit sector. These tools contribute to the improvement of fruit-producing crops such as pome and stone

M. S. Vieira · P. M. B. Fernandes (✉)
Núcleo de Biotecnologia, Universidade Federal do Espírito Santo, Vitória, ES, Brazil
e-mail: patricia.fernandes@ufes.br

G. F. Montovaneli · S. P. Rodrigues
Núcleo Multidisciplinar de Pesquisa em Biologia, Universidade Federal do Rio de Janeiro, Duque de Caxias, RJ, Brazil

M. Maurastoni
Department of Entomology and Plant Pathology, 27695, North Carolina State University, Raleigh, NC, USA

R. Kumar et al. (eds.), *Fruit Crops and Viral Pathogens Volume II*, Plant Pathology for Sustainable Agriculture, https://doi.org/10.1007/978-981-95-7261-8_22

fruits, citrus, papaya, banana, and grapevine, all of which are vital for domestic and international trade, but face significant challenges due to their susceptibility to viruses. This chapter presents the major viral diseases impacting these economically important crops and explores the forefront of development in biotechnology applied to perennial fruit-bearing plants.

Keywords

RNAi · Fruit crops · CRISPR-based gene editing · Viral pathogens · Biotechnology · Antiviral resistance

22.1 Perennial Fruit Crops: Biological Characteristics and Agricultural Importance

Perennial fruit crops play a crucial role in global agriculture due to their long life cycles and sustained productivity over multiple years (Pimentel et al. 2012). Unlike annual and biennial crops, which require yearly replanting and often occupy large areas under monoculture, perennials contribute significant ecological benefits, including soil conservation, carbon sequestration, and biodiversity enhancement (Kreitzman et al. 2020). Their stable yields and diverse products make them important sources of nutrition and income, especially in regions where agriculture is a key economic activity (Smith 2011; Zhang et al. 2011). Perenniality is characterized by the cyclical alternation between vegetative growth and reproductive phases (flowering and fruiting). The juvenile phase can last from months to years, and after flowering, the plant reverts to vegetative growth and adapts its meristems, making it capable of flowering repeatedly. Perennial reproductive cycles are influenced by seasonal and genetic factors, as explained by Friedman (2022) in his detailed review of perennial plants. In technical terms, a perennial plant can live for more than two years, even though unproductive periods, and is usually polycarpic; unlike annuals and biennials (Wang and Ding 2023). They can be separated into two groups: herbaceous, which do not have secondary growth and usually have a root/rhizome system that keeps the plant alive underground (Shefferson 2009) despite adversities like cold temperatures or the chopping of its leaves, like banana (*Musa* spp.) and ginger (*Zingiber officinale*); and woody perennials, which have primary and secondary growth and a high-lignin stem that keeps these plants above ground even during temperature changes (Khan et al. 2025), like apple (*Malus domestica*) and mango (*Mangifera indica* L.) trees. According to the Food and Agriculture Organization of the United Nations (FAO), of the 20 most produced fruit crops, 16 are considered perennials. However, crops such as pineapple and pepper are commonly cultivated on an annual or semi-annual basis, despite being botanically perennial. This practice is due to their high susceptibility to diseases, pests, and temperature fluctuations. The production classification for these perennial crops in 2023 is illustrated in Fig. 22.1.

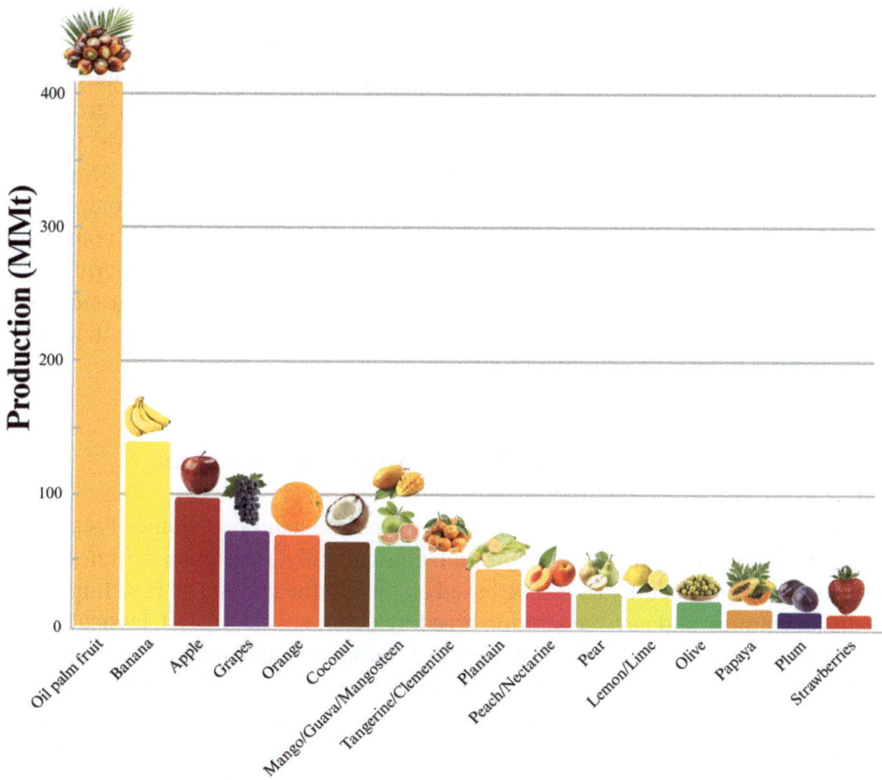

Fig. 22.1 World production of perennial fruit trees in 2023 (FAO 2023)

22.2 Viral Pathogens in Perennial Fructiferous Systems

Phytopathogens represent a central challenge to agricultural sustainability, affecting the productivity and commercial quality of crops. Disease management strategies are intrinsically linked to the host's biology, notably its life cycle (annual or perennial). In general, plant diseases cost US$220 billion per year to the economy (FAO 2019), and viruses are particularly significant, with economic losses in global agriculture estimated at up to US$30 billion annually (Sastry and Zitter 2014; Tatieni and Hein 2023). Viruses are obligate intracellular parasites, and plant viruses are composed mostly of RNA genomes, although species with DNA genomes also hold significant agricultural importance (Butkovic and González 2022). For primary infection to occur in nature, transmission relies on a vector that inoculates viral particles by causing an injury to the plant. The most common vectors are insects of the order Hemiptera, such as aphids, whiteflies, leafhoppers, and mealybugs (Catto et al. 2022). Once the infection is established in a susceptible host, the systemic infection stage begins. This is enabled by the virus's ability to synthesize movement proteins (MPs), allowing the traffic of virions or their genomes between adjacent cells through the plasmodesmata. For long-distance colonization, the plant's vascular system, primarily the phloem, is utilized. After an incubation period, symptoms manifest, whose severity can vary, requiring the implementation of control measures to mitigate damage. The longevity of perennial crops represents prolonged exposure to vectors and increases the risk of virus infections throughout the production season. As an example, the *C. papaya* L. plants typically require 9–15 months to mature from germination to harvest (Ming and Moore 2014). In contrast, the plum tree may begin flowering two years after germination, though it commonly takes up to four years (Neumüller 2011). This scenario is especially critical for crops of high commercial and social value fruits, such as pome and stone fruits, bananas, grapevines, citrus, and papayas. The most important viruses that affect these crops are detailed in Table 22.1.

22.3 Engineering Antiviral Resistance in Fruit Crops

Viral plasticity grants viruses the ability to overcome natural or induced resistance in plants, culminating in new variants and, consequently, new strains. Strategies for developing resistant cultivars can be based on three pillars: classical breeding, transgenic techniques, and next-generation breeding (Vieira et al. 2024). Classical breeding programs have been fundamental in the development and production of cultivars with high agronomic and economic quality, aiming to respond to the exponential population growth that demanded (and demands) global food security. Among breeding techniques, the cross between genetically distinct progenitors is termed hybridization. This process relies upon the selection of phenotypes of interest, such that the hybrid offspring exhibit the desired traits from both progenitors (Ribeiro and Costa 2018). The main bottleneck of conventional breeding lies in the time required: the market introduction of a new cultivar can take, on average, 8 to 12 years, which constrains the agricultural sector's ability to respond swiftly to emerging challenges, such as new pests and diseases (Wieczorek and Wright 2012). The breeding of perennial fruit trees encounters several challenges related to the biology of these organisms: high frequencies of self-incompatibility and heterozygosity, polyploidy, and the most significant challenge, the long juvenile phase, which can range from one to ten years in species like apple and citrus (Deng and Duan 2006; Callahan et al. 2016; Goulin et al. 2019). Within the context of next-generation breeding, this prolonged juvenility is one of the factors that reduces the efficiency of the CRISPR/Cas system (Wang et al. 2018). As a solution, CRISPR/Cas has been employed to induce early flowering by editing key repressor genes, notably TERMINAL FLOWER 1 (*TFL1*) Martín-Valmaseda et al. (2023) Charrier et al. (2019) validated the efficacy of this approach, and obtained early flowering in 93% of the apple and 9% of the pear specimens included in their study after editing the *TFL1* gene. In a similar breakthrough, editing genes of the CENTRORADIALIS (*CEN*) family in kiwifruit not only accelerated flowering and fruiting but also introduced the concept of "annualization" for these perennial species (Varkonyi-Gasic et al. 2018). These strategies reveal how gene editing technologies are essential not

Table 22.1 Perennial fructiferous crops and their main associated virus

Crop	Viruses	Yield losses	Disease control	Source
Pome fruits (*Malus domestica*; *Pyrus* spp.)	Apple mosaic virus (ApMV); apple stem grooving virus (ASGV); Apple chlorotic leaf spot virus (ACLSV)	ApMV–46% ASGV-30–40% ACLSV–30-40%	ApMV—Care on grafting trees, and roguing; ASGV/ACLSV—Virus certification before grafting	Cembali et al. (2003); Manzoor et al. (2023); Wu et al. (1997); Peter (2024)
Banana and plantain (*Musa* spp.)	Banana bunchy top virus (BBTV); Banana streak virus (BSV)	BBTV—Complete loss (of yield and plant) BSV—6%–15% (90% if not treated)	BBTV—Destroying diseased plants, controlling aphid population, and planting virus-free plants BSV—All the above, plus carefulness with management of tools used in diseased plants	Qazi (2016); Daniells et al. (2001); Wambulwa et al. (2013)
Citrus (*Citrus* spp.)	Citrus tristeza virus (CTV); citrus psorosis virus (CPsV)	CTV—Complete loss (of yield and plant) CPsV—35% loss	CTV—Quarantine and certification of budwood to be free of virus. Destruction of infected trees can delay or avoid a pandemic CPsV—Quarantine and certification of budwood to be free of virus	Moreno et al. (2008); Belabess et al. (2020); Moore and Nauer (1957)
Grapevine (*Vitis vinifera* L.)	Grapevine leafroll-associated virus (GLRaVs); Grapevine fanleaf virus (GFLV)	GLRaV-1—20-30% GLRaV-3—20-40% GFLV—Up to 80%	GLRaV—Roguing and planting virus-free plants GLFV—The virus and the nematode are necessary to induce the disease. Control methods are limited; nematocides and roguing are some of these, but they are not that effective. Some efforts on cross-breeding with GLFV- resistant strains but are still ineffective.	Andret-Link et al. (2004); Maree et al. (2013); Krebelj et al. (2015); Mannini and Digiaro (2017); Almeida et al. (2013)

(continued)

Table 22.1 (continued)

Crop	Viruses	Yield losses	Disease control	Source
Papaya (*Carica papaya* L.)	Papaya meleira virus (PMeV); Papaya ringspot virus (PRSV)	PMeV—Complete loss (of yield and plant); PRSV—complete loss (of yield and plant)	PMeV—Roguing in the early stages of the disease; PRSV—Roguing in the early stages of the disease, Breeding, conventional methods, transgenic and non-transgenic	Abreu et al. (2015); Gonsalves et al. (2010); Jyotika et al. (2024); Soares et al. (2017)
Stone fruits (*Prunus* spp.)	Plum pox virus (PPV); Prunus necrotic ringspot virus (PNRSV)	PPV—0.6 M m t/ year (apricot) 1.5 M m t/ year (plum) (90% yield loss per tree) 5% loss a year (peaches) PNRSV—20-60% depending on the host	PPV—Field surveys, control of prunus nurseries, use of diagnostics for virus identification, and eradication methods when the disease is present. PNRSV—Virus-free plants and rootstocks, and if it's known to have other plants with this disease in the cultivar, need to be chopped down because it can also disseminate via pollen.	Cambra et al. (2006); Chirkov et al. (2023); Pedrelli et al. (2024)

only for accelerating breeding cycles but also for engineering desirable fruit traits and ensuring durable resistance to viruses. This chapter will focus on transgenic and genome editing methodologies, as these approaches allow for the introduction or modification of traits with greater precision and in significantly shorter timeframes than conventional breeding. A clear distinction between these two strategies is essential. According to the World Health Organization (WHO), a Genetically Modified Organism (GMO) is an organism whose genetic material (DNA) has been altered in a way that does not occur naturally (WHO, 2014). In scientific contexts, the term GMO often specifically refers to transgenic organisms, which are created using recombinant DNA technology to insert exogenous DNA (a transgene) from a different species. In contrast, Genome Editing (GEd), primarily using tools like CRISPR, is considered a form of next-generation breeding. It works by making precise, targeted changes to the organism's own genome, typically without inserting foreign DNA. The resulting modifications are often indistinguishable from those that could potentially occur in nature or through conventional breeding, only achieved much more rapidly and efficiently. The application of these technologies is particularly crucial for controlling viral diseases due to their potential for rapid spread and significant economic impact. This is especially pertinent for long-lived perennial species, which require several years to begin fruiting; consequently, the loss of even a single tree represents a substantial loss of invested resources. On the other hand, studies with "HoneySweet" plum plants demonstrate that a transgenic tree, even after 12 years of cultivation under constant virus pressure and continuous aphid transmission, can maintain its health, and by the eighth year shows no symptoms of Sharka disease (Polák et al. 2017). Furthermore, genetic engineering and gene editing enable the utilization of resistance sources inaccessible through traditional breeding, thus expanding the range of available solutions to protect crops and ensure agricultural sustainability.

22.3.1 Developing Virus-Resistant Varieties via Transgenic RNAi

The GMOs gained momentum in the 1990s. This technology was the basis for the success of the "Rainbow", "SunUp", and "Huanong No.1" papaya plant varieties, the "HoneySweet" plum, and virus-resistant banana plants, through techniques such as RNA interference (RNAi) (Gonsalves 2006; Wu et al. 2018; Scorza et al. 1995; Shekhawat et al. 2012). RNAi, the mechanism through RNA silencing response, is the most widely used mechanism to confer virus resistance in genetically modified plants (Akbar et al. 2022; Fernandes et al. 2025). Besides the already approved and commercialized papaya and plum plant varieties available according to the International Service for the Acquisition of Agri-biotech Applications (ISAAA), research on other perennial fruit crops, such as banana plants resistant to the banana bunchy top virus (BBTV), also demonstrates the technique's effectiveness (Shekhawat et al. 2012). The application of this technology explores an innate plant defense mechanism, studied and characterized in the model organism *Arabidopsis thaliana* (Agius et al. 2012). Generally, the recognition of pathogen-derived double-stranded RNA (dsRNA) by the cell initiates the defense response (Fig. 22.2). The core molecular components participating in this pathway include small RNAs (sRNAs), categorized as small interfering (siRNAs, 21–25 nt), and microRNAs (21 nt), as well as proteins from the Dicer-like (DCL), Argonaute (AGO), and RNA-dependent RNA polymerase (RDRp) families (Agius et al. 2012). The presence of dsRNA acts as an alert to the host through a signal called virus-associated molecular pattern (VAMP) (Pumplin and Voinnet 2013). DCL proteins, specifically DCL4 and DCL2, are type III ribonucleases responsible for recognizing the VAMP signal and processing viral dsRNA and converting it into viral small interfering RNAs (vsiRNAs). In this process, DCLs are assisted by accessory double-stranded RNA-binding proteins (DRBs), ensuring a precise cleavage of the nucleic acid. The vsiRNAs are then recognized by endonucleases of the Argonaute family, such as AGO1 and AGO2 (Wang et al. 2011). The association of a vsiRNA with an AGO protein forms the RNA-induced silencing complex (RISC), which is guided to cleave the corresponding viral mRNA, thereby interrupting its translation. To ensure the process's efficiency, the RNA silencing signal is amplified by RDR proteins, which synthesize new dsRNA molecules from the viral RNA (Van-Esse et al. 2019; Zhao et al. 2019). The insertion of a transgene, such as the coat protein (*CP*) gene,

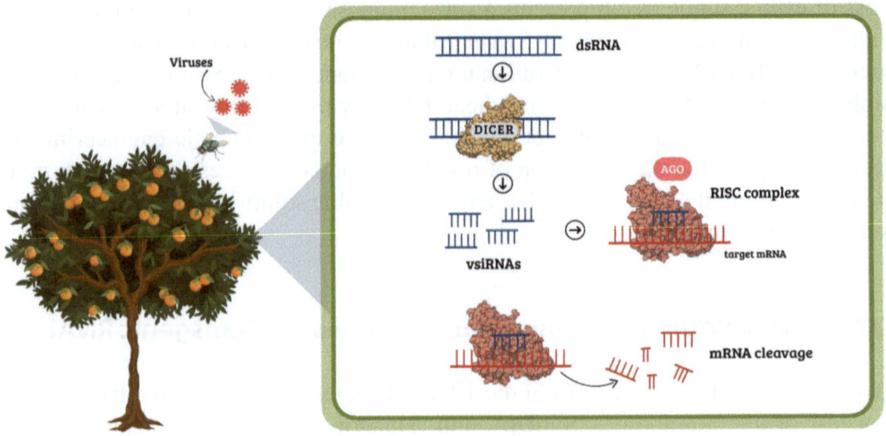

Fig. 22.2 The RNA interference (RNAi) pathway. Viral double-stranded RNA (dsRNA) is recognized and cleaved by the Dicer enzyme into viral small interfering RNAs (vsiRNAs). vsiRNA is incorporated into an Argonaute (AGO) protein, forming the RNA-induced silencing complex (RISC). The RISC complex then targets and degrades the complementary viral mRNA, silencing gene expression and inhibiting viral replication

confers protection by generating transcripts that form dsRNA, thus activating the plant's natural defense mechanism (Singh et al. 2019).

The constant evolution of organisms such as viruses leads them to develop ways to circumvent the plant's defense system through RNAi suppressors, known as viral suppressors of RNA silencing (VSRs) (Csorba et al. 2015). VSRs interrupt the silencing signal, as does the P3 protein of the citrus tristeza virus (CTV) (Ruiz-Ruiz et al. 2013). The strategies of VSRs target at least one of the pathway's steps, such as preventing siRNA production by DCLs, inhibiting the formation of the RISC complex, or suppressing signal amplification by RDRs (Burgyán and Havelda 2011). Despite these viral strategies, the ongoing study and refinement of RNAi for producing resistant fruit crops also currently focus on the topical application, such as spraying, of dsRNA produced and purified in the laboratory (Rank and Koch 2021). The spray acts as a trigger for silencing and, by not altering the plant's genome, represents a methodology that may have a simplified path to approval by regulatory agencies.

22.3.2 Gene Editing in the Context of Next-Generation Breeding

Next-generation breeding aims to produce genetically improved organisms using highly precise editing techniques that, as they may not incorporate exogenous genes, result in products similar to those obtained through natural processes. The clustered regularly interspaced short palindromic repeats (CRISPR) system, associated with a Cas enzyme (CRISPR-Cas), enables genome editing through the insertion, deletion, or substitution of a gene of interest, thereby creating the GEd organism. Although other editing techniques exist, such as zinc finger nucleases (ZFNs) and transcription activator-like effector nucleases (TALENs), Fernandes et al. (2023a) highlight CRISPR as a simpler, cheaper, and highly efficient research tool. Genetic engineering reached a new stage following the discovery of the CRISPR technique, which has allowed for more refined and targeted genome manipulation. Originally identified in the adaptive immune system of prokaryotes, it allows these organisms to store the information of exogenous DNA sequences that penetrate the cell, such as from bacteriophages or conjugative plasmids, to activate the necessary defense pathways later. The DNA fragments, ranging from 17 to 84 nucleotides, are stored as "spacers" in a CRISPR locus and are separated by palindromic repeats of 23 to 50 bases (Gostimskaya 2022). Associated with the CRISPR loci are enzymes with endonuclease characteristics, belonging to the Cas group. The CRISPR-Cas system is currently divided into two classes: class I utilizes a multi-protein complex, while class II, commonly used in biotechnology, consists of single effector proteins with multiple domains (Koonin et al. 2017). Currently, Cas9 (class II) is the protein of greatest scientific interest, as it is responsible for cleaving the target DNA. However, the cleavage is entirely dependent on the CRISPR RNAs (crRNAs) and the trans-activating RNA (tracrRNA), which direct the enzyme to the site of recognition and action (Prado et al. 2024). The precision of the cut requires a match between the crRNA and the target DNA, as well as the presence of a short sequence adjacent to the target site, known as the PAM (Protospacer Adjacent Motif) (Schiml and Puchta 2016). Moreover, enzymes of the Cas13a family (also designated C2c2) represent a new line of gene editing, as they target viral RNA. Cas13a is a class II enzyme, like Cas9, and has two nucleotide-binding domains with RNase function, which will be subsequently activated after recognition and binding of the target RNA to the crRNA (Kick et al. 2022). The work of Aman et al. (2018) and Shoeb et al. (2021) demonstrates that by expressing Cas13, plants become resistant by degrading the viral RNA. The precision of CRISPR-Cas allows for great versatility in genome editing through the artificial synthesis of single guide RNAs (sgRNAs), which guide the Cas complex to the gene of interest. The modification is consolidated by the cell's repair machinery, either by non-homologous end joining (NHEJ), which generally leads to random insertions or indels, or by homology-directed repair (HDR), which allows for the precise substitution or insertion of sequences, as shown in Fig. 22.3 (Jinek et al. 2012; Xu and Li 2020). In the effort to obtain virus-resistant or tolerant varieties, the main strategy consists of editing the host plant's genes that are crucial for the viral cycle, such as susceptibility factors necessary for viral replication or movement (Fernandes et al. 2025). Examples of perennial fruit crops under research for editing include citrus for citrus tristeza virus (CTV) (Sun et al. 2019), grapevine for grapevine virus A (GVA) (Spencer et al. 2023), and banana for banana streak virus (BSV) (Tripathi et al. 2019).

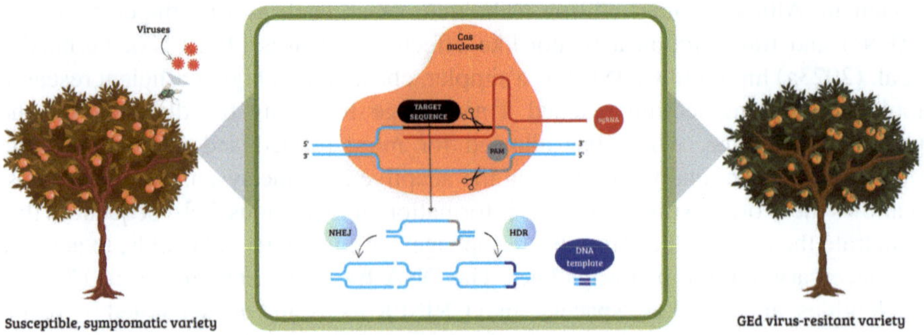

Fig. 22.3 Plant gene editing using the CRISPR/Cas system. A virus-susceptible plant variety (left panel) is selected, and the specific target sequence within its genome is identified for editing. A custom-designed single guide RNA (sgRNA) directs the Cas nuclease to create a precise double-strand break at the target site, with recognition facilitated by the protospacer adjacent motif (PAM) sequence. Post-cleavage, the cell employs one of two repair pathways: non-homologous end joining (NHEJ), which often introduces small insertions or deletions (indels) or direct rejoining; or homology-directed repair (HDR), which utilizes an exogenously supplied DNA template for highly precise sequence modification. The outcome is a genetically edited (GEd) plant with enhanced resistance to the pathogen (right panel)

22.4 Current Biotech Solutions for Virus Resistance: Insights from Transgenic Papaya and Plum Plants

Viral diseases pose the most significant threat to *C. papaya* L. cultivation. The papaya ringspot virus type P (PRSV-p) is particularly destructive, capable of causing total crop loss in the absence of effective management. Containment strategies currently depend on roguing, the eradication of symptomatic plants to disrupt the virus's transmission cycle (Galeano et al. 2022). However, this method, while critical, is the only viable control strategy, as no commercial varieties with effective PRSV-p resistance or tolerance are yet available in major producing countries like Brazil and Mexico. In this context, in the 1990s, transgenic *C. papaya* L. varieties resistant to PRSV-p were developed in Hawaii in response to the advance of the disease, which had already devastated production on the island of Oahu in the 1950s (Gonsalves 1998). The first control strategy explored was cross-protection, using the mild strain PRSV HA 5–1. However, despite promising results, Gonsalves (1998) reports that the technique did not gain acceptance from local producers due to factors such as adverse effects on fruit appearance, the associated management costs, and the farmers' distrust in intentionally inoculating their plants with a viral strain. Despite the lack of adoption, efforts continued, based on the strategy of pathogen-derived resistance (PDR), which had already demonstrated efficacy in other crops like tobacco. RNAi technology is embedded within the PDR concept, in which a gene from the pathogen itself is used to confer resistance to the host. Based on this principle, researchers in Hawaii used the PRSV HA 5-1 strain as a source to clone the gene encoding the coat protein (*CP*) into the *C. papaya* L. genome. The transgenic line that showed the greatest efficacy in the field was 55-1, from the

"Sunset" cultivar. However, this red-fleshed variety did not align with the local market's preference at the time, which preferred the yellow-fleshed "Kapoho". To solve this commercial impasse, the 55-1 line was crossed with the non-transgenic "Sunset" variety, yielding the homozygous cultivar "SunUp". Subsequently, "SunUp" was crossed with "Kapoho", resulting in the "Rainbow" hybrid (Gonsalves 2006). Field tests validated the efficacy of both cultivars, which were well-accepted in the market (Fig. 22.4). Regarding the resistance spectrum, "SunUp" showed broader resistance, whereas "Rainbow" exhibited restricted resistance to the Hawaiian variants of PRSV (Ferreira et al. 2002).

The commercial release of the transgenic Rainbow and SunUp papaya varieties revitalized Hawaii's agribusiness by effectively controlling PRSV-p. This success is a primary reason why the vast majority of papaya currently grown in Hawaii is genetically engineered. Despite this recovery, the papaya agribusiness's overall footprint has shrunk. According to the United States Department of Agriculture (USDA), the total land dedicated to papaya cultivation in Hawaii has declined significantly over the past 27 years, from 2180 acres in 1998 to just 650 acres in 2023. Yields have also seen a slight decrease from 18,800 to 15,800 pounds per acre over this period. However, the economic value of the crop has increased substantially. The price per pound of papaya has more than doubled, rising from \$0.31 in 1998 to \$0.68 in 2023, helping to sustain the industry's profitability on a smaller scale (HDOA 2023). In China, where *C. papaya* L. cultivation is concentrated in the southern part of the country, growers also faced massive losses due to PRSV infection (Wu et al. 2018). Motivated by the widespread distribution of the virus and the success of Hawaiian transgenic varieties, researchers at South China Agricultural University initiated a project in 1990 to develop a resistant cultivar to local viral strains using PDR technology based on the *CP* gene (Ye and Li 2010). After a construct designed from the *CP* gene did not confer the desired level of protection, the replicase (*RP*) gene proved to be the ideal alternative, conferring durable and persistent resistance (Ye et al. 2003). Field trials demonstrated that the transgenic cultivar was more productive and, although it showed mild symptoms early in the season,

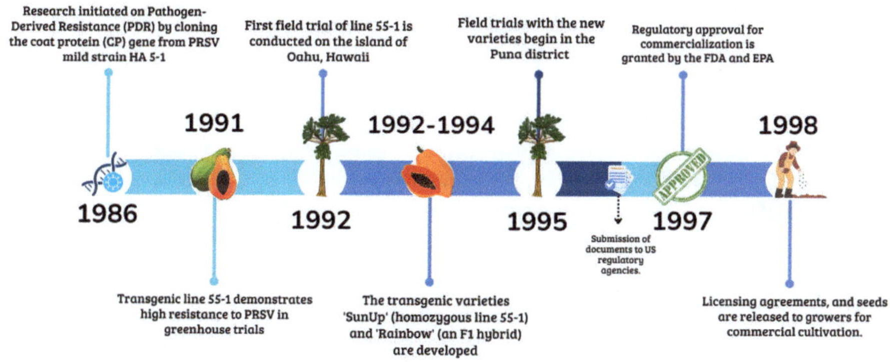

Fig. 22.4 Timeline of the development and commercialization of "Rainbow" and "SunUp" transgenic papayas

these had completely disappeared by the end of the cycle. The cultivar, named "Huanong No. 1", was deregulated for commercialization in Guangdong province in 2006 (Ye and Li 2010). Sharka disease, caused by plum pox virus (PPV), is a highly destructive pathogen of stone fruits that has spread globally since its initial identification in Bulgaria in the early twentieth century (García et al. 2024). It infects numerous Prunus species, including *P. domestica* (plum), *P. armeniaca* (apricot), *P. persica* (peach), and *P. avium* (cherry), and can cause cumulative production losses ranging from 48 to 100% over a ten-year period (Scorza et al. 2013). Research into natural resistance identified the German cultivar "Jojo", which exhibits a hypersensitive response (HR) conferring resistance to PPV. However, this resistance was determined to be strain-specific, and the multigenic nature of the HR can be associated with undesirable agronomic traits (Polák and Jarošová 2012).

"HoneySweet" plum represents a landmark achievement in applying genetic engineering to combat PPV. Its resistance was also based on RNAi, achieved by also inserting the PPV coat protein (*CP*) gene. This triggers a post-transcriptional gene silencing (PTGS) mechanism, characterized by the production of the siRNAs that confer highly effective and durable resistance without producing the CP protein itself (Scorza et al. 1995). Extensive international field trials conducted over two decades in endemic regions demonstrated that "HoneySweet" is completely resistant to natural aphid transmission of PPV. Even under intense graft-inoculation pressure, viral movement is confined and systemic infection does not occur; the resistance is stable, heritable as a dominant trait, and effective against all major PPV strains (Scorza et al. 2013). A comprehensive risk assessment confirmed its safety, finding no impact on non-target insects, no change in PPV population diversity, no generation of viral recombinants, and minimal gene flow (Scorza 2004). Furthermore, compositional analyses confirmed the fruit is not materially different from conventional plums. "HoneySweet" completed the US regulatory process, becoming the first genetically engineered woody perennial fruit tree to receive approvals from the USDA, FDA, and EPA (Fig. 22.5). Its development validates RNAi technology as a

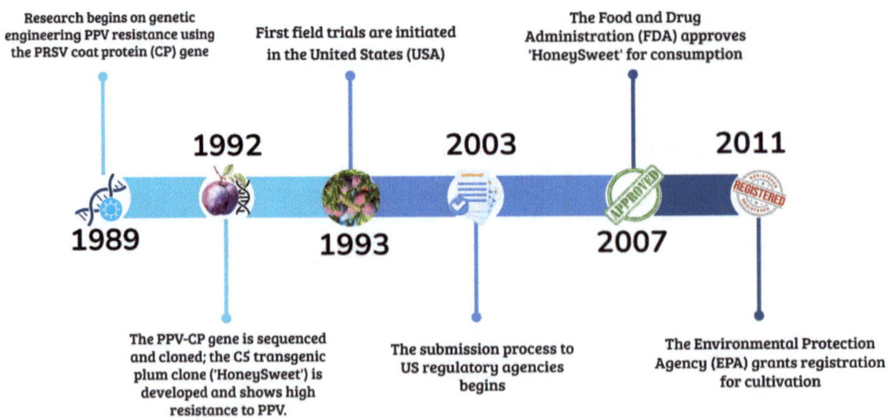

Fig. 22.5 Timeline of the development and deregulation of the "HoneySweet" transgenic plum

powerful and safe tool for creating sustainable virus resistance in perennial crops, providing a model for future biotechnological applications in agriculture.

22.5 Harnessing the Microbiome and AI as Emerging Frontiers in Virus Resistance

The plant microbiome is seen as having biotechnological potential, as its microorganisms can act as a line of defense against pathogens, a mechanism known as bioprotection (Maksimov et al. 2019; Trojak-Goluch 2024; Fernandes et al. 2025). The core of the bioprotection is to use microbiological biocontrol agents (MBCAs, such as bacteria, fungi, and even other viruses), which act through different mechanisms to induce resistance or tolerance in the hosts (Kohl et al. 2019). One of the bioprotection mechanisms is induced systemic resistance (ISR), in which beneficial microorganisms, such as plant growth-promoting microorganisms (PGPMs), colonize the rhizosphere or the plant's internal tissues (as endophytes) (Maksimov et al. 2015). This colonization results in the activated expression of pathogenesis-related proteins (PR proteins) with antiviral activity, leading to a rapid and robust defense response against an infection, as observed in plants colonized by rhizobacteria (Guevara-Morato et al. 2010; Manjunatha et al. 2022). Another mode of action is the direct activity of the microorganism or its metabolites on the pathogen or its vector. An example is the *Bacillus subtilis*, which can produce RNases capable of degrading viral RNA (Lehnik-Habrink et al. 2012). Indirectly, other microorganisms produce compounds such as antibiotics, lipopeptides, or enzymes that control insect vectors of viruses. The use of microbial consortia has also proven effective in controlling important viruses, such as tobacco mosaic virus (TMV) and potato virus Y (PVY) (Trojak-Goluch 2024). Tomato (*Solanum lycopersicum*) stands out as a crop with promising outcomes for the application of MCBAs against key viral pathogens. Vitti et al. (2016) demonstrated that soil application of *Trichoderma harzianum* strain T22 (Trianum G), 7 days after viral inoculation, reduced foliar symptoms of cucumber mosaic virus (CMV) by up to 82% compared to untreated controls. Similarly, Gupta et al. (2021) observed that tomato plants treated with *Pseudomonas putida* strain P1f showed a 60% reduction in the incidence of tobacco mosaic virus (TMV) 28 days post-inoculation. Even in the face of excellent results, data on the efficacy and durability of biocontrol agents for viruses in field trials are still scarce. As highlighted by He et al. (2021), a primary challenge is the viability and performance of these microorganisms, which are highly susceptible to fluctuating environmental conditions such as temperature and soil texture, factors that can significantly impact their consistent performance. In future perspectives for perennial fruit crops, biotechnology offers the opportunity to refine this approach further. A key strategy is the genetic modification of beneficial microorganisms to enhance their protective potential (Fernandes et al. 2025), and the integration of RNAi with the sustainability of bioprotection (Goodfellow et al. 2019). In this way, science can deliver eco-friendly and highly effective solutions, contributing to a more sustainable agriculture. On the same path, the advancement of artificial intelligence (AI) has also had a strong impact on the sustainability of farms that adopt this technology. By integrating data from the field, AI can determine the exact amounts of water, fertilizers, and pesticides needed, avoiding waste and, in the case of pesticides, reducing environmental contamination (Rejeb et al. 2022). AI is also an important tool for shortening the time spent in engineering and breeding plants. It accelerates selection during breeding through phenomics, variant calling models, gene discovery, genomic selection, and gene editing (Farooq et al. 2024). Traditionally, these processes were carried out empirically, by trial and error, often taking more than five years to develop a single potential plant variety (Nester et al. 2002) or even more for some woody perennials. Another challenge that AI can address is the risk and performance assessment of GM and GE plants, which often represents a roadblock for their approval (Mmbando 2024). Empirical tests with these plants are usually performed under specific and narrow conditions, such as exposure to UV-B stress (Mmbando et al. 2021) or infection by a given pathogen (Vigne et al. 2004), often in controlled laboratory environments and within limited time frames. AI, empowered by machine learning (ML) and deep learning (DL), can overcome these limitations by processing large datasets that include images of control and transgenic plants, weather and soil conditions, and genomic information. From these, AI can identify patterns that may be translated into possible risks, which can then be tested empirically, reducing the dependency on time-consuming trial and error approaches (Mmbando 2024). In genome editing, the integration of CRISPR/Cas9 with large language models (LLMs) opens new horizons. LLMs such as AlphaFold3, specialized in protein structure prediction, and OpenCRISPR-1, trained on over one million CRISPR operons, work similarly to Chatbot AI but are tailored to molecular biology tasks. These tools can considerably speed up the process by improving efficiency, specificity, and protein engineering capabilities (Farooq et al. 2024; Li et al. 2025). Although the impact of AI in genome editing is already remarkable, its full potential still faces challenges, such as the shortage of reliable training data and the inherent complexity of biological systems. Nevertheless, the prospects are highly promising, particularly regarding the possibility of designing novel proteins specifically engineered to protect plants against pathogens, opening exciting new directions for sustainable agriculture.

22.5.1 Development of Other Modifications in Long-Term Fruit Crops

Other popular fruits from perennial plants, like grapes and citrus, have yet to have approval status as GM or GE, even though these crops are highly vulnerable to viral infections. Citrus in particular is severely affected by CTV (Moreno et al. 2008). Attempts have been made to create resistant grapefruit (*Citrus paradisi*) using an insert of the *CP* gene of CTV, generating siRNA that mediates the degradation of viral RNA, in the PTGS process (Febres et al. 2008). Muniz et al. (2012) also attempted this method of immunization with the orange tree (*Citrus sinensis*), using three distinct constructs. These included *pCTV-CP*, containing the CTV *CP* gene;

pCTV-dsCP, which arranged the same capsid region into a hairpin design; and *pCTV-CS*, constructed from a conserved region of the CTV genome. When challenged with a mild strain of CTV, every *C. sinensis* clone containing these three constructs became infected. However, superior performance and tolerance were observed in plants carrying the *pCTV-dsCP* gene. In these particular plants, the viral titer was exceptionally low and comparable to that of healthy, non-inoculated plants. From these results, the authors concluded that the efficiency of transcription was directly related to the type of construct delivered to the host. As for grapes, resistance to GFLV has also been tested since 1995. While this strategy led to the successful generation of transgenic lines, it also produced considerable variability in transgene expression, with inconsistent mRNA levels observed across different individuals (Mauro et al. 1995; Tsvetkov et al. 2000; Gambino et al. 2005). Despite this, a single lineage was particularly promising for future field evaluation, a result attributed to its consistent protein accumulation and the consequent partial resistance to viral replication (Maghuly et al. 2006). As for the bananas, they improved poorly genetically speaking, because of the polyploidy, heterozygosity, and parthenocarpic fruits that become obstacles, but progress has been made in attempts to generate transgenicity of this species (Ganapathi et al. 2021). There were two events in 2011 that used RNAi to provide resistance to create two new transgenic plants resistant to BBTV, one derived from the *CP* gene of BBTV (Ismail et al. 2011) and the other utilizing the replicase-associated protein (*REP*) of BBTV (Borth et al. 2011). There were also other GM and GE of banana, constructed in response to fungal diseases like Fusarium wilt disease (Panama Disease) (Ghag et al. 2014; Dale et al. 2017), and bacterial diseases caused by *Xanthomonas campestris* (Tripathi et al. 2010, 2016). Recently, as of 2024, the governments of Australia and New Zealand approved a new transgenic variety, a Cavendish cv. Grand Nain banana resistant to *Fusarium oxysporum,* named event "QCAV-4" (Harding et al. 2025), which was made by introducing *M. acuminata* ssp. *malaccensis* resistance gene analog 2 *(RGA2)*, into the genome (Dale et al. 2017). As a successful new transgenic enters the market with approval status, it opens the possibility that other *Musa* spp. GM or GE resistant to other pathogens, like viruses, to also pass through the bureaucratic path that is the regulations of this plant.

22.6 Future Directions in the Development of GM and GE Fruit Trees

The current landscape of GM and GE fruit crops reveals a market predominantly characterized by varieties featuring herbicide tolerance and insect resistance (Klümper and Qaim 2014; Cheng et al. 2024) and promising nutritional profiles, such as the Golden Banana (Paul et al. 2017) and the Pinkglow® Pineapple, which represent important innovations. This reality prompts a critical question: given that viruses are a significant threat to the fruit sector, why is antiviral resistance not among the most widely deployed traits? The primary challenge lies in the rapid evolutionary capacity of viruses. As relatively simple organisms prone to genetic events like mutation and reassortment, they can generate new variants capable of overcoming the lines of defense, a phenomenon known as resistance breakdown. These resistance-breaking (RB) variants can exhibit altered host ranges and increased virulence (Moreno-Pérez et al. 2016). This is well-documented: a reassortant tomato spotted wilt virus was able to break the *tsw* gene-conferred resistance in pepper plants (Kwon et al. 2021), and RB strains were able to infect tomato resistant varieties designed with the *Sw-5* resistance gene (Lahre et al. 2023). Cases with the PRSV resistance varieties are also known. The remarkable Hawaiian "Rainbow" papaya plants are not completely immune; Ferreira et al. (2002) reported that despite their high degree of resistance, some individuals became infected during field tests and that young seedlings remained susceptible, serving as a cautionary sign of emerging RB variants. This is not an isolated case, as a similar situation was observed with the Chinese transgenic variety "Huanong No.1". After a massive sample collection from 2012 to 2016, Wu et al. (2018) also reported new RB variants that were genetically diverse and distinct from those previously seen in 2006. Therefore, in order to achieve resilient resistance strategies, it is necessary to have a deeper understanding of the host plant's genetic background, as the efficacy of any engineered trait is highly dependent on it (Gallois et al. 2018). Eukaryotic translation initiation factors (eIFs) serve as key targets for conferring broad-spectrum resistance to potyviruses, and CRISPR-mediated modification of these factors has already achieved success in important crops like cassava (Gomez et al. 2018), rice (Macovei et al. 2018), and cucumber (Chandrasekaran et al. 2016). Ultimately, genetic modification and gene editing technologies stand as essential tools for sustainable fruticulture, which must provide benefits not only to large agricultural companies but also to organic farming and the end consumer (Fernandes et al. 2023a). The continued growth in the adoption of transgenic and genetically modified crops demonstrates their expanding reach, reinforcing the strength of the agricultural biotechnology sector (ISAAA 2024). Future progress will therefore depend on the dual ability to engineer genetic solutions against phytopathogens that constantly challenge science, while simultaneously fostering global public trust through transparent communication.

References

Abreu PM, Antunes F, Magaña-Álvarez A, Pérez-Brito D et al (2015) A current overview of the Papaya meleira virus, an unusual plant virus. Viruses 7:1853–1870. https://doi.org/10.3390/v7041853

Agius C, Eamens AL, Millar AA et al (2012) RNA silencing and antiviral defense in plants. In: Watson JM, Wang MB (eds) Antiviral resistance in plants. Humana Totowa, New Jersey, pp 17–38. https://doi.org/10.1007/978-1-61779-882-5_2

Akbar S, Wei Y, Zhang MQ (2022) RNA interference: promising approach to combat plant viruses. Int J Mol Sci 23:5312. https://doi.org/10.3390/ijms23105312

Almeida RP, Daane KM, Bell VA, Blaisdell GK et al (2013) Ecology and management of grapevine leafroll disease. Front Microbiol 4:94. https://doi.org/10.3389/fmicb.2013.00094

Aman R, Ali Z, Butt H et al (2018) RNA virus interference via CRISPR/Cas13a system in plants. Genome Biol 19:1. https://doi.org/10.1186/s13059-017-1381-1

Andret-Link P, Laporte C, Valat L, Ritzenthaler C et al (2004) Grapevine fanleaf virus: still a major threat to the grapevine industry. J Plant Pathol 86:183–195. http://www.jstor.org/stable/41992424

Belabess Z, Sagouti T, Rhallabi N, Tahiri A et al (2020) Citrus Psorosis virus: current insights on a still poorly understood Ophiovirus. Microorganisms 8:1197. https://doi.org/10.3390/microorganisms8081197

Borth W, Perez E, Cheah K, Chen Y et al (2011) Transgenic banana plants resistant to *banana bunchy top virus* infection. In: Van den Bergh I, Smith M, Swennen R, Hermanto C (eds) V international symposium on Banana: ISHS-ProMusa symposium on global perspectives on Asian challenges. ISHS Acta Horticulturae, Guangzhou, pp 449–457. https://doi.org/10.17660/ActaHortic.2011.897.61

Burgyán J, Havelda Z (2011) Viral suppressors of RNA silencing. Trends Plant Sci 16:265–272. https://doi.org/10.1016/j.tplants.2011.02.010

Butkovic A, González R (2022) A brief view of factors that affect plant virus evolution. Front Virol 2:994057. https://doi.org/10.3389/fviro.2022.994057

Callahan AM, Srinivasan C, Dardick C, Scorza R (2016) Rapid cycle breeding: application of transgenic early flowering for perennial trees. In: Janick J (ed) Plant breeding reviews, 1st edn. Wiley, New Jersey, pp 299–334. https://doi.org/10.1038/hortres.2014.22

Cambra M, Capote N, Myrta A, Llácer G (2006) *Plum pox virus* and the estimated costs associated with sharka disease. EPPO Bull 36:202–204. https://doi.org/10.1111/j.1365-2338.2006.01027.x

Catto M, Mugerwa H, Myers B et al (2022) A review on transcriptional responses of interactions between insect vectors and plant viruses. Cells 11:693. https://doi.org/10.3390/cells11040693

Cembali T, Folwell RJ, Wandschneider P et al (2003) Economic implications of a virus prevention program in deciduous tree fruits in the US. Crop Prot 22:1149–1156. https://doi.org/10.1016/S0261-2194(03)00156-X

Chandrasekaran J, Brumin M, Wolf D et al (2016) Development of broad virus resistance in non-transgenic cucumber using CRISPR/Cas9 technology. Mol Plant Pathol 17:1140–1153. https://doi.org/10.1111/mpp.12375

Charrier A, Vergne E, Dousset N et al (2019) Efficient targeted mutagenesis in apple and first time edition of pear using the CRISPR-Cas9 system. Front Plant Sci 10:40. https://doi.org/10.3389/fpls.2019.00040

Cheng X, Li H, Tang Q et al (2024) Trends in the global commercialization of genetically modified crops in 2023. J Integr Agric 23:3943–3952. https://doi.org/10.1016/j.jia.2024.09.012

Chirkov S, Sheveleva A, Tsygankova S, Slobodova N et al (2023) Whole genome characterization of Prunus necrotic ringspot virus and prune Dwarf virus infecting stone fruits in Russia. Horticulturae 9:941. https://doi.org/10.3390/horticulturae9080941

Csorba T, Kontra L, Burgyán J (2015) Viral silencing suppressors: tools forged to fine-tune host-pathogen coexistence. Virology 479-480:85–103. https://doi.org/10.1016/j.virol.2015.02.028

Dale J, James A, Paul JY et al (2017) Transgenic Cavendish bananas with resistance to Fusarium wilt tropical race 4. Nat Commun 8:1496. https://doi.org/10.1038/s41467-017-01670-6

Daniells JW, Geering AD, Bryde NJ, Thomas JE (2001) The effect of Banana streak virus on the growth and yield of dessert bananas in tropical Australia. Ann Appl Biol 139:51–60. https://doi.org/10.1111/j.1744-7348.2001.tb00130.x

Deng X, Duan Y (2006) Modification of perennial fruit trees. In: Fladung M, Ewald D (eds) Tree transgenesis: recent developments, 1st edn. Springer Berlin, Heidelberg, pp 44–66. https://doi.org/10.1007/3-540-32199-3_3

Farooq MA, Gao S, Hassan MA, Huang Z et al (2024) Artificial intelligence in plant breeding. Trends Genet 40:891–908. https://doi.org/10.1016/j.tig.2024.07.001

Febres VJ, Lee RF, Moore GA (2008) Transgenic resistance to Citrus tristeza virus in grapefruit. Plant Cell Rep 27:93–104. https://doi.org/10.1007/s00299-007-0445-1

Fernandes AAR, Favaratto L, Fernandes PMB et al (2023a) Técnicas inovadoras de melhoramento de precisão para produção de organismos geneticamente editados. ILSI Brasil International Life Science do Brasil, São Paulo. Available at: https://ilsibrasil.org/publication/tecnicas-inovadoras-de-melhoramento-de-precisao-para-producao-de-organismos-geneticamente--editados/

Fernandes PMB, Fernandes AAR, Maurastoni M, Rodrigues SP (2025) Lab legends and field phantoms: the tale of virus-resistant plants. Annu Rev Virol 12:23–42. https://doi.org/10.1146/annurev-virology-092623-101850

Ferreira SA, Pitz KU, Manshardt R et al (2002) Virus coat protein transgenic papaya provides practical control of papaya ringspot virus in Hawaii. Plant Dis 86(2). https://doi.org/10.1094/PDIS.2002.86.2.101

Food and Agriculture Organization of the United Nations (FAO) – FAOSTAT (2023). Available at: https://www.fao.org/faostat/en/#data/QCL/visualize

Food and Agriculture Organization of the United Nations (FAO) – Plant Production and Protection (2019). Available at: https://www.fao.org/plant-production-protection/en

Friedman J (2022) The evolution of annual and perennial plant life histories: ecological correlates and genetic mechanisms. Annu Rev Ecol Evol Syst 51:461–481. https://doi.org/10.1146/annurev-ecolsys-110218-024638

Galeano EAV, Martins DS, Barros FLS et al (2022) Cadeia produtiva do mamão no Espírito Santo. Incaper, Vitória. Available at: https://biblioteca.incaper.es.gov.br/digital/handle/item/4206

Gallois JL, Moury B, German-Retana S (2018) Role of the genetic background in resistance to plant viruses. Int J Mol Sci 19:2856. https://doi.org/10.3390/ijms19102856

Gambino G, Gribaudo I, Leopold S, Schartl A et al (2005) Molecular characterization of grapevine plants transformed with GFLV resistance genes: I. Plant Cell Rep 24:655–662. https://doi.org/10.1007/s00299-005-0006-4

Ganapathi TR, Negi S, Tak H, Bapat VA (2021) Transgenic Banana: current status, opportunities and challenges. In: Kavi Kishor PB, Rajam MV, Pullaiah T (eds) Genetically modified crops. Springer, Singapore, pp 111–128. https://doi.org/10.1007/978-981-15-5932-7_5

García JA, Rodamilans B, Martínez-Turiño S et al (2024) Plum pox virus: an overview of the potyvirus behind sharka, a harmful stone fruit disease. Ann Appl Biol 186:49–75. https://doi.org/10.1111/aab.12958

Ghag SB, Shekhawat UKS, Ganapathi TR (2014) Transgenic banana plants expressing a *Stellaria media defensin* gene (*Sm*-AMP-D1) demonstrate improved resistance to *Fusarium oxysporum*. Plant Cell Tissue Organ Cult 119:247–255. https://doi.org/10.1007/s11240-014-0529-x

Gomez MA, Lin ZD, Moll T et al (2018) Simultaneous CRISPR/Cas9-mediated editing of cassava *eIF4E* isoforms *nCBP-1* and *nCBP-2* reduces cassava brown streak disease symptom severity and incidence. Plant Biotechnol J 17:421–434. https://doi.org/10.1111/pbi.12987

Gonsalves D (1998) Control of papaya ringspot virus in papaya: a case study. Annu Rev Phytopathol 36:415–437. https://doi.org/10.1146/annurev.phyto.36.1.415

Gonsalves D (2006) Transgenic papaya: development, release, impact and challenges. Adv Virus Res 67:317–354. https://doi.org/10.1016/S0065-3527(06)67009-7

Gonsalves D, Tripathi S, Carr JB, Suzuki JY (2010) Papaya Ringspot Virus. The Plant Health Instructor. Available at: https://www.apsnet.org/edcenter/pdlessons/Pages/PapayaRingspotvirus.aspx

Goodfellow S, Zhang D, Wang MB, Zhang R (2019) Bacterium-mediated RNA interference: potential application in plant protection. Plants 8:572. https://doi.org/10.3390/plants8120572

Gostimskaya I (2022) CRISPR–Cas9: a history of its discovery and ethical considerations of its use in genome editing. Biochemistry (Mosc) 87:777–788. https://doi.org/10.1134/S0006297922080090

Goulin EH, Galdeano DM, Granato LM et al (2019) RNA interference and CRISPR: promising approaches to better understand and control citrus pathogens. Microbiol Res 226:1–9. https://doi.org/10.1016/j.micres.2019.03.006

Guevara-Morato MA, de Lacoba MG, García-Luque I, Serra MT (2010) Characterization of a pathogenesis-related protein 4 (PR-4) induced in *Capsicum chinense* L. plants with dual RNase and DNase activities. J Exp Bot 61:3259–3271. https://doi.org/10.1093/jxb/erq148

Gupta AK, Verma J, Srivastava A et al (2021) A comparison of induced antiviral resistance by the phytoprotein CAP-34 and isolate P1f of the rhizobacterium *Pseudomonas putida*. 3 Biotech 11:509. https://doi.org/10.1007/s13205-021-03057-3

Harding R, Paul JY, James A, Smith M (2025) QCAV-4, the first genetically modified Cavendish (cv. Grand Nain) banana resistant to Fusarium wilt tropical race 4 approved for commercial production and consumption. Plant Biotechnol J 23:3628–3637. https://doi.org/10.1111/pbi.70178

Hawaii Department of Agriculture (HDOA) – Statistics of Papaya (2023) State of Hawaii, 2017–2023. Available at: https://dab.hawaii.gov/add/files/2024/05/Papaya-Stats-2023_SOH_05.16.24.pdf

He DC, He MH, Amalin DM et al (2021) Biological control of plant diseases: an evolutionary and eco-economic consideration. Pathogens 10:1311. https://doi.org/10.3390/pathogens10101311

International Service for the Acquisition of Agri-biotech Applications (ISAAA) – GM Crop Events List (2024). Available at: https://www.isaaa.org/gmapprovaldatabase/eventslist/default.asp

International Service for the Acquisition of Agri-biotech Applications (ISAAA) – GM Crop Market Projected to Reach USD 36 Billion by 2031. Available at: https://www.isaaa.org/kc/cropbiotechupdate/article/default.asp?ID=21124

Ismail RM, El-Domyati FM, Wagih EE, Sadik AS et al (2011) Construction of banana bunchy top nanovirus-DNA-3 encoding the coat protein gene and its introducing into banana plants cv. Williams. J Genet Eng Biotechnol 9:35–41. https://doi.org/10.1016/j.jgeb.2011.05.012

Jinek M, Chylinski K, Fonfara I et al (2012) A programmable dual-RNA–guided DNA endonuclease in adaptive bacterial immunity. Science 337:816–821. https://doi.org/10.1126/science.1225829

Jyotika RK, Harish S, Karthikeyan G et al (2024) Molecular approaches for the management of papaya ringspot virus infecting papaya: a comprehensive review. Mol Biol Rep 51:981. https://doi.org/10.1007/s11033-024-09920-9

Khan FS, Considine MJ, Can W, Xu Z et al (2025) From morphology to epigenetics: current understanding of the floral regulation in woody perennials. Plant Physiol Biochem 227:110192. https://doi.org/10.1016/j.plaphy.2025.110192

Kick LM, Wrisberg MKV, Runtsch LS, Schneider S (2022) Structure and mechanism of the RNA dependent RNase Cas13a from *Rhodobacter capsulatus*. Commun Biol 5:71. https://doi.org/10.1038/s42003-022-03025-4

Klümper W, Qaim M (2014) A meta-analysis of the impacts of genetically modified crops. PLoS One 9:e111629. https://doi.org/10.1371/journal.pone.0111629

Kohl J, Kolnaar R, Ravensberg WJ (2019) Mode of action of microbial biological control agents against plant diseases: relevance beyond efficacy. Front Plant Sci 10:845. https://doi.org/10.3389/fpls.2019.00845

Koonin EV, Makarova KS, Zhang F (2017) Diversity, classification and evolution of CRISPR-Cas systems. Curr Opin Microbiol 37:67–78. https://doi.org/10.1016/j.mib.2017.05.008

Krebelj AJ, Čepin U, Ravnikar M et al (2015) Spatio-temporal distribution of *Grapevine fanleaf virus* (GFLV) in grapevine. Eur J Plant Pathol 142:159–171. https://doi.org/10.1007/s10658-015-0600-4

Kreitzman M, Toensmeier E, Chan KMA et al (2020) Perennial staple crops: yields, distribution, and nutrition in the global food system. Front Sustain Food Syst 4:588988. https://doi.org/10.3389/fsufs.2020.588988

Kwon SJ, Cho YE, Kwon OH et al (2021) Resistance-breaking tomato spotted wilt virus variant that recently occurred in pepper in South Korea is a genetic Reassortant. Plant Dis 105:2771–2775. https://doi.org/10.1094/PDIS-01-21-0205-SC

Lahre K, Shekasteband R, Meadows I et al (2023) First report of resistance-breaking variants of tomato spotted wilt virus (TSWV) infecting tomatoes with the *Sw-5* resistance gene in North Carolina. Plant Dis 107:1968–2271. https://doi.org/10.1094/PDIS-11-22-2637-PDN

Lehnik-Habrink M, Lewis RJ, Mader U, Stulke J (2012) RNA degradation in *Bacillus subtilis*: an interplay of essential endo- and exoribonucleases. Mol Microbiol 84:1005–1017. https://doi.org/10.1111/j.1365-2958.2012.08072.x

Li G, An L, Yang W et al (2025) Integrated biotechnological and AI innovations for crop improvement. Nature 643:925–937. https://doi.org/10.1038/s41586-025-09122-8

Macovei A, Sevilla NR, Cantos C et al (2018) Novel alleles of rice *eIF4G* generated by CRISPR/Cas9-targeted mutagenesis confer resistance to *Rice tungro spherical virus*. Plant Biotechnol J 16:1918–1927. https://doi.org/10.1111/pbi.12927

Maghuly F, Leopold S, Machado AC et al (2006) Molecular characterization of grapevine plants transformed with GFLV resistance genes: II. Plant Cell Rep 25:546–553. https://doi.org/10.1007/s00299-005-0087-0

Maksimov IV, Veselova SV, Nuzhnaya TV (2015) Plant growth-promoting bacteria in regulation of plant resistance to stress factors. Russ J Plant Physiol 62:715–726. https://doi.org/10.1134/S1021443715060114

Maksimov IV, Sorokan AV, Burkhanova GF et al (2019) Mechanisms of plant tolerance to RNA viruses induced by plant-growth-promoting microorganisms. Plants 8:575. https://doi.org/10.3390/plants8120575

Manjunatha L, Rajashekara H, Uppala LS et al (2022) Mechanisms of microbial plant protection and control of plant viruses. Plants 11:3449. https://doi.org/10.3390/plants11243449

Mannini F, Digiaro M (2017) The effects of viruses and viral diseases on grapes and wine. In: Meng B, Martelli G, Golino D, Fuchs M (eds) Grapevine viruses: molecular biology, diagnostics and management. Springer, Cham, pp 453–482. https://doi.org/10.1007/978-3-319-57706-7_23

Manzoor S, Nabi SU, Baranwal VK, Verma MK et al (2023) Overview on century progress in research on mosaic disease of apple (*Malus domestica* Borkh) incited by apple mosaic virus/apple necrotic mosaic virus. Virology 587:109846. https://doi.org/10.1016/j.virol.2023.109846

Maree HJ, Almeida RP, Bester R, Chooi KM et al (2013) Grapevine leafroll-associated virus 3. Front Microbiol 4:82. https://doi.org/10.3389/fmicb.2013.00082

Martín-Valmaseda M, Devin SR, Ortuño-Hernández G et al (2023) CRISPR/Cas as a genome-editing technique in fruit tree breeding. Int J Mol Sci 24:16656. https://doi.org/10.3390/ijms242316656

Mauro MC, Toutain S, Walter B, Pinck L, Otten L et al (1995) High efficiency regeneration of grapevine plants transformed with the GFLV coat protein gene. Plant Sci 112:97–106. https://doi.org/10.1016/0168-9452(95)04246-Q

Ming R, Moore PH (eds) (2014) Genetics and genomics of papaya. Springer, New York, pp 3–16. https://doi.org/10.1007/978-1-4614-8087-7

Mmbando GS (2024) The use of artificial intelligence in the production of genetically modified (GM) crops: a recent promising strategy for enhancing the acceptability of GM products. Discov Appl Sci 6:581. https://doi.org/10.1007/s42452-024-06212-6

Mmbando GS, Teranishi M, Hidema J (2021) Transgenic rice *Oryza glaberrima* with higher CPD photolyase activity alleviates UVB-caused growth inhibition. GM Crops Food 12:435–448. https://doi.org/10.1080/21645698.2021.1977068

Moore PW, Nauer E (1957) Scaly bark disease of citrus: nine-year study of seven older Orange orchards indicates advance of Psorosis may be faster than is generally believed. Calif Agric 11:11–15. Available at: https://californiaagriculture.org/article/113933-scaly-bark-disease--of-citrus-nine-year-study-of-seven-older-orange-orchards-indicates-advance-of-psorosis-may-be-faster-than-is-generally-believed

Moreno P, Ambrós S, Albiach-Martí MR et al (2008) *Citrus tristeza virus*: a pathogen that changed the course of the citrus industry. Mol Plant Pathol 9:251–268. https://doi.org/10.1111/j.1364-3703.2007.00455.x

Moreno-Pérez MG, García-Luque I, Fraile A, García-Arenal F (2016) Mutations that determine resistance breaking in a plant RNA virus have pleiotropic effects on its fitness that depend on the host environment and on the type, single or mixed, of infection. J Virol 90:9128–9137. https://doi.org/10.1128/JVI.00737-16

Muniz FR, De Souza AJ, Stipp LCL, Schinor E et al (2012) Genetic transformation of *Citrus sinensis* with Citrus tristeza virus (CTV) derived sequences and reaction of transgenic lines to CTV infection. Biol Plant 56:162–166. https://doi.org/10.1007/s10535-012-0035-0

Nester EW, Thomashow LS, Metz M et al (2002) 100 years of *Bacillus thuringiensis*: a critical scientific assessment. American Society for Microbiology, Washington, DC. Available at: https://www.ncbi.nlm.nih.gov/books/NBK559445/

Neumüller M (2011) Fundamental and applied aspects of plum (*Prunus domestica*) breeding. Fruit Veg Cereal Sci Biotech 5:139–156. Available at: http://globalsciencebooks.info/Online/GSBOnline/images/2011/FVCSB_5(SI1)/FVCSB_5(SI1)139-156o.pdf

Paul JY, Khanna H, Kleidon J et al (2017) Golden bananas in the field: elevated fruit pro-vitamin A from the expression of a single banana transgene. Plant Biotechnol 15:520–532. https://doi.org/10.1111/pbi.12650

Pedrelli A, Panattoni A, Cotrozzi L (2024) The sharka disease on stone fruits in Italy: a review, with a focus on Tuscany. Eur J Plant Pathol 169:287–300. https://doi.org/10.1007/s10658-024-02827-y

Peter KA (2024) Apple Viruses. PennState Extension. Available at: https://extension.psu.edu/apple-viruses

Pimentel D, Cerasale D, Stanley RC et al (2012) Annual vs. perennial grain production. Agric Ecosyst Environ 161:1–9. https://doi.org/10.1016/j.agee.2012.05.025

Polák J, Jarošová J (2012) Susceptibility of plum trees cv. 'Jojo' to a Czech isolate of *Plum pox virus* strain D. Can J Plant Pathol 34:283–287. https://doi.org/10.1080/07060661.2011.640711

Polák J, Kundu JK, Krška B et al (2017) Transgenic plum *Prunus domestica* L., clone C5 (cv. HoneySweet) for protection against sharka disease. J Integr Agric 16:516–522. https://doi.org/10.1016/S2095-3119(16)61491-0

Prado GS, Rocha DC, Santos LN et al (2024) CRISPR technology towards genome editing of the perennial and semi-perennial crops citrus, coffee and sugarcane. Front Plant Sci 14:1331258. https://doi.org/10.3389/fpls.2023.1331258

Pumplin N, Voinnet O (2013) RNA silencing suppression by plant pathogens: defence, counter-defence and counter-counter-defence. Nat Rev Microbiol 11:745–760. https://doi.org/10.1038/nrmicro3120

Qazi J (2016) Banana bunchy top virus and the bunchy top disease. J Gen Plant Pathol 82:2–11. https://doi.org/10.1007/s10327-015-0642-7

Rank AP, Koch A (2021) Lab-to-field transition of RNA spray applications—how far are we? Front Plant Sci 12:755203. https://doi.org/10.3389/fpls.2021.755203

Rejeb A, Rejeb K, Zailani S, Keogh JG et al (2022) Examining the interplay between artificial intelligence and the agri-food industry. Artificial Intelligence in Agriculture 6:111–128. https://doi.org/10.1016/j.aiia.2022.08.002

Ribeiro FE, Costa EFN (2018) Melhoramento Genético. In: Sobral LF (ed) Coco: o produtor pergunta, a Embrapa responde. Embrapa, Brasília, pp 149–158. Available at: https://www.infoteca.cnptia.embrapa.br/infoteca/handle/doc/1118898

Ruiz-Ruiz S, Soler N, Sánchez-Navarro J et al (2013) *Citrus tristeza virus* p23: determinants for nucleolar localization and their influence on suppression of RNA silencing and pathogenesis. Mol Plant-Microbe Interact 26:306–318. https://doi.org/10.1094/MPMI-08-12-0201-R

Sastry KS, Zitter TA (2014) Management of virus and viroid diseases of crops in the tropics. In: Plant virus and viroid diseases in the tropics. Springer, Dordrecht, pp 149–480. https://doi.org/10.1007/978-94-007-7820-7_2

Schiml S, Puchta H (2016) Revolutionizing plant biology: multiple ways of genome engineering by CRISPR/Cas. Plant Methods 12:8. https://doi.org/10.1186/s13007-016-0103-0

Scorza R (2004) Application for determination of non-regulatory status for C5 ('HoneySweet') plum (*Prunus domestica* L.) resistant to *Plum Pox* virus. Available at: https://www.aphis.usda.gov/sites/default/files/04_26401p.pdf

Scorza R, Ravelonandro M, Callahan AM et al (1995) Transgenic plums (*Prunus domestica* L.) express the plum pox virus coat protein gene. Plant Cell Rep 14:18–22. https://doi.org/10.1007/BF00233291

Scorza R, Callahan A, Dardick C et al (2013) Genetic engineering of *Plum pox virus* resistance: 'HoneySweet' plum—from concept to product. Plant Cell Tissue Organ Cult 115:1–12. https://doi.org/10.1007/s11240-013-0339-6

Shefferson RP (2009) The evolutionary ecology of vegetative dormancy in mature herbaceous perennial plants. J Ecol 97:1000–1009. https://doi.org/10.1111/j.1365-2745.2009.01525.x

Shekhawat UKS, Ganapathi TR, Hadapad AB (2012) Transgenic banana plants expressing small interfering RNAs targeted against viral replication initiation gene display high-level resistance to banana bunchy top virus infection. J Gen Virol 93:1804–1813. https://doi.org/10.1099/vir.0.041871-0

Shoeb E, Badar U, Venkataraman S, Hefferon K (2021) CRISPR/Cas9 and Cas13a systems: a promising tool for plant breeding and plant defence. In: Abd-elsalam KA, Lim K (eds) CRISPR and RNAi systems: nanobiotechnology approaches to plant breeding and protection. Elsevier, Amsterdam, pp 211–231. https://doi.org/10.1016/B978-0-12-821910-2.00002-3

Singh K, Dardick C, Kundu JK (2019) RNAi-mediated resistance against viruses in perennial fruit plants. Plants 8:359. https://doi.org/10.3390/plants8100359

Smith J (2011) The history of temperate agroforestry. The Organic Research Centre, Bershire. Avaliable at: https://orgprints.org/id/eprint/18173/

Soares EDA, Werth EG, Madroñero LJ (2017) Label-free quantitative proteomic analysis of pre-flowering PMeV-infected *Carica papaya* L. J Proteome 151:275–283. https://doi.org/10.1016/j.jprot.2016.06.025

Spencer KP, Burger JT, Campa M (2023) CRISPR-based resistance to grapevine virus A. Front Plant Sci 14:1296251. https://doi.org/10.3389/fpls.2023.1296251

Sun L, Nasrullah, Ke F et al (2019) Citrus genetic engineering for disease resistance: past, present and future. Int J Mol Sci 20:5256. https://doi.org/10.3390/ijms20215256

Tatieni S, Hein GL (2023) Plant viruses of agricultural importance: current and future perspectives of virus disease management strategies. Phytopathology 113:117–141. https://doi.org/10.1094/PHYTO-05-22-0167-RVW

Tripathi L, Mwaka H, Tripathi JN, Tushemereirwe WK (2010) Expression of sweet pepper *Hrap* gene in banana enhances resistance to *Xanthomonas campestris* pv. musacearum. Mol Plant Pathol 11:721–731. https://doi.org/10.1111/j.1364-3703.2010.00639.x

Tripathi S, Patil BL, Verma R (2016) Viral diseases of bananas and their management. In: Gaur RK, Petrov NM, Patil BL, Stoyanova MI (eds) Plant viruses: evolution and management. Springer Singapore, Singapore, pp 289–308. https://doi.org/10.1007/978-981-10-1406-2

Tripathi JN, Ntui VO, Ron M et al (2019) CRISPR/Cas9 editing of endogenous *banana streak virus* in the B genome of *Musa* spp. overcomes a major challenge in banana breeding. Commun Biol 2:46. https://doi.org/10.1038/s42003-019-0288-7

Trojak-Goluch A (2024) The use of bacteria, Actinomycetes and fungi in the bioprotection of Solanaceous crops against Tobacco Mosaic Virus (TMV). Agriculture 14:1220. https://doi.org/10.3390/agriculture14081220

Tsvetkov IJ, Atanassov AI, Tsolova VM (2000) Gene transfer for stress resistance in grapes. Acta Hortic 528:389–396. https://doi.org/10.17660/ActaHortic.2000.528.56

United States Department of Agriculture (USDA) – Fruit and Tree Nuts Data – Fruit and Tree Nuts Yearbook Tables. Available at: https://www.ers.usda.gov/data-products/fruit-and-tree--nuts-data/fruit-and-tree-nuts-yearbook-tables

Van-Esse HP, Reuber TL, Does DVD (2019) Genetic modification to improve disease resistance in crops. New Phytol 225:70–86. https://doi.org/10.1111/nph.15967

Varkonyi-Gasic E, Wang T, Voogd C et al (2018) Mutagenesis of kiwifruit *CENTRORADIALIS*-like genes transforms a climbing woody perennial with long juvenility and axillary flowering into a compact plant with rapid terminal flowering. Plant Biotechnol J 17:869–880. https://doi.org/10.1111/pbi.13021

Vieira MS, Cabral RLR, Favaratto L et al (2024) Tropical fruit virus resistance in the era of next-generation plant breeding. SynBio 2:267–284. https://doi.org/10.3390/synbio2030016

Vigne E, Komar V, Fuchs M (2004) Field safety assessment of recombination in transgenic grapevines expressing the coat protein gene of *Grapevine fanleaf virus*. Transgenic Res 13:165–179. https://doi.org/10.1023/b:trag.0000026075.79097.c9

Vitti A, Pellegrini E, Nali C et al (2016) *Trichoderma harzianum* T-22 induces systemic resistance in tomato infected by cucumber mosaic virus. Front Plant Sci 7:1520. https://doi.org/10.3389/fpls.2016.01520

Wambulwa MC, Wachira FN, Karanja LS, Kiarie SM et al (2013) The influence of host and pathogen genotypes on symptom severity in banana streak disease. Afr J Biotechnol 12:27–31. https://doi.org/10.5897/AJB12.2536

Wang J, Ding J (2023) Molecular mechanisms of flowering phenology in trees. For Res (Fayettev) 3:2. https://doi.org/10.48130/FR-2023-0002

Wang XB, Jovel J, Udomporn P et al (2011) The 21-nucleotide, but not 22-nucleotide, viral secondary small interfering RNAs direct potent antiviral defense by two cooperative Argonautes in *Arabidopsis thaliana*. Plant Cell 23:1625–1638. https://doi.org/10.1105/tpc.110.082305

Wang X, Tu M, Li Z et al (2018) Current Progress and future prospects for the Clustered Regularly Interspaced Short Palindromic Repeats (CRISPR) genome editing technology in fruit tree breeding. CRC Crit Rev Plant Sci 37:233–258. https://doi.org/10.1080/07352689.2018.1517457

Wieczorek AM, Wright MG (2012) History of agricultural biotechnology: how crop development has evolved. Nature Education Knowledge 3:9. Available at: https://www.nature.com/scitable/knowledge/library/history-of-agricultural-biotechnology-how-crop-development-25885295/#:-~:text=Traditional%20breeding%20programs%20are%20time,produce%20a%20new%20crop%20variety

World Health Organization (WHO) – Food, Genetically Modified (2014). Available at: https://www.who.int/news-room/questions-and-answers/item/food-genetically-modified

Wu YQ, Zhang DM, Chen SY, Wang XF, Wang WH (1997) Comparison of three ELISA methods for the detection of apple chlorotic leaf spot virus and apple stem grooving virus. In: Hadidi A (ed) XVII international symposium virus and virus-like diseases of temperate fruit crops, vol 472. ISHS Acta Horticulturae, Bethesda, pp 55–60. https://doi.org/10.17660/ActaHortic.1998.472.3

Wu Z, Mo C, Zhang S, Li H (2018) Characterization of *Papaya ringspot virus* isolates infecting transgenic papaya 'Huanong no.1' in South China. Sci Rep 8:8206. https://doi.org/10.1038/s41598-018-26596-x

Xu Y, Li Z (2020) CRISPR-Cas systems: overview, innovations and applications in human disease research and gene therapy. Comput Struct Biotechnol J 18:2401–2415. https://doi.org/10.1016/j.csbj.2020.08.031

Ye C, Li H (2010) 20 years of transgenic research in China for resistance to *Papaya ringspot virus*. Transgenic Plant J 4:5863. Available at: http://www.globalsciencebooks.info/Online/GSBOnline/images/2010/TPJ_4(SI1)/TPJ_4(SI1)58-63o.pdf

Ye CM, Wei XD, Chen DH, Lan CY, Chu LM (2003) Analyses of virus resistance and transgenes for transgenic papaya. Hereditas (Beijing) 25, 181–184

Zhang Y, Li Y, Jiang L et al (2011) Potential of perennial crop on environmental sustainability of agriculture. Procedia Environ Sci 10:1141–1147. https://doi.org/10.1016/j.proenv.2011.09.182

Zhao Y, Yang X, Zhou G, Zhang T (2019) Engineering plant virus resistance: from RNA silencing to genome editing strategies. Plant Biotechnol J 18:328–336. https://doi.org/10.1111/pbi.13278

MIX
Papier aus verantwortungsvollen Quellen
Paper from responsible sources
FSC® C105338

If you have any concerns about our products,
you can contact us on
ProductSafety@springernature.com

In case Publisher is established outside the EU,
the EU authorized representative is:
Springer Nature Customer Service Center GmbH
Europaplatz 3, 69115 Heidelberg, Germany

Printed by Libri Plureos GmbH
in Hamburg, Germany